Advances in Thermal Design of Heat Exchangers

Related Titles

Combined Power and Process – An Exergy Approach	F J Barclay	1 86058 129 3
Optical Methods and Data Processing in Heat and Fluid Flow	Edited by C Greated, J Cosgrove, and J M Buick	1 86058 281 8
Axial-Flow Compessors – A Strategy for Aerodynamic Design and Analysis	R H Aungier	1 86058 422 5

Advances in Thermal Design of Heat Exchangers

A Numerical Approach: Direct-sizing, step-wise rating, and transients

Eric M Smith

John Wiley & Sons, Ltd

Professional Engineering Publishing

Published by John Wiley & Sons Ltd, The Atrium, Southern Gate, Chichester, West Sussex PO19 8SQ, England

Telephone (+44) 1243 779777

Email (for orders and customer service enquiries): cs-books@wiley.co.uk
Visit our Home Page on www.wiley.com

Other Wiley Editorial Offices

John Wiley & Sons Inc., 111 River Street, Hoboken, NJ 07030, USA

Jossey-Bass, 989 Market Street, San Francisco, CA 94103-1741, USA

Wiley-VCH Verlag GmbH, Boschstr. 12, D-69469 Weinheim, Germany

John Wiley & Sons Australia Ltd, 33 Park Road, Milton, Queensland 4064, Australia

John Wiley & Sons (Asia) Pte Ltd, 2 Clementi Loop #02-01, Jin Xing Distripark, Singapore 129809

John Wiley & Sons Canada Ltd, 22 Worcester Road, Etobicoke, Ontario, Canada M9W 1L1

Wiley also publishes its books in a variety of electronic formats. Some content that appears in print may not be available in electronic books.

British Library Cataloguing in Publication Data

A catalogue record for this book is available from the British Library

ISBN 1-86058-461-6

Typeset by Techset Composition Limited, Salisbury, Wiltshire
Printed and bound in Great Britain by Antony Rowe, Ltd, Chippenham, Wiltshire
This book is printed on acid-free paper responsibly manufactured from sustainable forestry in which at least two trees are planted for each one used for paper production.

This volume is dedicated to Dorothy my wife
for her unfailing kindness and understanding,
and to my three sons for their consistent support.

'If you can build hotter or colder than anyone else,
If you can build higher or faster than anyone else,
If you can build deeper or stronger than anyone else,
If . . .
Then, in principle, you can solve all the other problems in between.'

(*Attributed to Sir Monty Finniston, FRS*)

Contents

	Preface	**xxiii**
Chapter 1	**Classification**	**1**
1.1	Class definition	1
1.2	Exclusions and extensions	1
1.3	Helical-tube, multi-start coil	3
1.4	Plate–fin exchangers	5
1.5	RODbaffle	6
1.6	Helically twisted flattened tube	7
1.7	Spirally wire-wrapped	7
1.8	Bayonet tube	8
1.9	Wire-woven heat exchangers	9
1.10	Porous matrix heat exchangers	9
1.11	Some possible applications	10
Chapter 2	**Fundamentals**	**19**
2.1	Simple temperature distributions	19
2.2	Log mean temperature difference	21
2.3	LMTD–*Ntu* rating problem	23
2.4	LMTD–*Ntu* sizing problem	25
2.5	Link between *Ntu* values and LMTD	26
2.6	The 'theta' methods	26
2.7	Effectiveness and number of transfer units	27
2.8	ε–*Ntu* rating problem	31
2.9	ε–*Ntu* sizing problem	32
2.10	Comparison of LMTD–*Ntu* and ε–*Ntu* approaches	33
2.11	Sizing when Q is not specified	34
2.12	Optimum temperature profiles in contraflow	35
2.13	Optimum pressure losses in contraflow	40
2.14	Compactness and performance	42
2.15	Required values of *Ntu* in cryogenics	42
2.16	To dig deeper	45
2.17	Dimensionless groups	47
Chapter 3	**Steady-State Temperature Profiles**	**59**
3.1	Linear temperature profiles in contraflow	59
3.2	General cases of contraflow and parallel flow	61

3.3	Condensation and evaporation	66
3.4	Longitudinal conduction in contraflow	67
3.5	Mean temperature difference in unmixed crossflow	74
3.6	Extension to two-pass unmixed crossflow	79
3.7	Involute-curved plate–fin exchangers	82
3.8	Longitudinal conduction in one pass unmixed crossflow	83
3.9	Determined and undetermined crossflow	90
3.10	Possible optimization criteria	92
3.11	Cautionary remark about core pressure loss	92
3.12	Mean temperature difference in complex arrangements	93
3.13	Exergy destruction	94
Chapter 4	**Direct-Sizing of Plate–Fin Exchangers**	**99**
4.1	Exchanger lay-up	99
4.2	Plate–fin surface geometries	101
4.3	Flow-friction and heat-transfer correlations	103
4.4	Rating and direct-sizing design software	103
4.5	Direct-sizing of an unmixed crossflow exchanger	106
4.6	Concept of direct-sizing in contraflow	110
4.7	Direct-sizing of a contraflow exchanger	113
4.8	Best of rectangular and triangular ducts	120
4.9	Best small, plain rectangular duct	125
4.10	Fine-tuning of ROSF surfaces	127
4.11	Overview of surface performance	127
4.12	Headers and flow distribution	130
4.13	Multi-stream design (cryogenics)	130
4.14	Buffer zone or leakage plate 'sandwich'	130
4.15	Consistency in design methods	132
4.16	Geometry of rectangular offset strip fins	133
4.17	Compact fin surfaces generally	138
4.18	Conclusions	138
Chapter 5	**Direct-Sizing of Helical-Tube Exchangers**	**143**
5.1	Design framework	143
5.2	Consistent geometry	145
5.3	Simplified geometry	151
5.4	Thermal design	153
5.5	Completion of the design	159
5.6	Thermal design results for $t/d = 1.346$	162
5.7	Fine tuning	163
5.8	Design for curved tubes	168
5.9	Discussion	172
5.10	Part-load operation with by-pass control	174
5.11	Conclusions	174

Chapter 6	**Direct-Sizing of Bayonet-Tube Exchangers**	**177**
6.1	Isothermal shell-side conditions	177
6.2	Evaporation	178
6.3	Condensation	189
6.4	Design illustration	190
6.5	Non-isothermal shell-side conditions	191
6.6	Special explicit case	194
6.7	Explicit solution	196
6.8	General numerical solutions	199
6.9	Pressure loss	201
6.10	Conclusions	204
Chapter 7	**Direct-Sizing of RODbaffle Exchangers**	**207**
7.1	Design framework	207
7.2	Configuration of the RODbaffle exchanger	208
7.3	Approach to direct-sizing	208
7.4	Flow areas	209
7.5	Characteristic dimensions	209
7.6	Design correlations	210
7.7	Reynolds numbers	211
7.8	Heat transfer	211
7.9	Pressure loss tube-side	213
7.10	Pressure loss shell-side	214
7.11	Direct-sizing	215
7.12	Tube-bundle diameter	217
7.13	Practical design	217
7.14	Generalized correlations	220
7.15	Recommendations	222
7.16	Other shell-and-tube designs	222
7.17	Conclusions	224
Chapter 8	**Exergy Loss and Pressure Loss**	**229**
	Exergy loss	**229**
8.1	Objective	229
8.2	Historical development	230
8.3	Exergy change for any flow process	231
8.4	Exergy loss for any heat exchangers	233
8.5	Contraflow exchangers	234
8.6	Dependence of exergy loss number on absolute temperature level	236
8.7	Performance of cryogenic plant	238
8.8	Allowing for leakage	240
8.9	Commercial considerations	242
8.10	Conclusions	242

	Pressure loss	**243**
8.11	Control of flow distribution	243
8.12	Header design	244
8.13	Minimizing effects of flow maldistribution	250
8.14	Embedded heat exchangers	251
8.15	Pumping power	253
Chapter 9	**Transients in Heat Exchangers**	**257**
9.1	Review of solution methods – contraflow	257
9.2	Contraflow with finite differences	259
9.3	Further considerations	265
9.4	Engineering applications – contraflow	266
9.5	Review of solution methods – crossflow	267
9.6	Engineering applications – crossflow	268
Chapter 10	**Single-Blow Test Methods**	**275**
10.1	Features of the test method	275
10.2	Choice of theoretical model	276
10.3	Analytical and physical assumptions	277
10.4	Simple theory	278
10.5	Relative accuracy of outlet response curves in experimentation	284
10.6	Conclusions on test methods	287
10.7	Practical considerations	287
10.8	Solution by finite differences	289
10.9	Regenerators	290
Chapter 11	**Heat Exchangers in Cryogenic Plant**	**297**
11.1	Background	297
11.2	Liquefaction concepts and components	298
11.3	Liquefaction of nitrogen	307
11.4	Hydrogen liquefaction plant	313
11.5	Preliminary direct-sizing of multi-stream heat exchangers	314
11.6	Step-wise rating of multi-stream heat exchangers	317
11.7	Future commercial applications	321
11.8	Conclusions	322
Chapter 12	**Heat Transfer and Flow Friction in Two-Phase Flow**	**325**
12.1	With and without phase change	325
12.2	Two-phase flow regimes	326
12.3	Two-phase pressure loss	327

12.4	Two-phase heat-transfer correlations	331
12.5	Two-phase design of a double-tube exchanger	333
12.6	Discussion	336
12.7	Aspects of air conditioning	340
12.8	Rate processes	343
Appendix A	**Transient Equations with Longitudinal Conduction and Wall Thermal Storage**	**349**
A.1	Mass flow and temperature transients in contraflow	349
A.2	Summarized development of transient equations for contraflow	352
A.3	Computational approach	355
Appendix B	**Algorithms And Schematic Source Listings**	**361**
B.1	Algorithms for mean temperature distribution in one-pass unmixed crossflow	361
B.2	Schematic source listing for direct-sizing of compact one-pass crossflow exchanger	364
B.3	Schematic source listing for direct-sizing of compact contraflow exchanger	365
B.4	Parameters for rectangular offset strip fins	366
B.5	Longitudinal conduction in contraflow	370
B.6	Spline-fitting of data	375
B.7	Extrapolation of data	376
B.8	Finite-difference solution schemes for transients	377
Supplement to Appendix B – Transient Algorithms		**383**
Appendix C	**Optimization of Rectangular Offset Strip, Plate–Fin Surfaces**	**405**
C.1	Fine-tuning of rectangular offset strip fins	405
C.2	Trend curves	407
C.3	Optimization graphs	408
C.4	Manglik & Bergles correlations	409
Appendix D	**Performance Data for RODbaffle Exchangers**	**411**
D.1	Further heat-transfer and flow-friction data	411
D.2	Baffle-ring by-pass	414
Appendix E	**Proving the Single-Blow Test Method – Theory and Experimentation**	**419**
E.1	Analytical approach using Laplace transforms	419

E.2	Numerical evaluation of Laplace outlet response	420
E.3	Experimental test equipment	423
Appendix F	**Most Efficient Temperature Difference in Contraflow**	**425**
F.1	Calculus of variations	425
F.2	Optimum temperature profiles	426
Appendix G	**Physical Properties of Materials and Fluids**	**429**
G.1	Sources of data	429
G.2	Fluids	429
G.3	Solids	431
Appendix H	**Source Books on Heat Exchangers**	**433**
H.1	Texts in chronological order	433
H.2	Exchanger types not already covered	439
H.3	Fouling – some recent literature	442
Appendix I	**Creep Life of Thick Tubes**	**443**
I.1	Applications	443
I.2	Fundamental equations	443
I.3	Early work on thick tubes	445
I.4	Equivalence of stress systems	446
I.5	Fail-safe and safe-life	447
I.6	Constitutive equations for creep	447
I.7	Clarke's creep curves	449
I.8	Further and recent developments	451
I.9	Acknowledgements	451
Appendix J	**Compact Surface Selection for Sizing Optimization**	**455**
J.1	Acceptable flow velocities	455
J.2	Overview of surface performance	455
J.3	Design problem	458
J.4	Exchanger optimization	466
J.5	Possible surface geometries	467
Appendix K	**Continuum Equations**	**469**
K.1	Laws of continuum mechanics	469
K.2	Coupled continuum theory	473
K.3	De-coupling the balance of energy equation	474
Appendix L	**Suggested Further Research**	**477**
L.1	Sinusoidal–lenticular surfaces	477
L.2	Steady-state crossflow	478

L.3	Header design	478
L.4	Transients in contraflow	479
Appendix M	**Conversion Factors**	**483**
Notation		**487**
Commentary		487
Chapter 2	Fundamentals	488
Chapter 3	Steady-state temperature profiles	489
Chapter 4	Direct-sizing of plate–fin exchangers	490
Chapter 5	Direct-sizing of helical-tube exchangers	491
Chapter 6	Direct-sizing of bayonet-tube exchangers	493
Chapter 7	Direct-sizing of RODbaffle exchangers	494
Chapter 8	Exergy loss and pressure loss	495
Chapter 9	Transients in heat exchangers	496
Chapter 10	Single-blow test methods	497
Chapter 11	Heat exchangers in cryogenic plant	498
Chapter 12	Heat transfer and flow friction in two-phase flow	499
Appendix A	Transient equations with longitudinal conduction and wall thermal storage	500
Appendix I	Creep life of thick tubes	501
Index		**503**

THERMAL DESIGN ROADMAP
(outline guide for contraflow)

DIRECT-SIZING
(minimum input data required)

INPUT DATA
contraflow

$$\left\{ \begin{matrix} & \dot{Q}duty & & \\ \dot{m}_h & T_{h1} & p_{h1} & \Delta p_h \\ \dot{m}_c & T_{c2} & p_{c2} & \Delta p_c \end{matrix} \right\}$$

OPTIMAL TEMPERATURE DISTRIBUTION
Grassman & Kopp

$$\left\{ \begin{matrix} exergy\ constraint & \dfrac{\Delta T}{T_h} = const. \\ check & \dfrac{\dot{m}_h C_h}{\dot{m}_c C_c} = 1 - \dfrac{\Delta T}{T_h} \end{matrix} \right\}$$

Ntu VALUES
$\{find \quad T_{h2} \quad T_{c1}\}$

LMTD–*Ntu* approach $\{N_h \, N_c\}$ | ε–*Ntu* approach $\{Ntu \, W\}$

EXCHANGER TYPE
Plate–fin
Helical-tube
RODbaffle

MEAN PHYSICAL PROPERTIES
specific heats
absolute viscosities
thermal conductivities

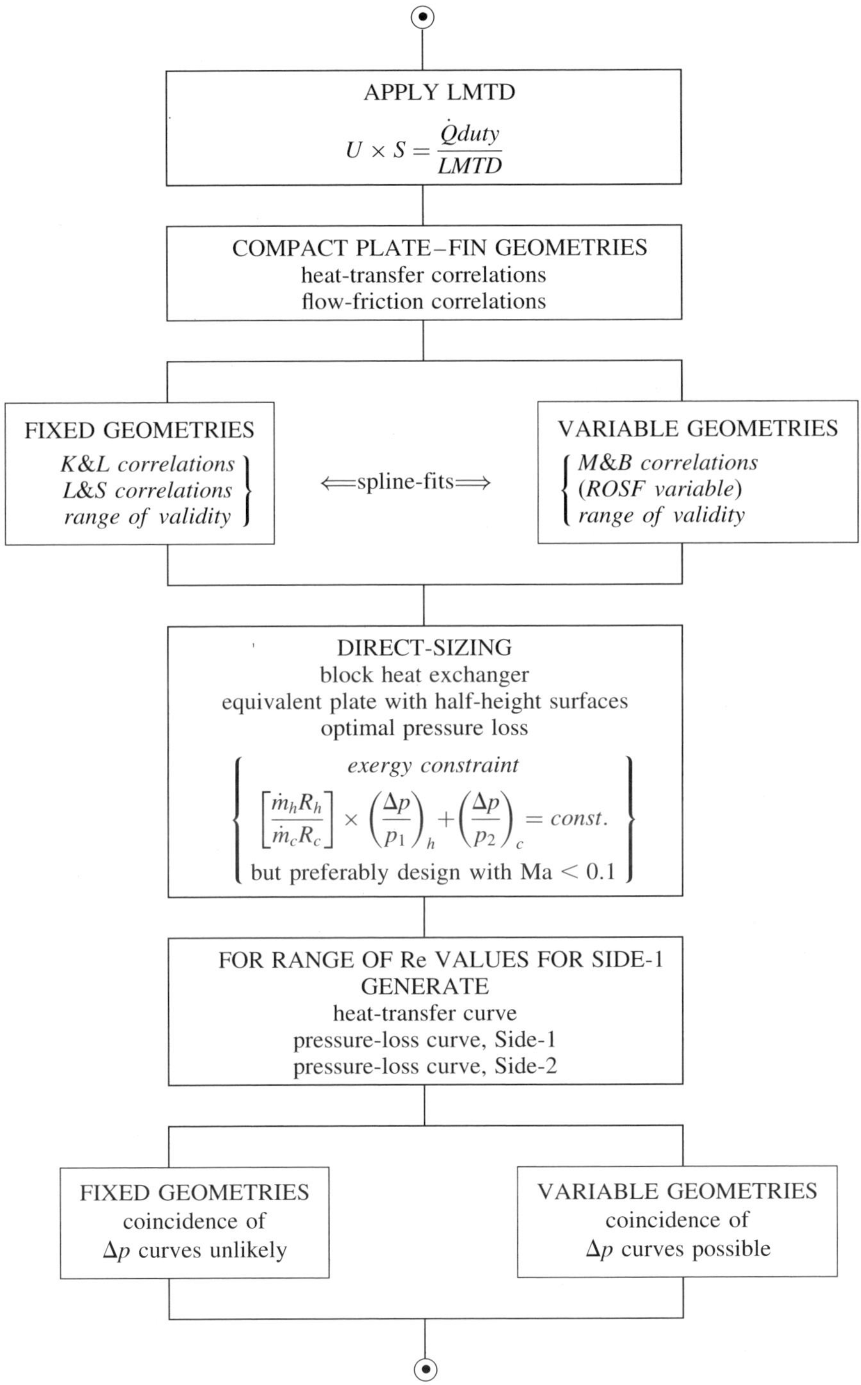
APPLY LMTD
$U \times S = \frac{\dot{Q}duty}{LMTD}$
COMPACT PLATE–FIN GEOMETRIES
heat-transfer correlations
flow-friction correlations
FIXED GEOMETRIES
K&L correlations
L&S correlations
range of validity
⟸spline-fits⟹
VARIABLE GEOMETRIES
M&B correlations
(ROSF variable)
range of validity
DIRECT-SIZING
block heat exchanger
equivalent plate with half-height surfaces
optimal pressure loss
exergy constraint
$\left[\frac{\dot{m}_h R_h}{\dot{m}_c R_c}\right] \times \left(\frac{\Delta p}{p_1}\right)_h + \left(\frac{\Delta p}{p_2}\right)_c = const.$
but preferably design with $Ma < 0.1$
FOR RANGE OF Re VALUES FOR SIDE-1
GENERATE
heat-transfer curve
pressure-loss curve, Side-1
pressure-loss curve, Side-2
FIXED GEOMETRIES
coincidence of
Δp curves unlikely
VARIABLE GEOMETRIES
coincidence of
Δp curves possible

NEAR-OPTIMUM EXCHANGER
estimated cross-section and length

LONGITUDINAL CONDUCTION
(reduced performance in most exchangers)

STEADY-STATE TEMPERATURE PROFILES
three simultaneous partial differential equations

{ *hot fluid ... balance of energy*
solid wall ... balance of energy
cold fluid ... balance of energy }

LMTD REDUCTION
(allowing for longitudinal conduction)

KROEGER SOLUTION
equal water equivalents
{ *analytical solution*
greatest conduction }

GENERAL SOLUTION
unequal water equivalents
{ *numerical solution*
Crank–Nicholson }

APPLY CORRECTION TO DIRECT-SIZING LMTD
leading to
CONSERVATIVE DIRECT-SIZING DESIGN
with mean thermophysical values
accurate cross-section and length

OPTIMIZED DESIGN
vary local surface geometries
until Δp curves coincide

STEP-WISE RATING
(using cross-section from direct-sizing)

AWKWARD CONDITIONS
arbitrary temperature profiles
physical properties varying along length

FOR TEMPERATURE RANGE OF EXCHANGER
spline-fit thermophysical data
for interpolation

SECTION-WISE DESIGN
assume equal temperature intervals for one fluid
and use enthalpy balance to calculate
corresponding temperatures for other fluid

LMTD AND MEAN TEMPERATURES AT EACH SECTION
thermophysical properties for each section
using spline-fitted data

CALCULATE SURFACE AREA AND LENGTH
for each section

CALCULATE PRESSURE LOSS FOR EACH FLUID
for each section

SUM LENGTHS AND PRESSURE LOSSES
to obtain final step-wise design

MULTI-STREAM DESIGN
(refer to specialist papers for cross-conduction and near-optimization)

TRANSIENTS
(for known steady-state design)

COMBINED MASS FLOW AND TEMPERATURE TRANSIENTS
temperature-dependent physical properties
seven simultaneous partial differential equations

hot fluid
- *balance of mass*
- *balance of linear momentum*
- *balance of energy*

solid wall – balance of energy

cold fluid
- *balance of energy*
- *balance of linear momentum*
- *balance of mass*

FINITE-DIFFERENCE SOLUTION
solve sequentially by finite differences
two pairs of outer equations for density and velocity
three simultaneous central equations for temperature field

DELAYED ENTRY TO INDIVIDUAL CHANNELS
AND CROSS-CONDUCTION EFFECTS
(refer to specialist papers)

In general, finite-difference schemes were preferred for both steady-state analysis, and for transients, as temperature-dependent physical properties could then most easily be accommodated.

About the Author

Eric Smith is a Fellow of both the Institution of Mechanical Engineers and of the American Society of Mechanical Engineers, and is also a Member of the Institute of Refrigeration. He received his BSc and PhD degrees from the University of Glasgow. His early research on the strength of high-temperature materials was complemented by an interest in heat transfer.

His early career included a total of 5 years in civil nuclear engineering research at C.A. Parsons & Co. Ltd of Newcastle upon Tyne and at the Institut für Reaktor Bauelemente, Kernreaktor Kernforschungszentrum, Karlsruhe. The next 20 years involved teaching and research to post-graduate level in Mechanical Engineering at the University of Newcastle upon Tyne.

Dr Smith has published with IMechE, ASME, and ASTM, and has presented papers at international and national level. He has represented the UK at a Nato AGARD Special Technical Meeting in Washington DC, and was retained as an expert witness by Norton Rose of London on behalf of shipbuilders Harland and Wolff of Belfast.

Following short periods in Defence Consultancy in the UK, and teaching engineering in Hong Kong, he returned to the UK to pursue his interests in long-range engineering.

The Book

All material presented in this volume has been computed from scratch by the author, and a number of new points of understanding have been uncovered. Specifically and illustratively:

- Alternative Effectiveness–*Ntu* and LMTD–*Ntu* approaches to design are shown to be equivalent in finding terminal temperatures or *Ntu* values.
- Recommended *Ntu* performance limits for parallel flow, crossflow, and contraflow designs have been established, permitting appropriate choice for duty.
- With the LMTD–*Ntu* approach, longitudinal conduction can be approximated by calculating the LMTD reduction factor in contraflow sizing.
- The application of direct-sizing to three different types of heat exchanger is presented in some detail, and matching of local surface geometries prior to direct sizing is explored.
- An unambiguous measure of specific thermal performance is defined, applicable to all types of exchanger.
- Exergy loss number is defined, and its relation to quality of heat exchange and absolute temperature level of operation developed. Elimination of pressure losses in headers, and the proper way of evaluating pumping power are presented.
- Methods are set out for predicting full transients in contraflow allowing for temperature-dependent physical properties. The single-blow method for determining heat-transfer and flow-friction correlations is outlined.
- Aspects of design for cryogenic and two-phase flow problems are examined.
- Numerical methods are emphasized throughout, starting from the controlling differential equations and building towards understanding of thermal design at every level.

Preface

'I would like to extend the way in which you may think about the design of heat exchangers ... '

(author, 10th International Heat Transfer Conference, Brighton 1994)

Purpose of this work

The primary objective in any engineering design process has to be the elimination of uncertainties. In thermal design of heat exchangers there are presently many stages in which assumptions in mathematical solution of the design problem are being made. Accumulation of these assumptions (e.g. use of mean values) may introduce variations in design as large as the uncertainties introduced in heat-transfer and flow-friction correlations. The designer needs to understand where these inaccuracies may arise, and strive to eliminate as many sources of error as possible by choosing design configurations that avoid such problems at source.

This book is set at research graduate and professional level in clean technologies, and is designed as a reference text. Theory is explained simply, so that the reader can develop his/her own approach to solution of problems. The text is not intended as a collection of heat-transfer and pressure-loss correlations, although a fair amount of such material has been included.

Historical development of the subject

Up until the early 1940s virtually all papers employ 'mean temperature difference' as the design parameter, a good collection being found in the two reference volumes by Jakob (1949, 1957). Around 1942 the method of designing contra-parallel-flow heat exchangers was effectively changed by London & Seban (1980) from using LMTD to using the ε–*Ntu* approach, partly on the grounds that the LMTD approach did not give explicit results in some elementary cases, and partly because the concept of 'effectiveness' provided a measure of the approach to ultimate performance of the exchanger.

The consequence has been that since 1942 many important papers have concentrated on expressing results in terms of effectiveness in preference to mean temperature difference, which in this author's view has not been entirely beneficial, particularly in the case of variable thermophysical properties (Soyars, 1992), and in the case of crossflow. The separate concepts of mean temperature difference and of effectiveness both have useful roles to play in assessing the performance of heat exchangers and should be used in combination.

The present treatment shows that:

- The LMTD–*Ntu* approach is fully explicit in finding terminal temperatures in contraflow and parallel flow, and contains expressions for 'energy' and 'rate' processes (... iteration cannot be avoided when only inlet temperatures and LMTD are known and outlet temperatures are required).
- Effectiveness may be a measure of performance for entropy loss, but exergy loss number is essential in cryogenics. Aiming for minimum entropy generation in contraflow leads to temperature profiles with a pinch point at the hot end, while aiming for least exergy loss in contraflow leads to temperature profiles with a pinch point at the cold end.
- Exchanger comparisons are best made using the specific performance parameter

$$\frac{\dot{Q}_{duty}}{V\Delta\theta_{mean}} \equiv \frac{U\,S_{ref}}{V} \quad \text{kW/(m}^3\text{K)}$$

where S_{ref} is the reference surface.

The case of unmixed–unmixed two-pass crossflow is examined in some detail. The very comprehensive analytical paper by Baclic (1990) which examines 72 possible configurations for two-pass crossflow, concentrates on presenting results in terms of effectiveness alone. However, understanding has been lost in not computing temperature sheets, and for accurate design the assumption of equal mass flowrate at inlet to each channel should be replaced by the assumption of equal pressure loss in each flow channel. Dow's (1950) approach for designing headers with zero pressure loss allows concentration on core pressure loss.

Direct-sizing methods

'Sizing' methods have traditionally posed more problems than 'rating' methods. Guessing one principal dimension of the exchanger may be necessary before the performance of the core can be compared with design requirements. For the class of heat exchangers in which 'local' geometry of the heat-transfer surface is fully representative of the whole geometry, guessing is no longer necessary. Methods of direct-sizing go straight to the dimensions of the heat exchanger core, while satisfying all thermal performance constraints.

Design approaches for contraflow

Data	*Rating*	*Direct sizing*
Given	block size $L \times W \times H$	thermal duty $\dot{Q},\ dP_h,\ dP_c$
Inlet values	$\dot{m}_h,\ C_h,\ T_{h1}$ $\dot{m}_c,\ C_c,\ T_{c2}$	$\dot{m}_h,\ C_h,\ T_{h1}$ $\dot{m}_c,\ C_c,\ T_{c2}$
Find	thermal duty	block size

In direct-sizing, the design approach is limited to that class of heat exchangers in which 'local geometry' is fully representative of the complete heat exchange surface and which provide core layouts which eliminate/minimize parasitic losses due to flow leakage and by-pass flows, typically those described by Tinker[1] (1951, Part I, Fig. 6).

The concept can be applied to such different designs as:

- compact plate–fin exchangers
- helical-tube, multi-start coil exchangers
- platen-type heat exchangers
- RODbaffle shell-and-tube exchangers
- lamella heat exchangers
- flattened and helically twisted tubes
- printed-circuit heat exchangers
- HELIXCHANGERs

As all terminal temperatures may be determined in advance of direct-sizing, the necessary input data for complete sizing take the following form:

- exchanger duty ($\dot{Q}$)
- mean temperature difference for heat exchange ($\Delta\theta_m$)
- 'local' geometry on both sides
- mass flowrates of both fluids ($\dot{m}$)
- physical properties for both fluids at mean bulk temperature (Pr, C, η, λ)
- allowable pressure loss data (ΔP, P, T_{bulk}, R_{gas})
- physical properties of material of construction (λ, ρ, C)

For the selected geometry, the standard procedure is to evaluate heat-transfer performance over the range of valid Reynolds numbers for both sides of the exchanger. This provides a heat-transfer curve. Pressure-loss performance is similarly evaluated for both sides over the same range of valid Reynolds numbers, providing two separate pressure-loss curves. Both pressure-loss curves will intersect the heat-transfer curve, and the intersection furthest to the right provides the initial design point.

In cases where heat-transfer and pressure-loss correlations are suitable a fully algebraic solution may be possible – as with the helical-tube, multi-start coil heat exchanger. More often a numerical approach is preferred because the interpolating cubic spline-fit provides more accurate temperature-dependent physical properties plus heat-transfer and flow-friction coefficients, with the special advantage that spline-fits cannot be extrapolated outside the range of their validity.

Longitudinal and cross-conduction

Techniques for estimating longitudinal conduction effects in both contraflow and crossflow exchangers are described. Longitudinal conduction reduces exchanger performance, and the design approach is to calculate and apply the reduction in mean temperature difference.

With multi-stream and crossflow exchangers, an additional problem exists, caused by fluids flowing in the same direction having different temperature profiles along the length of the exchanger. Cross-conduction effects may then have to be taken into account.

Step-wise rating

When changes in thermophysical properties are significant, and it is not desired to go to full transient analysis, it may be appropriate to design by step-wise rating. To start this process an initial cross-section of the exchanger is required, and direct-sizing can be helpful in providing the start-up information. By using temperature-dependent physical properties, and evaluating LMTD for each section, the design can be made more accurate. Step-wise rating is the intermediate stage between design assessment using mean values and full numerical prediction of transient performance of a design.

Transient response

In contraflow exchangers which experience transient temperature disturbances, longitudinal conduction terms appear in the set of seven simultaneous partial differential equations. As the Mach number in heat exchangers is normally less than ($Ma = 0.1$) it becomes practicable to separate the problem into solution of mass flowrate and temperature-field disturbances.

A full numerical study of transients in a two-stream contraflow exchanger involves preparation of interpolating cubic spline-fits for both heat-transfer and flow-friction data against Reynolds number, and for all temperature-dependent physical parameters. At least 50 stations along the length of a contraflow exchanger are desirable.

Single-blow testing

To measure heat-transfer and flow-friction performance of heat exchanger surfaces the contraflow transient equations are simplified to the point where they become the Single-Blow Transient Test Equations. This method of obtaining data for heat-transfer and flow-friction correlations is well established as reliable, providing that assumptions in the mathematical analysis are correctly matched to the experimental method.

Arbitrary inlet temperature disturbances and longitudinal conduction effects in a matrix involve deeper analytical theory than presented in this text. A subset of the full transient equations, combined with a numerical approach is one way forward.

The extraction of dimensionless numbers from governing differential equations is covered in outline, to reveal the extent of their limitations when incorporated in heat-transfer and flow-friction correlations. Rayleigh's empirical approach and Buckingham's π-method are not included as they are readily available elsewhere.

---

[1]Tinker's work on baffle losses in shell-and-tube exchangers has been reproduced in textbooks and papers since McAdams (1954).

Design involving phase change

For fluids that experience changes in thermal capacity, e.g. phase-change applications, the exchanger can sometimes be split into two or more sections in which single-phase behaviour can be assumed, or in which different stages of two-phase behaviour can be shown to exist. The two final chapters outline some considerations in both step wise rating and variable fluid properties

Optimization in direct-sizing

First stages in optimization of compact exchangers may involve:

- limitation of exergy loss in contraflow due to pressure losses, Section 2.13
- selection of appropriate local surface geometries, viz.
  - choice between rectangular and triangular ducts, Section 4.8
  - preference of rectangular ducts over square ducts, Section 4.9
  - overview of plain duct performance, Section 4.11
  - closer choice of starting values for compact surfaces, Appendix J
- study of directions for improvement in rectangular offset strip-fin (ROSF) surfaces, looking for minimum core volume, minimum frontal area, etc., Appendix C

Subsequently, for a two-stream contraflow exchanger with single-phase fluids, direct-sizing involves construction of three design curves, one for heat transfer and two pressure-loss curves, one for each fluid. Design is achieved when each pressure-loss curve cuts the heat-transfer curve at the same point.

The refined approach involves independent optimization of surface geometry for each side in turn (Appendix J). This is done by setting the pressure loss on the other side as high as permitted to reduce its influence on design to a minimum. Once the two independently optimized pressure-loss curves have been found, the design is recalculated with the pair of optimized surfaces, and final slight adjustments made to have the pressure-loss curves coincide at the design point. With such an approach it is more easily seen when the use of primary surfaces becomes advantageous.

Notation

The international standards for nomenclature are adopted, with a few exceptions to improve clarity. Hewitt *et al.* (1993) provide good argument for accepting new notation in a Preface to their handbook on process heat transfer. Full listings for each chapter are provided at the end of the appendices.

Much of the difficulty that arises in reading the literature on heat exchangers stems from the way in which temperatures are labelled at the ends of the exchanger. For a heat exchanger under steady-state conditions two possibilities exist, viz.:

- label one end with subscript 1 and the other end with subscript 2
- label the exchanger to give all inlet temperatures subscript 1 and all outlet temperatures subscript 2

The second option leads to confusion as one is always referring back to ascertain if the analysis has been correctly assembled. The first option is to be preferred and is used in this text.

References

The reader will find that some references listed at the end of each chapter are not directly mentioned in the text. These are papers that may indicate possible further directions of development, and it is worth scanning the references for interesting titles. Appendix H contains a list of useful textbooks together with other publications of interest to designers.

Applications

The possible applications for exchangers suitable for direct-sizing are quite wide, including aerospace, marine propulsion systems, land-based power plant, chemical engineering plant, etc. It is possible to go directly to the optimum exchanger core and minimize the choice of core volume, core mass, core frontal area, etc. Multi-stream exchangers for cryogenic duty must usually be sized by step-wise rating.

Stressing of exchanger tubes, shells, and other components is specifically excluded from this volume. However, the author found the solution to the problem of creep in thick tubes which may be of use in extreme temperature/pressure conditions. A brief summary is given in Appendix I.

Computer software

All computation for this volume was developed in HP-Pascal 3.0 on an 8 MHz machine with 1 Mbyte of RAM and no hard disk. A dual floppy drive for 1 Mbyte disks was the only storage used. This platform is now obsolete, but the software remains valid. Any future development on a modern platform would require conversion from Macintosh WordPerfect (where listings now reside) to Mac OS-X, Unix, and/or Linux. Most of the software will be acceptable to free Pascal Compilers, requiring only minor changes.

Much of the original software was designed to be 'pipelined' – output from one package becoming input for the next package. Some of the original (extended) procedure bodies would require a rebuild, retaining the same names for compatibility.

Acknowledgements

In the author's early days, Sir Monty Finniston, Director of International Research and Development of C.A. Parsons & Co. Ltd, inspired research engineers to achieve of their best through his leadership. Professor Albert Green, late of the University of Newcastle upon Tyne, was kind enough to permit a young engineering lecturer to attend the post-graduate mathematics course in Continuum Mechanics, and over many years of discussion, encouragement was provided by Professor Alan Jeffrey of the Department of Engineering Mathematics at the University of Newcastle upon Tyne. Professor Aubrey Burstall, late head of the Department of Mechanical Engineering at the University of Newcastle upon Tyne supported initiation of both experimental and theoretical work by the author over a range of topics, including investigative research on the single-blow test method. Both teaching and research continued successfully under Professor Leonard Maunder. Outstanding facilities were provided by Professor Ewan Page and staff of the university Computing Laboratory. These involved mainframe computers that were nationally state of the art, together with matching technical and software support. Elizabeth Barraclough, Deputy Director of the Computing Laboratory, allowed a mechanical engineer occasional overnight access to the machine room. Chris Woodford on the staff of the same laboratory developed the original spline-fitting algorithm since used extensively over many years.

Special thanks to my close colleagues Tom Frost and Attila Fogarasy in the Department of Mechanical Engineering, University of Newcastle upon Tyne, whose patience and talents were outstanding, both professionally and personally. Particularly so when Tom took the leadership with an inexperienced partner in rock climbing and hilarious hill-walking, and when Attila explained the advantage of transforming differential equations into Riemannian space before solving explicitly along a geodesic.

From 1983, apart from less than 2 years of formal employment, the author chose to work privately on material for this book while undertaking a modest amount of industrial consultancy to help defray the cost.

The author particularly wishes to thank the referees for the care taken in assessing this final manuscript. Thoughtful and constructive suggestions have been made by many experienced engineers in the preparation of this text. If any material in this text has been included without proper recognition the author would be pleased to have this drawn to his attention.

Special acknowledgements are due to Sheril Leitch (Commissioning Editor) and Lou Attwood (Co-ordinating Editor) of Professional Engineering Publishing, Ltd, for their consideration, contributions and patience with the author in arriving at the near-final text. Late changes to the manuscript occurred and these made the final version through the kind assistance of Martin Tribe (Executive Project Editor) at John Wiley & Sons, Ltd.

Material from the author's published papers is included with the permission and/or acknowledgement of the Institution of Chemical Engineers, the Institution of Mechanical Engineers, the American Society of Mechanical Engineers, and the American Society for Testing and Materials. Clarification of some points in the design of ROD-baffle exchangers was kindly provided by C. C. Gentry of Philips Petroleum.

Assurance

Serious preparation of the first edition of this text began around 1994. All information used in production of this present edition was sourced from the open literature, suggested by others, or developed from scratch by the author.

Eric M. Smith
St Andrews, UK

References

Baclic, B.S. (1990) ε-Ntu analysis of complicated flow arrangements. *Compact Heat Exchangers – A Festschrift for A.L. London.* (Eds, R.K. Shah, A.D. Kraus, and D. Metzger), Hemisphere, New York, pp. 31–90.

Dow, M.W. (1950) The uniform distribution of a fluid flowing through a perforated pipe. *Trans. ASME, J. Appl. Mechanics*, December, 431–438.

Hewitt, G.F., **Shires, G.L.**, and **Bott, T.R.** (1993) *Process Heat Transfer*, CRC Press, Florida.

Jakob, M. (1949, 1957) *Heat Transfer*, vol. I (1949) and vol. II (1957), John Wiley.

London, A.L. and **Seban, R.A.** (1980) (Release of unpublished 1942 paper) A generalisation of the methods of heat exchanger analysis. *Int. J. Heat Mass Transfer*, **23**, 5–16.

McAdams, W.H. (1954) *Heat Transmission*, 3rd edn, McGraw-Hill, New York, London, p. 278.

Soyars, W.M. (1992) The applicability of constant property analyses in cryogenic helium heat exchangers. *Advances in Cryogenic Engineering*, vol. 37, Part A, Plenum Press, pp. 217–223.

Tinker, T. (1951) Shell-side characteristics of shell and tube heat exchangers, Parts I, II and III. In *Proceedings of General Discussion on Heat Transfer,* 11–13 September 1951, Institution of Mechanical Engineers, London, and American Society of Mechanical Engineers, New York, pp. 89–96, 97–109, 110–116.

CHAPTER 1
Classification

Consistent core geometry in heat exchangers

1.1 Class definition

Direct-sizing is concerned with members of the class of heat exchangers that have consistent geometry throughout the exchanger core, such that local geometry is fully representative of the whole surface. The following configurations are included in that class and are discussed further in this chapter, but the list is short and illustrative only, namely:

- Helical-tube, multi-start coil
- Plate–fin
- RODbaffle
- Helically twisted, flattened tube
- Spirally wire-wrapped
- Bayonet tube
- Wire-woven tubes
- Porous matrix heat exchanger

Illustrations of many types of exchanger are included in the following recent texts:

- Hewitt *et al.* (1994), Chapter 4
- Hesselgreaves (2001), Chapter 2
- Shah & Sekulic (2003), Chapter 1

1.2 Exclusions and extensions

Exclusions

Not every heat exchanger design is considered in this textbook, for the main objective is to study thermal design of contraflow exchangers proceeding via steady-state direct-sizing, through optimization, to the study of transients.

Most automotive heat exchangers operate in crossflow, and have a relatively small flow length on the air-side. They may be constructed of tubes inserted in corrugated plate–fins, or made up from welded channels with corrugated fins. The small air flow length rather marks them out as a special design case and the subject deserves separate attention. It is not covered in this text.

Segmentally baffled shell-and-tube designs

Segmentally baffled and disc-and-doughnut baffled shell-and-tube designs are not specifically included because the exchanger core may not have sufficiently regular flow geometry. However, there have been some attempts to develop a direct-sizing approach for these exchangers, plus helically baffled shell-and-tube exchangers which are referenced in Chapter 7.

Single-spiral radial flow

Also excluded is the single-spiral heat exchanger with inward and outward spiral (pseudo-radial) flow. Papers analysing performance of this exchanger design have been published by Bes & Roetzel (1991, 1992, 1993). The omission of this design is not a criticism of its usefulness, for in the right application such exchangers may be more economic, or more suitable for corrosive or fouling service.

Extensions

Exchangers that may be suitable for direct-sizing include:

Single-spiral axial design

The single-spiral exchanger with axial flow has been realized and is a candidate for direct-sizing using the thermal design approach outlined in Chapter 4 (Oswald *et al.*, 1999).

Plate–frame designs

The plate-and-frame heat exchanger is not specifically considered, because steady-state design follows standard contraflow or parallel-flow procedures. It is only necessary to source sets of heat-transfer and flow-friction correlations before proceeding.

Plate-and-frame designs can be similar in flow arrangement to plate-fin designs, but there is restriction on the headering geometry. Optimization may proceed in a similar way as for compact plate–fin heat exchangers, but is likely to be less comprehensive until universal correlations for the best plate–panel corrugations become available. The text by Hewitt *et al.* (1994) provides an introduction to steady-state design using plates with standard corrugations, and provides further references. The paper by Focke (1985) considers asymmetrically corrugated plates.

Inlet and return headering for plate-and-frame designs, and the same arrangement for plate–fin designs, may add a phase shift to the outlet transient response following an inlet disturbance. Effects of this headering arrangement have been considered by Das & Roetzel (1995). Faster response is obtained with U-type headering than with Z-type headering, and the choice of U-type headering is evident in the paper by Crisalli & Parker (1993) describing a recuperated gas-turbine plant using plate–fin heat exchangers. However, the reader should consider Dow's (1950) approach to the design of headers in Chapter 8 of this text.

Printed-circuit heat exchangers

These are constructed first by taking a suitable flat plate, then printing a chemically resistant photographic image of material between desired flow channels on to the plate, and then etching the plate to a depth not exceeding 2.0 mm. For the second fluid a further plate with similar etched channels, but probably of different design, is placed on top of the first plate, and the stacking process repeated until a desired stack height is reached. The stack of plates is then diffusion bonded together to form the single core of an exchanger.

Two-stream and multi-stream exchangers may be constructed in this way. It is important that the best geometry of flow channel is selected for each fluid stream, and that proper consideration is given to inlet and outlet headers so as not to create an exchanger with mixed crossflow and contraflow features, as it then becomes problematic to calculate correct temperature profiles.

Depending on geometry and availability of appropriate heat-transfer and flow-friction correlations, thermal design can be approached in the same way as for plate–fin exchangers.

Lamella heat exchangers

Flat tube ducts are fitted inside a tubular shell, leaving equal spacing for shell-side flow between the flat tube ducts. The geometry offers a very flexible surface arrangement, with good means for header connections to shell- and tube-side flow.

Rapid prototyping (but real) designs

The technique of producing rapid prototypes of complex components has now been extended to include construction of complete heat exchangers (see UK Patent GB2338293). The technique involves slicing the finished concept drawings into flat shapes which then may be either cut from meta sheet by laser, or stamped out. These metal sections are then stacked and diffusion bonded to recover the final exchanger. Small ligaments may be required to locate otherwise unsupported parts of a slice in place. If adjacent slices also require support, then ligaments are staggered to preserve flow paths past the ligaments. This approach has already been successful in creating a small and well-designed shell-and-tube heat exchanger, in which baffle passes are repeated to minimize the number of slices required.

Porous metal developments

New interest has been noted in the use of porous, foamed metal fillings inside tubes, and sometimes as external fins. Potential advantages which can be identified include greater metal/fluid surface area for heat transfer, and the possibility of using the porous substrate for mounting catalysts.

1.3 Helical-tube, multi-start coil

This design shown in Fig. 1.1 has no internal baffle leakage problems, it permits uninterrupted crossflow through the tube bank for high heat-transfer coefficients, and provides advantageous counterflow terminal temperature distribution in the

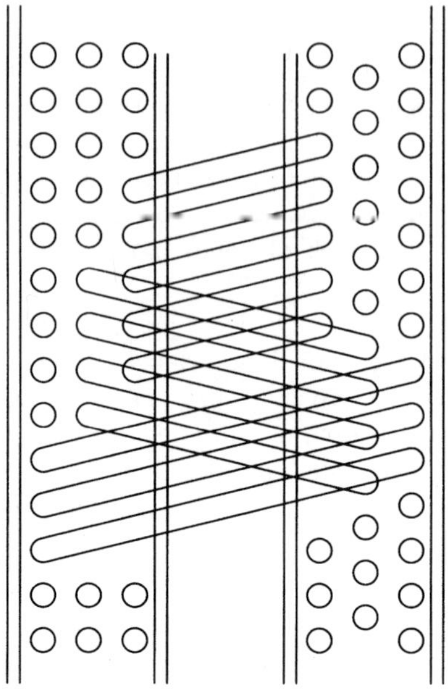

Fig.1.1 Helical-tube multi-start coil exchanger

whole exchanger. Some modification to the log mean temperature difference (LMTD) is necessary when the number of tube turns is less than about ten and this analysis has been provided by Hausen (1950, 1983) in both his German and his English texts.

Although exchangers of this type had been in use since the first patents by Hampson (1895) and L'Air Liquide (1934), consistent geometry in the coiled tube bundle does not seem to have been known before Smith (1960). Since that time programmes of work on helical-coil tube bundles have appeared (Gilli, 1965; Smith & Coombs, 1972; Smith & King, 1978; Gill *et al.*, 1983), and a method of direct-sizing has been obtained by Smith (1986) which is further reported in this text.

Cryogenic heat exchangers to this design have been built by Linde AG and are illustrated in both editions of Hausen (1950, 1983), further examples being found in the papers by Abadzic & Scholz (1972), Bourguet (1972) and Weimer & Hartzog (1972). High-temperature nuclear heat exchangers have been constructed in very large multiple units by Babcock Power Ltd for two AGR reactors (Perrin, 1976), and by Sulzer and others for several HTGR reactors (Kalin, 1969; Profos, 1970; Bachmann, 1975; Chen, 1978; Anon, 1979). A single unit may exceed 18 m in length and 25 tonnes in mass with a rating of 125 MWt.

The pressurized-water reactor (PWR) nuclear ship Otto Hahn was provided with a helical-coil integral boiler built by Deutsche Babcock (Ulken, 1971). For LNG applications, Weimer & Hartzog (1972) report that coiled heat exchangers are preferred for reduced sensitivity to flow maldistribution. Not all of the above heat exchangers have consistent geometry within the tube bundle.

1.4 Plate–fin exchangers

The compact plate–fin exchanger is now well known due to the work of Kays & London (1964), London & Shah (1968), and many others. It is manufactured in several countries, and its principal use has been in cryogenics and in aerospace where high performance with low mass and volume are important. Constructional materials include aluminium alloys, nickel, stainless steel, and titanium. The lay-up is a stack of plates and finned surfaces which are either brazed or diffusion bonded together. Flat plates separate the two fluids, to which the finned surfaces are attached. The finned surfaces are generally made from folded and cut sheet and serve both as spacers separating adjacent plates, and as providers of channels in which the fluids may flow [Fig 1.2(a)].

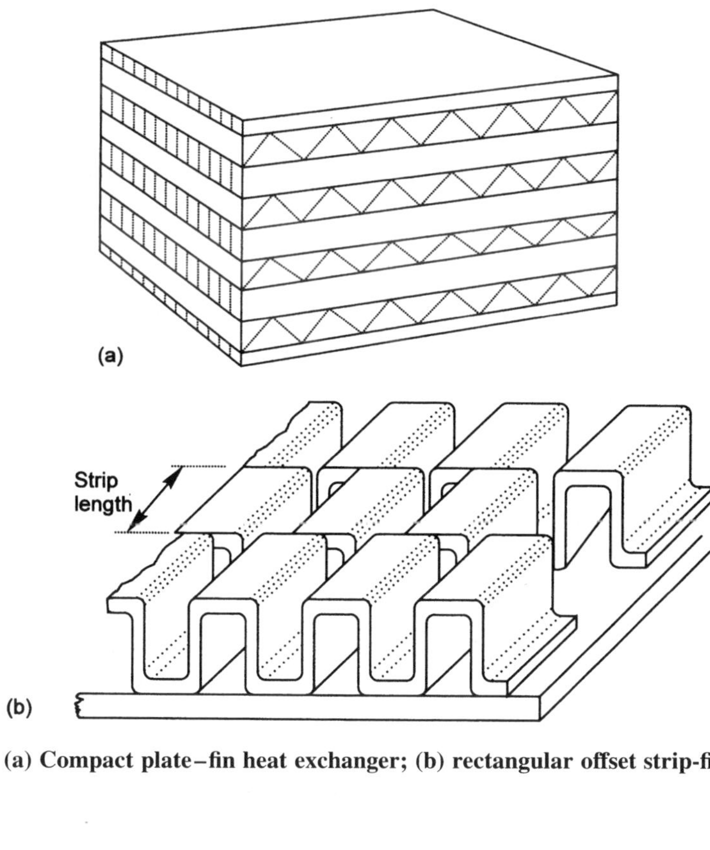

Fig.1.2 (a) Compact plate–fin heat exchanger; (b) rectangular offset strip-fin surface

Many types of finned surface have been tested, see e.g. Kays & London (1964) and Fig. 1.2(b) shows an example of a rectangular offset strip-fin surface which is one of the best-performing geometries. The objective is to obtain high heat-transfer coefficients without correspondingly increased pressure-loss penalties. As the strip-fins act as flat plates in the flowing fluid, each new edge starts a new boundary layer which is very thin, thus high heat-transfer coefficients are obtained.

1.5 RODbaffle

The RODbaffle exchanger is essentially a shell-and-tube exchanger with conventional plate-baffles (segmental or disc-and-doughnut) replaced by grids of rods. Unlike plate-baffles, RODbaffle sections extend over the full transverse cross-section of the exchanger.

Originally the design was produced to eliminate tube failure due to transverse vortex-shedding-induced vibration of unsupported tubes in crossflow (Eilers & Small, 1973), but the new configuration also provided enhanced performance and has been developed further by Gentry (1990) and others.

Square pitching of the tube bundle is considered the most practicable with RODbaffles, and circular rods are placed between alternate tubes to maintain spacing. To

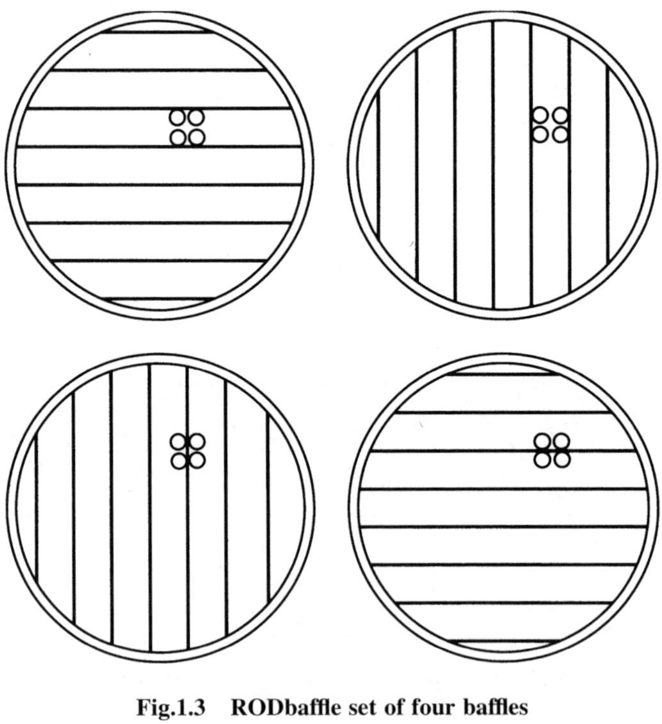

Fig.1.3 RODbaffle set of four baffles

minimize blockage, one set of vertical rods in a baffle section is placed between every second row of tubes. At the next baffle section the vertical rods are placed in the alternate gaps between tubes not previously filled at the first baffle section. The next two baffle sections have horizontal rod spacers, similarly arranged. Thus each tube in the bank receives support along its length.

It might be argued that the RODbaffle geometry is not completely consistent throughout its shell-side, and that it should not therefore be included in this study. However, the spacing rods in the shell-side fluid were found to be shedding von Karman vortex streets *longitudinally* which persist up to the next baffle rod. Thus as far as the shell-side fluid is concerned there is consistent geometry in the exchanger even though the RODbaffles themselves are placed 150 mm apart.

Tube counts are possible for square pitching using the Phadke (1984) approach. Figure 1.3 illustrates arrangement of baffles in the RODbaffle design.

1.6 Helically twisted flattened tube

This compact shell-and-tube design was developed by Dzyubenko *et al.* (1990) for aerospace use, and it complies with the requirement of consistent local geometry in every respect when triangular pitching is used. The outside of the tube bundle requires a shield to ensure correct shell-side flow geometry, and the space between the exchanger pressure shell and the shield can be filled with internal insulating material. The design is illustrated in Fig. 1.4, and the performance of this design is discussed thoroughly in the recent textbook by Dzyubenko *et al.* (1990), although the title of the book is somewhat misleading. Tube counts on triangular pitching are possible using the Phadke (1984) approach.

1.7 Spirally wire-wrapped

A further shell-and-tube concept is based on providing spiral wire-wraps to plain tubes – a concept used with nuclear fuel rods. With triangular pitching it is possible

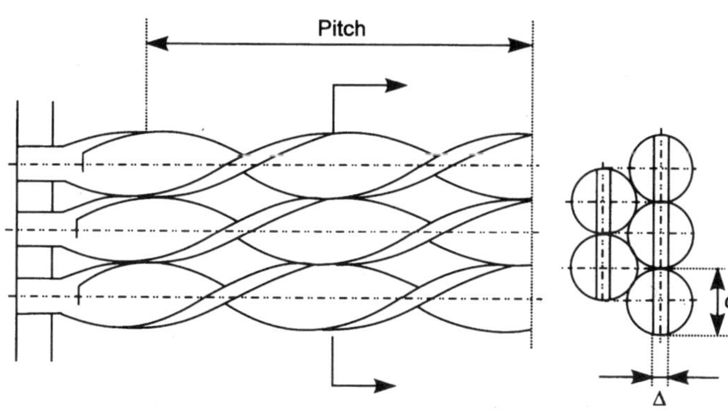

Fig.1.4 Helically twisted flattened tube

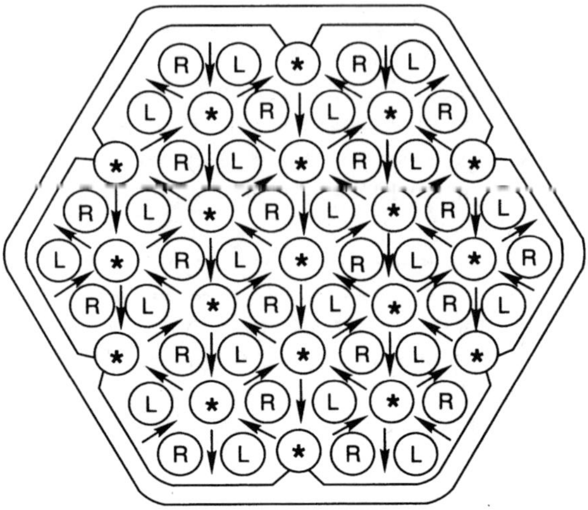

Fig.1.5 Cross-section of R–0–L spirally wire-wrapped layout

to arrange a mixture of right-hand (R), plain (0) and left-hand (L) wire-wraps so as to reinforce mixing in the shell-side fluid. This concept has not been tested for heat exchangers, and it does not quite fulfil the requirements of consistent local geometry, as the plain tubes lack the finning effect of the wire-wrap. The cross-section of a tube bundle is shown in Fig. 1.5, and the wire-wraps extend for about the central 90 per cent of the tube length.

The most common spiral wire-wrap configuration is to have *all* nuclear fuel rods with the same-handed spiral. This leads to opposing streams at the point of closest approach of rods, and swirling in the truncated triangular cusped flow principal flow channels. The spiral wrap is slow, being of the order 12–18° to the longitudinal axis of the rod. In several nuclear fuel rod geometries the arrangement of rods does not follow a regular triangular pattern, and correlations need to be assessed accordingly.

The R–0–L configuration provides even shell-side fluid distribution and mixing. The Phadke (1984) tube-count method will apply to triangular pitching.

1.8 Bayonet tube

Both bayonet-tube and double-pipe heat exchangers satisfy the concept of consistent shell-side and tube-side geometry, both have been discussed in other works, e.g. Martin (1992). Hurd (1946) appears to be the first to have analysed the performance of the bayonet-tube heat exchanger, but his analysis was not complete and further results are reported in the present text. The upper diagram in Fig. 1.6 show a typical exchanger. Practical uses include heating of batch processing tanks, sometimes with vertical bayonet tubes with condensation of steam in the annuli

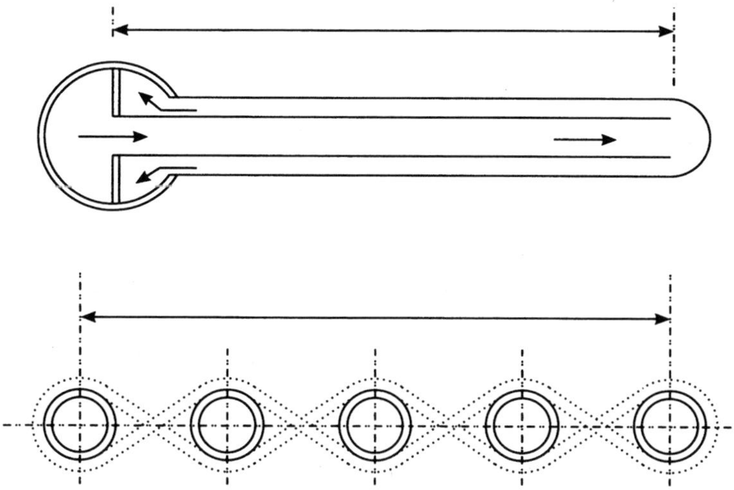

Fig.1.6 Bayonet-tube exchanger (upper diagram). Wire-woven tubes (lower diagram)

(Holger, 1992), freezing of ground, and cooling of cryogenic storage tanks, and high-temperature recuperators using silicon carbide tubes.

Residence time of the fluid in the annulus may be extended by adding a spiral wire-wrap to the outside of the inner tube, thus forcing fluid in the annulus to follow a helical path. When this is combined with insulating the inner tubes, improved external heat transfer will result.

1.9 Wire-woven heat exchangers

The concept of fine tubes woven with wire threads into a flat sheet is a recent proposal by Echigo *et al.* (1992). Given the right layout this arrangement could easily qualify for direct-sizing. The lower diagram in Fig. 1.6 shows the arrangement.

1.10 Porous matrix heat exchangers

The surface of the porous matrix heat exchanger described by Hesselgreaves (1995, 1997) is built up from flattened sections of perforated plate, or flattened expanded mesh metal, stacked so that each section is offset half a pitch from its immediate neighbours (Fig. 1.7). The fluid flows in and out of the plane of the fins in its passage through the exchanger, coupled with diverging and converging flow, thus creating a three-dimensional flow field in the matrix. Individual plate thicknesses are much thinner than with conventional plate–fin geometries, presently ranging from 0.137 to 0.38 mm.

The new geometry offers an increased number of 'flat plate' edges to the flow stream, plus greater cross-sectional area for heat to flow towards the channel

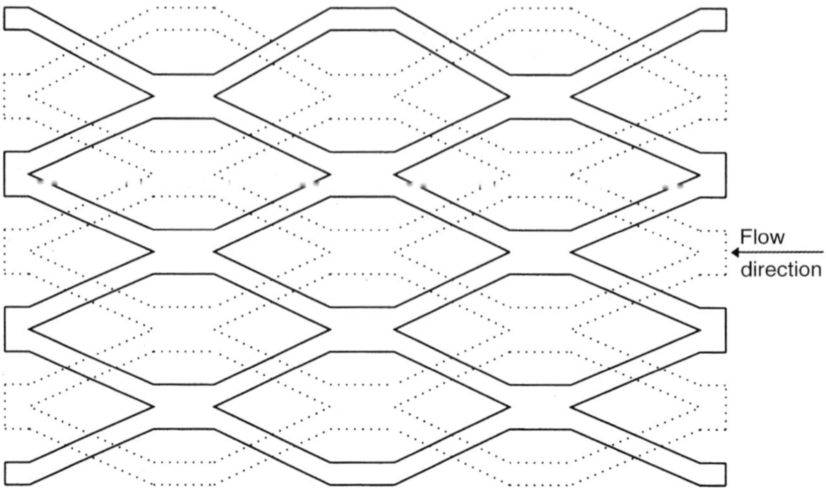

Fig.1.7 Stacked plates of porous matrix heat exchanger

separating plates. The layout is thus better configured for heat transfer than conventional plate–fin geometries. An infinite number of geometries are possible, with the possibility of changing mesh size along the length of the exchanger. Presently only preliminary test results are available, but there is every indication that the pressure loss will be lower, and the heat transfer higher, than for plate–fin designs.

So far, the flattened expanded mesh plates have been diffusion bonded together in packs of from 6 to 15 layers, with separating plates between streams, to form a very strong exchanger. As such a construction seems amenable to forming plate-packs with involute curvature, as illustrated in Fig. 1.8, the prospect of constructing a completely bonded two-pass annular flow exchanger exists. This arrangement could prove suitable for the vehicular gas-turbine application shown in Fig. 1.9.

Sufficient examples of exchangers with a recognizable 'local geometry' have now been given to allow the reader to recognize new types of exchanger which conform to requirements for 'direct-sizing'.

1.11 Some possible applications

At this stage it is only possible to indicate some applications for the heat exchanger configurations described earlier. Not all of this technology is yet in service, or indeed constructed, and the reader is simply asked to appreciate some possible applications which exist for the new direct-sizing designs described.

Propulsion systems

Intercooled and recuperated gas turbine cycles for marine propulsion are presently being developed for the considerable fuel savings that are possible (Cownie, 1993;

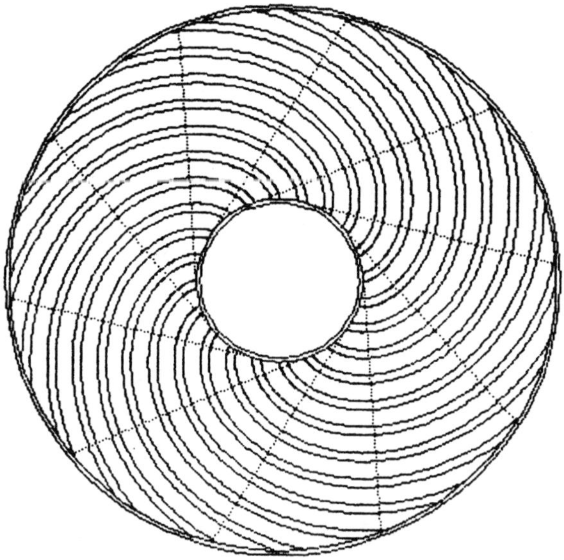

Fig.1.8 Cross-section of involute-curved plate–fin heat exchanger

Crisalli & Parker, 1993). In every case a contraflow heat exchanger arrangement is the natural first choice as it provides more energy recovery than multi-pass cross-flow, but the practicalities of inlet and outlet ducting have also to be considered.

Small recuperators

A recuperator of two-pass involute design has been developed for military tank propulsion (Ward & Holman, 1992), (Fig. 1.8). In the compact vehicle propulsion system shown in Fig. 1.9 (after Collinge, 1994), the power turbine exhaust flows outwards through the exchanger core while high-pressure combustion air flows axially through the exchanger in two passes. Swirling exhaust gases can be directed by an outlet scroll before entering the exhaust stack. Some development work would be required to realize the involute-curved plate–fin exchanger. Thermal sizing is identical to that for the compact flat-plate design. Plate spacing on the high-pressure cold air side is narrow while the spacing on the low-pressure hot gas side is wide.

For a single-pass contraflow design some thought would be required in the arrangement of headers. A further problem with the involute exchanger is the difficulty of cleaning curved channels. Wilson (1995) believes that a rotating ceramic regenerator should be preferred, as it could be more easily cleaned, but it introduces the problem of sliding seals.

Large recuperators

For the recuperator of a larger gas turbine a plate-and-frame design with U-type headering was developed for marine propulsion (Valenti, 1995). Crisalli & Parker

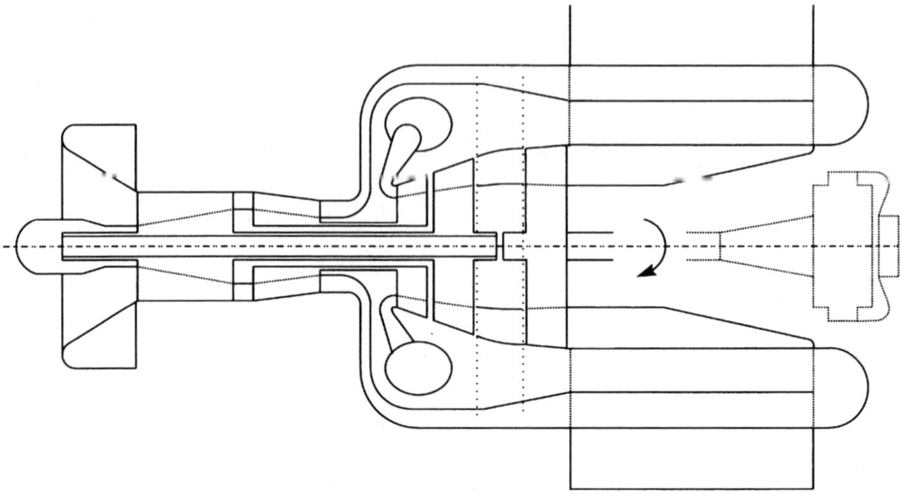

Fig.1.9 Schematic arrangement of two-spool power gas turbine with two-pass crossflow exhaust exchanger

(1993) and Shepard *et al.* (1995) report on development of the WR-21 intercooled and recuperated marine gas turbine. In the latest Rolls-Royce design of recuperator, compact plate–fin-type surfaces were used.

Intercoolers

An intercooler can also be fitted between low-pressure and high-pressure compressors of large marine and land-based systems (Crisalli & Parker, 1993; Bannister *et al.*, 1994). This has to be a contraflow plate–fin design for compactness, with fresh-water/glycol supply and return from external annular header pipes. The exchanger can be segmented for ease of maintenance. With this arrangement the gas turbine would not become exposed to sea-water leaking from a damaged intercooler. Pressure in the closed-loop fresh-water/glycol system can be adjusted to suit the desired operating conditions.

Liquid hydrogen propulsion

Related to both the helical-tube, multi-start coil heat exchanger and the involute-curved plate–fin exchanger is the involute-curved, serpentine-tube panel design used by Pratt & Whitney in one of their experimental engines powered by liquid hydrogen (Mulready, 1979). The tube panel is a single serpentine tube arranged such that the shell-side fluid may flow transversely over the tube as in crossflow over rows of tubes, Fig. 1.10(a).

Each panel is given an involute curve and is placed together with others in an annular pattern as in Fig. 1.10(b), the pitch between adjacent panels is constant with radius and the shell-side fluid sees the same geometry everywhere. Thermal

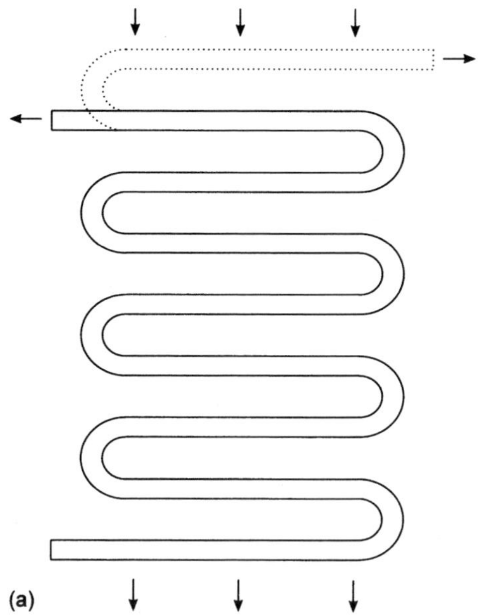

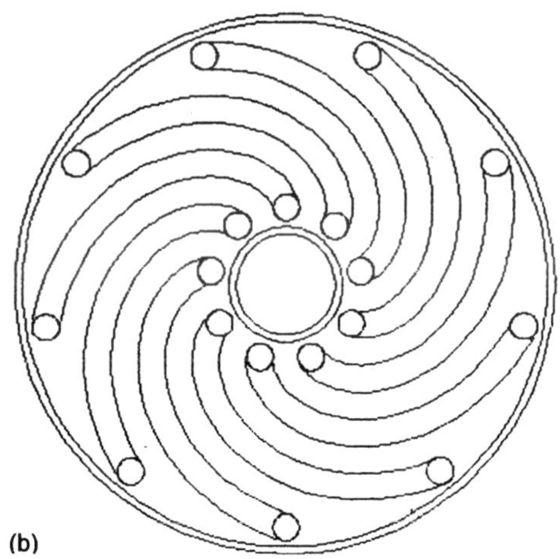

Fig.1.10 (a) Serpentine tube panel; (b) involute pattern

performance of flat serpentine panels has been discussed by Hausen (1950), and will apply to involute-curved serpentine panels also. The only difference is a secondary effect due to involute curvature, which may affect tube-side heat transfer and pressure loss hardly at all. The incentive to go to this design must be high, to make pressure loss in repeated bends acceptable.

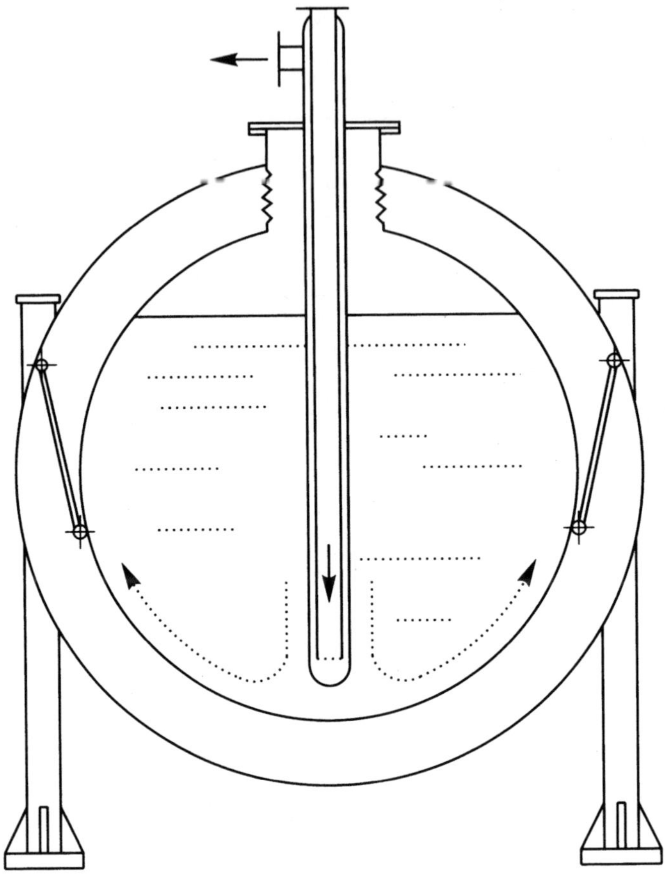

Fig.1.11 Cryogenic storage tank with bayonet-tube exchanger (schematic)

Cryogenic plant

Cryogenic heat exchangers

The sizing of cryogenic heat exchangers is discussed in Chapter 11. A step-wise rating approach is required, especially when near to the critical point, and once that technique has been developed it is straightforward to apply the same methods to cryogenic designs in which the LMTD concept would otherwise be perfectly viable.

Cryogenic storage tank

One problem that has troubled cryogenic and petrochemical industries is that of 'roll-over' in cryogenic storage tanks. The liquid cryogen is at a higher pressure at the bottom of the tank compared with the surface, due to liquid density. The saturation pressure is thus higher at the bottom of the tank. If there is external heat leak into the tank then it is possible for the liquid at the bottom of the tank to be at a temperature higher than that at the top. If conditions are such that this liquid travels to the top of the tank by convection then the massive evaporation which ensues at the lower pressure can be sufficient to rupture the tank.

Presently a mixing propeller on a shaft is used to circulate liquid within the tank to keep the contents close to isothermal.

The bayonet-tube heat exchanger is a design that requires only single penetration of a pressure vessel. If such an exchanger is fitted to the top of a cryogenic storage tank and a cryogen used in this exchanger to cool the contents of the tank, then controlled circulation may be set up in the tank with colder fluid at the centre falling to the bottom and warmer fluid at the side walls rising to the free surface (Fig. 1.11.) An extension of this would be to provide the mixing propeller with a hollow shaft incorporating a bayonet-tube exchanger. The possible effectiveness of the bayonet-tube exchanger in inhibiting 'roll-over' seems worthy of investigation.

It is hoped that the above long-range concepts may stimulate the reader to consider other arrangements for heat exchangers that can be directly sized.

References

Abadzic, E.E. and **Scholz, H.W.** (1972) Coiled tubular heat exchangers. In *Proceedings of the 1972 Cryogenic Engineering Conference, Advances in Cryogenic Engineering*, Plenum Press, Paper B-1, vol. 18, pp. 42–51.

Anon. (1979) The high temperature reactor and process applications. In *Proceedings of an International Conference*, London, 26–28 November 1979, British Nuclear Energy Society. See also IAEA Specialist meeting, Julich 1979, in *Nuclear Engng Int.*, March 1980, 24.

Bachmann, U. (1975) Steam generators for the 300-MWe power station with a thorium high-temperature reactor, *Sulzer Tech. Rev.*, **57**(4), 189–194.

Bannister, R.L., Cheruvu, N.S., Little, D.A., and **McQuiggan, G.** (1994) Turbines for the turn of the century. *Mech. Engng*, **116**(6), June, 68–75.

Bes, Th. and **Roetzel, W.** (1991) Approximate theory of the spiral heat exchanger. *Design and Operation of Heat Exchangers* (Eds. W. Roetzel, P.J. Heggs, and D. Butterworth), *Proceedings of the EUROTHERM Seminar no. 18*, Hamburg, 27 February–1 March 1991, Springer Verlag, Berlin, pp. 223–232.

Bes, Th. and **Roetzel, W.** (1992) Distribution of heat flux density in spiral heat exchangers. *Int. J. Heat Mass Transfer*, **35**(6), 1331–1347.

Bes, Th. and **Roetzel, W.** (1993) Thermal theory of the spiral heat exchanger. *Int. J. Heat Mass Transfer*, **36**(3), 765–773.

Bourguet, J.M. (1972) Cryogenics technology and scaleup problems of very large LNG plants. In *Proceedings of the 1972 Cryogenic Engineering Conference, Advances in Cryogenic Engineering*, vol. 18, Plenum Press, Paper A-2, pp. 9–26.

Chen, Y.N. (1978) General behaviour of flow induced vibrations in helical tube bundle heat exchangers. *Sulzer Tech. Rev.*, Special Number 'NUCLEX 78', pp. 59–68.

Collinge, K. (1994) Lycoming AGT1500 powers the M1 Abrams. *IGTI Global Gas Turbine News*, ASME, May, 4–7.

Cownie, J. (1993) Aerospace 90 years on. *Prof. Engng*, **6**(11), December, 17–19.

Crisalli, A.J. and **Parker, M.L.** (1993) Overview of the WR-21 intercooled recuperated gas turbine engine system. A modern engine for a modern fleet. In *International Gas Turbine and Aeroengine Congress and Exposition*, Cincinnati, Ohio, 24–27 May 1993, ASME Paper 93-GT-231.

Das, S.K. and **Roetzel, W.** (1995) Dynamic analysis of plate heat exchangers with dispersion in both fluids (plate-and-frame exchanger). *Int. J. Heat Mass Transfer*, **38**(6), 1127–1140.

Dow, M.W. (1950) The uniform distribution of a fluid flowing through a perforated pipe. *Trans. ASME J. Appl. Mech.*, December, 431–438.

Dzyubenko, B.V., **Dreitser, G.A.**, and **Ashmantas, L.-V.A.** (1990) *Unsteady Heat and Mass Transfer in Helical Tube Bundles*, Hemisphere, New York.

Echigo, R., **Yoshida, H.**, **Hanamura, K.**, and **Mori, H.** (1992) Fine-tube heat exchanger woven with threads. *Int. J. Heat Mass Transfer*, **35**(3), 711–717.

Eilers, J.F. and **Small, W.M.** (1973) Tube vibration in a thermosiphon reboiler. *Chem. Engng Prog.*, **69**(7), 57–61.

Focke, W.W. (1985) Asymmetrically corrugated plate heat exchanger plates (plate-and-frame exchanger). *Int. Comm. Heat Mass Transfer*, **12**(1), 67–77.

Gentry, C.C. (1990) RODbaffle heat exchanger technology. *Chem. Engng Prog.*, **86**(7), July, 48–57.

Gill, G.M., **Harrison, G.S.**, and **Walker, M.A.** (1983) Full scale modelling of a helical boiler tube. In *International Conference on Physical Modelling of Multi-Phase Flow*, BHRA Fluid Engineering Conference, April 1983, Paper K4, pp. 481–500.

Gilli, P.V. (1965) Heat transfer and pressure drop for crossflow through banks of multistart helical tubes with uniform inclinations and uniform longitudinal pitches. *Nuclear Sci. Engng*, **22**, 298–314.

Hampson, W. (1895) Improvements relating to the progressive refrigeration of gases. British Patent 10165.

Hausen, H. (1950) Kreuzstrom in Verbindung mit Parallelstrom im Kreuzgegenstromer. *Warmeubertragung im Gegenstrom, Gleichstrom und Kreuzstrom*, Technische Physik no. 8, 1st (German) edn, Springer, Berlin, pp. 213–228.

Hausen, H. (1983) Crossflow combined with parallel flow in the cross-counterflow heat exchanger. *Heat Transfer in Counter Flow, Parallel Flow and Cross Flow*, 2nd (English) edn, McGraw-Hill, New York, pp. 232–248.

Hesselgreaves, J. (1995) Concept proving of a novel compact heat exchanger surface. In *Proceedings of the 4th National Conference on Heat Transfer*, Manchester, 26–27 September 1995, IMechE Paper C5109/082/95, pp. 479–486.

Hesselgreaves, J.F. (1997) Single phase and boiling performance of a novel highly compact heat exchanger surface. In *5th UK National Conference on Heat Transfer*, Imperial College of Science & Technology, London, 17–18 September 1997, Institution of Chemical Engineers.

Hesselgreaves, J.F. (2001) *Compact Heat Exchangers*, Pergamon.

Hewitt, G.F., **Shires, G.L.**, and **Bott, T.R.** (1994) *Process Heat Transfer*, CRC Press, Florida.

Holger, M. (1992) *Heat Exchangers*, Hemisphere, New York.

Hurd, N.L. (1946) Mean temperature difference in the field or bayonet tube. *Ind. Engng Chem.*, **38**(12), December, 1266–1271.

Kalin, W. (1969) The steam generators of Fort St. Vrain nuclear power plant, USA. *Sulzer Tech. Rev.*, **52**(3), 167–174, Special number 'NUCLEX 1969', 12–34.

Kays, W.M. and **London, A.L.** (1964) *Compact Heat Exchangers*, 2nd edn (3rd edn exists), McGraw-Hill, New York.

L'Air Liquide (1934) Improvements relating to the progressive refrigeration of gases. British Patent 416,096.

London, A.L. and **Shah, R.K.** (1968) Offset rectangular plate-fin surfaces – heat transfer and flow-friction characteristics. *ASME J. Engng Power*, **90**, 218–228.

Martin, H. (1992) *Heat Exchangers*, Hemisphere, New York.

Mulready, R.C. (1979) Liquid hydrogen engines. In *DFVLR International Symposium: Hydrogen in Air Transportation*, Stuttgart, 11–14 September 1979.

Oswald, J.I., **Dawson, D.A.**, and **Clawley, L.A.** (1999) A new durable gas turbine recuperator. In *International Gas Turbine and Aeroengine Congress and Exhibition*, Indianapolis, Indiana, 7–10 June 1999, ASME Paper 99-GT-369.

Perrin, A.J. (1976) Hartlepool and Heysham pod boilers. *Nuclear Engng Int.*, **21**(239), February, 48–51.

Phadke, P.S. (1984) Determining tube counts for shell-and-tube exchangers. *Chem. Engng*, September, 65–68.

Profos, O. (1970) The new steam generating system for the French nuclear power station EL-4. *Sulzer Tech. Rev.*, **53**(2), 69–83.

Shah, R.K. and **Sekulic, D.P.** (2003) *Fundamentals of Heat Exchanger Design*, John Wiley.

Shepard, S.B., **Bowen, T.L.**, and **Chiprich, J.M.** (1995) Design and development of the WR-21 intercooled recuperated (ICR) marine gas turbine. *ASME J. Engng Gas Turbines Power*, **117**, July, 557–562.

Smith, E.M. (1960) The geometry of multi-start helical coil heat exchangers. Unpublished report.

Smith, E.M. (1986) Design of helical-tube multi-start coil heat exchangers. In *Advances in Heat Exchanger Design, ASME Winter Annual Meeting*, Anaheim, California, 7–12 December 1986, ASME Publication HTD-Vol. 66, pp. 95–104.

Smith, E.M. and **Coombs, B.P.** (1972) Thermal performance of cross-inclined tube bundles measured by a transient technique. *J. Mech. Engng Sci.*, **14**(3), 205–220.

Smith, E.M. and **King, J.L.** (1978) Thermal performance of further cross-inclined in-line and staggered tube banks. In *6th International Heat Transfer Conference*, Toronto, Paper HX-14, pp. 276–272.

Ulken, D. (1971) N.S. Otto Hahn. *Trans Inst. Marine Engs*, **83**, 65–83.

Valenti, M. (1995) A turbine for tomorrow's Navy (recuperated WR-21). *Mech. Engng*, **117**(9), September, 70–73.

Ward, M.E. and **Holman, L.** (1992) Primary surface recuperator for high performance prime movers. *SAE International Congress and Exposition*, Detroit, Michigan, 24–28 February 1994, SAE Paper 920150.

Weimer, R.F. and **Hartzog, D.G.** (1972) Effects of maldistribution on the performance of multistream multi-passage heat exchangers. *Proceedings of the 1972 Cryogenic Engineering Conference, Advances in Cryogenic Engineering*, vol. 18, Plenum Press, New York, Paper B-2, pp. 52–64.

Wilson, D.G. (1995) Automotive gas turbines: government funding and the way ahead. *IGTI Global Gas Turbine News*, **35**(4), 17–21.

Bibliography

Adderley, C. and **Hallgren, L.H.** (1994) The design and manufacture of diffusion bonded plate-fin heat exchangers. In *The Industrial Sessions Papers, 10th International Heat Transfer Conference*, Brighton, 14–18 August 1994, Institution of Chemical Engineers.

Devois, J.F. and **Duranstanti, J.F.** (1993) Modelling of spiral plate heat exchangers by a finite-difference method. In *Conference: New Developments in Heat Exchangers*, Lisbon, Portugal, (Eds. N. Afgan, M.G. Caravello, A Bar-Cohen, D. Butterworth, and W. Roetzel), Gordon & Breach, pp. 317–386.

McDonald, C.F. (1994) Heat exchanger ubiquity in advanced gas turbine cycles. In *ASME Cogen Turbo Power, 8th Congress and Exposition on Gas Turbines in Cogeneration and Utility, Industrial and Independent Power Generation*, ASME/IGTI Portland, Oregon, 25–27 October 1994, vol. 9, pp. 681–703.

McDonald, C.F. (1996) Compact buffer zone plate-fin IHX – the key component for high-temperature nuclear process heat realisation with advanced MHR. *Appl. Thermal Engng*, **16**(1), 3–32.

McDonald, C.F. and **Wilson, D.G.** (1996) The utilisation of recuperated and regenerated engine cycles for high efficiency gas turbines in the 21st century. *Appl. Thermal Engng*, **16**(8/9), 635–653.

Ogorkiewicz, R.M. (1991) *Technology of Tanks*, 2 vols, Chapter 11, Section 11.9, Jane's Information Group, Gas Turbines, pp. 258–262.

CHAPTER 2
Fundamentals

Insights

2.1 Simple temperature distributions

In considering overall heat-transfer coefficients, an elementary length of surface is taken across which the temperature difference is assumed to remain constant. In general, temperatures of the fluids change over the length of the surface, and a principal method of classifying heat exchangers is according to the directions of fluid flow on each 'side' and the effect upon the temperatures in the system.

The first design problem is to determine the mean temperature difference for heat transfer between the two fluids, so as to be able to use the design equation

$$\dot{Q} = US\Delta\theta_m \tag{2.1}$$

where $\Delta\theta_m$ is the mean temperature difference between fluids.

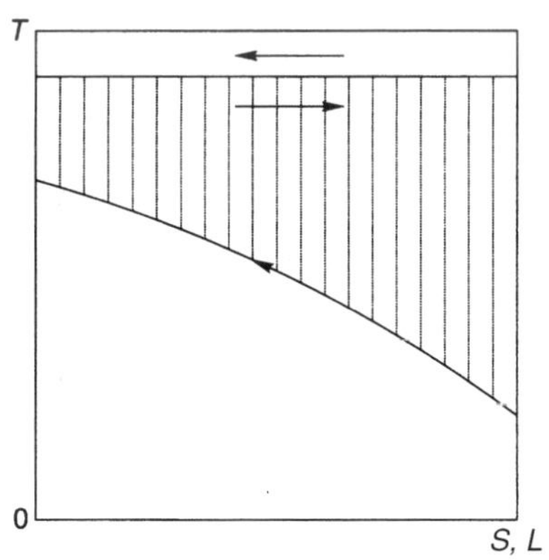

Fig.2.1 Temperature profiles. Condenser: Fluid at constant temperature gives up heat to a colder fluid whose temperature increases

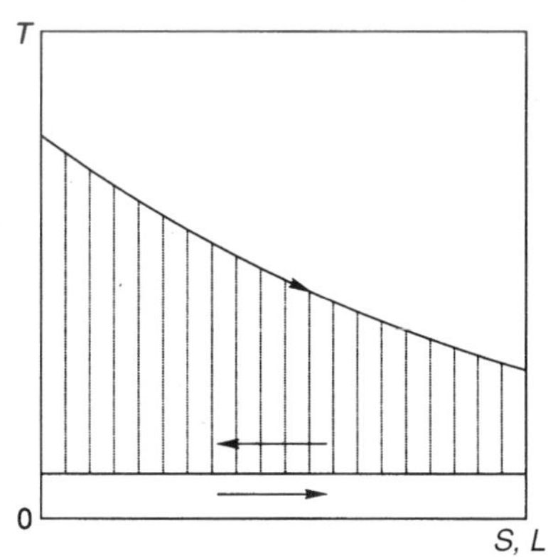

Fig.2.2 Temperature profiles. Evaporator: Fluid at constant temperature receives heat from a hotter fluid whose temperature decreases

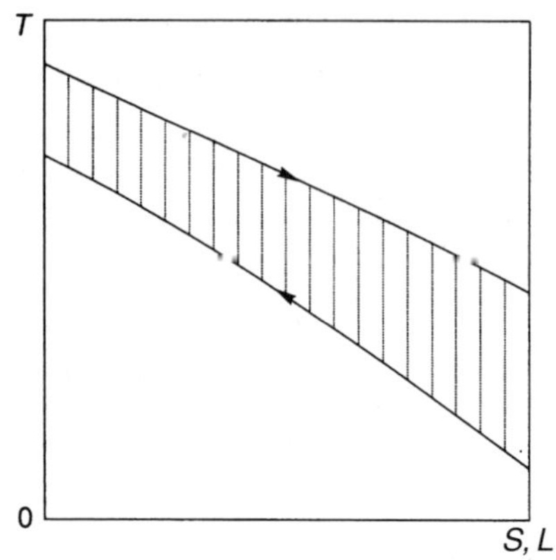

Fig.2.3 Temperature profiles. Contra-flow: Fluids flow in opposite directions. More efficient use of temperature difference allows colder fluid to exit at higher temperature than the hotter fluid

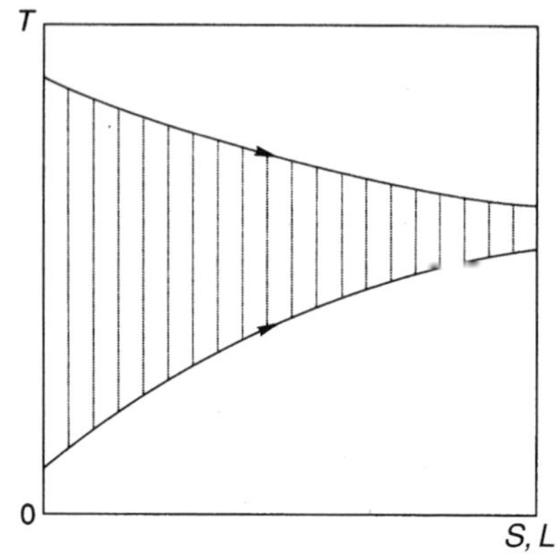

Fig.2.4 Temperature profiles. Parallel flow: Both fluids flow in the same direction, one increasing in temperature, the other decreasing in temperature

It is essential to watch out for *phase changes*, or situations in which *enthalpy changes* of single-phase fluids do not change linearly with temperature (as in some cryogenic heat exchangers). In the desuperheating *feed-heater* below, the 'internal' terminal temperatures must first be found, and separate values of LMTD calculated for each section. Temperature cross-over is a fictitious temperature profile situation which can be assumed unconsciously if 'external' terminal temperatures are used to produce a single LMTD for the design of the whole exchanger.

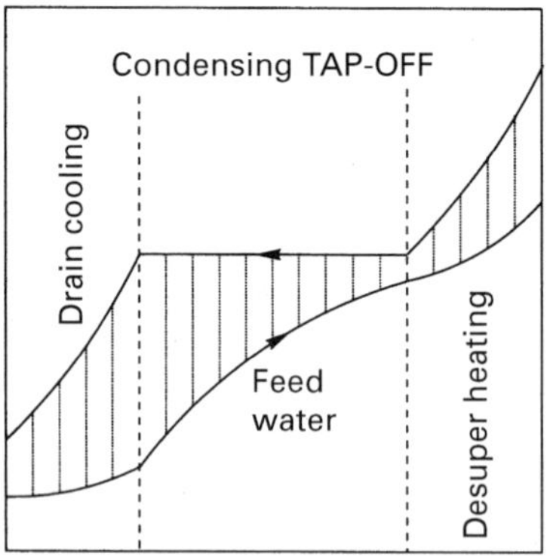

Fig.2.5 Desuperheating feed-heater with phase changes

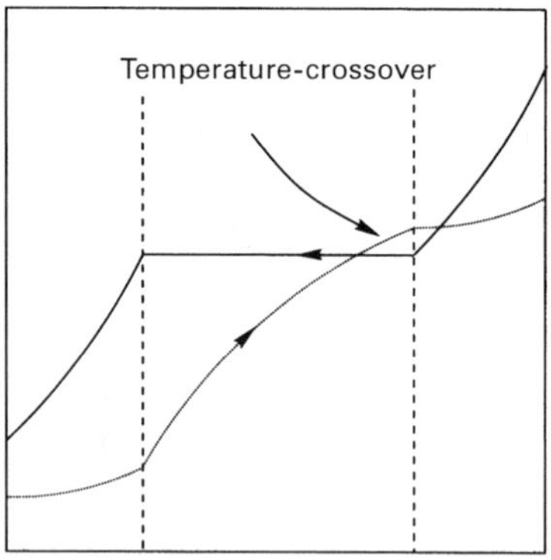

Fig.2.6 Temperature cross-over in feed-heater

Similarly with *cryogenic recuperators*, in which one fluid at a pressure just above its critical value is being cooled around its critical temperature, against the flow of a colder fluid at lower pressure well away from its critical point.

The contraflow feed-heater may be properly designed as three separate exchangers. The cryogenic exchanger may require incremental design along its length. For the simplest exchangers the problem reduces either to '*rating*' an existing design, or '*sizing*' a new exchanger.

Rating	*Sizing*
Given: *geometry* $\dot{m}_h$, C_h, T_{h1}, Δp_h $\dot{m}_c$, C_c, T_{c2}, Δp_c	Given: $\dot{Q}$ *(duty)* $\dot{m}_h$, C_h, T_{h1}, Δp_h $\dot{m}_c$, C_c, T_{c2}, Δp_c
Find: $\dot{Q}$ *(duty)*	Find: *geometry*

Note: The allowable pressure losses (Δp_h, Δp_c) have a role to play in design which will be explored in later chapters.

Two design approaches which can be used are the LMTD–*Ntu* method and the ε–*Ntu* method. Notation is explained in the sections which follow.

2.2 Log mean temperature difference

Consider parallel and contraflow heat exchangers assuming:

- fluids with steady mass flowrates ($\dot{m}_h$, $\dot{m}_c$)
- constant overall heat-transfer coefficient (U)
- constant specific heats (C_h, C_c)
- negligible heat loss to surroundings

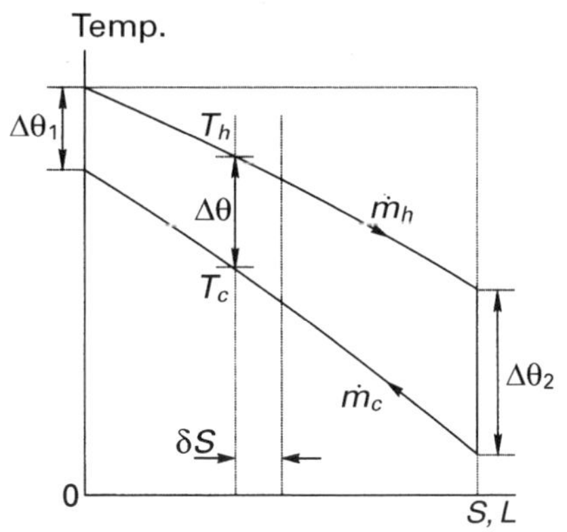

Fig.2.7 Contraflow

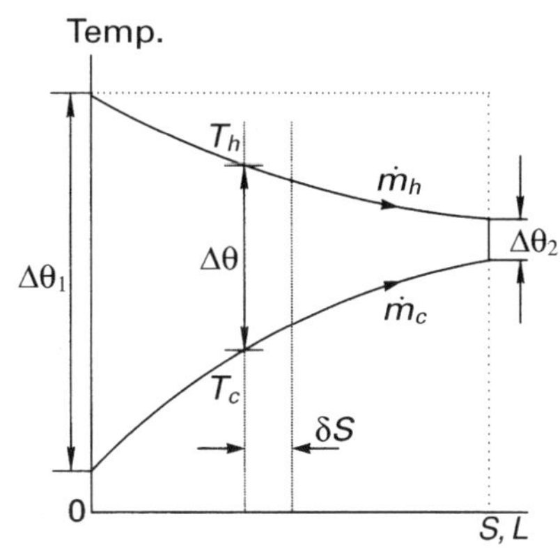

Fig.2.8 Parallel flow

For an element of area δS, then

$$\delta\dot{Q} = U\delta S\Delta\theta_m \tag{2.2}$$

Also, considering each fluid

$$\delta\dot{Q} = \dot{m}_h C_h(-\delta T_h) = \dot{m}_c C_c(\pm\delta T_c) \tag{2.3}$$

Now the increment in temperature difference $\Delta\theta$ is

$$\begin{aligned}\delta(\Delta\theta) &= \delta(T_h - T_c) = (\delta T_h - \delta T_c)\\ &= \left[\frac{-1}{\dot{m}_h C_h} - \frac{\pm 1}{\dot{m}_c C_c}\right]\delta Q = \left[\frac{-1}{\dot{m}_h C_h} - \frac{\pm 1}{\dot{m}_c C_c}\right]U\delta S\Delta\theta\end{aligned} \tag{2.4}$$

Thus

$$\frac{\delta(\Delta\theta)}{\Delta\theta} = \left[\frac{-1}{\dot{m}_h C_h} - \frac{\pm 1}{\dot{m}_c C_c}\right]U\delta S$$

Integrating between stations 1 and 2

$$\ln\frac{\Delta\theta_2}{\Delta\theta_1} = \left[\frac{-1}{\dot{m}_h C_h} - \frac{\pm 1}{\dot{m}_c C_c}\right]US \tag{2.5}$$

also integrating equation (2.4) between stations 1 and 2

$$\Delta\theta_2 - \Delta\theta_1 = \left[\frac{-1}{\dot{m}_h C_h} - \frac{\pm 1}{\dot{m}_c C_c}\right]Q \tag{2.6}$$

Eliminating the square bracket term between equations (2.5) and (2.6)

$$\dot{Q} = US\frac{(\Delta\theta_2 - \Delta\theta_1)}{\ln(\Delta\theta_2/\Delta\theta_1)}$$

and comparing this expression with equation (2.1), $\Delta\theta_m$ can be written as the *logarithmic mean temperature difference*, $\Delta\theta_{lmtd}$

$$\Delta\theta_{lmtd} = \frac{(\Delta\theta_2 - \Delta\theta_1)}{\ln(\Delta\theta_2/\Delta\theta_1)} \tag{2.7}$$

In real exchangers the mean temperature difference $\Delta\theta_m$ may not necessarily equal the mathematical LMTD expression based on four terminal temperatures. This is true, for example, when specific heats vary along the length of an exchanger, then mean temperature difference as defined below can always be employed

$$\Delta\theta_m = \left(\sum_1^n \Delta\theta\right)/n$$

where n is the number of stations along the length of the exchanger at which temperature differences $\Delta\theta$ are known.

Equation (2.7) is correct also for condensers and evaporators. In the special case of contraflow with equal water equivalents ($\dot{m}_h C_h = \dot{m}_c C_c$), equation (2.7) is indeterminate because $\Delta\theta_1 = \Delta\theta_2$, as temperature profiles are parallel straight lines.

$$\Delta\theta_{lmtd} = \Delta\theta_1 = \Delta\theta_2 \tag{2.8}$$

2.3 LMTD–*Ntu* rating problem

Contraflow (LMTD–Ntu) rating	*Parallel flow (LMTD–Ntu) rating*
Given: *geometry* $\dot{m}_h$, C_h, T_{h1} $\dot{m}_c$, C_c, T_{c2}	Given: *geometry* $\dot{m}_h$, C_h, T_{h1} $\dot{m}_c$, C_c, T_{c1}
Find: $\dot{Q}$ *(duty)*	Find: $\dot{Q}$ *(duty)*

$$\dot{Q} = US\Delta\theta_{lmtd} = \dot{m}_h C_h (T_{h1} - T_{h2}) = \dot{m}_c C_c (T_{c1} - T_{c2}) \tag{2.9}$$

$$\dot{Q} = US\Delta\theta_{lmtd} = \dot{m}_h C_h (T_{h1} - T_{h2}) = \dot{m}_c C_c (T_{c2} - T_{c1}) \tag{2.9}$$

Since the geometry is defined, the surface area S is known. Also hot-side, solid-wall, and cold-side heat-transfer coefficients α_h, α_c, and $\alpha_w = (\lambda_w/t_w)$ may be evaluated from heat-transfer correlations and physical properties, giving the overall

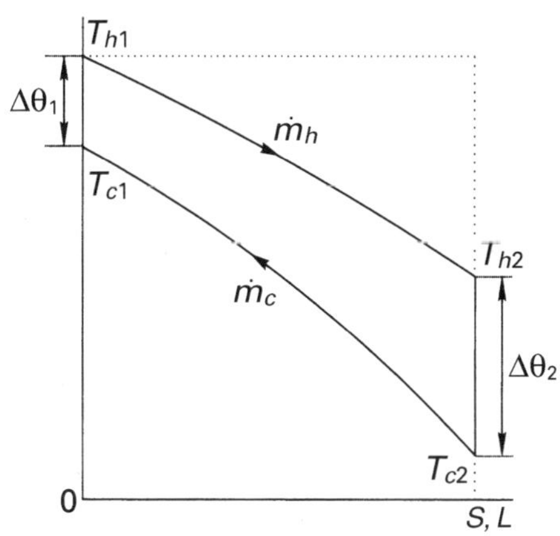

Fig.2.9 Contraflow profiles

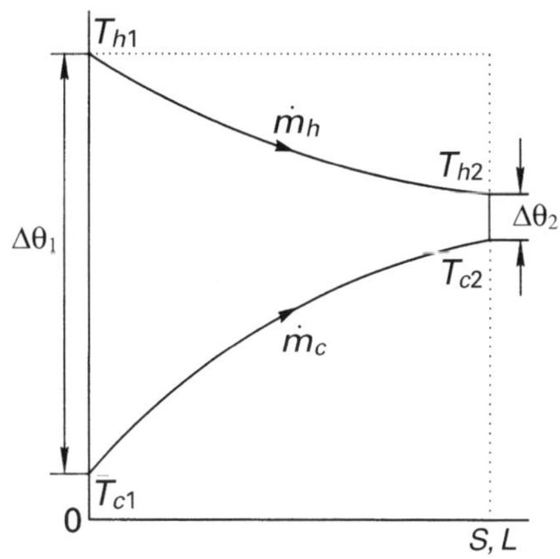

Fig.2.10 Parallel flow profiles

heat-transfer coefficient U.

$$\frac{1}{U} = \frac{1}{\alpha_h} + \frac{1}{\alpha_w} + \frac{1}{\alpha_c}$$

all coefficients being referred to the same reference surface area S.

For duty $\dot{Q}$, proceed as follows:

From equation (2.9)

$$\Delta\theta_{lmtd} = \frac{T_{h1} - T_{h2}}{N_h} = \frac{T_{c1} - T_{c2}}{N_c} \quad (2.10)$$

where

$$N_h = \frac{US}{\dot{m}_h C_h} \quad \text{and} \quad N_c = \frac{US}{\dot{m}_c C_c}$$

Then

$$T_{c1} = \frac{N_c}{N_h}(T_{h1} - T_{h2}) + T_{c2} \quad (2.11)$$

$$\Delta\theta_{lmtd} = \frac{\Delta\theta_1 - \Delta\theta_2}{\ln(\Delta\theta_1/\Delta\theta_2)} = \frac{[T_{h1} - T_{c1}] - [T_{h2} - T_{c2}]}{\ln(\Delta\theta_1 - \Delta\theta_2)}$$

The first bracket in the numerator may be written

$$\left[T_{h1} - \frac{N_c}{N_h}(T_{h1} - T_{h2}) - T_{c2}\right]$$

hence

$$\Delta\theta_{lmtd} = \frac{[T_{h1} - T_{h2}](1 - N_c/N_h)}{\ln(\Delta\theta_1/\Delta\theta_2)} \quad (2.12)$$

Equating equations (2.10) and (2.12)

$$\frac{[T_{h1} - T_{h2}]}{N_h} = \frac{[T_{h1} - T_{h2}](1 - N_c/N_h)}{\ln(\Delta\theta_1/\Delta\theta_2)}$$

$$\ln\left[\frac{\Delta\theta_1}{\Delta\theta_2}\right] = (N_h - N_c) \quad (2.13)$$

From equation (2.9)

$$\Delta\theta_{lmtd} = \frac{T_{h1} - T_{h2}}{N_h} = \frac{T_{c2} - T_{c1}}{N_c} \quad (2.10)$$

where

$$N_h = \frac{US}{\dot{m}_h C_h} \quad \text{and} \quad N_c = \frac{US}{\dot{m}_c C_c}$$

Then

$$T_{c2} = \frac{N_c}{N_h}(T_{h1} - T_{h2}) + T_{c1} \quad (2.11)$$

$$\Delta\theta_{lmtd} = \frac{\Delta\theta_1 - \Delta\theta_2}{\ln(\Delta\theta_1/\Delta\theta_2)} = \frac{[T_{h1} - T_{c1}] - [T_{h2} - T_{c2}]}{\ln(\Delta\theta_1/\Delta\theta_2)}$$

The second bracket in the numerator may be written

$$\left[T_{h2} - \frac{N_c}{N_h}(T_{h1} - T_{h2}) - T_{c1}\right]$$

hence

$$\Delta\theta_{lmtd} = \frac{[T_{h1} - T_{h2}](1 + N_c/N_h)}{\ln(\Delta\theta_1/\Delta\theta_2)} \quad (2.12)$$

Equating equations (2.10) and (2.12)

$$\frac{[T_{h1} - T_{h2}]}{N_h} = \frac{[T_{h1} - T_{h2}](1 + N_c/N_h)}{\ln(\Delta\theta_1/\Delta\theta_2)}$$

$$\ln\left[\frac{\Delta\theta_1}{\Delta\theta_2}\right] = (N_h + N_c) \quad (2.13)$$

From equations (2.10) and (2.13) the LMTD–*Ntu* equation pairs for contraflow and parallel flow are

Contraflow		Parallel flow
$\dfrac{\Delta\theta_1}{\Delta\theta_2} = \dfrac{T_{h1} - T_{c1}}{T_{h2} - T_{c2}} = \exp[N_h - N_c]$	'rate'	$\dfrac{\Delta\theta_1}{\Delta\theta_2} = \dfrac{T_{h1} - T_{c1}}{T_{h2} - T_{c2}} = \exp(N_h + N_c)$
$\dfrac{\Delta T_c}{\Delta T_h} = \dfrac{T_{c1} - T_{c2}}{T_{h1} - T_{h2}} = \dfrac{N_c}{N_h}$	'energy'	$\dfrac{\Delta T_c}{\Delta T_h} = \dfrac{T_{c2} - T_{c1}}{T_{h1} - T_{h2}} = \dfrac{N_c}{N_h}$

Given (N_h, T_{h1}) and (N_c, T_{c2}) for contraflow, or (N_h, T_{h1}) and (N_c, T_{c1}) for parallel flow, numerical values may be inserted in the simultaneous LMTD–*Ntu* equations to find the two unknown temperatures. [In a slightly different form, these equations were anticipated by Clayton (1984), whose paper was not seen until the first edition of this text was in its final editing stage.]

Explicit algebraic solutions for the unknowns in contraflow can be written

$$T_{h2} = \frac{(a-1)T_{h1} - (b-1)T_{c2}}{(a-1)-(b-1)} \qquad T_{c1} = \frac{a(b-1)T_{h1} - b(a-1)T_{c2}}{(b-1)-(a-1)}$$

where $a = N_c/N_h$, and $b = \exp(N_h - N_c)$, with similar expressions for parallel flow.

Exchanger duty can now be determined from $\dot{Q} = US\Delta\theta_{lmtd}$.

2.4 LMTD–*Ntu* sizing problem

Contraflow LMTD–Ntu sizing	*Parallel flow LMTD–Ntu sizing*
Given: $\dot{Q}$ *(duty)* $\dot{m}_h$, C_h, T_{h1} $\dot{m}_c$, C_c, T_{c2}	Given: $\dot{Q}$ *(duty)* $\dot{m}_h$, C_h, T_{h1} $\dot{m}_c$, C_c, T_{c1}
Find: *geometry*	Find: *geometry*

Writing energy balances

$$\dot{Q} = US\Delta\theta_{lmtd} = \dot{m}_h C_h (T_{h1} - T_{h2}) = \dot{m}_c C_c (T_{c1} - T_{c2})$$

$$\dot{Q} = US\Delta\theta_{lmtd} = \dot{m}_h C_h (T_{h1} - T_{h2}) = \dot{m}_c C_c (T_{c2} - T_{c1})$$

In sizing, the unknown temperatures are found from the energy balances allowing the LMTD to be calculated directly. The product *US* can then be evaluated and used in direct-sizing methods of design described in later sections.

2.5 Link between *Ntu* values and LMTD

For the contraflow heat exchanger

$$\dot{Q} = US\Delta\theta_{lmtd} = \dot{m}_h C_h (T_{h1} - T_{h2}) = \dot{m}_c C_c (T_{c1} - T_{c2})$$

$$\Delta\theta_{lmtd} = \frac{\dot{m}_h C_h}{US}(T_{h1} - T_{h2}) - \frac{\dot{m}_c C_c}{US}(T_{c1} - T_{c2})$$

$$\Delta\theta_{lmtd} = \frac{T_{h1} - T_{h2}}{N_h} = \frac{T_{c1} - T_{c2}}{N_c} \tag{2.14}$$

Since

$$\frac{a}{b} = \frac{c}{d} = \frac{a-c}{b-d} \left(\frac{a+c}{b+d} \quad \text{for parallel flow} \right)$$

then

$$\Delta\theta_{lmtd} = \frac{(T_{h1} - T_{h2}) - (T_{c1} - T_{c2})}{N_h - N_c} = \frac{(T_{h1} - T_{c1}) - (T_{h2} - T_{c2})}{N_h - N_c}$$

$$= \frac{\Delta\theta_1 - \Delta\theta_2}{N_h - N_c} \tag{2.15}$$

but

$$\Delta\theta_{lmtd} = \frac{\Delta\theta_1 - \Delta\theta_2}{\ln(\Delta\theta_1/\Delta\theta_2)} \tag{2.16}$$

Combining equations (2.15) and (2.16) there results

$$\boxed{\Delta\theta_{lmtd} = \frac{T_{h1} - T_{h2}}{N_h} = \frac{T_{c1} - T_{c2}}{N_c} = \frac{\Delta\theta_1 - \Delta\theta_2}{N_h - N_c} = \frac{\Delta\theta_1 - \Delta\theta_2}{\ln(\Delta\theta_1/\Delta\theta_2)}}$$

which includes the relationship $N_h - N_c = \ln(\Delta\theta_1/\Delta\theta_2)$. Thus explicit expressions for LMTD may be written in the form given by Spalding (1990)

$\Delta\theta_{lmtd} = \left\lvert \frac{\Delta\theta_1 - \Delta\theta_2}{N_h - N_c} \right\rvert$	$\Delta\theta_{lmtd} = \left\lvert \frac{\Delta\theta_1 - \Delta\theta_2}{N_h + N_c} \right\rvert$
Contraflow	*Parallel flow*

2.6 The 'theta' methods

Alternative methods of representing the performance of heat exchangers exist, and reference to these may be found in Hewitt *et al.* (1994). The basic parameter 'theta' was devised by Taborek (1983), where in the notation of this text

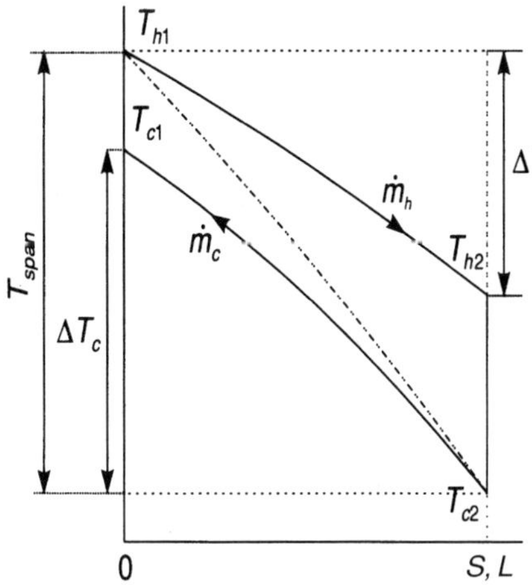

Fig.2.11 Contraflow

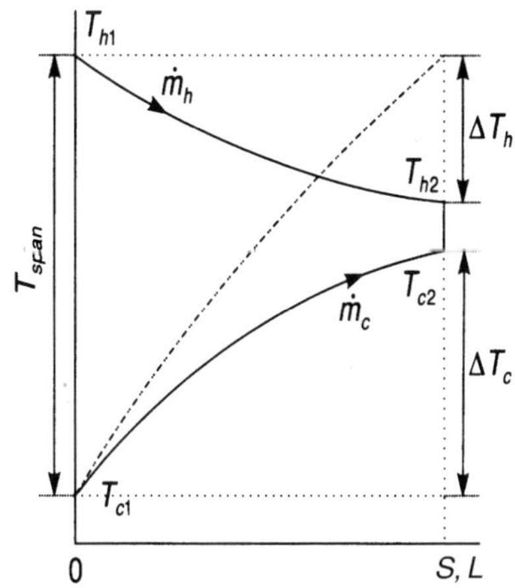

Fig.2.12 Parallel flow

(Figs 2.11 and 2.12)

$$\text{‘}\Theta\text{’} = \frac{\Delta\theta_m}{Tspan}$$

where $\Delta\theta_m$ is mean temperature difference.

The 'theta' method is related to associated 'E' and 'P' methods by the expressions

$$\text{‘}\Theta\text{’} = \frac{\Delta\theta_m}{Tspan} = \frac{\text{‘}P\text{’}}{Ntu_{cold}} = \frac{\text{‘}E\text{’}}{Ntu_{\min}}$$

and the relationships between parameters are often presented in graphical form. However, they all depend on finding $\Delta\theta_m$ or $\Delta\theta_{lmtd}$.

2.7 Effectiveness and number of transfer units

Considering contraflow and parallel-flow exchangers, the assumptions remain the same as in Section 2.2.
Define

$$\varepsilon = \frac{\Delta T_c}{Tspan} \quad \text{or} \quad \varepsilon = \frac{\Delta T_h}{Tspan} \qquad \text{whichever is the greater,}$$

and assume $(\dot{m}_c C_c < \dot{m}_h C_h) \ldots$ then $(N_c > N_h)$.

$$\Delta T_c = T_{c1} - T_{c2}$$
$$Tspan = T_{h1} - T_{c2}$$
$$\varepsilon = \frac{\Delta T_c}{Tspan} = \frac{T_{c1} - T_{c2}}{T_{h1} - T_{c2}} \tag{2.17}$$

Equation (2.17) may be written

$$\exp[N_h - N_c]$$
$$= \frac{(T_{h1} - T_{c2}) - (T_{c1} - T_{c2})}{(T_{h1} - T_{c2}) - (T_{h1} - T_{h2})}$$
$$= \frac{1 - \varepsilon}{1 - \left(\dfrac{T_{h1} - T_{h2}}{T_{h1} - T_{c2}}\right)}$$
$$= \frac{1 - \varepsilon}{1 - \varepsilon(N_h/N_c)}$$

$$\Delta T_c = T_{c2} - T_{c1}$$
$$Tspan = T_{h1} - T_{c1}$$
$$\varepsilon = \frac{\Delta T_c}{Tspan} = \frac{T_{c2} - T_{c1}}{T_{h1} - T_{c1}} \tag{2.17}$$

Equation (2.17) may be written

$$\exp[N_h + N_c]$$
$$= \frac{(T_{h1} - T_{c1})}{(T_{h1} - T_{c1}) - (T_{c2} - T_{c1}) - (T_{h1} - T_{h2})}$$
$$= \frac{1}{1 - \varepsilon - \left(\dfrac{T_{h1} - T_{h2}}{T_{h1} - T_{c1}}\right)}$$
$$= \frac{1}{1 - \varepsilon(1 + N_h/N_c)}$$

Solving for effectiveness

$\varepsilon = \dfrac{1 - \exp(N_h - N_c)}{1 - \dfrac{N_h}{N_c}\exp(N_h - N_c)}$ *Contraflow*	$\varepsilon = \dfrac{1 - \exp[-(N_h + N_c)]}{1 + \dfrac{N_h}{N_c}}$ *Parallel flow*

(2.18) (2.18)

These equations may be expressed in alternative form by writing

$$W = \frac{\dot{m}_c C_c}{\dot{m}_h C_h} = \frac{N_h}{N_c} \quad \text{(it is necessary to have } W < 1)$$

and write $N_c = Ntu$, then

$\varepsilon = \dfrac{1 - \exp[-Ntu(1 - W)]}{1 - W\exp[-Ntu(1 - w)]}$ *Contraflow*	$\varepsilon = \dfrac{1 - \exp[-Ntu(1 + W)]}{1 + W}$ *Parallel flow*

(2.19) (2.19)

The exponentials are written with negative exponents so that limiting values for effectiveness may be obtained

As $S \to \infty$ $Ntu \to \infty$ $\varepsilon_{\lim} \to 1$ (2.20)	As $S \to \infty$ $Ntu \to \infty$ $\varepsilon_{\lim} \to 1/(1+W)$ (2.20)

A special case occurs with *contraflow* when the water equivalents of the two fluids are equal. Then $\dot{m}_c C_c = \dot{m}_h C_h$ and $W = 1$, and the temperature profiles are parallel straight lines (proof in Section 3.1).

Directly $\Delta\theta_m = T_{h1} - T_{c1}$

and $\varepsilon = \dfrac{\Delta T_c}{Tspan} = \dfrac{T_{c1} - T_{c2}}{T_{h1} - T_{c2}}$

$$Ntu = \frac{US}{\dot{m}_c C_c} = \frac{T_{c1} - T_{c2}}{\Delta\theta_m} = \frac{T_{c1} - T_{c2}}{T_{h1} - T_{c1}}$$

$$Ntu + 1 = \frac{T_{h1} - T_{c2}}{T_{h1} - T_{c1}} \quad \text{thus}$$

$$\boxed{\varepsilon = \frac{Ntu}{1 + Ntu} \quad \text{with} \quad Ntu = N_h = N_c \quad \textit{Special case}} \tag{2.21}$$

For *condensation* the condition $\dot{m}_c C_c < \dot{m}_h C_h$ holds for the definition of *Ntu*, and the effectiveness solution for either contraflow or parallel flow applies.

For *evaporation* the condition $\dot{m}_h C_h < \dot{m}_c C_c$ holds for the definition of *Ntu*, and the same effectiveness solution for either contraflow or parallel flow applies.

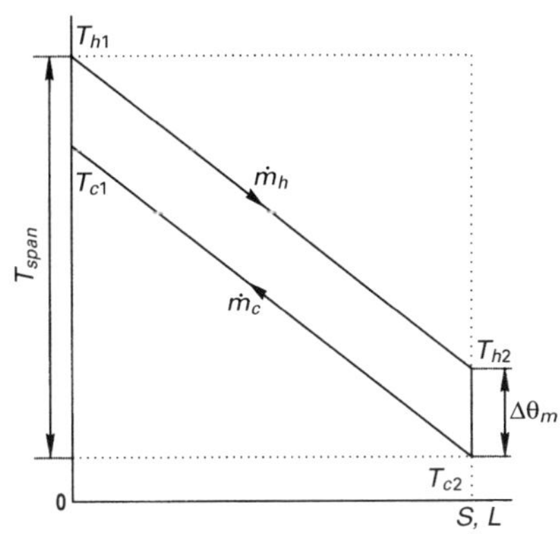

Fig.2.13 Contraflow

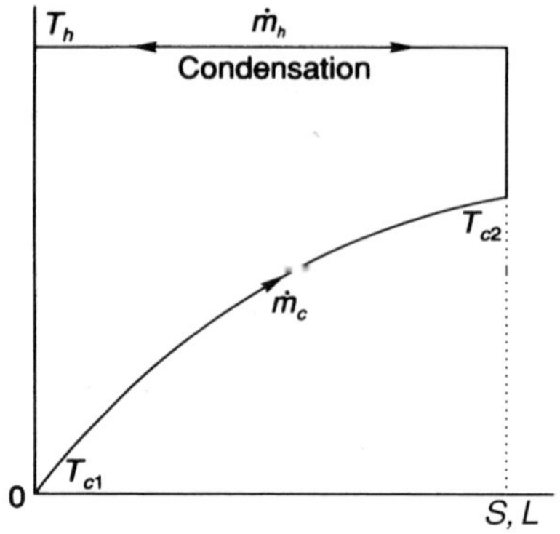

Fig.2.14 Condensation

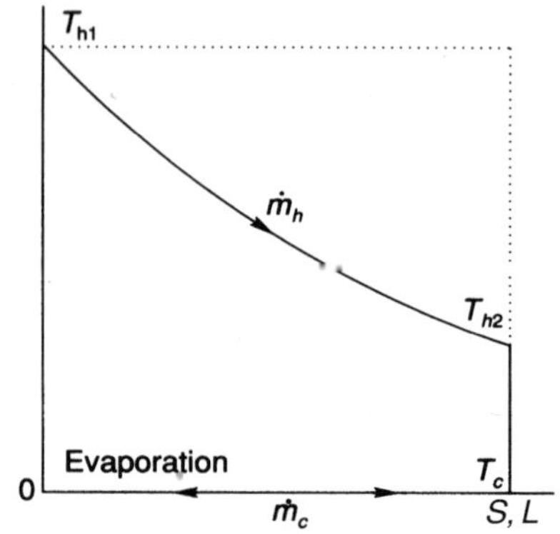

Fig.2.15 Evaporation

For both cases assume first that a small temperature difference exists in the phase-change fluid. This is not far from the truth because pressure loss always exists in fluid flow. Then

$$\dot{m}_c C_c (T_{c2} - T_{c1}) = \dot{m}_h C_h \delta T_h$$
$$= \dot{m}_h \left(\frac{\partial h_h}{\partial T_h} \right)_p \delta T_h$$

as

$$\delta T_h \to 0, \; C_h \to \infty$$

and

$$\dot{m}_c C_c (T_{c2} - T_{c1}) = \dot{m}_h \times h_{fg_h}$$

thus

$$W = \frac{\dot{m}_c C_c}{\dot{m}_h C_h} \to 0 \quad \text{and}$$
$$\varepsilon = 1 - e^{-Ntu} \tag{2.22}$$

$$\dot{m}_h C_h (T_{h1} - T_{h2}) = \dot{m}_c C_c \delta T_c$$
$$= \dot{m}_c \left(\frac{\partial h_c}{\partial T_c} \right)_p \delta T_c$$

as

$$\delta T_c \to 0, \quad C_c \to \infty$$

and

$$\dot{m}_h C_h (T_{h1} - T_{h2}) = \dot{m}_c \times h_{fg_c}$$

thus

$$W = \frac{\dot{m}_h C_h}{\dot{m}_c C_c} \to 0 \quad \text{and}$$
$$\varepsilon = 1 - e^{-Ntu} \tag{2.22}$$

Equations (2.22) for effectiveness hold equally for condensation or evaporation.

Results of equations (2.18), (2.19), and (2.22) are presented graphically in Fig. 2.16. To maintain reasonable terminal temperature differences the value $N_h + N_c$ for parallel flow should not exceed about 4.0, thus for equal values, $N_h = N_c = 2.0$. In a later section it will be shown that for crossflow with both fluids unmixed, equal values of Ntu should not exceed about $N_h = N_c = 4.0$. For contraflow, comfortable equal values of Ntu lie well above 10.0.

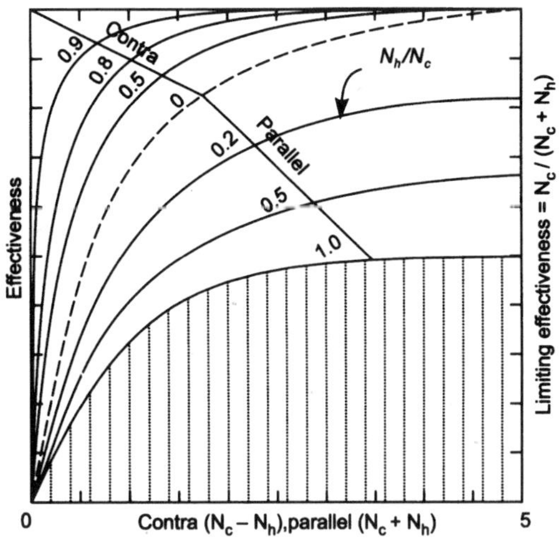

Fig.2.16 Generalized effectiveness plot with parameter N_h/N_c for contraflow and parallel-flow heat exchangers

Effectiveness versus Ntu *plots*

Many texts present curves of effectiveness (ε) versus number of transfer units (Ntu) over a range of capacity–rate ratios for a variety of flow arrangements, and apply these curves in design. Examining definitions of these axes in contraflow

$$Ntu = \frac{US}{\dot{m}C} = \frac{\Delta T_{max}}{\Delta \theta_{mean}}$$

allows effectiveness as

$$\varepsilon = \frac{\Delta T_{max}}{Tspan} = Ntu\left(\frac{\Delta \theta_{max}}{Tspan}\right)$$

and equally we might have plotted $(\Delta\theta_{max}/Tspan)$ versus Ntu. The parameter pair (ε, Ntu) thus says nothing about the physical size of the exchanger core (see Section 2.14).

2.8 ε–*Ntu* rating problem

Contraflow ε–*Ntu rating*	*Parallel flow* ε–*Ntu rating*
Given: *geometry* $\dot{m}_h$, C_h, T_{h1} $\dot{m}_c$, C_c, T_{c2}	Given: *geometry* $\dot{m}_h$, C_h, T_{h1} $\dot{m}_c$, C_c, T_{c1}
Find: $\dot{Q}$ *(duty)*	Find: $\dot{Q}$ *(duty)*

Referring to the figures in Section 2.4, then for $\dot{m}_c C_c < \dot{m}_h C_h$ the values $Ntu = US/(\dot{m}_c C_c)$ and $W = \dot{m}_c C_c / \dot{m}_h C_h$ apply, thus

Contraflow	Parallel flow
$\varepsilon = \dfrac{1 - e^{-Ntu(1-W)}}{1 - We^{-Ntu(1-W)}}$	$\varepsilon = \dfrac{1 - e^{-Ntu(1+W)}}{1 + W}$
$= \dfrac{\Delta T_c}{Tspan}$ for $\dot{m}_c C_c < \dot{m}_h C_h$	$= \dfrac{\Delta T_c}{Tspan}$ for $\dot{m}_c C_c < \dot{m}_h C_h$
$= \dfrac{T_{c1} - T_{c2}}{T_{h1} - T_{c2}}$	$= \dfrac{T_{c2} - T_{c1}}{T_{h1} - T_{c1}}$
Hence T_{c1} may be found, then T_{h2}, so	Hence T_{c2} may be found, then T_{h2}, so
$\dot{Q} = \dot{m}_c C_c (T_{c1} - T_{c2})$	$\dot{Q} = \dot{m}_c C_c (T_{c2} - T_{c1})$

2.9 ε–*Ntu* sizing problem

Contraflow *ε–Ntu sizing*	*Parallel flow* *ε–Ntu sizing*
Given: $\dot{Q}$ (duty) $\dot{m}_h$, C_h, T_{h1} $\dot{m}_c$, C_c, T_{c2}	Given: $\dot{Q}$ (duty) $\dot{m}_h$, C_h, T_{h1} $\dot{m}_c$, C_c, T_{c1}
Find: *geometry*	Find: *geometry*

For $\dot{m}_c C_c < \dot{m}_h C_h$, $W = \dot{m}_c C_c / (\dot{m}_h C_h)$ and unknown temperatures may be found immediately:

Contraflow	Parallel flow
$\dot{Q} = \dot{m}_h C_h (T_{h1} - T_{h2})$	$\dot{Q} = \dot{m}_h C_h (T_{h1} - T_{h2})$
$= \dot{m}_c C_c (T_{c2} - T_{c1})$	$= \dot{m}_c C_c (T_{c1} - T_{c2})$
$\varepsilon = \dfrac{\Delta T_c}{Tspan}$ for $\dot{m}_c \dot{C}_c < \dot{m}_h C_h$	$\varepsilon = \dfrac{\Delta T_c}{Tspan}$ for $\dot{m}_c \dot{C}_c < \dot{m}_h C_h$
$= \dfrac{T_{c1} - T_{c2}}{T_{h1} - T_{c2}}$	$= \dfrac{T_{c2} - T_{c1}}{T_{h1} - T_{c1}}$
Then solve for *Ntu* from	Then solve for *Ntu* from
$\varepsilon = \dfrac{1 - \exp(-Ntu(1 - W)]}{1 - W\exp[-Ntu(1 - W)]}$	$\varepsilon = \dfrac{1 - \exp(-Ntu(1 + W)]}{1 + W}$

Since $\dot{m}_c C_c < \dot{m}_h C_h$, the value of $Ntu = US/(\dot{m}_c C_c)$ gives the product US, leaving the geometry to be determined as in Section 2.4.

2.10 Comparison of LMTD–*Ntu* and ε–*Ntu* approaches

Contraflow rating problem
Given: *geometry* $\dot{m}_h,\ C_h,\ T_{h1}$ $\dot{m}_c,\ C_c,\ T_{c2}$
Find: $\dot{Q}$ *(duty)*

Consider the problem of rating a contraflow exchanger with surface area $S = 40.0\ \text{m}^2$, and for which the overall heat-transfer coefficient $U = 100\ \text{J}/(\text{m}_s^2\ \text{K})$ has been estimated from appropriate correlations by taking physical properties at the assumed mean bulk temperature $(T_{h1} + T_{c2})/2$, then with

Mass flowrates (kg/s)	*Inlet temperatures (°C)*	*Specific heats (J/kg K)*
$\dot{m}_h = 1.60$	$T_{h1} = 250$	$C_h = 1000$
$\dot{m}_c = 1.00$	$T_{c2} = 50$	$C_c = 4000$

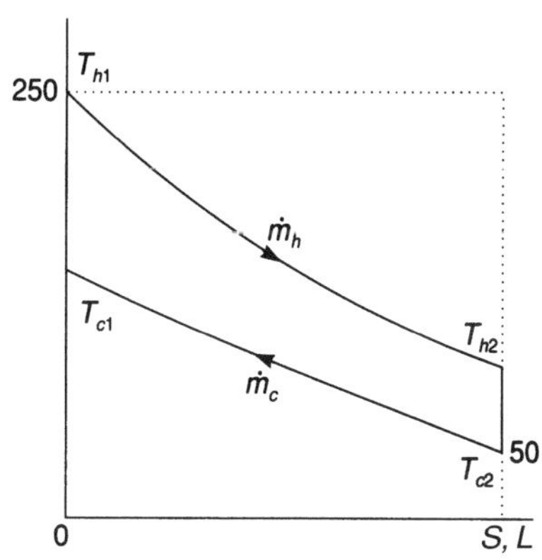

Fig.2.17 Contraflow profiles

LMTD–Ntu approach

$$N_h = \frac{US}{\dot{m}_h C_h} = \frac{100 \times 40}{1.6 \times 1000} = 2.5$$

$$N_c = \frac{US}{\dot{m}_c C_c} = \frac{100 \times 40}{1.0 \times 4000} = 1.0$$

$$\frac{\Delta\theta_1}{\Delta\theta_2} = \frac{T_{h1} - T_{c1}}{T_{h2} - T_{c2}} = \exp(N_h - N_c)$$

$$\frac{250 - T_{c1}}{T_{h2} - 50} = \exp(2.5 - 1.0)$$

$$= 4.4817$$

$$T_{c1} - 4.4817 \times T_{h2} = 474.08 \qquad \text{(a)}$$

$$\frac{\Delta T_c}{\Delta T_h} = \frac{T_{c1} - T_{c2}}{T_{h1} - T_{h2}} = \frac{N_c}{N_h}$$

$$\frac{T_{c1} - 50}{250 - T_{h2}} = \frac{1.0}{2.5} = 0.4$$

$$T_{c1} + 0.4 \times T_{h2} = 150.0 \qquad \text{(b)}$$

Subtracting equations (a) from (b)

$$4.0817 \times T_{h2} = 324.084$$

ε–Ntu approach

$$Ntu = \frac{US}{\dot{m}_h C_h} = \frac{100 \times 40}{1.6 \times 1000} = 2.5$$

$$W - \frac{\dot{m}_h C_h}{\dot{m}_c C_c} - 0.4$$

$$\varepsilon = \frac{1 - \exp[-Ntu(1 - W)]}{1 - W\exp[-Ntu(1 - W)]}$$

$$\varepsilon = \frac{1 - \exp[-2.5(1 - 0.4)]}{1 - 0.4\exp[-2.5(1 - 0.4)]}$$

$$\varepsilon = 0.853$$

$$\varepsilon = \frac{\Delta T_h}{Tspan} = \frac{T_{h1} - T_{h2}}{T_{h1} - T_{c2}}$$

$$0.853 = \frac{250 - T_{h2}}{250 - 50}$$

$$T_{h2} = 79.4\,^\circ\mathrm{C}$$

$$\dot{m}_c C_c(T_{c1} - T_{c2}) = \dot{m}_h C_h(T_{h1} - T_{h2})$$

$$4000(T_{c1} - 50) = 1600(250 - 79.4)$$

$$T_{c1} = 118.24\,^\circ\mathrm{C}$$
$$T_{h2} = 79.40\,^\circ\mathrm{C}$$

2.11 Sizing when $\dot{Q}$ is not specified

Occasionally a sizing problem may arise when $\dot{Q}$ is not known but 'best practicable recuperation' is desired. The normal procedure is to determine the end of the exchanger at which the temperature pinch point occurs, and give $\Delta\theta$ an appropriate value, say 4 °C. With two known terminal temperatures it is then possible to solve the heat balance equations for the unknown temperatures directly in five out of six possible cases.

The remaining case arises when the pinch point is at the same end as the two known temperatures. When an exchanger of this type is required, there is usually a limiting temperature for one of the unknowns. If this is not a restriction, then an alternative approach may be to determine the limiting value of

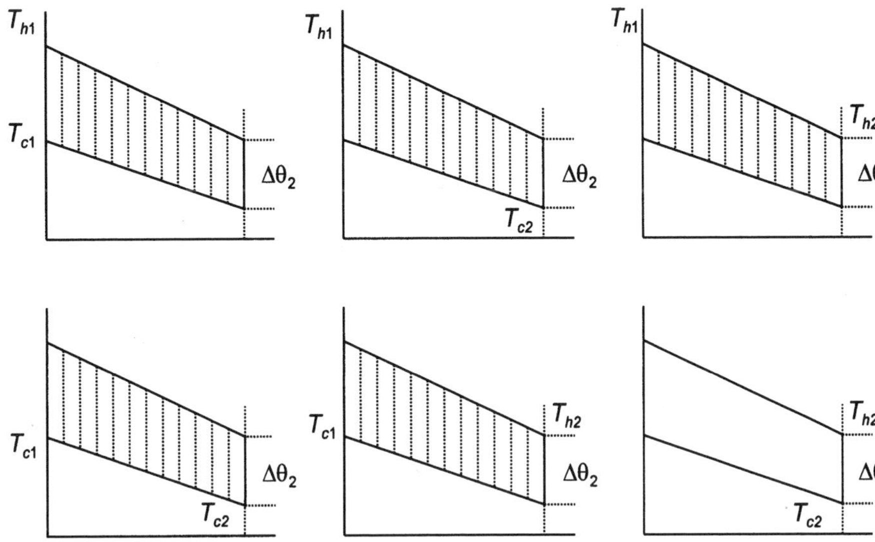

Fig.2.18 Six possible cases when Q is unknown

effectiveness, e.g.

$$\text{Contraflow,} \quad \varepsilon_{\text{lim}} = 1 \qquad \text{Parallel flow,} \quad \varepsilon_{\text{lim}} = \frac{1}{1+W}$$

and obtain the actual effectiveness from $\varepsilon = f * \varepsilon_{\text{lim}}$ where from practical experience $(0.7 < f < 0.9)$ approximately. Alternatively, and preferably, it is possible to proceed along lines suggested by Grassmann & Kopp (1957), as discussed in Section 2.12.

2.12 Optimum temperature profiles in contraflow

The paper by Grassmann & Kopp (1957) presented an analysis based on minimizing exergy loss for temperatures below the dead state T_0. This original work is now adapted and extended.

In the optimization of power plant the objective is to get the maximum amount of work out of a given amount of supplied heat. Usable work (or exergy) can be lost if the design of a heat exchanger is poor. In the optimization of cryogenic plant the objective is to minimize the compression work required to operate the system by minimizing exergy loss in the heat exchangers.

Minimum exergy loss targets may differ from some targets in chemical plant design, where a particular temperature level has to be achieved for a process to occur, but in general the overall target is still to minimize the work required to operate the plant.

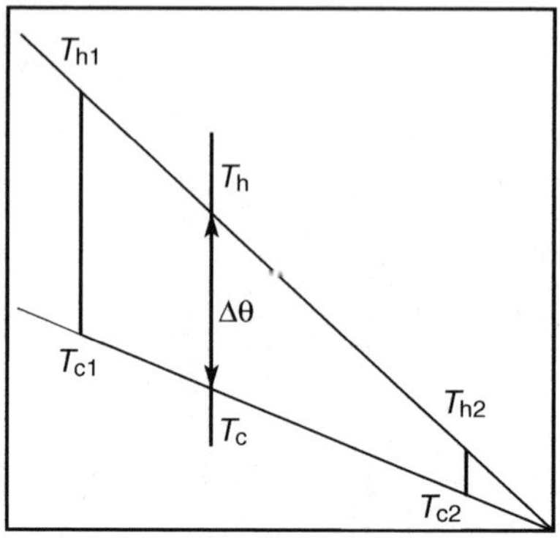

Fig.2.19 Schematic temperature profiles

In a contraflow heat exchanger the differential exergy rate changes of the hot and cold fluid are

$$\delta \dot{E}_h = \left(\frac{T_h T_0}{T_h}\right)(-\delta \dot{q})$$

Total exergy rate change is thus

$$\delta \dot{E}_h + \delta \dot{E}_c = \delta \dot{q} T_0 \left(\frac{T_c - T_h}{T_c T_h}\right)$$

Write $\Delta\theta = (T_h - T_c)$ then

$$\delta \dot{E}_{total} = \delta \dot{q} T_0 \left[\frac{-\Delta\theta}{(T_h - \Delta\theta) T_h}\right]$$

Over the whole exchanger

$$\Delta \dot{E}_{total} = -T_0 \int_{T_{h_1}}^{T_{h2}} \frac{\Delta\theta}{(T_h - \Delta\theta) T_h} d\dot{q}$$

since $\delta \dot{q} = \dot{m}_h C_h \delta T_h = \dot{m}_c C_c \delta T_c$ then

$$\Delta \dot{E}_{total} = \dot{m}_h T_0 \int_{T_{h2}}^{T_{h1}} \frac{C_h \Delta\theta}{(T_h - \Delta\theta) T_h} dT_h$$

$$\delta \dot{E}_c = \left(\frac{T_c - T_0}{T_c}\right)(+)\delta \dot{q}$$

$$\delta \dot{E}_{total} = \delta \dot{q} T_0 \left[\frac{-\Delta\theta}{(T_c + \Delta\theta) T_c}\right]$$

$$\Delta \dot{E}_{total} = -T_0 \int_{T_{c1}}^{T_{c_2}} \frac{\Delta\theta}{(T_c + \Delta\theta) T_c} d\dot{q}$$

$$\Delta \dot{E}_{total} = \dot{m}_h T_0 \int_{T_{c2}}^{T_{c1}} \frac{C_c \Delta\theta}{(T_c + \Delta\theta) T_c} dT_c \qquad (2.23)$$

and the problem reduces to the variational problem 'find a function $\Delta\theta = \phi(T)$ such that the integral in these last equations is a minimum, *subject to the constraint that the surface area S has a fixed value*'. For constant specific heats, generally

$$\Delta\dot{E}_{total} = \dot{m}_h C_h T_0 \int_{T_{h2}}^{T_{h1}} \frac{\Delta\theta}{(T_h - \Delta\theta)T_h} dT_h = \dot{m}_h C_h T_0 \left[\int_{T_{h2}}^{T_{h1}} \frac{dT_h}{(T_h - \Delta\theta)} - \int_{T_{h2}}^{T_{h1}} \frac{dT_h}{T_h} \right]$$

$$\Delta\dot{E}_{total} = \dot{m}_c C_c T_0 \int_{T_{c2}}^{T_{c1}} \frac{\Delta\theta}{(T_c + \Delta\theta)T_c} dT_c = \dot{m}_c C_c T_0 \left[\int_{T_{c2}}^{T_{c1}} \frac{dT_c}{(T_c + \Delta\theta)} - \int_{T_{c2}}^{T_{c1}} \frac{dT_c}{T_c} \right] \quad (2.24)$$

and it becomes evident that $\Delta\theta = \phi(T)$ should be as small as practicable, but not zero which would be for zero heat exchange.

Values of $\Delta\theta$ must also satisfy the energy balance equation

$$\delta\dot{Q} = -U\delta S(T_h - T_c) = -U\delta S\Delta\theta = \dot{m}_h C_h \delta T_h = \dot{m}_c C_c \delta T_c \quad (2.25)$$

where the negative sign is present because T_h is reducing as S is increasing.

With overall heat-transfer coefficient U and surface area S, constant equations (2.25) may be integrated to give the energy equations

$$N_h = \frac{US}{\dot{m}_h C_h} = +\int_{T_{h2}}^{T_{h1}} \frac{dT_h}{\Delta\theta} \qquad N_c = \frac{US}{\dot{m}_c C_c} = +\int_{T_{c2}}^{T_{c1}} \frac{dT_c}{\Delta\theta} \quad (2.26)$$

Equations (2.24) are to be solved, subject to the constraint of equation (2.26)

Using variational calculus (see Appendix F), by inspection, the equations so far developed are seen to match those given in that appendix, viz.

Integral

$$\Delta\dot{E}_{total} = \dot{m}_h C_h T_0 \int_{T_{h2}}^{T_{h1}} \frac{\Delta\theta}{(T_h - \Delta\theta)T_h} dT_h \qquad \Delta\dot{E}_{total} = \dot{m}_c C_c T_0 \int_{T_{c2}}^{T_{c1}} \frac{\Delta\theta}{(T_c + \Delta\theta)T_c} dT_c$$

from which

$$F = \frac{\Delta\theta}{(T_h - \Delta\theta)T_h} = \frac{1}{(T_h - \Delta\theta)} - \frac{1}{T_h} \qquad F = \frac{\Delta\theta}{(T_c + \Delta\theta)T_c} = \frac{-1}{(T_c + \Delta\theta)} + \frac{1}{T_c}$$

Constraint

$$N_h = \frac{US}{\dot{m}_h C_h} = \int_{T_{h2}}^{T_{h1}} \frac{dT_h}{\Delta\theta} \qquad N_c = \frac{US}{\dot{m}_c C_c} = \int_{T_{c2}}^{T_{c1}} \frac{dT_c}{\Delta\theta}$$

from which $G = \dfrac{1}{\Delta\theta}$ | from which $G = \dfrac{1}{\Delta\theta}$

End conditions

$$\begin{cases} \Delta\theta(T_{h1}) = \Delta\theta_1 \\ \Delta\theta(T_{h2}) = \Delta\theta_2 \end{cases}$$

thus

$$H = F + \lambda G$$
$$= \left[\frac{1}{(T_h - \Delta\theta)} - \frac{1}{T_h} + \frac{\lambda}{\Delta\theta}\right]$$

hence

$$\frac{\partial H}{\partial(\Delta\theta)} = \left[\frac{1}{(T_h - \Delta\theta)^2} - \frac{\lambda}{(\Delta\theta)^2}\right]$$

and

$$\frac{\partial H}{\partial(\Delta\theta')} = 0$$

The Euler equation becomes

$$\frac{1}{(T_h - \Delta\theta)^2} - \frac{\lambda}{(\Delta\theta)^2} = 0$$

$$\left(\frac{\Delta\theta}{T_h - \Delta\theta}\right)^2 = \lambda$$

Only the +ve square root will do

$$\left(\frac{\Delta\theta}{T_h - \Delta\theta}\right) = +\sqrt{\lambda}$$

with solution

$$\frac{\Delta\theta}{T_h} = \frac{\sqrt{\lambda}}{1 + \sqrt{\lambda}}$$

giving

$$\frac{\Delta\theta}{T_h} = \frac{a}{1+a} \quad \text{and} \quad \frac{\Delta\theta}{T_c} = a$$

$$\begin{cases} \Delta\theta(T_{c1}) = \Delta\theta_1 \\ \Delta\theta(T_{c2}) = \Delta\theta_2 \end{cases}$$

$$H = F + \lambda G$$
$$= \left[\frac{1}{T_c} - \frac{1}{(T_c + \Delta\theta)} + \frac{\lambda}{\Delta\theta}\right]$$

$$\frac{\partial H}{\partial(\Delta\theta)} = \left[\frac{1}{(T_c - \Delta\theta)^2} - \frac{\lambda}{(\Delta\theta)^2}\right]$$

$$\frac{\partial H}{\partial(\Delta\theta')} = 0$$

$$\frac{1}{(T_c - \Delta\theta)^2} - \frac{\lambda}{(\Delta\theta)^2} = 0$$

$$\left(\frac{\Delta\theta}{T_c - \Delta\theta}\right)^2 = \lambda$$

$$\left(\frac{\Delta\theta}{T_c - \Delta\theta}\right) = +\sqrt{\lambda}$$

$$\frac{\Delta\theta}{T_c} = \frac{\sqrt{\lambda}}{1 - \sqrt{\lambda}}$$

$$\frac{\Delta\theta}{T_c} = \frac{b}{1-b} \quad \text{and} \quad \frac{\Delta\theta}{T_h} = b$$

The pairs of solutions are equivalent, only depending on whether the initial reckoning of $\Delta\theta$ is to be from T_h or T_c.

Consider the 'a' form.

$$\frac{T_h}{T_c} = \left(\frac{T_h}{\Delta\theta}\right)\left(\frac{\Delta\theta}{T_c}\right) = 1 + a \qquad \text{thus} \qquad \frac{T_{h1}}{T_{c1}} = \frac{T_{h2}}{T_{c2}} \qquad \text{hence}$$

$$\frac{T_{h1}}{T_{h2}} - 1 = \frac{T_{c1}}{T_{c2}} - 1 \qquad \text{thus} \qquad \frac{T_{h1} - T_{h2}}{T_{h2}} = \frac{T_{c1} - T_{c2}}{T_{c2}} \qquad \text{hence}$$

$$\frac{T_{h1} - T_{h2}}{T_{c1} - T_{c2}} = \frac{T_{h2}}{T_{c2}} = \frac{T_{h1}}{T_{c1}} \qquad \text{giving}$$

$$\left(\frac{\dot{m}_c C_c}{\dot{m}_h C_h}\right) = \frac{T_{h1} - T_{h2}}{T_{c1} - T_{c2}} = \frac{T_{h2}}{T_{c2}} = \frac{T_{h1}}{T_{c1}} = 1 + a \tag{2.27}$$

At the optimum it follows that

$$\underset{\left\{\begin{array}{c}\text{product of}\\ \text{inlet temps}\end{array}\right\}}{T_{h1} \times T_{c2}} = \underset{\left\{\begin{array}{c}\text{product of}\\ \text{outlet temps}\end{array}\right\}}{T_{h2} \times T_{c1}} \tag{2.28}$$

When Grassmannn & Kopp made the heat-transfer coefficient a function of temperature also, the influence of temperature was found to be very weak, and could be neglected.

Generally we may thus write

$$\boxed{\Delta\theta = T/const.} \tag{2.29}$$

For cryogenic design, equation (2.29) might be written as $\Delta\theta = T/20$, while for near-ambient conditions the expression might be written $\Delta\theta = T/30$. The constant is chosen to produce an acceptable temperature difference between streams. Three valid expressions exist for equation (2.29), viz.

$$\frac{\Delta\theta}{T_h} \quad \text{or} \quad \frac{\Delta\theta}{T_c} \quad \text{or} \quad \frac{\Delta\theta}{T_m}$$

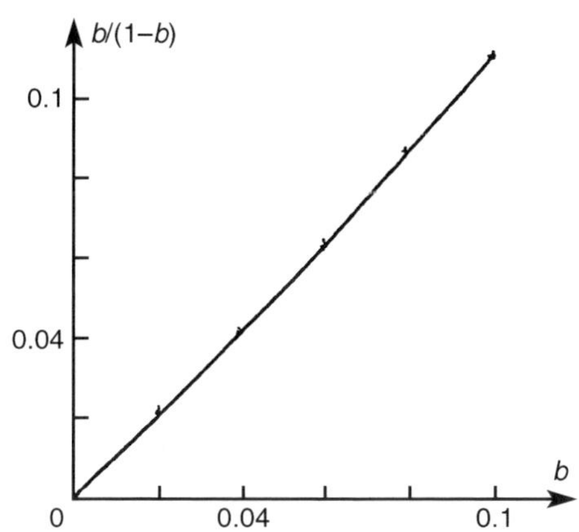

Fig.2.20 Variation of $b/(1-b)$ versus b [and a versus $a/(1+a)$]

where

$$T_m = \frac{1}{2}(T_h + T_c)$$

but the same constant cannot be used in each form.

The optimum temperature profiles can be found analytically, and the derivations are given in Appendix F.2

$$\text{Hot fluid profile} => \frac{T_h}{T_{h1}} = \left(\frac{T_{h2}}{T_{h1}}\right)^{\frac{x}{L}}$$

$$\text{Cold fluid profile} => \frac{T_c}{T_{c1}} = \left(\frac{T_{c2}}{T_{c1}}\right)^{\frac{x}{L}}$$

2.13 Optimum pressure losses in contraflow

So far, no account has been taken of pressure-loss optimization.

Exergy approach

Starting from the exergy change for a single fluid

$$\Delta b = (h_1 - T_0 s_1) - (h_2 - T_0 s_2) \tag{2.30}$$

the rate of entropy generation for a contraflow exchanger transferring heat between two fluids may be found as

$$\dot{S} = \left(\frac{\dot{m}_h \Delta b_h + \dot{m}_c \Delta b_c}{T_0}\right) = -\dot{m}_h(s_{h1} - s_{h2}) + \dot{m}_c(s_{c1} - s_{c2})$$

For perfect gases

$$\dot{S} = \dot{m}_h\left[C_h \ln\left(\frac{T_{h2}}{T_{h1}}\right) - R_h \ln\left(\frac{p_{h2}}{p_{h1}}\right)\right] + \dot{m}_c\left[C_c \ln\left(\frac{T_{c1}}{T_{c2}}\right) - R_c \ln\left(\frac{p_{c1}}{p_{c2}}\right)\right]$$

which may be written

$$\dot{S} = \left[-\dot{m}_h C_h \ln\left(\frac{T_{h1}}{T_{h2}}\right) + \dot{m}_c C_c \ln\left(\frac{T_{c1}}{T_{c2}}\right)\right] + \left[-\dot{m}_h R_h \ln\left(\frac{p_{h2}}{p_{h1}}\right) - \dot{m}_c R_c \ln\left(\frac{p_{c1}}{p_{c2}}\right)\right]$$

or

$$\dot{S} = [\dot{S}_{temp}] + [\dot{S}_{press}] \tag{2.31}$$

Considering pressure terms alone

$$\dot{S}_{press} = +\dot{m}_h R_h \ln\left(\frac{p_{h1}}{p_{h2}}\right) + \dot{m}_c R_c \ln\left(\frac{p_{c2}}{p_{c1}}\right) \tag{2.32}$$

or

$$\dot{S}_{press} = -\dot{m}_h R_h \ln\left(1 - \frac{\Delta p_h}{p_{h1}}\right) - \dot{m}_c R_c \ln\left(1 - \frac{\Delta p_c}{p_{c2}}\right)$$

Further simplification is possible using the expansion for $\ln(1 - x)$ and neglecting higher-order terms, viz.

$$\dot{S}_{press} = +\dot{m}_h R_h \left(\frac{\Delta p_h}{p_{h1}}\right) + \dot{m}_c R_c \left(\frac{\Delta p_c}{p_{c2}}\right) \tag{2.33}$$

but it is safer to remain with equation (2.32) whenever possible.

Once a core has been sized, inlet and outlet pressure losses, acceleration losses, and header pressure losses, have to be added to the core pressure losses.

Equation (2.33) can be rewritten as

$$\left[\frac{\dot{m}_h R_h}{\dot{m}_c R_c}\right] \times \left(\frac{\Delta p_h}{p_{h1}}\right) + \left(\frac{\Delta p_c}{p_{c2}}\right) = \left[\frac{\dot{S}_{press}}{\dot{m}_c R_c}\right] \tag{2.34}$$

If the exergy loss rate for pressure loss ($\dot{S}_{press}$) is required to be fixed, then equation (2.34) provides some allowance for redistribution of pressure loss between the two streams.

Mach number approach

It is natural to seek a relationship between pressures and pressure losses such as equation (2.34), as an aid to minimizing the effort to find the optimum exchanger with minimum core volume. However, it is not necessary to specify both pressure losses, because the use of direct-sizing software always produces a 'controlling' pressure loss, and the other pressure-loss curve must always fall on the same design point found in direct-sizing (see Chapter 4).

In many cases the allowable pressure-loss level is related to Mach number in the flow channels. Unless there is a special requirement it is not desirable to move out of the incompressible flow region, and Mach numbers should not ever exceed 0.1, while the better limit would be 0.05 to permit use of most experimental correlations for flow friction.

For near-ideal gases the maximum choice of flow velocity is then given by

$$u = \mathrm{Ma}\sqrt{\gamma R T} \quad \text{(m/s)}$$

where γ is the ratio of specific heats, and R is the gas constant.

The 'controlling' pressure loss

In any two-stream exchanger without phase change, the 'controlling' pressure loss is nearly always associated with the lower pressure level, i.e. the hotter gas stream in recuperator design, and the colder gas stream in cryogenic design (but see Appendix J).

The strategy for initial optimization might follow the undernoted sequential scheme:

1. First fix both surface geometries, and vary the 'controlling' Δp until acceptable performance is found (the other Δp arrives automatically in direct-sizing).
2. Now fix both Δp values, and systematically vary both surface geometries until appropriate performance is found.
3. Check the Mach numbers.

In all of these stages it is wise to find the LMTD reduction factor for longitudinal conduction and apply this in design, for very short heat exchangers may sometimes be produced.

For full optimization there remains the problem of 'best guessing' both surface geometries, and it is easier to do this logically than to guess pressure losses. For rectangular offset strip-fin geometries, a guide will be found in Appendix C suggesting appropriate directions in which changes might be made. The summary of fully developed laminar flow solutions reported by Bhatti & Shah (1987) and by Webb (1994) also provide useful guide information. See also Appendix J.

2.14 Compactness and performance

Specific performance

For comparison of different heat exchanger designs some authors use the parameter

$$\frac{S}{V}\left(\frac{m^2}{m^3}\right)$$

This is unsatisfactory because surface area S may or may not include secondary surfaces. A safer measure is obtained by starting from $\dot{Q} = US\Delta\theta_m$ which allows the expression for specific performance

$$\frac{\dot{Q}}{V\Delta\theta_m}\left(=\frac{US}{V}\right)\left(\frac{\mathrm{kW}}{\mathrm{m^3\,K}}\right) \tag{2.35}$$

as this is now independent of choice of surface area. Allowances for longitudinal conduction may be made in the value for $\Delta\theta_m$, see Chapter 3.

Performance comparison

The dimensionless exergy loss number provides universal performance comparison for all heat exchangers, see Chapter 8.

$$N_x = \left[-\left(\frac{s_1 - s_2}{h_1 - h_2}\right)_{hot} + \left(\frac{s_1 - s_2}{h_1 - h_2}\right)_{cold}\right] \times T_0 \tag{2.36}$$

2.15 Required values of *Ntu* in cryogenics

The problem of lifting energy at cryogenic temperatures is best illustrated by Fig. 2.21 which makes use of the basic expressions for Carnot efficiency above

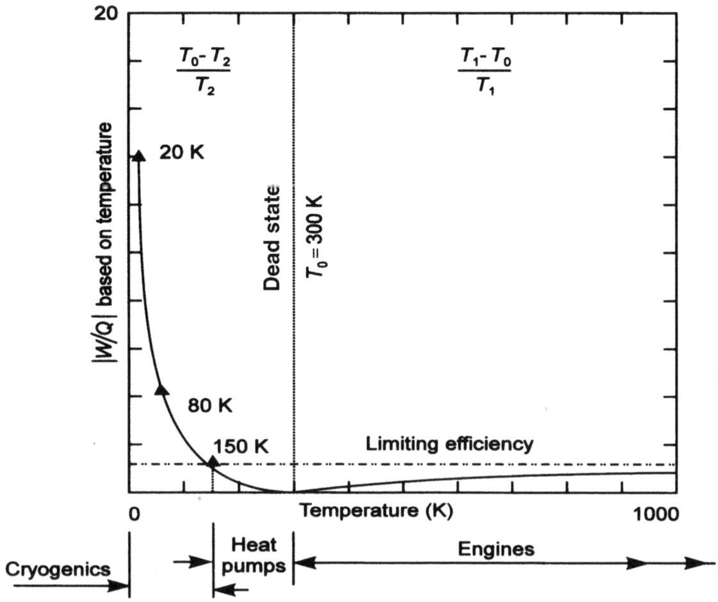

Fig.2.21 Carnot efficiency above and below dead state T_0

and below the dead state to define the regions named 'cryogenics', 'heat pumps', and 'engines'. The work required to lift heat from cryogenic temperatures and reject it to the dead state places a premium on achieving best possible heat exchange conditions.

A thermophysically efficient means of creating a refrigerating stream is to expand high-pressure gas in an inward radial flow cryo-turbine. The temperature reduction so caused depends on the isentropic index ($\gamma = C_p/C_v$) of the expanded gas. For the cryogenic gases of interest, viz. H_2, He, Ne, N_2, O_2, and Ar, γ is around 1.66 for monatomic gases and around 1.4 for diatomic gases (see Chapter 11).

To avoid loss-producing shock waves a limiting Mach number of 1 in the inlet nozzles of the cryo-turbine produces limiting pressure ratios of around 10 : 1 and 6 : 1 for monatomic and diatomic gases, respectively; these ratios correspond to maximum temperature reduction ratios of around 1 : 2 and 2 : 3, respectively. In cryogenic practice desirable pressure ratios are about one-third of the limiting values.

This information, together with the result obtained by Grassmann & Kopp that the temperature difference in cryogenic heat exchange should be proportional to absolute temperature, e.g. $\Delta\theta = T_h/20$ allows evaluation of the levels of *Ntu* required in cryogenic heat exchange.

Let us assume cooling of high-pressure product stream of nitrogen by a second refrigerating stream of nitrogen first expanded in a cryo-turbine, both streams being sufficiently far away from the critical point so that linearization of the $h-T$ curve for each fluid is a reasonable assumption (see Section 11.3). The outlet stream from the cryo-expander is at 100 K and is rewarmed to 150 K while cooling the product stream.

Inlet and outlet temperatures are 157.5 and 105 K for the product stream being cooled, providing terminal temperature differences of 7.5 and 5.0 K.

Applying the 'rate' equation

$$\frac{\Delta\theta_1}{\Delta\theta_2} = \frac{T_{h1} - T_{c1}}{T_{h2} - T_{c2}} = \exp(N_h - N_c)$$
$$N_h - N_c = 0.405\ 465 \tag{2.37}$$

Applying the 'energy' equation

$$\frac{\Delta T_c}{\Delta T_h} = \frac{T_{c1} - T_{c2}}{T_{h1} - T_{h2}} = \frac{N_c}{N_h}$$
$$N_c = 0.952\ 381 \times \mathrm{N_h} \tag{2.38}$$

Solving simultaneous equations (2.37) and (2.38) we obtain $N_h = 8.5148$ and $N_c = 8.1093$, and these values have to be equalled or exceeded. With these overall values of *Ntu*, the resulting temperature profiles would be as shown in Fig. 2.22, providing the specific heats of both fluids remain constant.

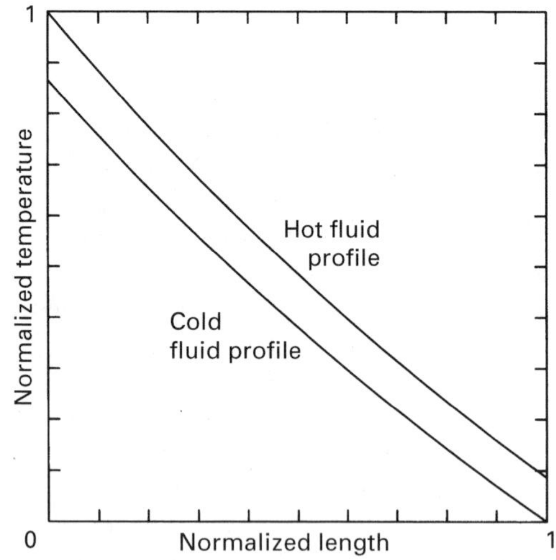

Fig.2.22 Normalized temperature profiles for contraflow with optimized temperature profiles

2.16 To dig deeper

It is useful to think a little more carefully about Figs 2.1 to 2.4 before proceeding to the rest of the text, as a number of features may be seen by careful observation.

1. With the condenser and the evaporator, it does not matter whether the constant-temperature fluid flows to the right or to the left. In a real exchanger, e.g. one of shell-and-tube design, this also allows the condensing or evaporating fluid to flow at right angles to the tubes and to the flow directions shown above, without changing the temperature field.
2. By comparing Figs 2.1 and 2.3 (the condenser and contraflow pair of diagrams), it becomes obvious that as the hotter fluid in the contraflow exchanger increases in water equivalent beyond all bounds ($\dot{m}C \rightarrow \infty$), the condenser is a limiting-case of contraflow. The same is true for the parallel-flow exchanger (Fig. 2.4 can be flipped about its vertical axis).

   Similarly by comparing Figs 2.2 and 2.4 the evaporator is a limiting-case of parallel flow. Thus all four exchanger configurations are closely related, and this observation is expressed formally in the 'generalized effectiveness plot' (Fig. 2.16). The reader may like to think about where the condenser and the evaporator might fit into this diagram.
3. The curved temperature profiles in Figs 2.1 and 2.2 can be flipped about their vertical axes without changing the concept. Shifting the origin can be helpful in simplifying mathematics, cf. Figs 3.9 and 3.10 on condensation and evaporation later in the text. In Chapter 6 on bayonet-tube exchangers, shifting the origin from one end of the exchanger to the other greatly simplified the mathematics for the isothermal and non-isothermal cases.
4. Anything to do with temperatures and temperature differences involves rate processes which are usually governed by exponentials. Exponentials should be expected in the solutions to most of the cases examined in this text. Whenever possible the final expressions are expressed as dimensionless ratios for neatness.
5. Figures 2.1 to 2.4 can be drawn with the vertical scales corresponding to real temperatures and the horizontal scales corresponding to either exchanger length or surface area (the class of exchanger examined here has this constraint). But these figures can be redrawn so that the maximum dimension in each direction is unity. This 'normalization' does not change the relation of the curves to each other, but simplifies the mathematics. However, normalized results must be converted back to engineering dimensions before they can be applied.

Engineers may find that full normalization of the mathematics sometimes takes away too much from the solution. A good example of this is to be found in Chapter 3, Section 3.2, where full normalization would produce the following canonical equation pair (Nusselt equations) at the expense of obscuring the problem.

$$\frac{\partial T_h}{\partial X} = -(T_h - T_c)$$
$$\frac{\partial T_c}{\partial Y} = +(T_h - T_c)$$

The effectiveness concept

Effectiveness is a measure of how closely the temperature of the fluid with the least water equivalent approaches the maximum possible temperature rise *Tspan* in the exchanger. For the contraflow arrangement, and to some extent for the crossflow arrangement, this corresponds to seeking the closest temperature approach between fluids. When care is taken to keep the temperature approach as small as practicable, then good effectiveness values should be achieved without the need to address the effectiveness issue specifically in design.

It is possible to think of two contraflow exchangers with the same effectiveness, viz.

$$\left(\varepsilon = \frac{\Delta T_c}{Tspan}\right) \quad \text{and} \quad \left(\varepsilon = \frac{\Delta T_h}{Tspan}\right)$$

only one of which has minimum entropy generation.

Thinking is different for parallel-flow arrangements, because parallel-flow applications are usually more concerned with limiting the maximum temperature of the cold fluid being heated, or to controlling the drop in temperature of the hot fluid being cooled, while recovering energy. Here the closest temperature approach in the exchanger is related to temperatures of fluids at the same end, and the actual value of effectiveness achieved can usefully be compared to its limiting value.

Units in differential equations

Throughout this text SI units are used. It is perhaps not always realized that ordinary and partial differential equations have units, and checking these units is a valuable way of confirming that the equation has been correctly formulated.

Consider the symbols x and t representing distance (metres) and time (seconds). It is familiar territory to recognize velocity and acceleration, respectively, as

$$\dot{x} = \frac{dx}{dt} \quad \text{and} \quad \ddot{x} = \frac{d^2x}{dt^2}$$

and but a short step to recognize that the units 'go' as the back end of the differential expressions: for velocity and acceleration, respectively

$$\dot{x} = \frac{d}{d}\left(\frac{x}{t}\right) => \left(\frac{\mathrm{m}}{\mathrm{s}}\right) \quad \text{and} \quad \ddot{x} = \frac{d^2}{d}\left(\frac{x}{t^2}\right) => \left(\frac{\mathrm{m}}{\mathrm{s}^2}\right)$$

Where differential terms are themselves raised to powers, then units are obtained as

$$\dot{x}^2 = \left(\frac{dx}{dt}\right)^2 = \left[\frac{d}{d}\left(\frac{x}{t}\right)\right]^2 => \left(\frac{\mathrm{m}}{\mathrm{s}}\right)^2$$

The central partial differential equation of a set of three given as equation (A.1) of Appendix A is given below, and the individual terms must have identical dimensions for the equation to make sense, viz.

$$\frac{\partial T_w}{\partial t} - \kappa_x \frac{\partial^2 T_w}{\partial x^2} = + \left(\frac{\alpha_h S}{M_w C_w}\right)(T_h - T_2) - \left(\frac{\alpha_c S}{M_w C_w}\right)(T_w - T_c)$$

$$\left(\frac{\mathrm{K}}{\mathrm{s}}\right) - \left(\frac{\mathrm{m}^2}{\mathrm{s}}\right)\left(\frac{\mathrm{K}}{\mathrm{m}^2}\right) = + \left(\frac{\mathrm{J}}{\mathrm{m}^2\,\mathrm{s}\,\mathrm{K}}\right)\left(\frac{\mathrm{m}^2}{\mathrm{kg}}\right)\left(\frac{\mathrm{kg}\,\mathrm{K}}{\mathrm{J}}\right)\mathrm{K} - \left(\frac{\mathrm{J}}{\mathrm{m}^2\,\mathrm{s}\,\mathrm{K}}\right)\left(\frac{\mathrm{m}^2}{\mathrm{kg}}\right)\left(\frac{\mathrm{kg}\,\mathrm{K}}{\mathrm{J}}\right)\mathrm{K}$$

2.17 Dimensionless groups

It would not be proper to proceed further without some discussion of dimensionless groups which arise in both heat-transfer and flow-friction correlations used in the design of heat exchangers. This subject may require deeper study in other texts, as here it has been simplified as far as seems practicable without destroying fundamental concepts of dimensional analysis of linear systems.

Rayleigh's method and Buckingham's π-theorem

The reader may come across one or both of these algebraic approaches used in finding dimensionless groups. There is some merit in examining both methods, for situations do exist where the form of the differential equations governing the phenomena under consideration may not be known.

With both these approaches it is necessary first to intelligently 'guess' the number of independent variables involved in a problem. If too many are guessed then the number of dimensionless groups may become over-large. If too few are guessed then valid groups will still be produced, but they will be unfamiliar and difficult to apply.

When the governing differential equations are known in advance, the exact number of dimensionless groups can be extracted from them quite naturally. This is the approach adopted below.

Fundamental approach via differential equations

A differential equation is a mathematical model of a whole class of phenomena (Luikov, 1966). To obtain one particular solution from the multitude of possible solutions we must provide additional information – the conditions of single-valuedness. Into these conditions enter:

(a) *geometrical properties* of the system
(b) *physical properties* of the bodies involved in the phenomena under consideration

(c) *initial conditions* describing the state of the system at the first instant
(d) *boundary conditions* giving the interaction of the system with its surroundings

Two conditions are similar if they are described by one and the same system of differential equations and have similar conditions of single-valuedness.

We recognize the concepts of:

- a class of phenomena – partial differential equations
- a group of phenomena – similarity
- a single phenomenon – partial differential equations plus conditions of single-valuedness

Similarity in transient thermal conduction

Examine the case without internal heat generation. There is no increased difficulty with heat generation, but it introduces another parameter. Consider the Cartesian form of the 'energy balance + Fourier constitutive' differential equation with constant physical properties.

$$\rho C\frac{\partial T}{\partial t} = \lambda\left(\frac{\partial^2 T}{\partial x^2} + \frac{\partial^2 T}{\partial y^2} + \frac{\partial^2 T}{\partial z^2}\right)$$

For body 1 this becomes

$$\frac{\partial T_1}{\partial t_1} = \kappa_1\left(\frac{\partial^2 T_1}{\partial x_1^2} + \frac{\partial^2 T_1}{\partial y_1^2} + \frac{\partial^2 T_1}{\partial z_1^2}\right) \qquad \text{where} \quad \kappa_1 = \frac{\lambda_1}{\rho_1 C_1}$$

If the surroundings are at T_0 then

$$\frac{\partial \theta_1}{\partial t_1} = \kappa_1\left(\frac{\partial^2 \theta_1}{\partial x_1^2} + \frac{\partial^2 \theta_1}{\partial y_1^2} + \frac{\partial^2 \theta_1}{\partial z_1^2}\right) \qquad \text{where} \quad \theta_1 = T_1 - T_0 \tag{2.39}$$

For body 2, the corresponding equation is

$$\frac{\partial \theta_2}{\partial t_2} = \kappa_2\left(\frac{\partial^2 \theta_2}{\partial x_2^2} + \frac{\partial^2 \theta_2}{\partial y_2^2} + \frac{\partial^2 \theta_2}{\partial z_2^2}\right) \tag{2.40}$$

Let the quantities referring to body 2 be related everywhere and for all times to the corresponding quantities of the first body where the Γ values are constants of proportionality, then

$$\begin{aligned} x_2 &= (\Gamma_x)x_1 & t_2 &= (\Gamma_t)t_1 \\ y_2 &= (\Gamma_y)y_1 & \kappa_2 &= (\Gamma_\kappa)\kappa_1 \\ z_2 &= (\Gamma_z)z_1 & \theta_2 &= (\Gamma_\theta)\theta_1 \end{aligned}$$

In equation (2.40) we can therefore write

$$\frac{\Gamma_\theta}{\Gamma_t}\frac{\partial \theta_1}{\partial t_1} = \Gamma_\kappa \kappa_1\left(\frac{\Gamma_\theta \partial^2 \theta_1}{\Gamma_x^2 \partial x_1^2} + \frac{\Gamma_\theta}{\Gamma_y^2}\frac{\partial^2 \theta_1}{\partial y_1^2} + \frac{\Gamma_\theta}{\Gamma_z^2}\frac{\partial^2 \theta_1}{\partial z_1^2}\right) \tag{2.41}$$

Equation (2.41) will be identical to equation (2.39), and thus the heat flow in the two bodies similar, providing

$$\frac{\Gamma_\theta}{\Gamma_t} = 1 \tag{2.42}$$

$$\frac{\Gamma_\kappa \Gamma_\theta}{\Gamma_x^2} = \frac{\Gamma_\kappa \Gamma_\theta}{\Gamma_y^2} = \frac{\Gamma_\kappa \Gamma_\theta}{\Gamma_z^2} = 1 \tag{2.43}$$

First, from equation (2.43) it follows that

$$\Gamma_x = \Gamma_y = \Gamma_z \qquad (= \Gamma_\ell, \text{ say}) \tag{2.44}$$

Thus

$$\frac{x_1}{x_2} = \frac{y_1}{y_2} = \frac{z_1}{z_2} = \frac{\ell_1}{\ell_2} \tag{2.45}$$

where ℓ_1 and ℓ_2 are characteristic (or reference) lengths similarly defined in the two bodies. In other words, equation (2.45) implies that the bodies must be geometrically similar.

Second, from equations (2.42) and (2.43) it follows that

$$\frac{\Gamma_\theta}{\Gamma_t} = \frac{\Gamma_\kappa \Gamma_\theta}{\Gamma_x^2}, \qquad \text{i.e.} \quad \frac{\Gamma_\kappa \Gamma_t}{\Gamma_x^2} = 1$$

or, where ℓ is any characteristic dimension, it follows that

$$\frac{\kappa_1 t_1}{\ell_1^2} = \frac{\kappa_2 t_2}{\ell_2} = \text{Fo} \quad \text{(Fourier number)} \tag{2.46}$$

The Fourier number, which includes the physical constants, is in a sense 'generalized time', and must be dimensionless because the Γ values are dimensionless.

$$\text{Fo} = \frac{\lambda}{\rho C}\frac{t}{\ell^2} \qquad \frac{\text{J}}{(\text{m s K})}\frac{\text{m}^3}{\text{kg}}\frac{(\text{kg K})}{\text{J}}\frac{\text{s}}{\text{m}^2}$$

If equations (2.45) and (2.46) are satisfied, then the temperature distribution in the two bodies will be similar, provided the boundary conditions and the initial conditions are also similar.

Simple boundary conditions are illustrated in Fig. 2.23, and these relate to equation (2.47). T_{bulk} is the temperature of the flowing stream and T_s is the surface temperature. The difference, $T_s - T_{bulk} = \theta_s$, is the temperature difference across the boundary layer, and θ is the temperature in the solid wall.

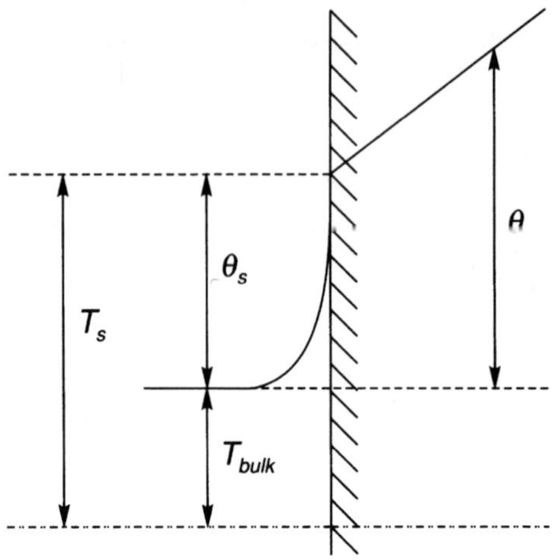

Fig.2.23 Surface temperature profiles

At the boundary

$$\left\{\begin{matrix}\text{heat tranported across} \\ \text{boundary surface}\end{matrix}\right\} = \left\{\begin{matrix}\text{heat flowing in} \\ \text{body at surface}\end{matrix}\right\}$$

$$\alpha\theta_s = -\lambda\left(\frac{d\theta}{d\ell}\right)_s \tag{2.47}$$

where

α = surface heat transfer coefficient (J/m^2 s K)
θ_s = temperature excess of surface above reference (K)
ℓ = dimension normal to the surface (m)

Then by the same argument as before

$$\alpha_2 = (\Gamma_\alpha)\alpha_1, \qquad \theta_2 = (\Gamma_\theta)\theta_1$$
$$\lambda_2 = (\Gamma_\lambda)\lambda_1, \qquad \ell_2 = (\Gamma_\ell)\ell_1$$

thus

$$(\Gamma_\alpha\Gamma_\theta)\alpha_1\theta_{b_1} = -\left(\frac{\Gamma_\lambda\Gamma_\theta}{\Gamma_\ell}\right)\lambda_1\left(\frac{d\theta_1}{d\ell_1}\right)_b$$

and the further condition is required that

$$\Gamma\alpha = \frac{\Gamma_\lambda}{\Gamma_t}, \qquad \text{i.e. } \frac{\Gamma_\alpha\Gamma_t}{\Gamma_\lambda} = 1$$

or where ℓ is a characteristic dimension

$$\frac{\alpha_1 \ell_1}{\lambda_1} = \frac{\alpha_2 \ell_2}{\lambda_2} = \text{Bi (Biot number)} \tag{2.48}$$

The Biot number differs from the Nusselt number in that λ refers to the solid, and not to the fluid surrounding the body.

The condition that the ratio of the temperatures θ at any point in the bodies to their surface temperatures θ_s is constant must also apply. This gives the condition of similarity of temperature distribution throughout the bodies at all times, including similarity at the start, i.e. of the initial conditions.

Thus the relationships

$$\frac{x}{\ell} = \text{const.,} \qquad \text{Fo} = \text{const.,} \qquad \text{Bi} = \text{const.,} \qquad \frac{\theta}{\theta_s} = \text{const.}$$

define the conditions for similarity of heat conduction in a solid body. Transient heat flow is therefore characterized by relations of the form

$$f\left(\text{Fo,Bi,} \frac{\theta}{\theta_s}, \frac{x}{\ell}\right) = 0 \qquad \text{or} \qquad \frac{\theta}{\theta_s} = \phi\left(\text{Fo,Bi,} \frac{x}{\ell}\right)$$

Comparison with analytical solution

To illustrate the connection between analytical solutions and conditions of similarity, consider the problem of a wall of finite thickness ℓ, heated on both sides in such a way that the surface temperatures are suddenly raised and maintained constant at temperature T_s.

The basic 'energy balance + Fourier constitutive' differential equation governing this problem is

$$\rho.C\frac{\partial T}{\partial t} = \lambda \frac{\partial^2 T}{\partial x^2} \tag{2.49}$$

with initial conditions $T = 0$ at $0 < x < L$ and surface conditions $T = T_s$ at $x = 0$ and $x = L$.

The analytical solution to this problem is given in terms of a Fourier series, which converges in about five terms

$$\frac{\theta}{\theta_s} = \frac{2}{\pi}\sum_{1}^{n}\left[\{1 - \cos(n\pi)\}\frac{1}{n}\exp\left\{-(n\pi)^2\left(\frac{\kappa t}{\ell^2}\right)\right\}\sin\left\{n\pi\left(\frac{x}{\ell}\right)\right\}\right] \tag{2.50}$$

where the term $\{1 - \cos(n\pi)\}$ is either 2 or 0, thus

$$\frac{\theta}{\theta_s} = f\left[\left(\frac{\kappa t}{\ell^2}\right), \left(\frac{x}{\ell}\right)\right] = f\left[\text{Fo,} \left(\frac{x}{\ell}\right)\right]$$

where

$$\frac{\theta}{\theta_s} = \frac{(T - T_0)}{(T_s - T_0)}$$

The Biot number does not enter into this solution because surface temperatures were *specified* in fixing the boundary conditions. If surface heat-transfer coefficients had also been involved, the relationship would be of the form

$$\frac{\theta}{\theta_s} = f\left[\text{Fo,Bi,}\left(\frac{x}{\ell}\right)\right]$$

Williamson & Adams (1919) developed analytical solutions for the history of centre-line temperatures for a number of shapes whose surface temperatures were suddenly changed to a new value. The shapes considered were:

- infinitely wide slab
- cylinder with length equal to diameter
- cylinder with infinite length
- cube
- square bar
- sphere

Of these, the infinitely wide slab is the simplest case as heat flows along one axis only (thickness). A graphical plot prepared for this analytical solutions is presented in Fig. 2.24, illustrating the use of dimensional groups as coordinates.

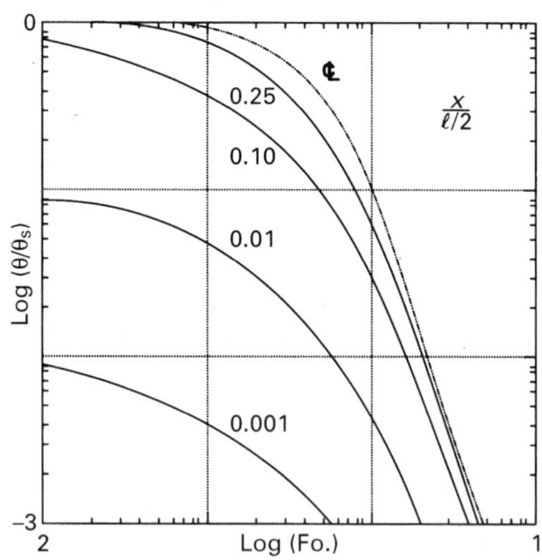

Fig.2.24 Dimensionless plot of temperature history of an infinite flat plate with step change at the surface

Convective heat transfer

Turning to convective heat transfer, the basic differential equations become extremely complex, and analytical solutions exist only for very simple physical situations. To appreciate this, it would be necessary to solve simultaneously:

(a) the single balance of mass equation

$$\frac{\partial \rho}{\partial t} + \frac{\partial (\rho u)}{\partial x} + \frac{\partial (\rho v)}{\partial y} + \frac{\partial (\rho w)}{\partial z} = 0$$

(b) the three Navier–Stokes field equations (momentum balance + Newtonian constitutive), describe the velocity components of a Newtonian fluid at each point in the fluid and at each instant of time. Only the x-direction equation is given below

$$\rho\left(u\frac{\partial u}{\partial x} + v\frac{\partial u}{\partial y} + w\frac{\partial u}{\partial z} + \frac{\partial u}{\partial t}\right) = \rho X - \frac{\partial p}{\partial x} + \frac{1}{3}\eta\frac{\partial}{\partial x}\left(\frac{\partial u}{\partial x} + \frac{\partial v}{\partial y} + \frac{\partial w}{\partial z}\right) + \eta\nabla^2 u$$

(c) the single-field equation (energy balance + Newtonian constitutive), gives the temperature distribution throughout the fluid with respect to (w.r.t.) location of each point in space w.r.t. time

$$\rho C\left(u\frac{\partial T}{\partial x} + v\frac{\partial T}{\partial y} + w\frac{\partial T}{\partial z} + \frac{\partial T}{\partial t}\right) = \left(u\frac{\partial p}{\partial x} + v\frac{\partial p}{\partial y} + w\frac{\partial p}{\partial z} + \frac{\partial p}{\partial t}\right) + \lambda\nabla^2 T + \eta\Phi$$

where Φ is the Rayleigh dissipation function

$$\Phi = 2\left[\left(\frac{\partial u}{\partial x}\right)^2 + \left(\frac{\partial v}{\partial y}\right)^2 + \left(\frac{\partial w}{\partial z}\right)^2\right] + \left[\left(\frac{\partial u}{\partial y} + \frac{\partial v}{\partial x}\right)^2 + \left(\frac{\partial v}{\partial z} + \frac{\partial w}{\partial y}\right)^2 + \left(\frac{\partial w}{\partial x} + \frac{\partial u}{\partial z}\right)^2\right] - \frac{2}{3}\left[\frac{\partial u}{\partial x} + \frac{\partial v}{\partial y} + \frac{\partial w}{\partial z}\right]^2$$

These equations must be solved in association with:

- boundary conditions (velocity and temperature conditions at the surface)
- initial conditions (velocity and temperature conditions at time zero)
- temperature-dependent physical properties

Referring back to the very simple conduction equation (2.49) and its more complex analytical solution (2.50), it is not surprising that general solutions for the simultaneous linear differential equations describing fluid flow have not been found.

In forced turbulent convection simple experimental correlations for fluids and gases flowing through pipes, may be written

$$\mathrm{Nu} = 0.023(\mathrm{Re})^{0.8}(\mathrm{Pr})^{0.4} \tag{2.51}$$

Comparing this with equation (2.50) it is easy to induce that it might be better written as a more complicated series expansion

$$\mathrm{Nu} = a_1(\mathrm{Re})^{b_1}(\mathrm{Pr})^{c_1} + a_2(\mathrm{Re})^{b_2}(\mathrm{Pr})^{c_2} + \cdots$$

This comparison suggests simply that empirical correlations are at best an approximation to what is actually happening, that they should be used with caution, and that their range of applicability must always be known.

Dimensionless groups in heat transfer and fluid flow

It is straightforward to set about extracting dimensionless groups from the Navier–Stokes and Newtonian energy balance equations. This is explained in Schlichting (1960), and in other engineering texts.

The extraction will not be repeated here, and it will suffice to provide some physical interpretation of the dimensionless groups which may be encountered in experimental correlations for heat transfer and fluid flow.

The dimensionless groups involved would include the following:

From similarity of the velocity fields (Navier–Stokes)

$$\text{Reynolds number, Re} = \frac{\ell G}{\eta} = \frac{\text{inertia force}}{\text{viscous force}}$$

$$\text{Grashof number, Gr} = \frac{\mathrm{Ra}}{\mathrm{Pr}} = \frac{(\beta g)\rho^2\ell^3\theta}{\eta^2} = \frac{\text{buoyancy force}}{\text{viscous force}}$$

$$\text{Euler number, Eu} = \frac{\Delta p}{\rho u^2} = \frac{\text{pressure force}}{\text{inertia force}}$$

From similarity of the temperature fields (Newtonian energy balance)

$$\text{Eckert number,}\quad \mathrm{Ec} = \frac{u^2}{C.\theta} = \frac{2 \times \text{temperature increase at stagnation}}{\text{temperature difference between wall and fluid}}$$

$$\text{perhaps to be understood from } \frac{mu^2/2}{mC\theta} = \frac{\text{kinetic energy}}{\text{thermal energy}}$$

$$\text{Mach number,}\quad \mathrm{Ma} = \sqrt{\mathrm{Ec}/(\gamma - 1)}$$

$$= \frac{u}{a} = \frac{\text{velocity of fluid flow}}{\text{speed of sound in fluid}} \text{ for perfect gas}$$

$$\text{Peclet number,} \quad \text{Pe} = \text{Pr Re} = \frac{u\ell}{\kappa} = \frac{\text{heat transfer by convection}}{\text{heat transfer by conduction}}$$

$$\text{Prandtl number,} \quad \text{Pr} = \frac{\text{Pe}}{\text{Re}} = \frac{C\eta}{\lambda} = \frac{\eta/\rho}{\kappa} = \frac{\text{momentum diffusivity}}{\text{thermal diffusivity}}$$

From similarity at the boundary

$$\text{Nusselt number, Nu} = \frac{\alpha\ell}{\lambda} = \frac{\text{total heat transfer}}{\text{conductive heat transfer of fluid}}$$

From geometric similarity

One, two, or three lengths as appropriate, e.g. (d/ℓ) as one length ratio.

A general function obtained from governing equations for convective heat transfer may look like

$$\text{Nu} = f\left(\text{Re, Pr, Gr, Ec, } \frac{d}{\ell}\right)$$

Whether the Eckert number need be present may be determined by the Mach number, which is a measure of whether heating effects caused by compressibility are likely to be important. If $\text{Ma} < 0.30$, then the fluid may be regarded as incompressible and the Eckert number can be omitted.

It is not usual to have both Reynolds number and Grashof number present together, because Reynolds number applies to forced convection, while Grashof number applies to natural convection. Only when the magnitude of the two effects are of similar order will it be necessary to include both numbers. The Peclet number is adequately represented by the (Pr, Re) groups and need not be explicitly present. The Stanton number may be used to replace the Nusselt number in some correlations, viz.

$$\text{St} = \frac{\text{Nu}}{\text{Re Pr}}$$

Typical heat-transfer correlations are

$$\text{Nu} = 0.027(\text{Re})^{0.8}(\text{Pr})^{0.33}\left(\frac{\eta_b}{\eta_w}\right)^{0.14}$$

Forced turbulent convection inside a tube

$$\text{Nu} = \left\{0.6 + \frac{0.387\text{Ra}^{1/6}}{[1 + (0.559/\text{Pr})^{9/16}]^{8/27}}\right\}^2$$

Natural convection air over horizontal pipes

where the Rayleigh number Ra is the product of Grashof and Prandtl numbers.

The Euler number provides a pressure-loss coefficient for flow, $\text{Eu} = \phi(\text{Re})$. At the elementary level used in heat transfer, the friction factor (f) provides the link. For forced turbulent convection inside a tube,

$$f = 0.046(\text{Re})^{-0.2} \qquad \Delta p = \frac{4fG^2}{2\rho}\left(\frac{\ell}{d}\right)$$

Flow drag expressions in natural convection may be more complicated.

Coupling between the equation for heat transfer and the equation for pressure loss is through the Reynolds number, and these effects are separable because the Eckert number is small, i.e. thermal effects due to friction are small. When the Eckert number is large, thermal effects due to compressibility become significant, e.g. aerodynamic heating in high-speed aircraft.

The reader is referred to Bejan (1995) for an up to date treatment of correlations.

Applicability of dimensionless groups

There are many applications where dimensional analysis provides information which would not otherwise be easily seen, e.g. Obot *et al.* (1991) and Obot (1993) who extend flow similarity concepts to include transition to turbulent flow for different channel geometries.

Similarity can also be applied to mechanical structures (see e.g. Lessen, 1953; Dugundji & Calligeros, 1962; Hovanesian & Kowalski, 1967; Jones, 1974). However the principal applications have been in the field of fluid mechanics and heat transfer, illustrated by the papers by Boucher & Alves (1959), Klockzien & Shannon (1969), and Morrison (1969).

The reader may be impressed by the number of dimensionless groups listed by Catchpole & Fulford (1966, 1968).

In using heat-transfer and flow-friction correlations it is not essential to have a correlation expressed in mathematical form, e.g.

$$\mathrm{Nu} = 0.023(\mathrm{Re})^{0.8}(\mathrm{Pr})^{0.4}$$

This equation is simply a mathematical 'best' fit to a graph of experimental data, and frequently a better fit can be produced employing an interpolating cubic spline-fit which allows for individual experimental errors at each data point. Some recommendations on spline-fitting procedures are given in Appendix B.6, at the end of the book.

References

Bejan, A. (1995) *Convective Heat Transfer*, 2nd edn, Wiley Interscience.

Bhatti, M.S. and **Shah, R.K.** (1987) Laminar convective heat transfer in ducts. *Handbook of Single-phase Heat Transfer*, Chapter 3 (Eds. S. Kakaç, R.K. Shah, and W. Aung), John Wiley, New York.

Boucher, D.F. and **Alves, G.E.** (1959) Dimensionless numbers for fluid mechanics, heat transfer, mass transfer and chemical reaction. *Chem. Engng Progress*, **55**(9), September, 55–83.

Catchpole, J.P. and **Fulford, G.** (1966) Dimensionless groups. *Ind. Engng Chem.*, **58**(3), March, 46–60.

Catchpole, J.P. and **Fulford, G.** (1968) Dimensionless groups. *Ind. Engng Chem.*, **60**(3), March, 71–78.

Clayton, D.G. (1984) Increasing the power of the LMTD method for heat exchangers. *Int. J. Mech. Engng Education*, **13**(3), 183–190.

Dugundji, J. and **Calligeros, J.M.** (1962) Similarity laws for aerothermoelastic testing. *J. Aerospace Sci.*, **29**, August, 935–950.

Grassmann, P. and **Kopp, J.** (1957) Zur gunstigen Wahl der Temperaturdifferenz und der Wärmeübergangszahl in Wärmeaustauchern. *Kaltetechnik*, **9**(10), 306–308.

Hewitt, G.F., **Shires, G.L.**, and **Bott, T.R.** (1994) *Process Heat Transfer*, CRC Press, Florida.

Hovanesian, J.D. and **Kowalski, H.C.** (1967) Similarity in elasticity. *Exp. Mechanics*, **7**, February, 82–84.

Jones, N. (1974) Similarity principles in structural mechanics. *Int. J. Mech. Engng Education*, **2**(2), 1–10.

Klockzien, V.G. and **Shannon, R.L.** (1969) Thermal scale modelling of spacecraft. *Soc. Automotive Engineers*, 13–17 January 1969, Paper 690196.

Lessen, M. (1953) On similarity in thermal stresses in bodies. *J. Aerospace Sci.*, **20**(10), 716–717.

Luikov, A.V. (1966) *Heat and Mass Transfer in Capillary-Porous Bodies* (English Translation by P.W.B. Harrison and W.M. Pun), Pergamon.

Morrison, F.A. (1969) Generalised dimensional analysis and similarity analyses. *Bull. Mech. Engng Education*, **8**, 289–300.

Obot, N.T. (1993) The frictional law of corresponding states: its origin and applications. *Trans. Inst. Chem. Engineers*, **71**(A), January, 3–10.

Obot, N.T., **Jendrzejczyk, J.A.**, and **Wambsganss, M.W.** (1991) Direct determination of the onset of transition to turbulence in flow passages. *Trans. ASME, J. Fluids Engng*, **113**, 602–607.

Schlichting, H. (1960) *Boundary Layer Theory*, 4th edn, McGraw-Hill, New York.

Spalding, D.B. (1990) Analytical solutions. *Hemisphere Handbook of Heat Exchanger Design* (Ed. G.F. Hewitt), Section 1.3.1-1, Hemisphere, New York.

Taborek, J. (1983) *Heat Exchanger Design Handbook*, vol. 1, Section 1.5, Hemisphere, New York.

Webb, R.L. (1994) *Principles of Enhanced Heat Transfer*, Table 2.2, John Wiley, p. 43.

Williamson, E.D. and **Adams, L.H.** (1919) Temperature distribution in solids during heating or cooling. *Physical Rev.*, **14**, 99–114.

Bibliography

Herbein, D.S. (1987) Comparison of entropy generation and conventional design methods for heat exchangers. *MS Thesis: Massachusetts Institute of Technology*, June 1987. (The author is grateful to Captain David Herbein for making a copy of this thesis available.)

Paterson, W.R. (1984) A replacement for the logarithmic mean. *Chem. Engng Sci.*, **39**(11), 1635–1636.

Underwood, A.J.V. (1933) Graphical computation of logarithmic mean temperature difference. *Ind. Chemist*, May, 167–170.

CHAPTER 3
Steady-State Temperature Profiles

Mostly ordinary differential equations

3.1 Linear temperature profiles in contraflow

This is a special case of contraflow that is of interest for heat exchangers in idealized recuperated gas turbine plant. Proof of linear temperature profiles requires a simple introduction to the development of the differential equations that govern temperature distributions (see Fig. 3.1).

Taking differential energy balances.

Hot fluid

$$\left\{\begin{matrix}\text{energy entering}\\ \text{with hot fluid}\end{matrix}\right\} - \left\{\begin{matrix}\text{energy leaving}\\ \text{with hot fluid}\end{matrix}\right\} - \left\{\begin{matrix}\text{heat transferred}\\ \text{to cold fluid}\end{matrix}\right\} = \left\{\begin{matrix}\text{energy stored}\\ \text{in hot fluid}\end{matrix}\right\}$$

$$\dot{m}_h C_h T_h - \dot{m}_h C_h \left(T_h + \frac{dT_h}{dx}\delta x \right) - U\left(S\frac{\delta x}{L} \right)(T_h - T_c) = 0$$

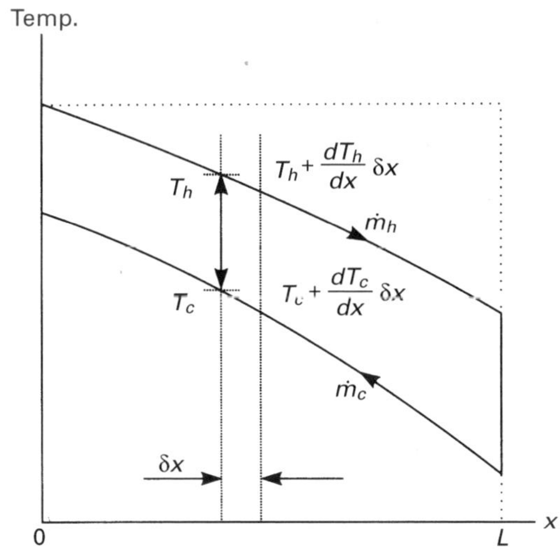

Fig.3.1 Arbitrary temperature profiles

giving

$$\frac{dT_h}{dx} = -\frac{US}{\dot{m}_h C_h} \cdot \frac{1}{L}(T_h - T_c) \tag{3.1}$$

Cold fluid

$$\dot{m}_c C_c \left(T_c + \frac{dT_c}{dx} \delta x \right) + U \left(S \frac{\delta x}{L} \right) (T_h - T_c) - \dot{m}_c C_c T_c = 0$$

giving

$$\frac{dT_c}{dx} = -\frac{US}{\dot{m}_c C_c} \cdot \frac{1}{L}(T_h - T_c) \tag{3.2}$$

Writing overall values of *Ntu* as

$$N_h = \frac{US}{\dot{m}_h C_h} \qquad \text{and} \qquad N_c = \frac{US}{\dot{m}_c C_c}$$

the coupled equations (3.1) and (3.2) become

$$\boxed{\begin{aligned} \frac{dT_h}{dx} &= -\frac{N_h}{L}(T_h - T_c) \\ \frac{dT_c}{dx} &= -\frac{N_c}{L}(T_h - T_c) \end{aligned}} \tag{3.3}$$

For equal water equivalents $N_h = N_c$, and it follows that $dT_h/dx = dT_c/dx$, which shows that the gradients are the same at any 'x'. It is a necessary but not sufficient condition for straight and parallel temperature profiles. It only remains to show that one temperature profile is linear.

From equation (3.3)

$$T_h - T_c = -\left(\frac{L}{N_h} \right) \frac{dT_h}{dx}$$

and differentiating

$$\frac{dT_h}{dx} - \frac{dT_c}{dx} = -\left(\frac{L}{N_h} \right) \frac{d^2 T_h}{dx^2}$$

but

$$\frac{dT_h}{dx} = \frac{dT_c}{dx} \qquad \text{thus} \qquad \frac{d^2 T_h}{dx^2} = 0$$

hence a linear profile exists (Fig. 3.2).

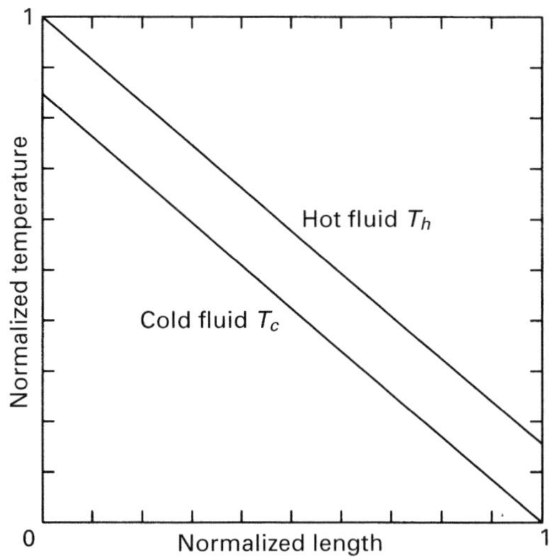

Fig.3.2 Normalized temperature profiles with $N_h = N_c = 5.0$

3.2 General cases of contraflow and parallel flow

In the treatments shown in Figs 3.3 and 3.4 there is no longitudinal conduction in the wall, no energy storage in the fluids or the wall (transients), no heat generation in the fluids or the wall, and no external losses.

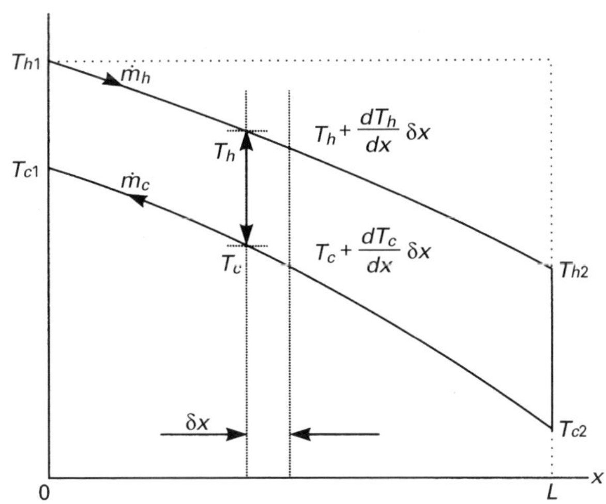

Fig.3.3 Contraflow

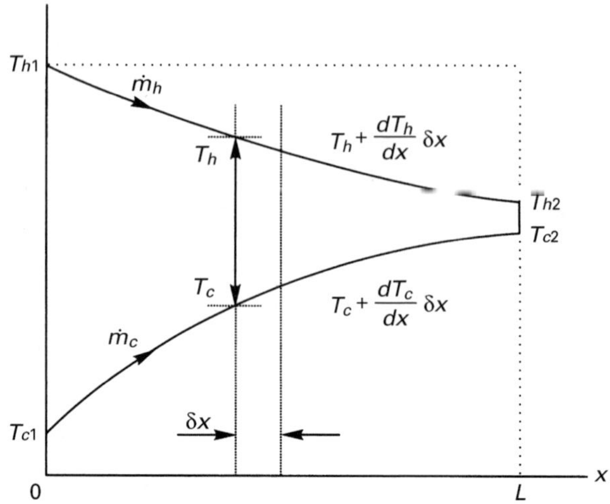

Fig.3.4 Parallel flow

The analysis for each of the above heat exchanger flow configurations is practically identical, thus only the contraflow exchanger will be considered. Similarly, the treatment of hot and cold fluids is virtually identical.

Hot fluid

$$\begin{Bmatrix}\text{energy entering}\\\text{with hot fluid}\end{Bmatrix} - \begin{Bmatrix}\text{energy leaving}\\\text{with hot fluid}\end{Bmatrix} - \begin{Bmatrix}\text{heat transferred}\\\text{to cold fluid}\end{Bmatrix} = \begin{Bmatrix}\text{energy stored}\\\text{in hot fluid}\end{Bmatrix}$$

$$\dot{m}_h C_h T_h - \dot{m}_h C_h \left(T_h + \frac{dT_h}{dx}\delta x \right) - U\left(S\frac{\delta x}{L} \right)(T_h - T_c) = 0$$

$$\frac{dT_h}{dx} = -\frac{US}{\dot{m}_h C_h} \cdot \frac{1}{L}(T_h - T_c) = -\frac{N_h}{L}(T_h - T_c) \tag{3.4}$$

Cold fluid

$$\frac{dT_c}{dx} = -\frac{US}{\dot{m}_c C_c} \cdot \frac{1}{L}(T_h - T_c) = -\frac{N_c}{L}(T_h - T_c) \tag{3.5}$$

Scaling of length x is possible by writing

$$\xi = \frac{x}{L} \cdot N_h \qquad \text{giving} \qquad \frac{d\xi}{dx} = \frac{N_h}{L}$$

so that with

$$\frac{dT}{dx} = \frac{dT}{d\xi} \cdot \frac{d\xi}{dx} = \frac{dT}{d\xi} \cdot \frac{N_h}{L}$$

equations (3.4) and (3.5) become for $0 < \xi < 1$

$$\boxed{\begin{aligned} \frac{dT_h}{d\xi} &= -(T_h - T_c) \\ \frac{dT_c}{d\xi} &= -\frac{N_c}{N_h}(T_h - T_c) \end{aligned}}$$

but including N_h in the independent variable takes too much information away from the engineering and it is better to write $\xi = x/L$. Then reverting back to the original notation

$$\boxed{\begin{aligned} \frac{dT_h}{dx} &= -N_h(T_h - T_c) \\ \frac{dT_c}{dx} &= -N_c(T_h - T_c) \end{aligned}}$$

Scaling and normalization will be useful later when compact notation is helpful, but for the present we can proceed more directly from equations (3.4) and (3.5).

From equation (3.4)

$$T_c = \frac{L}{N_h} \cdot \frac{dT_h}{dx} + T_h$$

Differentiating

$$\frac{dT_c}{dx} = \frac{L}{N_h} \cdot \frac{d^2T_h}{dx^2} + \frac{dT_h}{dx}$$

In equation (3.5)

$$\frac{L}{N_c}\left[\frac{L}{N_h} \cdot \frac{d^2T_h}{dx^2} + \frac{dT_h}{dx}\right] = -T_h + \left[\frac{L}{N_h} \cdot \frac{dT_h}{dx} + T_h\right]$$

$$\frac{d^2T_h}{dx^2} + \left[\frac{N_h - N_c}{L}\right] \cdot \frac{dT_h}{dx} = 0 \tag{3.6}$$

Similarly for the cold fluid

$$\frac{d^2T_c}{dx^2} + \left[\frac{N_h - N_c}{L}\right] \cdot \frac{dT_c}{dx} = 0 \tag{3.7}$$

The solution to equation (3.6) $T_h = A.e^{-(N_h - N_c)x/L} + B$ with boundary conditions

$$\left.\begin{aligned} x &= 0 \\ T_h &= T_{h1} \end{aligned}\right\} \qquad \left.\begin{aligned} x &= L \\ T_h &= T_{h2} \end{aligned}\right\}$$

from which the following dimensionless (and normalized) result is obtained, valid for both hot fluid and cold fluid in *contraflow*

$$\boxed{\frac{T_{h1} - T_h}{T_{h1} - T_{h2}} = \frac{1 - \exp\left[-(N_h - N_c)\dfrac{x}{L}\right]}{1 - \exp[-(N_h - N_c)]} = \frac{T_{c1} - T_c}{T_{c1} - T_{c2}}} \tag{3.8}$$

A similar analysis for the *parallel-flow* case produces

$$\boxed{\frac{T_{h1} - T_h}{T_{h1} - T_{h2}} = \frac{1 - \exp\left[-(N_h + N_c)\dfrac{x}{L}\right]}{1 - \exp[-(N_h + N_c)]} = \frac{T_{c1} - T_c}{T_{c1} - T_{c2}}} \tag{3.9}$$

Expressions $(N_h - N_c)$ and $(N_h + N_c)$ relate back to both the common effectiveness diagram, and 'rate' and 'energy' equations derived earlier in the LMTD–*Ntu* approach in Section 2.3.

The following dimensionless expressions control the temperature profiles, and provide means for assessing the useful length of the exchanger, i.e. the condition when the driving temperature differences in equations (3.8) and (3.9) cease to be effective. They are perhaps more fundamental than the concept of effectiveness itself, in that they come directly from the differential equations.

$$\frac{1 - \exp\left[-(N_h - N_c)\dfrac{x}{L}\right]}{1 - \exp[-(N_h - N_c)]} \quad \text{and} \quad \frac{1 - \exp\left[-(N_h + N_c)\dfrac{x}{L}\right]}{1 - \exp[-(N_h + N_c)]}$$

Figures 3.5 to 3.8 provide normalized temperature profiles for two cases of contraflow and two cases of parallel flow, respectively. The chain-dotted lines generated

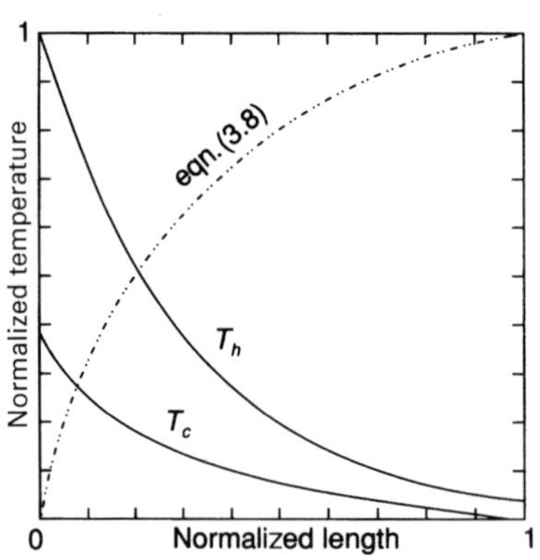

Fig.3.5 Contraflow. Normalized temperature profiles with $N_h = 5.0$, $N_c = 2.0$

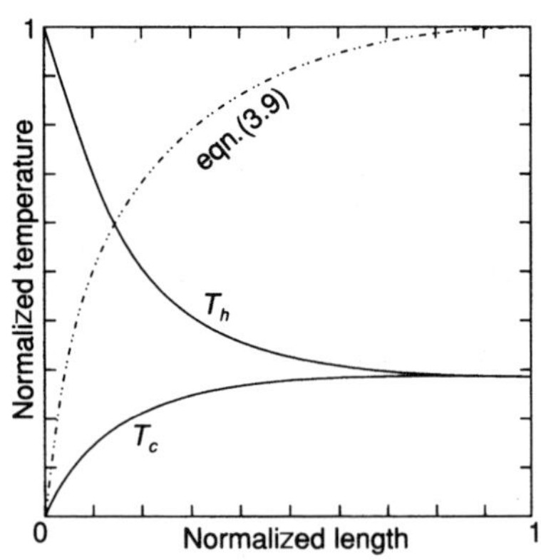

Fig.3.6 Parallel flow. Normalized temperature profiles with $N_h = 5.0$, $N_c = 2.0$

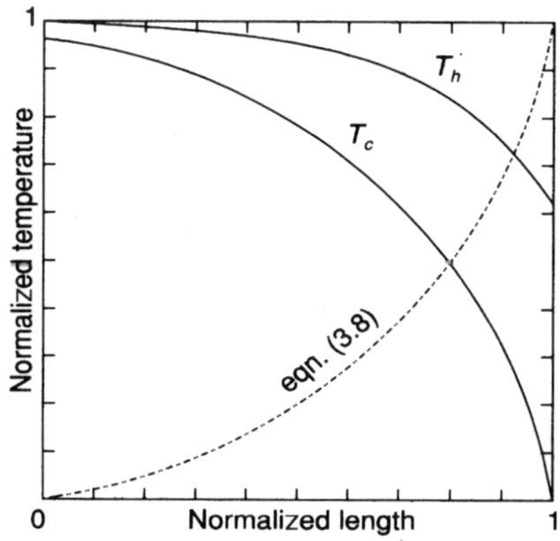

Fig.3.7 Contraflow. Normalized temperature profiles with $N_h = 2.0$, $N_c = 5.0$

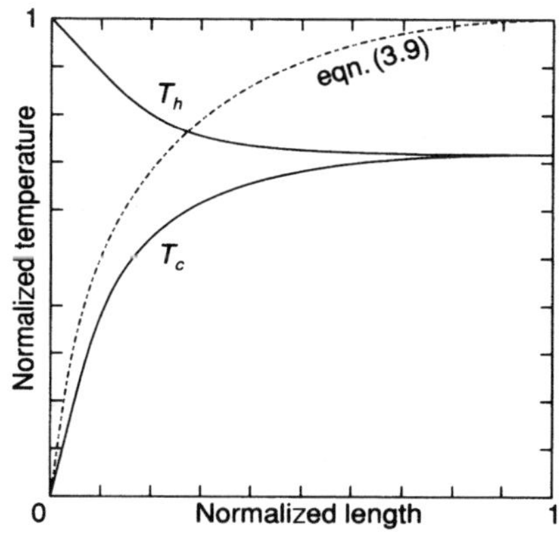

Fig.3.8 Parallel flow. Normalized temperature profiles with $N_h = 2.0$, $N_c = 5.0$

from the above dimensionless exponential ratios allow assessment of useful length of exchanger for effective heat transfer in practical cases.

Intermediate wall temperature

In steady-state heat exchange it is possible to obtain approximate expressions for wall temperature without formally introducing a differential equation for the wall. In such circumstances the wall has no stored energy (no transients), and there is no longitudinal conduction along the wall, so that wall temperature gradients at the ends are not correctly represented. The representation is, however, quite good. Wall temperatures are required whenever a heat-transfer correlation contains a correction factor of the form $(\eta_b/\eta_w)^{0.14}$.

At any intermediate point in the exchanger, differential heat flow and energy expressions may be written

$$\left.\begin{aligned} d\dot{Q} &= \dot{m}_h C_h dT_h = UdS(T_h - T_c) = \dot{m}_c C_c dT_c \\ &= \alpha_h\, dS(T_h - T_w) \qquad\qquad = \alpha_c\, dS(T_w - T_c) \end{aligned}\right\} \tag{3.10}$$

To confirm the consistency of these equations, write

$$T_h - T_w = \frac{dT_h}{dS} \cdot \frac{\dot{m}_h C_h}{\alpha_h} = \frac{U(T_h - T_c)}{\dot{m}_h C_h} \cdot \frac{\dot{m}_h C_h}{\alpha_h}$$

$$T_w - T_c = \frac{dT_c}{dS} \cdot \frac{\dot{m}_c C_c}{\alpha_c} = \frac{U(T_h - T_c)}{\dot{m}_c C_c} \cdot \frac{\dot{m}_c C_c}{\alpha_c}$$

Adding

$$(T_h - T_c) = (T_h - T_c)U\left[\frac{1}{\alpha_h} + \frac{1}{\alpha_c}\right]$$

which is only true if

$$\frac{1}{U} = \frac{1}{\alpha_h} + \frac{1}{\alpha_c}$$

seen to be correct for the case when no wall equation is present.

Again from the above derived equation

$$U(T_h - T_c) = \alpha_h(T_h - T_w) \quad \text{and} \quad U(T_h - T_c) = \alpha_c(T_w - T_c)$$

giving

$$\boxed{\frac{T_h - T_w}{T_h - T_c} = \frac{U}{\alpha_h} \quad \text{and} \quad \frac{T_w - T_c}{T_h - T_c} = \frac{U}{\alpha_c}} \tag{3.11}$$

In practical cases α_h and α_c will have been evaluated. It is not possible to obtain T_w from knowledge of N_h and N_c alone.

The results in equation (3.11) will apply *locally* to the case of simple cross-flow with both fluids unmixed and, by inspection, may apply to other cases as well.

By measuring two inlet and two outlet temperatures in parallel-flow and contra-flow exchangers it is practicable to calculate values of (N_h, N_c) using the LMTD-*Ntu* equations developed in Section 2.10. This provides a measure of the overall heat-transfer coefficient, which could be used to assess whether exchanger performance is deteriorating due to fouling. However, equations (3.11) above show that it would not be possible to determine which side of the exchanger is fouling without additional information.

3.3 Condensation and evaporation

Two special cases in which the temperature of one fluid remains constant require separate consideration. The analysis can be simplified by considering normalized temperatures from the start.

Although these cases are simple, their solutions also provide the inlet temperature distributions for the case of unmixed crossflow with constant inlet conditions, and thus are worthy of consideration.

Consider the case of condensation.

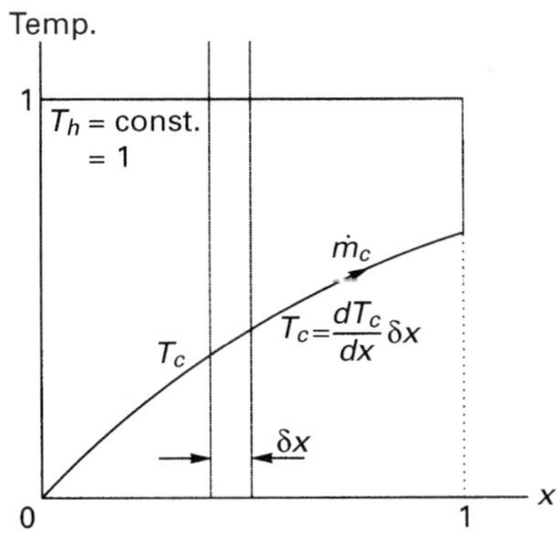

Fig.3.9 Condensation

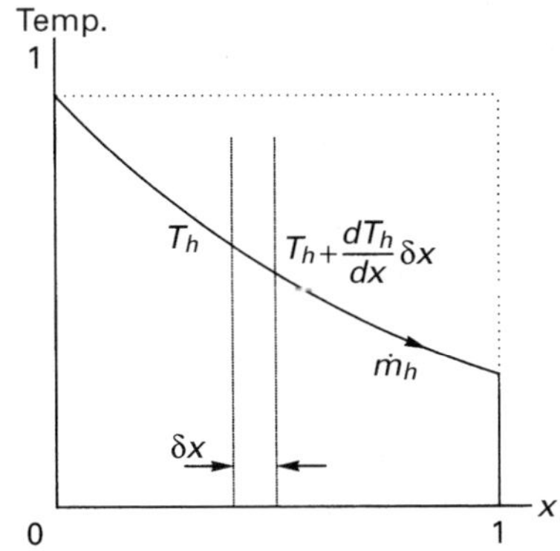

Fig.3.10 Evaporation

Cold fluid

$$\left\{\begin{matrix}\text{energy entering}\\ \text{with cold fluid}\end{matrix}\right\} - \left\{\begin{matrix}\text{energy leaving}\\ \text{with cold fluid}\end{matrix}\right\} + \left\{\begin{matrix}\text{heat transferred}\\ \text{from hot fluid}\end{matrix}\right\} = \left\{\begin{matrix}\text{energy stored}\\ \text{in cold fluid}\end{matrix}\right\}$$

$$\dot{m}_c C_c T_c - \dot{m}_c C_c\left(T_c + \frac{dT_c}{dx}\delta x\right) + U\left(S\frac{\delta x}{L}\right)(1 - T_c) = 0$$

$$\frac{US}{L}(1 - T_c) = \dot{m}_c C_c \frac{dT_c}{dx}$$

$$\frac{dT_c}{dx} + \frac{N_c}{L}\cdot T_c - \frac{N_c}{L} = 0$$

which provides the solution

$$\boxed{\begin{matrix}\textit{Condensation}\\ T_c = 1 - e^{-N_c(x/L)}\end{matrix}} \tag{3.12}$$

Similarly for the case of evaporation

$$\boxed{\begin{matrix}\textit{Evaporation}\\ T_h = e^{-N_h(x/L)}\end{matrix}} \tag{3.13}$$

3.4 Longitudinal conduction in contraflow

Longitudinal conduction in the direction of falling temperature is a problem in high-effectiveness contraflow heat exchangers, particularly in the design of cryogenic plant. Conduction will occur in each fluid, in the wall separating the two fluids and in the shell of the exchanger. For gases, conduction effects in the fluid can usually be neglected,[1]

[1]Thermal conductivity in the supercritical liquid region starts to become important.

and with plate–fin heat exchangers shell conduction losses are also small, leaving only longitudinal conduction in the wall as a significant effect.

Equal water equivalents

The following analysis is based on the paper by Kroeger (1966) which includes a closed-form solution for the case of equal water equivalents – identical to that for the case of most severe deterioration in heat exchanger performance. Given the terminal temperatures of a contraflow heat exchanger, the problem is to find the corrected LMTD to use in design when longitudinal conduction effects are present.

In sizing a heat exchanger, a first design pass is necessary to determine the approximate size of the exchanger and obtain the wall cross-section involved in longitudinal conduction. The reduction in LMTD is then obtained using Kroeger's method, and this reduced value of LMTD is applied in a second design pass to obtain the final size of the exchanger.

This avoids the necessity of solving three simultaneous partial differential equations for each individual case. The approach is approximate, but as maximum deterioration in performance has been incorporated, the final design should be conservative.

The balance of energy equations for the steady state when longitudinal conduction is present is obtained from Appendix A.1, equations (A.1), by omitting

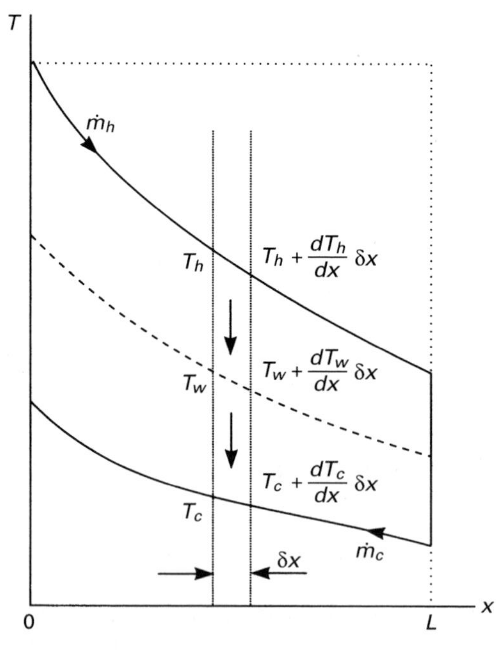

Fig.3.11 Schematic temperature profiles including the wall

time-dependent terms, and writing the effective thermal diffusivity as

$$\hat{\kappa}_w = \left(\frac{A_w L}{V_w}\right)\kappa_w = \left(\frac{A_w L}{V_w}\right)\frac{\lambda_w}{\rho_w C_w} = \frac{\lambda_w A_w L}{M_w C_w} \tag{3.14}$$

$$\begin{aligned}
+u_h\frac{\partial T_h}{\partial x} &= -\frac{u_h}{L}\left(\frac{\alpha_h S}{\dot{m}_h C_h}\right)(T_h - T_w) \\
-\hat{\kappa}_w\frac{\partial^2 T_w}{\partial x^2} &= +\left(\frac{\alpha_h S}{M_w C_w}\right)(T_h - T_w) - \left(\frac{\alpha_c S}{M_w C_w}\right)(T_w - T_c) \\
-u_c\frac{\partial T_c}{\partial x} &= +\frac{u_c}{L}\left(\frac{\alpha_c S}{\dot{m}_c C_c}\right)(T_w - T_c)
\end{aligned} \tag{3.15}$$

The central equation of equations (3.15) may be written as

$$+\left(\frac{\lambda AL}{M_w C_w}\right)\frac{d^2 T_w}{dx^2} = -\left(\frac{\alpha_h S}{M_w C_w}\right)(T_h - T_w) + \left(\frac{\alpha_c S}{M_w C_w}\right)(T_w - T_c) \tag{3.16}$$

Using the first and last of equations (3.15) and equation (3.14), the central equation may be written as

$$\left(\frac{\lambda_w A_w L}{M_w C_w}\right)\frac{d^2 T_w}{dx^2} = +\left(\frac{\dot{m}_h C_h L}{M_w C_w}\right)\frac{dT_h}{dx} - \left(\frac{\dot{m}_c C_c L}{M_w C_w}\right)\frac{dT_c}{dx}$$

and for the special case of equal water equivalents ($\dot{m}_h C_h = \dot{m}_c C_c = \dot{m}C$), and understanding that (λ, A) refer to the wall, we may write

$$\left(\frac{\lambda A}{\dot{m}C}\right)\frac{d^2 T_w}{dx^2} = \frac{dT_h}{dx} - \frac{dT_c}{dx} \tag{3.17}$$

Writing

$$n_h = \left(\frac{\alpha_h S}{\dot{m}_h C_h}\right) \quad \text{and} \quad n_c = \left(\frac{\alpha_c S}{\dot{m}_c C_c}\right)$$

as local *Ntu* values, the set of equations (3.15) now becomes

$$\begin{aligned}
\left(\frac{d}{dx} + \frac{n_h}{L}\right)T_h \qquad\qquad\qquad - \left(\frac{n_h}{L}\right)T_w &= 0 \\
-\frac{dT_h}{dx} \qquad + \frac{dT_c}{dx} + \left(\frac{\lambda A}{\dot{m}C}\right)\frac{d^2 T_w}{dx^2} &= 0 \\
+\left(\frac{d}{dx} - \frac{n_c}{L}\right)T_c \qquad + \left(\frac{n_c}{L}\right)T_w &= 0
\end{aligned} \tag{3.18}$$

Writing

$$\beta_1 = \frac{n_h}{L}, \qquad \beta_2 = \frac{n_c}{L}, \qquad \sigma = \frac{\lambda A}{\dot{m}C}, \qquad \mu = \frac{\dot{m}_h C_h}{\dot{m}_c C_c} = 1$$

and normalizing the length of the exchanger by putting $X = x/L$, then

$$\boxed{\begin{aligned} &\left(\frac{d}{dX} + \beta_1\right)T_h && && - \beta_1 T_w = 0 \\ &-\frac{dT_h}{dX} && +\frac{dT_c}{dX} && + \left(\frac{\sigma}{\mu}\right)\frac{d^2T_w}{dX^2} = 0 \\ & && +\left(\frac{d}{dX} - \beta_2\right)T_c && + \beta_2 T_w = 0 \end{aligned}}$$

corresponding with Kroeger's equations, which have the solutions

$$\boxed{\begin{aligned} T_h &= A_0 - \frac{A_1}{\beta_1} + A_1 X + \left(\frac{\beta_1}{\beta_1 - r_2}\right)e^{-r_2 X}A_2 + \left(\frac{\beta_1}{\beta_1 - r_3}\right)e^{-r_3 X}A_3 \\ T_w &= A_0 \qquad + A_1 X \qquad + e^{-r_2 X}A_2 \qquad + e^{-r_3 X}A_3 \\ T_c &= A_0 + \frac{A_1}{\beta_2} + A_1 X + \left(\frac{\beta_2}{\beta_2 + r_2}\right)e^{-r_2 X}A_2 + \left(\frac{\beta_2}{\beta_2 + r_3}\right)e^{-r_3 X}A_3 \end{aligned}} \tag{3.19}$$

where

$$r_2 = (p + q), \qquad r_3 = (p - q), \qquad p = \frac{(\beta_1 - \beta_2)}{2}$$

$$q = \sqrt{\frac{(\beta_1 + \beta_2)^2}{4} + \frac{1}{\sigma}(\beta_1 + \beta_2)}$$

The boundary conditions are

$$\left.\begin{aligned} X &= 0 \\ T_h &= 1 \\ dT_w/dX &= 0 \end{aligned}\right\} \qquad \left.\begin{aligned} X &= 0 \\ T_c &= 0 \\ dT_w/dX &= 0 \end{aligned}\right\}$$

Substituting in the first and last of equations (3.19)

$$1 = A_0 + \left(\frac{-1}{\beta_1}\right)A_1 + \left(\frac{\beta_1}{\beta_1 - r_2}\right)A_2 + \left(\frac{\beta_1}{\beta_1 - r_3}\right)A_3 \tag{3.20}$$

$$0 = A_0 + \left(\frac{1}{\beta_2} + 1\right)A_1 + \left[\left(\frac{\beta_2}{\beta_2 + r_2}\right)e^{-r_2}\right]A_2 + \left[\left(\frac{\beta_2}{\beta_2 + r_3}\right)e^{-r_3}\right]A_3 \tag{3.21}$$

Differentiating the central equation of equations (3.19)

$$\frac{dT_w}{dX} = A_1 + [-r_2 e^{-r_2 X}]A_2 + [-r_3 e^{-r_3 X}]A_3 \tag{3.22}$$

and substituting from the two boundary conditions given above

$$0 = A_1 + (-r_2)A_2 + (-r_3)A_3 \tag{3.23}$$

$$0 = A_1 + (-r_2 e^{-r_2})A_2 + (-r_3 e^{-r_3})A_3 \tag{3.24}$$

Subtracting equation (3.20) from equation (3.21)

$$\begin{aligned} -1 = {} & \left[\frac{1}{\beta_2} + 1 + \frac{1}{\beta_1}\right]A_1 + \left[\left(\frac{\beta_2}{\beta_2 + r_2}\right)e^{-r_2} - \left(\frac{\beta_1}{\beta_1 - r_2}\right)\right]A_2 \\ & + \left[\left(\frac{\beta_2}{\beta_2 + r_3}\right)e^{r_3} - \left(\frac{\beta_1}{\beta_1 - r_3}\right)\right]A_3 \end{aligned} \tag{3.25}$$

Simultaneous equations (3.23), (3.24), and (3.25) may be written

$$\left.\begin{aligned} -1 &= a_{11}A_1 + a_{12}A_2 + a_{13}A_3 \\ 0 &= a_{21}A_1 + a_{22}A_2 + a_{23}A_3 \\ 0 &= a_{31}A_1 + a_{32}A_2 + a_{33}A_3 \end{aligned}\right\}$$

and solved for A_1, A_2, and A_3. Equation (3.20) is then solved for A_0, allowing complete solution for temperature profiles (T_h, T_w, T_c) in equation (3.19).

Schematic temperature profiles

Analytical solution

The data in Table 3.1 have been selected to exaggerate the effect of longitudinal conduction on the performance of a contraflow exchanger.

Figure 3.12 applies to the case of equal water equivalents only, and parameters have been 'tweaked' to emphasize features. The group of curves marked 'Temperature profiles' have solid lines for actual temperature profiles and chain-dotted line for the wall temperature profile when longitudinal conduction is present. The

Table 3.1 Data for evaluating longitudinal conduction profiles

Parameter (units)	*Hot gas*	*Cold fluid*	*Solid wall*
Mass flowrate, kg/s	1.00	1.50	—
Specific heat, J/(kg K)	1200.0	800.0	—
Local transfer units, *Ntu*	5.00	4.00	—
Density, kg/s	—	—	1000.0
Cross-section, m^2	—	—	0.10
Exchanger length, m	—	—	0.50
Thermal conductivity, J/(m s K)	—	—	500.0

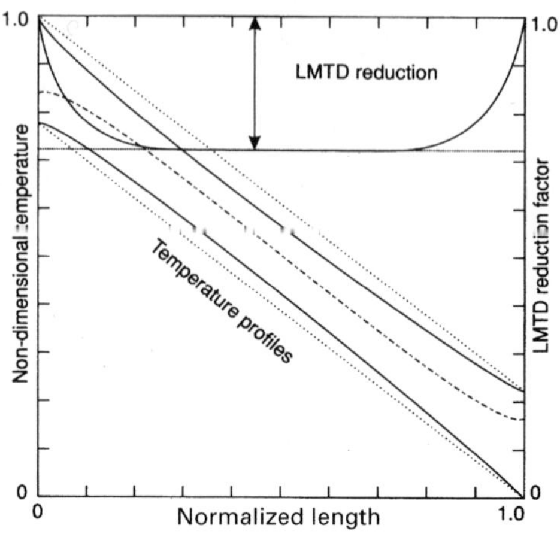

Fig.3.12 Temperature profiles with longitudinal conduction (equal water equivalents)

straight dotted lines in Fig. 3.12 show temperature profiles when longitudinal conduction is ignored. The LMTD reduction curve shows the difference in temperature of the two solid temperature profile lines.

Figure 3.14 applies to realistic profiles. Dotted lines have been omitted on the 'Temperature profiles' curves, and the LMTD reduction curves have been calculated numerically. The reduction to be applied to LMTD calculated from terminal temperatures is obtained numerically and is the average over 50 stations.

The computational results presented in Fig. 3.12 show fluid and solid wall temperatures deviating from expected linear temperature profiles. The reduced temperature difference along the exchanger and the LMTD reduction factor due to longitudinal conduction is also shown.

The solution for equal water equivalents ($\dot{m}_h C_h = \dot{m}_h C_h$) gives the maximum reduction in LMTD. The LMTD reduction factor is used later in design.

Unequal water equivalents

Kroeger's paper does not discuss evaluation of LMTD reduction for unequal water equivalents ($\dot{m}_h C_h \neq \dot{m}_c C_c$). We now proceed to obtain a numerical solution for the unequal case. The steady-state equations including longitudinal conduction are set out below. Here, to simplify notation, temperatures are replaced by their subscripts in upper case.

$$\begin{aligned} \frac{\partial H}{\partial x} &= -P(H-W) \\ -\frac{\partial^2 W}{\partial x^2} &= +Q(H-W) - R(W-C) \\ -\frac{\partial C}{\partial x} &= \qquad\qquad\quad + S(W-C) \end{aligned} \tag{3.25(a)}$$

where

$$P = \left(\frac{\alpha_h S}{\dot{m}_h C_h}\right)\frac{1}{L}, \qquad Q = \left(\frac{\alpha_h S}{M_w C_w}\right)\frac{1}{\hat{\kappa}_x}, \qquad R = \left(\frac{\alpha_c S}{M_w C_w}\right)\frac{1}{\hat{\kappa}_x},$$

$$S = \left(\frac{\alpha_c S}{\dot{m}_c C_c}\right)\frac{1}{L}$$

The numerical solution for unequal water equivalents depends on writing finite-difference expressions for wall temperatures only at internal positions (Fig. 3.13). The reason for this is examined at the beginning of Section 3.8. Matrices are most easily generated by writing out in full the equations for five finite-difference intervals, and then generalizing this to 'm' intervals.

Care is required when creating the forward finite-difference expressions to ensure that temperatures are always evaluated at the internal wall station. For the first equation,

$$\frac{\partial H}{\partial x} = P(H - W)$$

becomes

$$\frac{H_{j+1} - H_j}{\Delta x} = -P_j\left[\frac{H_{j+1} + H_j}{2} - W_j\right]$$

where P_j refers to the W_j position, thus

$$H_{j+1}\left[1 + \frac{\Delta x}{2}P_j\right] + H_j\left[-1 + \frac{\Delta x}{2}P_j\right] + W_j[-\Delta x P_j] = 0$$

At hot fluid inlet where inlet temperature (H_0) is known

$$H_1\left[1 + \frac{\Delta x}{2}P_0\right] + W_0[-\Delta x P_0] = H_0\left[1 - \frac{\Delta x}{2}P_0\right]$$

Similar expressions exist for the two remaining reduced equations, and the coefficients may be inserted in a $[3m, 3m + 1]$ matrix – see the example in Section 3.8

Schematic 5-cell layout

	known											
	H_0		H_1		H_2		H_3		H_4		H_5	
···	+	−	+	−	+	−	+	−	+	−	+	···
W_0	I	W_0	I	W_1	I	W_2	I	W_3	I	W_4	I	W_4
···	+	−	+	−	+	−	+	−	+	−	+	···
	C_0		C_1		C_2		C_3		C_4		C_5	
											known	

Fig.3.13 Notation for five finite-difference intervals

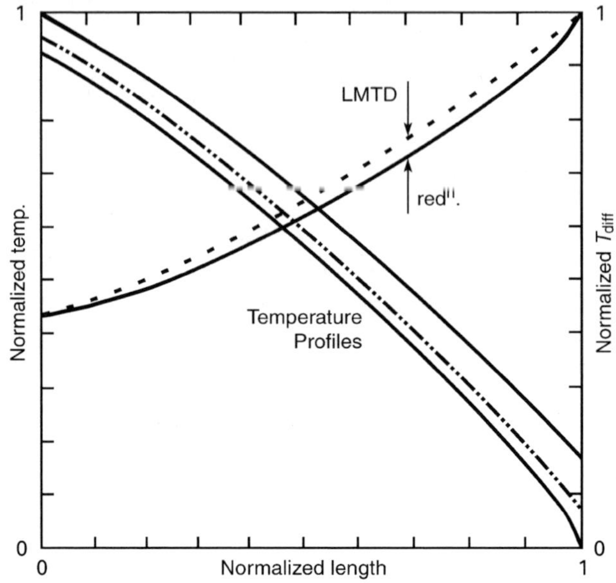

Fig.3.14 LMTD reduction for unequal water equivalents $(\dot{m}_h C_h) = 1.10 \times (\dot{m}_c C_c)$

for the approach. Solution of the matrix is by Gaussian inversion. The parameters (P, Q, R, S) may be taken as constants, or evaluated at each station when temperatures are known. Results for one computation are shown in Fig. 3.14.

The finite-difference computation does not work for equal water equivalents, but it does work so close to the balanced situation so as not to matter. This avoids having to proceed via the Kroeger analytical approach.

3.5 Mean temperature difference in unmixed crossflow

One-pass crossflow offers several possible arrangements, the fundamental configuration having both fluids unmixed (Fig. 3.15). A simple case is for one fluid mixed. When two and more passes are involved then the number of possible arrangements increases, for heat exchange in each pass can be both unmixed or with one fluid mixed, and between passes each fluid can be mixed or unmixed.

Here we shall consider the fundamental problem of both fluids unmixed in a one-pass exchanger and neglect effects of longitudinal conduction. A solution to the problem was obtained first by Nusselt (1911) in the form of analytical series expansions. Since that time many other workers have sought improved solutions, but it was left to Baclic & Heggs (1985) to show that all of these were mathematically equivalent. All of the expression obtained require numerical evaluation to be of use. We shall not travel down that path.

Instead the fundamental equations will first be obtained in canonical form, followed by direct numerical solution of these equations using finite differences.

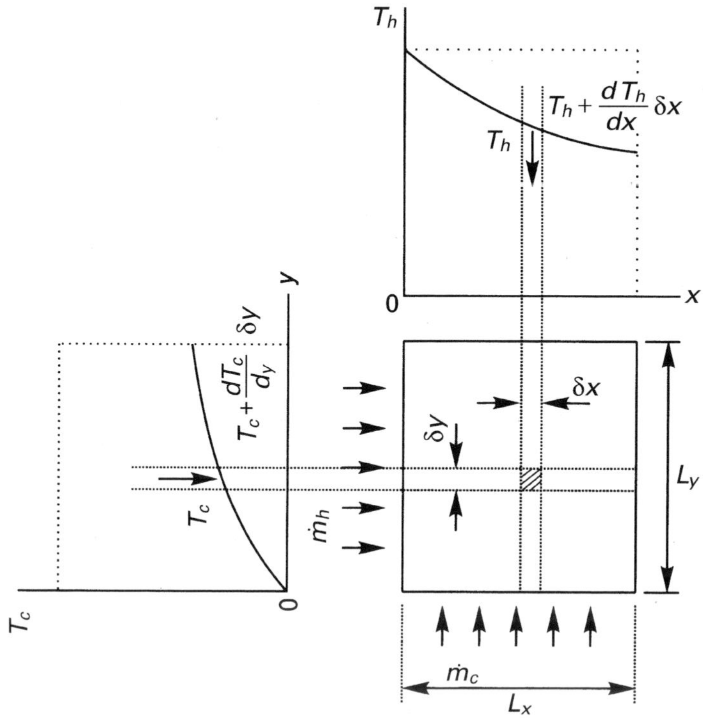

Fig.3.15 Differential temperatures for one-pass unmixed crossflow

This prepares the reader for later chapters in which transients are considered, where a numerical approach rather like solving the unmixed crossflow problem is required.

The mass flow in the x-direction entering the element of side δy is

$$\dot{m}_h \frac{\delta y}{L_y}$$

Striking energy balances for hot fluid and cold fluid together:

Hot fluid

$$\begin{Bmatrix}\text{energy entering}\\ \text{with hot fluid}\end{Bmatrix} - \begin{Bmatrix}\text{energy leaving}\\ \text{with hot fluid}\end{Bmatrix} - \begin{Bmatrix}\text{heat transferred}\\ \text{to cold fluid}\end{Bmatrix} = \begin{Bmatrix}\text{energy stored}\\ \text{is hot fluid}\end{Bmatrix}$$

$$\left(\dot{m}_h \frac{\delta y}{L_y}\right) C_h T_h - \left(\dot{m}_h \frac{\delta y}{L_y}\right) C_h \left[T_h + \frac{\partial T_h}{\partial x}\delta x\right] - U(\delta x \delta y)(T_h - T_c) = 0$$

Cold fluid

$$\left(\dot{m}_c \frac{\delta x}{L_x}\right) C_c T_c - \left(\dot{m}_c \frac{\delta x}{L_x}\right) C_c \left[T_c + \frac{\partial T_c}{\partial y}\delta y\right] + U(\delta x \delta y)(T_h - T_c) = 0$$

Thus

$$\frac{\partial T_h}{\partial x} = -\frac{UL_y}{\dot{m}_h C_h}(T_h - T_c) = -\frac{US}{\dot{m}_h C_h} \cdot \frac{1}{L_x}(T_h - T_c)$$

$$\frac{\partial T_c}{\partial y} = +\frac{UL_x}{\dot{m}_c C_c}(T_h - T_c) = +\frac{US}{\dot{m}_c C_c} \cdot \frac{1}{L_y}(T_h - T_c)$$

giving

$$\boxed{\begin{aligned} \frac{\partial T_h}{\partial x} &= -\frac{N_h}{L_x}(T_h - T_c) \\ \frac{\partial T_c}{\partial y} &= +\frac{N_c}{L_y}(T_h - T_c) \end{aligned}} \tag{3.26}$$

Normalizing the sides (L_x, L_y) of the exchanger and the temperatures on both sides with linear scaling

$$X = \frac{x}{L_x}, \qquad Y = \frac{y}{L_y}, \qquad \theta = \frac{T - T_{c1}}{T_{h1} - T_{c1}}$$

results in the equations

$$\boxed{\begin{aligned} \frac{\partial \theta_h}{\partial X} &= -N_h(\theta_h - \theta_c) \\ \frac{\partial \theta_c}{\partial Y} &= +N_c(\theta_h - \theta_c) \end{aligned}}$$

To keep the *Ntu* values visible, refrain from making the complete normalization

$$X = \frac{x}{L_x} N_h, \qquad Y = \frac{y}{L_y} N_c$$

which would produce the canonical pair of Nusselt equations (see Section 3.2). The present choice retains the values of *Ntu* to help with physical interpretation of the finite-difference solution, and from this point onwards reverting to the original notation for temperature for simplicity.

$$\boxed{\begin{aligned} \frac{\partial T_h}{\partial X} &= -N_h(T_h - T_c) \\ \frac{\partial T_c}{\partial Y} &= +N_c(T_h - T_c) \end{aligned}} \tag{3.27}$$

For temperature distributions in crossflow, symbolic logic software evaluation of one of the many equivalent mathematical series solutions obtained by

Baclic & Heggs (1985) may be less comprehensive than proceeding directly from the finite-difference numerical solution, which provides:

- mean temperature difference
- mean outlet temperatures
- outlet temperature profiles
- temperature sheets
- temperature difference sheet
- effectiveness
- self-checking heat balances

We notice that the independent variables occur in different equations. This allows a straightforward numerical solution similar in style to solving the transient contraflow problem.

It is straightforward to set up a finite-difference solution starting with equations (3.27), using a unit block (x, y, T) as a means of representing the temperature field. If both T_h and T_c are uniform at inlet, then the initial states $T_h = 1.0$ and $T_c = 0.0$ can be used to obtain the temperature responses for T_h at $x = 0$ and T_h at $y = 0$, which turn out to be the same temperature distributions as for condensation and evaporation, respectively, viz.

Condensation	*Evaporation*
$T_c = 1 - e^{-N_c(y/L_y)}$	$T_h = e^{-N_h(x/L_x)}$

Now consider the (x, y) face of the block as a 50×50 square grid, and from Nusselt slopes at $x = 0$ and $y = 0$ obtain

$$T_h[1, y] \text{ over } (1 \ldots y \ldots 50) \qquad \text{and} \qquad T_c[x, 1] \text{ over } (1 \ldots x \ldots 50)$$

Since both $T_h[1, 1]$ and $T_c[1, 1]$ are now known, use the modified Euler–Cauchy method values to generate

$$T_h[x, 1] \text{ over } (2 \ldots x \ldots 50) \qquad \text{and} \qquad T_c[1, y] \text{ over } (2 \ldots y \ldots 50)$$

The process is repeated until the temperature sheets are complete. All essential design parameters may now be found, and results for $N_h = N_c = 5.0$ are shown in Figs 3.16 and 3.17. The above approach is explicit finite difference, and accuracy is affected by error propagation, particularly from the steepest parts of the temperature field. For equally spaced intervals at least a 50×50 mesh should be used, and desirably correct analytical values used as input data.

With $N_h = N_c = 10.0$, the computed temperature difference sheet $(T_h - T_c)$ is near zero at the corners $(0, 1, T)$ and $(1, 0, T)$, providing no driving force for heat transfer. Clearly the crossflow exchanger is not an appropriate design selection when *Ntu* values close to 10.0 are involved. A better choice is for maximum values of *Ntu* around 4.0 for one-pass unmixed crossflow.

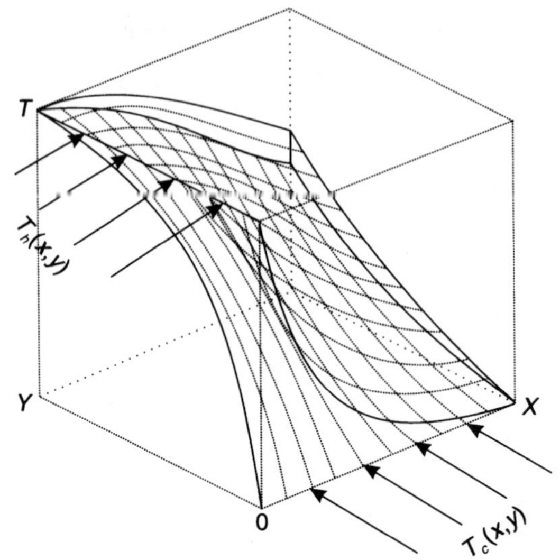

Fig.3.16 Temperature sheets T_h and T_c for $N_h = N_c = 5.0$

Schematic algorithms

Listings that allow construction of finite-difference algorithms in any source code are presented in Appendix B. The graphical results of Figs 3.16 and 3.17 were produced with these listings.

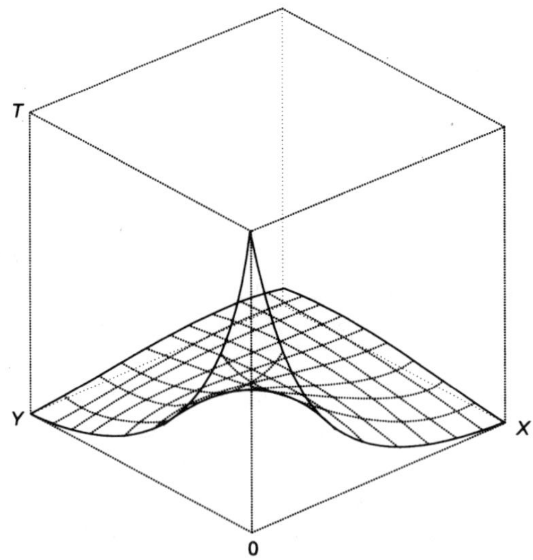

Fig.3.17 Temperature difference sheet ($T_h - T_c$) for $N_h = N_c = 5.0$

3.6 Extension to two-pass unmixed crossflow

Figure 3.18 illustrates one flow arrangement for a two-pass crossflow exchanger. Of the possible arrangements suggested by Stevens *et al.* (1957) and by Baclic (1990), this is possibly the most practicable configuration.

The exchanger can be considered as two single-pass exchangers of equal surface area, and temperature and temperature-difference distributions may be obtained by making use of the algorithm for the simple unmixed crossflow case. Values of *Ntu* employed in the solution now refer to half the total surface area, and the solution proceeds by solving the simple cases A and B successively until there is no change in intermediate temperature distributions for T_h and T_c (Fig. 3.19).

It is convenient to take the hot fluid inlet temperature as $T_h = 1$, and the cold fluid inlet temperature as $T_c = 0$. A solution to case A is first obtained by assuming that the intermediate cold inlet temperature distribution is $T_c = 0.5$ everywhere. An approximate intermediate warm outlet temperature distribution for T_h is calculated, which then becomes the input for case B. This produces a better estimate for the intermediate temperature distribution T_c, and the process continues until no significant changes in intermediate temperature distributions for T_h and T_c are obtained.

In this coupled solution, negative temperature differences may appear as profiles approach. This may indicate the need for one or more of:

- improved algorithms
- an increase in the number of finite-difference stations
- use of a computer with more accurate floating point capability

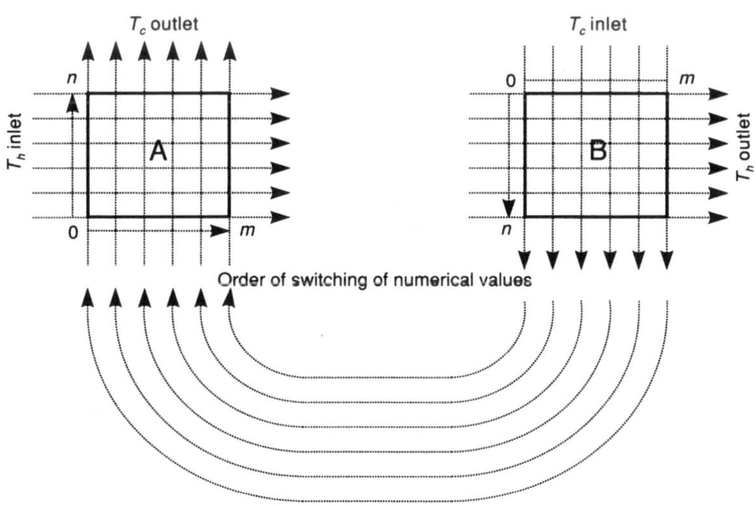

Fig.3.18 Computation of mean temperature difference for two-pass crossflow

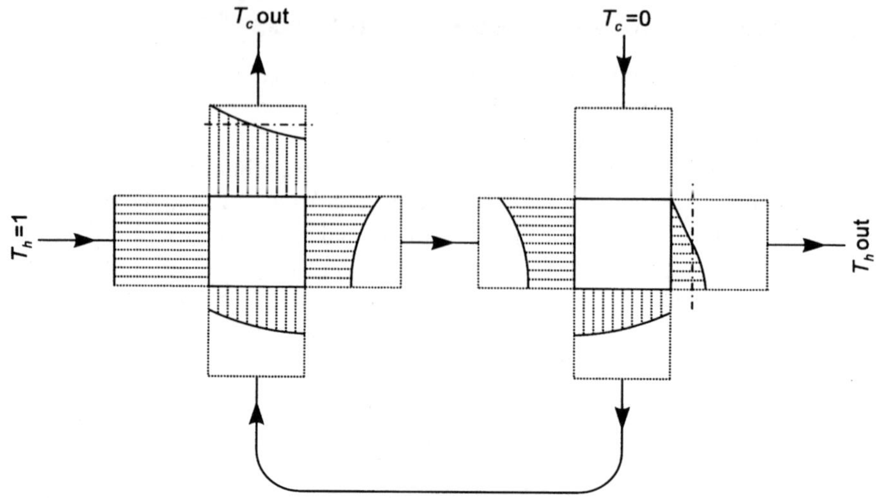

Fig.3.19 Terminal temperature profiles for $N_h = 7.0$, $N_c = 7.0$

As soon as a negative temperature difference appears an adjusted mean of the two temperatures was substituted according to the following scheme:

$$T_h = 1.010 \times (T_h + T_c)/2$$
$$T_c = 0.990 \times (T_h + T_c)/2$$

Mean outlet values for T_h and T_c can then be evaluated, together with all other information concerning temperature distributions.

In preparing the algorithm it is important to note the order of switching of numerical values for intermediate temperature distributions between cases A and B, and these are best studied on Fig. 3.18. Appropriate selection of order switching leads to solutions for different exchanger configurations.

Little difference in performance is evident from the ε–*Ntu* curves which are based on mixed outlet temperature value (Figs 3.20 and 3.21). However, it is also important to examine the outlet temperature profiles in Figs 3.22 and 3.23.

Final mixing involves external thermodynamic irreversibilities, and it is relevant to note the values of *Ntu* above which local temperature differences in the exchanger become negligible. Surface areas of such regions are inefficient in heat transfer. Figures 3.24 and 3.25 show mean outlet temperatures together with their associated temperature bands. Band limits should not get too close to the limiting values 1 and 0, indicating that one-pass unmixed crossflow is limited to *Ntu* values below 4.0, while two-pass unmixed crossflow might be used to *Ntu* values around 7.0 whenever exchanger mass is important, e.g. aerospace applications.

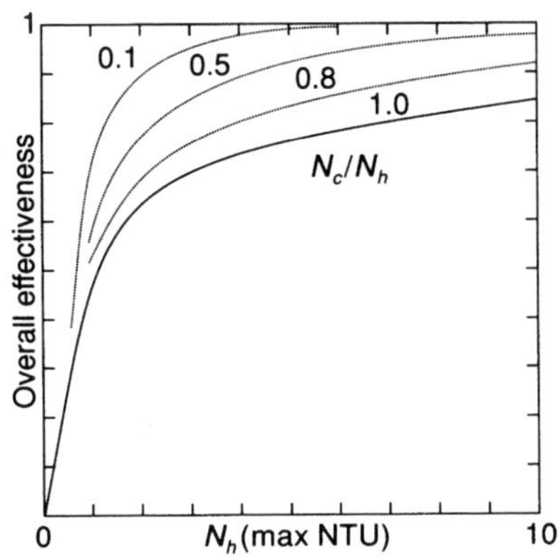

Fig.3.20 ε–*Ntu* for one-pass unmixed crossflow

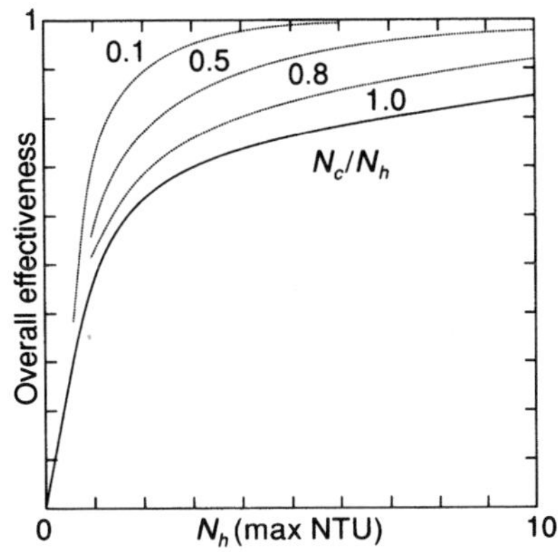

Fig.3.21 ε–*Ntu* for two-pass unmixed crossflow

There is no restriction in going to higher *Ntu* values – it just means that some of the surface is doing nothing while the remainder carries more duty, but the designer should be aware of the situation and decide when another pass is to be added. In the limit, the performance of a four-pass crossflow arrangement approaches that of a contraflow design.

For sizing A and B, there exists an alternative design philosophy to that of equal surface areas. It involves iterating until effectiveness values are equal for A and B. This results in different paired values (N_h, N_c) for each section, and the surface areas involved will thus be different.

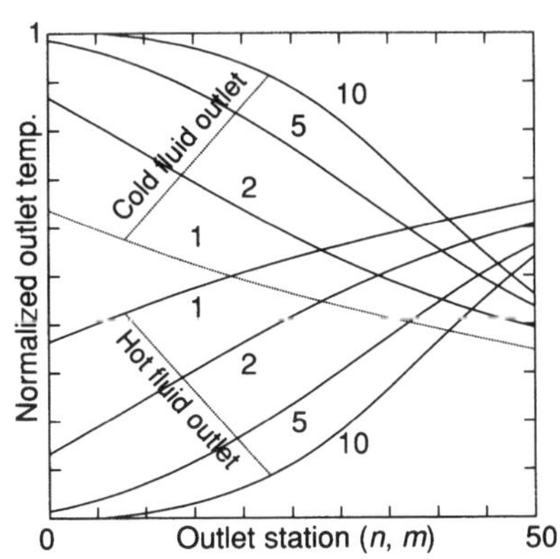

Fig.3.22 Outlet temperature profiles for one-pass unmixed cross-flow, $N_h = N_c = 1, 2, 5, 10$

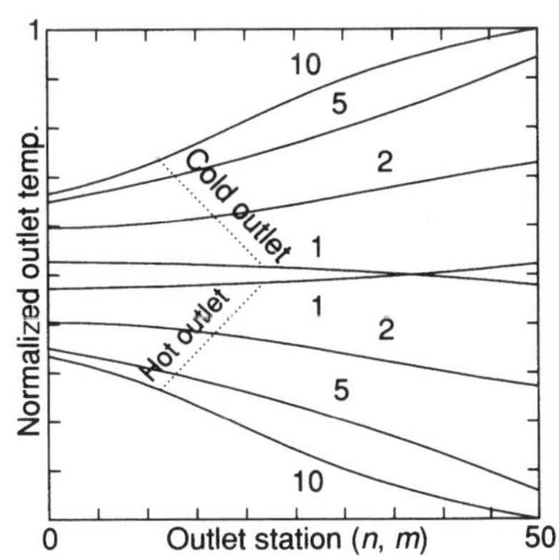

Fig.3.23 Outlet temperature profiles for two-pass unmixed cross-flow, $N_h = N_c = 1, 2, 5, 10$

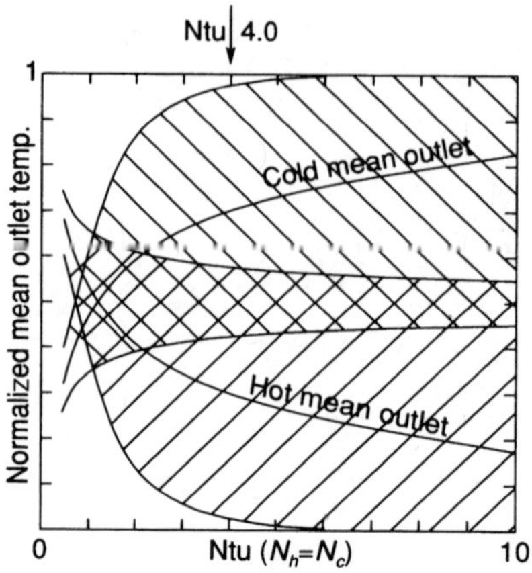

Fig.3.24 Mean outlet temperatures and temperature bands for one-pass unmixed crossflow, $N_h = N_c = 1, 2, 5, 10$

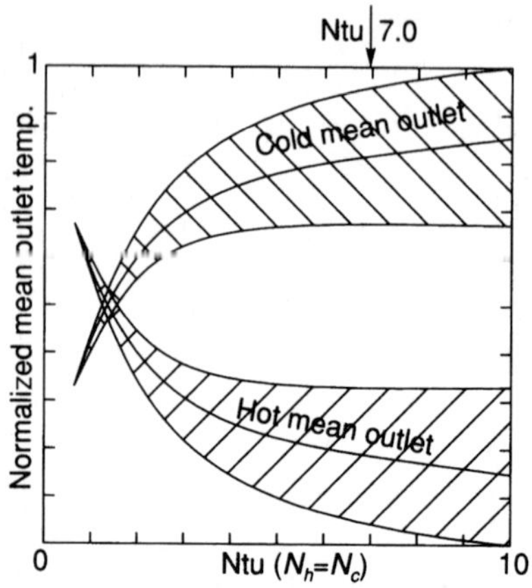

Fig.3.25 Mean outlet temperatures and temperature bands for two-pass unmixed crossflow, $N_h = N_c = 1, 2, 5, 10$

3.7 Involute-curved plate–fin exchangers

These may be designed as unmixed/unmixed single-pass contraflow exchangers using the same theory as developed for flat-plate compact contraflow exchangers in Chapter 4. It is only necessary to relate the curved length of an involute plate (s) to the inner (a) and outer (b) radii of the exchanger core.

Parametric representation of the involute curve depends on angle t measured on the base circle

$$\left.\begin{aligned} x &= a\cos(t) + at\sin(t) \\ y &= a\sin(t) - at\cos(t) \end{aligned}\right\} \tag{3.28}$$

Differential length of arc is given by

$$\frac{ds}{dx} = \sqrt{1 + \left(\frac{dy}{dx}\right)^2}$$

giving $ds = \sqrt{(dx)^2 + (dy)^2}$ and

$$\left(\frac{ds}{dt}\right)^2 = \left(\frac{dx}{dt}\right)^2 + \left(\frac{dy}{dt}\right)^2 \tag{3.29}$$

Using equations (3.26) in equations (3.27)

$$\begin{aligned} ds &= a.t.dt \\ s &= \int ds = \frac{a.t^2}{2} \end{aligned} \tag{3.30}$$

Outer radius b is given by

$$\begin{aligned} b^2 &= x^2 + y^2 \\ b &= a\sqrt{1+t^2} \end{aligned} \tag{3.31}$$

One- and two-pass crossflow arrangements may also be developed using the same theory developed for flat-plate compact crossflow exchangers in Chapter 4.

The unmixed one-pass crossflow arrangement has one fluid flowing axially and the other fluid flowing radially.

Two unmixed two-pass crossflow arrangements are possible. The first arrangement has one radial pass and two axial passes. The second arrangement has two radial passes and one axial pass. Insulation strips would be required in both cases.

Arrangements with higher numbers of passes are also possible but at the expense of greater complexity in manufacture.

3.8 Longitudinal conduction in one-pass unmixed crossflow

There is no clean analytical solution to this problem, or a numerical solution which can be coded without some effort, see Chiou (1978).

Chiou investigated performance deterioration for *Ntu* values in the range 1–100. Elsewhere in this text it is shown that surface area in the $(0, 1)$ and $(1, 0)$ corners of the unit block for one-pass unmixed crossflow tend to become ineffective at *Ntu* values greater than about 5.0. Thus Chiou's results are of practical interest perhaps only for *Ntu* values less than 5.0 for single-phase fluids.

The greatest deterioration occurred with equal water equivalents, as was the case with pure contraflow.

A simpler approach based on that of Chiou may be employed. The starting point is the set of three partial differential energy equations for steady-state crossflow – by comparison with those for contraflow in Appendix A.1, equations (A.1).

$$\boxed{\begin{array}{llcll} +u_h\dfrac{\partial T_h}{\partial x} & & = -\left(\dfrac{a_h S}{\dot{m}_h C_h}\right)\dfrac{u_h}{L_x}(T_h - T_w) & & \\ -\hat{k}_x\dfrac{\partial^2 T_w}{\partial x^2} & -\hat{k}_y\dfrac{\partial^2 T_w}{\partial y^2} & = +\left(\dfrac{a_h S}{M_w C_w}\right)(T_h - T_w) & -\left(\dfrac{a_c S}{M_w C_w}\right)(T_w - T_c) \\ & +u_c\dfrac{\partial T_c}{\partial y} & = & +\left(\dfrac{a_c S}{\dot{m}_c C_c}\right)\dfrac{u_c}{L_y}(T_w - T_c) \end{array}} \tag{3.32}$$

In considering how we might solve these coupled equations it is useful first to consider whether solutions can be found for the two flow-inlet faces of the crossflow exchanger. If these 'boundary' conditions become available, then the chances of finding a solution to the complete problem are much improved.

Equation pairs for the inlet faces

The governing equations for both hot fluid and cold fluid conditions may be extracted from the above set, and when normalized to unit length and unit temperature span these provide two separate sets of equations, viz.

Hot inlet face

$$\boxed{\begin{aligned} \frac{dT_h}{dX} &= -n_h(T_h - T_w) \\ -\frac{\hat{\kappa}_x}{L_x^2}\frac{d^2T_w}{dX^2} &= +\frac{n_h}{R_h\tau_h}(T_h - T_w) - \frac{n_c}{R_c\tau_c}(T_w - T_c) \end{aligned}} \tag{3.33}$$

where, for constant fluid velocities only

Residence mass $\tilde{m}_h = \dot{m}_h(L_x/u_h)$ giving parameter $R_h = M_wC_w/(\tilde{m}_hC_h)$
Residence mass $\tilde{m}_c = \dot{m}_c(L_y/u_c)$ giving parameter $R_c = M_wC_w/(\tilde{m}_cC_c)$

and thermal diffusivities are defined as

$$\hat{\kappa}_x = \left(\frac{A_xL_x}{V}\right)\kappa_x \quad \text{and} \quad \hat{\kappa}_y = \left(\frac{A_yL_y}{V}\right)\kappa_y$$

Cold inlet face

$$\boxed{\begin{aligned} \frac{dT_c}{dY} &= +n_c(T_w - T_c) \\ -\frac{\hat{\kappa}_y}{L_y^2}\cdot\frac{d^2T_w}{dY^2} &= +\frac{n_h}{R_h\tau_h}(T_h - T_w) - \frac{n_c}{R_c\tau_c}(T_w - T_c) \end{aligned}} \tag{3.34}$$

Solution methods

Analytical approach

Attempts were made first to solve these equations analytically, for the presence of T_w alone in the first equation of both sets means that explicit equations for T_w and its derivatives can be found for substitution in the second equation of each pair.

Boundary conditions are $T_h = 1$ for $x = 0$ and for all y, and $T_c = 0$ for $y = 0$ and for all x. Wall temperature at the origin might be obtained from knowledge of the local heat-transfer coefficients on each side of the exchanger $T_w = \alpha_h/(\alpha_h + \alpha_c)$ as there is no longitudinal conduction at this point, but as will be seen later this may be a questionable assumption when longitudinal conduction is present.

This results in a third-order ordinary differential equation in T_h or T_c which may be solved by assuming the usual exponential solution. A cubic equation which can be solved numerically permits an analytical solution for both temperature profiles on each face and the necessary boundary conditions are available for a complete solution.

However, when numerical values for a real exchanger were inserted, although the cubic could be solved, the roots so produced became arguments in exponentials, some of which could not be evaluated on the computer. It became apparent at this stage that the equations were very 'stiff', and some other method of solution would be required.

Explicit finite-difference approach

When an explicit finite-difference method was applied starting at the origin, again the solution 'blew up'. This may be because the requirement of zero wall temperature gradient at the origin is very swiftly followed by a point of inflexion in the wall profile, and also because the assumption of $T_w = \alpha_h/(\alpha_h + \alpha_c)$ at the origin is required. Another approach thus became necessary.

Implicit modified finite-difference approach

The method of Chiou (1978, 1980) provides a solution for the temperature field in crossflow when longitudinal conduction is present, but the equations presented by Chiou are quite complicated, as the author freely admitted in his paper.

Examining this solution it became apparent that it was entirely equivalent to writing simultaneous finite-difference expressions, then loading these in a matrix for direct solution. For equations (3.34) the trick is to avoid writing wall equations for $x = 0$ and $x = L_x$, and as we are only interested in the difference between hot and cold fluid temperatures in determining degradation in exchanger performance this seems quite acceptable.

The same remarks apply for equations (3.34) and the y-direction.

Formulation of the matrix

Matrices are most easily generated by writing out in full the equations for five finite-difference intervals, and then generalizing this to 'm' intervals. To simplify, notation temperatures are replaced by their subscripts in upper-case (Fig. 3.26).

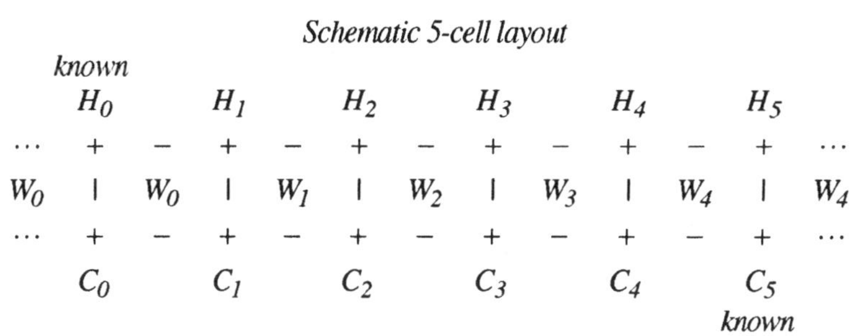

Fig.3.26 Notation for five finite-difference intervals

For five intervals on the 'hot' face (H), the first equation of equations (3.33) above may be written

$$j=0 \qquad \left[1+\frac{n_h\Delta X}{2}\right]H_1 \qquad -[n_h\Delta X]W_0 = +\left[1-\frac{n_h\Delta X}{2}\right]H_0$$

$$j=1,2,3 \qquad \left[1+\frac{n_h\Delta X}{2}\right]H_{j+1} - \left[1-\frac{n_h\Delta X}{2}\right]H_j - [n_h\Delta X]W_j = 0$$

$$j=4 \qquad \left[1+\frac{n_h\Delta X}{2}\right]H_5 - \left[1-\frac{n_h\Delta X}{2}\right]H_4 - [n_h\Delta X]W_4 = 0$$

The second equation of equations (3.33) may be written

$$j=0 \qquad \left[\frac{-\hat{\kappa}_x}{L_x^2\Delta X^2}\right]W_1 + \left[\frac{\hat{\kappa}_x}{L_x^2\Delta X^2}+\frac{n_h}{R_h\tau_h}+\frac{n_c}{R_c\tau_c}\right]W_0$$

$$-\left[\frac{n_h}{2R_h\tau_h}\right]H_1 = \left[\frac{n_h}{2R_h\tau_h}\right]H_0$$

$$j=1,2,3 \qquad \left[\frac{-\hat{\kappa}_x}{L_x^2\Delta X^2}\right](W_{j+1}+W_{j-1}) + \left[\frac{2\hat{\kappa}_x}{L_x^2\Delta X^2}+\frac{n_h}{R_h\tau_h}+\frac{n_c}{R_c\tau_c}\right]W_j$$

$$-\left[\frac{n_h}{2R_h\tau_h}\right](H_{j+1}+H_j) = 0$$

$$j=4 \qquad \left[\frac{-\hat{\kappa}_x}{L_x^2\Delta X^2}\right]W_3 + \left[\frac{\hat{\kappa}_x}{L_x^2\Delta X^2}+\frac{n_h}{R_h\tau_h}+\frac{n_c}{R_c\tau_c}\right]W_4$$

$$-\left[\frac{n_h}{2R_h\tau_h}\right](H_5+H_4) = 0$$

Writing the coefficients as follows

$$P = +\left[1+\frac{n_h\Delta X}{2}\right] \qquad S = -\left[\frac{n_h}{2R_h\tau_h}\right]$$

$$Q = +\left[1-\frac{n_h\Delta X}{2}\right] \qquad T = -\left[\frac{\hat{\kappa}_x}{L_x^2\Delta X^2}\right]$$

$$R = -[n_h\Delta X] \qquad U = +\left[\frac{2\hat{\kappa}_x}{L_x^2\Delta X^2}+\frac{n_h}{R_h\tau_h}+\frac{n_c}{R_c\tau_c}\right]$$

the 10×11 matrix to be solved is given in Table 3.2. Inversion of this matrix gives the values of T_h at $X = j.\Delta X$ ($j = 1, 2, 3, 4, 5$) and T_w at $X = (j+0.5).\Delta X$

Table 3.2 Matrix of coefficients

T *Eq.*	1 *H1*	2 *H2*	3 *H3*	4 *H4*	5 *H5*	6 *W0*	7 *W1*	8 *W2*	9 *W3*	10 *W4*	11 *RHS*
1	P					R					−Q
2	Q	P					R				
3		Q	P					R			
4			Q	P					R		
5				Q	P					R	
6	S					U + T	T				−S
7	S	S				T	U	T			
8		S	S				T	U	T		
9			S	S				T	U	T	
10				S	S				T	U + T	

RHS, right-hand side.

($j = 0$, 1, 2, 3, 4). The value of T_h at $X = 0$ is 1.0 and the value of T_w at $X = 0$ may be taken as $\alpha_h/(\alpha_h + \alpha_c)$.

Using the matrix in Table 3.2 as a model, the general PASCAL algorithms for a ($2m$, $2m + 1$) matrix may easily be constructed as follows:

```
           FOR j:=1 TO m DO
BEGIN      k:=j;
           A[j, k]:=P
END;
           FOR j:=2 TO m DO
BEGIN      k:= j-1;
           A[j,k]:=Q
END;
```

and so on. A similar matrix exists for equations (3.32).

Solution of the inlet problem

Data for the crossflow exchanger are due to Shah (1983). In equations (3.31) and (3.32) coefficients are evaluated in terms of $\hat{\kappa} = (AL/V)\kappa$, where A is the cross-section for axial conduction, L is length of the exchanger in the direction of conduction, V is the volume of solid material in the heat exchange surfaces, and κ is the thermal diffusivity of the material of construction. Evaluation of the cross-section for the Kays & London (1964) plain plate–fin surface 19.86 is eased if the centre-line of the geometry is first obtained. In this case it is approximately a sine curve (Fig. 3.27), and the arc length may be determined using standard methods, e.g. Wylie (1953).

Note that we are interested in only half-fin height associated with one plate in the exchanger. The half-height cross-section for rectangular offset strip-fin surface

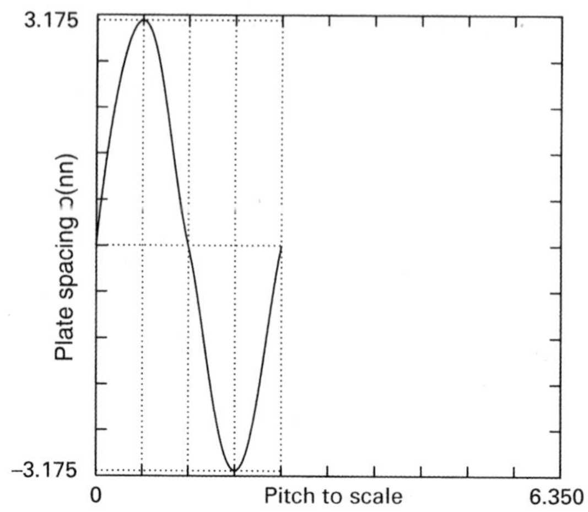

Fig.3.27 Centre-line of plain plate–fin surface 19.86 (approximately to scale)

104(S), designated 1/8-15.61 is more easily determined. This provides values of surface cross-sections useful in determining V which must also include the plate.

Cross-sections for longitudinal conduction may be obtained at the same time. In the case of plain plate–fin surface 19.86 the conduction section is the same as that for determining volume of material. In the case of rectangular offset strip-fin surface 104(S) the conduction path is effectively reduced to material attached directly to the plate, and this is only half the width of the plate.

A little more thought is required to evaluate the contribution of both surfaces in conduction at right angles to their flow lengths, which contributes to deterioration in performance on the other side of the plate. For the plain surface 19.86 this was taken as the cross-section of the material. For the rectangular offset strip-fin surface again half the cross-section of the material was assumed effective.

The reader may rightly enquire whether such detailed evaluation of the surfaces for longitudinal conduction is necessary, for we could more simply use κ instead of $\hat{\kappa}$ and still obtain a conservative design. We still have to find V to calculate the mass of material for evaluating R_h and R_c, but this simpler procedure is altogether less traumatic.

Normalized inlet conditions for both sides of the exchanger were computed using equations (3.31) and (3.32) and are presented graphically in Fig. 3.28. It was found possible to use 50 divisions along each face, the ends of the computed curve for wall temperature being marked by 'plus' symbols to show how distant they were from the end values.

Also shown as dotted lines are the steady-state values computed for the case without the wall equation, normalized wall temperature being evaluated by using the appropriate ratios of local heat transfer coefficients, $T_w = \alpha_h/(\alpha_h + \alpha_c)$.

For the exchanger examined, longitudinal conduction in the wall does affect performance. Reduction in mean temperature difference on the 'condenser' face was

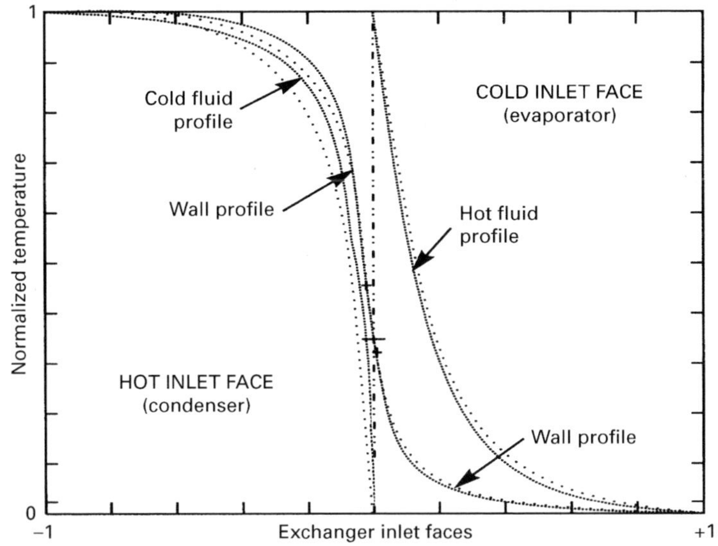

Fig.3.28 Schematic temperature profiles on inlet faces of a crossflow exchanger

evaluated as 0.992 58, and reduction in mean temperature on the 'evaporator' face of 0.992 51 was found.

Figures 3.16 and 3.17 show the temperature ridge starting at ($x = 1$, $y = 1$) finishing at ($x = 0$, $y = 0$) and as any hill climber knows the easiest route is along the ridge. It is likely that greatest longitudinal conduction effects are to be found at right angles to this ridge and near to the point ($x = 0$, $y = 0$). The two inlet faces provide an immediate approximation to the steepest slopes, and temperature profiles on these faces will be used to assess the amount of longitudinal conduction.

If we use the maximum reduction factor found, and divide the exchanger duty by this factor, then a second run of the program will produce a design that allows for the effect. Of course, if the mean temperature difference reduction factor for the whole exchanger is computed this would be better.

If the objective is to reduce longitudinal conduction to a minimum so as to be able to calculate transients, then the reduction factor obtained from temperature profiles on the inlet faces will suffice. This involves systematically changing individual surface geometries. It is beyond the scope of this book as the heat-transfer and flow-friction correlations for the plain sinusoidal gas-side surface geometry have not yet been reduced to universal correlations, but the approach should be clear.

The one reservation associated with Chiou's method is that it cannot properly evaluate $dT_w/dx = 0$ at $x = 0$ unless the increments Δx are extremely small. In this mathematical respect, we still do not have an exact solution to the problem (Fig. 3.29).

WARNING The reduction in mean temperature difference caused by longitudinal conduction in crossflow is very small, and probably may be neglected. However,

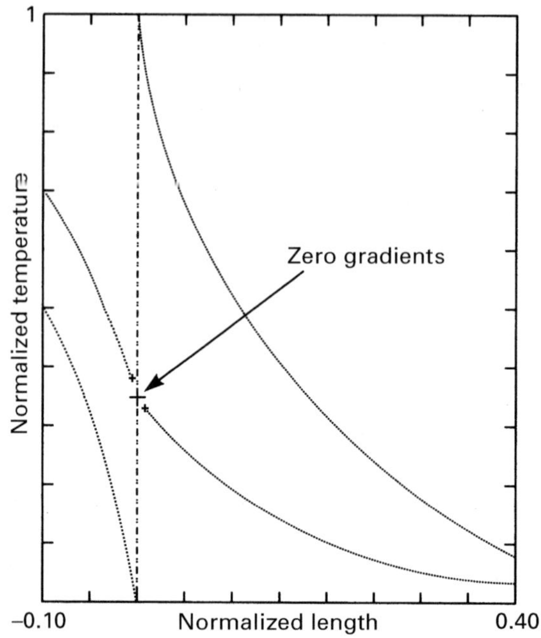

Fig.3.29 Temperature profiles near the origin for the 'evaporator'

a much more serious problem exists. Viscosity is temperature-dependent and the variation in temperature along each flow channel could invalidate the assumption of equal mass flowrates. See Appendix L, Section L.1.

3.9 Determined and undetermined crossflow

A principal difficulty in sizing crossflow heat exchangers arises when the configuration does not exactly match the mathematical requirements for mixed or unmixed crossflow, making the mean temperature difference uncertain. Permutation pairs of three flow arrangements – mixed, unmixed, and undetermined – are possible.

Figure 3.30 is an example of mixed/unmixed crossflow which is repeated in a simpler form in Fig. 3.32(a). However triangular tube pitching is better suited to estimating fin performance of the flat plates, as the performance ratio of circular disc fins may be used in place of the hexagonal shapes. Figure 3.32(b) is also mixed/unmixed.

Figure 3.31 is an example of unmixed/undetermined crossflow in which flat ducts are used in place of tubes. Once internal finning is introduced as shown in Fig. 3.32(c) the configuration becomes unmixed/unmixed. The performance of the fin plates may be handled as for straight rectangular fins, but the external finning lacks recreation of the boundary layers to obtain higher heat-transfer coefficients. The flat plates may be rippled transversely to the flow to improve heat transfer, but the overall design will still suffer from longitudinal conduction in the flat

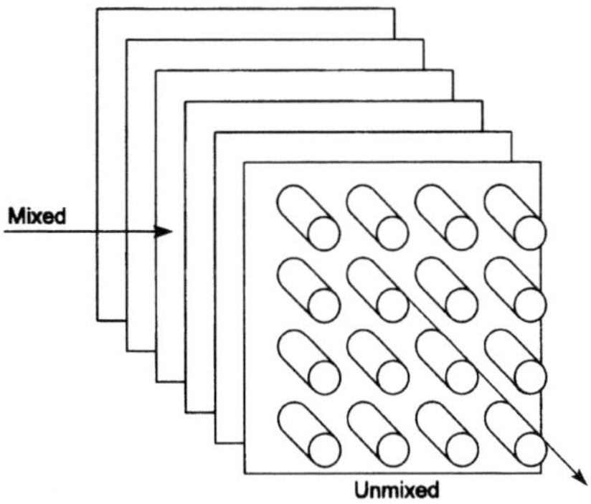

Fig.3.30 Fin and tube exchanger, mixed/unmixed crossflow

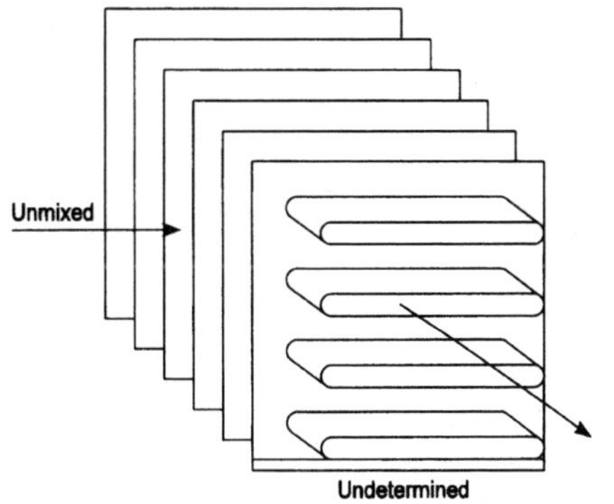

Fig.3.31 Fin and duct exchanger, unmixed/undetermined crossflow

plate fins. Slotted or louvred fins may then become necessary both to recreate boundary layers and to control longitudinal conduction.

Figure 3.32(d) breaks up the single, long flat ducts into many small staggered flat ducts or flattened tubes, approaching the unmixed/unmixed configuration without quite achieving it. With rippled plates this configuration begins to approximate to the rectangular offset-strip plate–fin surfaces used in compact heat exchangers,

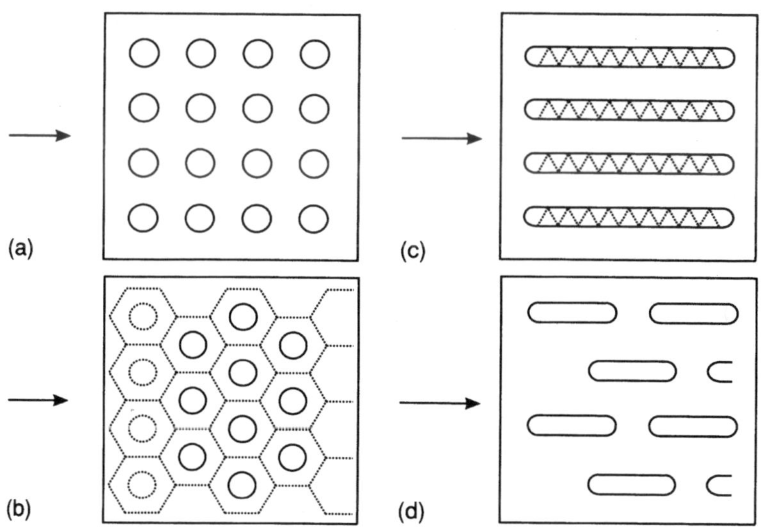

Fig.3.32 Some possible crossflow configurations: (a) plate–fin and tube with square pitching, mixed/unmixed crossflow; (b) plate–fin and tube with triangular pitching, mixed/unmixed crossflow; (c) plate–fin and flat ducts with internal finning, unmixed/unmixed crossflow; (d) plate–fin and flattened tubes, approach to unmixed/unmixed crossflow

but in that case the fins are so small and close that a good approximation to unmixed/unmixed flow exists. When the mean temperature difference cannot easily be determined then recourse to testing may be necessary, even with single-phase crossflow.

3.10 Possible optimization criteria

It is possible to optimize the performance of an exchanger by progressively changing the local geometry on one or both sides. Optimization may include minimization of any one of several criteria, including:

- core volume
- core mass
- frontal area etc.

We should not be surprised if the final result should differ from case to case – only one optimum at a time. However, a little thought reveals that when an exchanger is optimized so as to utilize all the allowable pressure loss on both sides, then this also corresponds to minimization of surface area, see Figs 5.8 and 5.9.

In cases where the *sum* of pressure losses is the constraint (see Section 3.13), it is straightforward to write a statement that calculates the allowable pressure loss on one side when that on the other side is known, and proceed with the normal optimization procedure.

Pinch technology

In a process plant with many heat exchangers, optimization of the whole system has been made thermodynamically simple by process integration technology developed by Linnhoff & Flower (1978) and others. The technique introduces a distinction between 'avoidable' and 'unavoidable' inefficiencies. By first constructing separate composite T–h curves for hot streams and cold streams, and by sliding the cold curve horizontally towards the hot curve on the T–h diagram, a minimum temperature difference or 'pinch point' is found somewhere in the operational field. This represents the best energy recovery configuration for the system that can be achieved with the chosen minimum temperature difference, with the rider that no heat should be transferred across the pinch point.

The second stage is to consider the heat exchangers making up the network. Process integration is best approached by first consulting the 'User Guide' prepared by Linhoff *et al.* (1982), the recent reference text by Hewitt *et al.* (1994), and the short overview paper by Terrill & Douglas (1987).

3.11 Cautionary remark about core pressure loss

For approximate core pressure loss, the simple expression for friction loss may be used, viz.

$$\Delta p = \frac{4fG^2}{2\rho_m}\left(\frac{L}{D}\right) \tag{3.35}$$

Loss due to friction is often 98 per cent or more of the total pressure loss given by Kays & London (1964) in their expression for pressure loss in an exchanger core, viz.

$$\Delta p = \frac{G^2 v_1}{2}\left[\underbrace{(Kc + 1 - \sigma^2)}_{\text{inlet loss}} + \underbrace{2\left(\frac{v_2}{v_1} - 1\right)}_{\text{flow acceleration}} + \underbrace{f\,\frac{Stotal}{Aflow}\frac{v_m}{v_1}}_{\text{core friction}} - \underbrace{(1 - \sigma^2 - Ke)\frac{v_2}{v_1}}_{\text{outlet loss}}\right] \tag{3.36}$$

See Shah (1983) for a detailed breakdown of equation (3.36).

Much grief can be avoided by using a consistent method for evaluating mean density (ρ_m) in equation (3.35) and mean specific volume (v_m) in equation (3.36). The mean bulk condition for gases may be approximated by first evaluating the mean pressure and temperature levels, *and only then* evaluating the mean density and mean specific volume. With this method it will be found that the reciprocal of mean density is equal to the mean specific volume.

This will not happen when the mean specific volume is taken as the mean of specific volumes at inlet and outlet conditions, and the mean density is similarly taken as the mean of densities at inlet and outlet conditions.

In both direct-sizing and rating design methods it is essential to be consistent in the treatment of variables, otherwise significant discrepancies between the two solutions of the same problem will arise.

3.12 Mean temperature difference in complex arrangements

The reader will notice that some arrangements of heat exchanger are not considered in this volume, firstly because some of these exchangers are simply more complex arrangements of those that are discussed, and secondly because their mathematical analyses are already well covered in other publications.

For completeness, it was decided to include the following list of solutions for mean temperature difference in other arrangements, together with the name of the author(s) who first published the solutions. Not all of the arrangements listed are appropriate for direct-sizing.

Mean temperature difference in multipass parallel flow

- one pass shell-side, two passes tube-side — Underwood (1934)
- one pass shell-side, four passes tube-side — Underwood (1934)
- one pass shell-side, three passes tube-side — Fischer (1938)
- one pass shell-side, infinite number of passes tube-side — Smith (1934)
- two passes shell-side, four passes tube-side — Underwood (1934)

Mean temperature difference in multiple crossflow

- single passes, both fluids unmixed — Nusselt (1911)
- single passes, both fluids mixed — Smith (1934)

- single passes, first fluid mixed, second fluid unmixed — Smith (1934)
- first fluid single-pass mixed, second fluid two passes unmixed, except between passes — Smith (1934)

All the above mathematical analyses (and references) are included in the textbook by Jakob (1957). Further analyses of single-, two- and three-pass crossflow arrangements are considered by Stevens *et al.* (1957). Four-pass arrangements tend to approach true crossflow. These results are suitable for use in LMTD–*Ntu* type solutions, while some of the equivalent ε–*Ntu* solutions are given by Shah (1982). A more recent paper by Spang & Roetzel (1995) provides an approximate equation with three or four empirical parameters for the calculation of LMTD correction factors for about 50 different flow configurations.

Comprehensive ε–*Ntu* type results for 36 two-pass cross-parallel flow arrangements and 36 two-pass cross-counterflow arrangements have been presented by Baclic (1990), which greatly extends the work of Stevens *et al.* (1957). Graphs are presented to help select the best flow arrangements which minimize irreversibilities due to mixing at exit and between passes. Earlier, explicit solutions were provided by Baclic & Gvozdenac (1981).

However, the correct requirement for every fluid in any crossflow arrangement is equal pressure loss in each channel as this controls flow distribution, and not the assumption of equal mass flowrate in each channel.

Three-fluid heat exchangers

Sekulic & Shah (1995) published a 105-page paper on thermal design of three-fluid heat exchangers, and readers are directed to this monograph-sized article. Yuan & Kou (1998) have considered the effects of longitudinal conduction in a three-fluid crossflow exchanger.

3.13 Exergy destruction

By this stage the reader should be aware that the traditional heat exchanger design is based on the *balance of energy* axiom, plus a degenerate form of the *balance of linear momentum* axiom for pressure loss. Only briefly, in Section 2.12, was there an appeal to the *growth of entropy* inequality (Second Law, or *destruction of exergy).*

Two possible approaches to reducing exergy destruction can be identified. The first is exergy analysis of the isolated heat exchanger. The second is exergy analysis of the exchanger already embedded in a thermodynamic cycle.

Cryogenic liquefaction plant provides one example of a system where design of the ‘isolated’ exchanger is practicable, because configuration of the plant allows for such optimization – or perhaps we should say that successful operation of the whole plant ‘demands’ such local optimization. This does not imply that exergy analysis of the complete plant can then be neglected.

Recuperated gas turbine plant provides one example of a system where the isolated exchanger cannot normally be optimized properly, because the temperature distribution produced by embedding requires the largest temperature difference to be at the cold end of the exchanger. Compromise in design of the exchanger is then necessary, and exergy optimization has rather to be for the complete plant.

Questions of exergy destruction should thus be settled at the plant layout stage, well before detailed heat exchanger design commences.

To obtain an overview of the role of thermodynamics in the design of heat exchangers (plus power and cryogenic cycles) the reader should study the excellent text by Bejan (1988), *Advanced Engineering Thermodynamics*, which presents thorough and readable analyses of many engineering systems of interest. Bejan's analysis of a heat exchanger showed that there exists a trade-off in performance between mean temperature difference and pressure loss, indicating that minimization of irreversibility (*entropy* generation) is possible. Sekulic (1986, 1990) emphasized the importance of including irreversibility due to pressure loss. Das & Roetzel (1995) have applied the approach to plate–fin exchangers.

Herbein & Rohsenow (1988) compared the entropy generation method of optimization of Bejan with conventional methods of optimizing a recuperator in a gas turbine cycle, and concluded that Bejan's entropy generation method did not necessarily find the best optimum. In concurrence with this observation, the reader should now study Chapter 8. *Exergy* destruction in heat exchangers also arises through 'heat transfer across a finite temperature difference' and by 'flow with friction'.

The value of the exergy destruction approach is that it usually reveals where best to make improvements.

References

Baclic, B.S. (1990) *ε–Ntu* analysis of complicated flow arrangements. *Compact Heat Exchangers – a festschrift for A. L. London* (Eds. R.K. Shah, A.D. Kraus, and D. Metzger), Hemisphere, New York, pp. 31–90.

Baclic, B.S. and **Gvozdenac, D.D.** (1981) Exact explicit equations for some two and three pass crossflow heat exchangers. *Heat Exchangers – Thermal Hydraulic Fundamentals and Design* (Eds. S. Kakaç, A.E. Bergles, and F. Mayinger), Hemisphere, New York, pp. 481–494.

Baclic, B.S. and **Heggs, P.J.** (1985) On the search for new solutions of the single-pass crossflow heat exchanger problem. *Int. J. Heat Mass Transfer*, **28**(10), 1965–1976.

Bejan, A. (1988) *Advanced Engineering Thermodynamics*, Wiley, New York.

Chiou, J.P. (1978) The effect of longitudinal heat conduction on crossflow exchanger. *ASME J. Heat Transfer*, **100**, May, 346–351.

Chiou, J.P. (1980) The advancement of compact heat exchanger theory considering the effects of longitudinal heat conduction and flow nonuniformity. In *Symposium on Compact Heat Exchangers–History, Technological Advancement and Mechanical Design Problems* (Eds. R.K. Shah, C.F. McDonald, and C.P. Howard), ASME book no. G00183, New York.

Das, S.K. and **Roetzel, W.** (1995) Exergetic analysis of plate heat exchanger in presence of axial dispersion in fluid. *Cryogenics* **35**(1), 3–8.

Fischer, K.F. (1938) Mean temperature difference correction in multipass exchangers. *Ind. Engng. Chem.*, **30**, 377–383.

Herbein, D.S. and **Rohsenow, W.M.** (1988) Comparison of entropy generation and conventional method of optimizing a gas turbine regenerator (Note: 'regenerator' is used in the sense of heat recovery in a thermodynamic cycle). *Int. J. Heat Mass Transfer*, **31**(2), February, 241–244.

Hewitt, G.F., **Shires, G.L.**, and **Bott, T.R.** (1994) *Process Heat Transfer*, CRC Press, Florida.

Jakob, M. (1949, 1957) *Heat Transfer*, vol. I (1949) and vol. II (1957), John Wiley.

Kays, W.M. and **London, A.L.** (1964) *Compact Heat Exchangers*, 2nd edn, McGraw-Hill, New York, pp. 250–257.

Kroeger, P.G. (1966) Performance deterioration in high effectiveness heat exchangers due to axial conduction effects. In *Proceedings of 1966 Cryogenic Engineering Conference, Advances in Cryogenic Engineering*, vol. 12, Plenum Press, Paper E-5, pp. 363–372.

Linnhoff, B. and **Flower, J.R.** (1978) Synthesis of heat exchanger networks. Part 1 – systematic generation of energy optimal networks. Part 2 – evolutionary generation of networks with various criteria of optimality. *AIChE J.* **24**, 633–642 and 642–654.

Linnhoff, B. Townsend, D.W., Boland, D., Hewitt, G.F., Thomas, B.E.A., Guy, A.R. and **Marsland, R.H.** (1982) *A User Guide on Process Integration for the Efficient Use of Energy*, Institution of Chemical Engineers, Rugby, UK.

Nusselt, W. (1911) Der Wärmeübergang im Kreuzstrom. *Zeitschrift Vereines deutscher Ingenieure*, **55**, 2021–2024.

Sekulic, D.P. (1986) Entropy generation in a heat exchanger. *Heat Transfer Engng*, **7**, 83–88.

Sekulic, D.P. (1990) A reconsideration of the definition of a heat exchanger. *Int. J. Heat Mass Transfer*, **33**(12), 2748–2750.

Sekulic, D.P. and **Shah, R.K.** (1995) Thermal design of three-fluid heat exchangers. *Adv. Heat Transfer*, **26**, 219–324.

Shah, R.K. (1983) Heat exchanger basic design methods. *Low Reynolds Number Flow Heat Exchangers* (Eds. S. Kakaç, R.K. Shah, and A.E. Bergles), Hemisphere, New York, pp. 21–72.

Smith, D.M. (1934) Mean temperature difference in crossflow. *Engineering*, **138**, November, 479–481 and 606–607.

Spang, B. and **Roetzel, W.** (1995) Neue Näherungsgleichung zur einheitlichen Berechnung von Wärmeübertragern (new approximate equation for uniform heat exchanger design). *Heat and Mass Transfer (Wärme und Stoffübertragung)*, **30**, 417–422.

Stevens, R.A., **Fernandez, J.** and **Woolf, J.R.** (1957) Mean temperature difference in one, two and three-pass crossflow heat exchangers. *Trans. ASME*, **79**, 287–297.

Terrill, D.L. and **Douglas, J.M.** (1987) A (T–h) method for heat exchanger network synthesis. *Ind. Engng Chem. Res.*, **26**, 175–179.

Underwood, A.J.V. (1934) The calculation of the mean temperature difference in industrial heat exchanger configuration. *J. Inst. Petroleum Technology*, **20**, 145–158.

Wylie, C.R. (1953) *Calculus*, McGraw-Hill, New York, pp. 283–289.

Yuan, P. and **Kou, H.-S.** (1998) Effect of longitudinal wall conduction in a three-fluid crossflow heat exchanger. *Numerical Heat Transfer, Part A. Applications*, **31**(2), 135–150.

Bibliography

Hausen, H. (1950) *Wärmeübertragung im Gegenstrom, Gleichstrom und Kreuzstrom*, Technische Physik in Einzeldarstellung, Band 8, Springer, Berlin.

Linnhoff, B. (1983) New concepts in thermodynamics for better chemical process design. *Proc. R. Soc., Series A*, 386 (1790), March 8th, 1–33. (Also *Chem. Engng Res. Design*, **61**(4), July, 207–223.

Linnhoff, B. and **Hindmarsh, E.** (1983) The pinch design method for heat exchanger networks. *Chem. Engng Sci.*, **38**(5) 745–763.

Manson, J.L. (1954) Heat transfer in crossflow. In *Proceedings of 2nd US National Applied Mechanics Conference*, pp. 801–803.

Nusselt, W. (1930) Eine neue Formel fur den Wärmedurchgang im Kreuzstrom. *Technische Mechanik und Thermodynamik*, **1**, 417–422.

Townsend, D.W. and **Linnhoff, B.** (1982) Designing total energy systems by systematic methods. *Chem. Engng*, **378**, March, 91–97.

CHAPTER 4

Direct-Sizing of Plate–Fin Exchangers

Practical design examples

4.1 Exchanger lay-up

Figures 4.1 to 4.4 show the lay-up of simple two-stream plate–fin heat exchangers made up of alternate layers of extended surface separated by flat plates. A complete block is usually of brazed construction, although diffusion bonding can be used with suitable materials. Careful design places half-height surfaces at the top and bottom of a well insulated block. The exchanger may then be considered as 'slices', made up of a flat plate with a half-height extended surface attached to each side.

For the *crossflow* design, thermal sizing is eased by first determining the heat exchange duty of one 'slice' (Fig. 4.2), the complete exchanger then being made up of an appropriate number of slices.

Flow-friction and heat-transfer correlations are based on the cell geometries of full-height surfaces. For correct pressure loss the task is to calculate the loss for an individual flow channel. For correct heat flow the task is to associate heat-transfer coefficients with half-height fin surfaces, and relate these to the single, flat plate surface, edge length × flow length ($E_1L_1 = E_2L_2$).

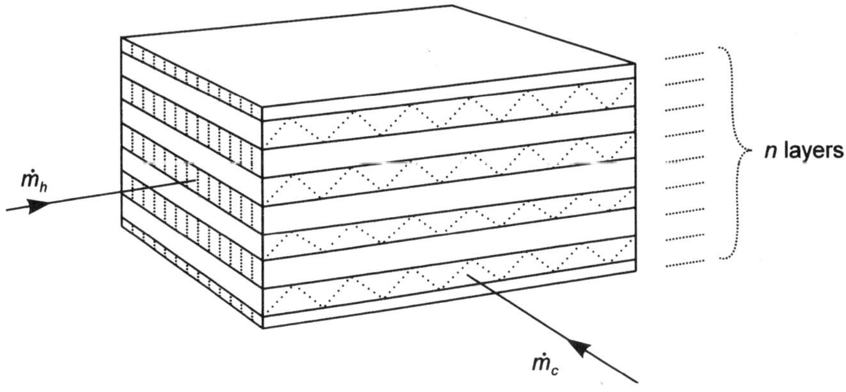

Fig.4.1 Crossflow plate–fin heat exchanger

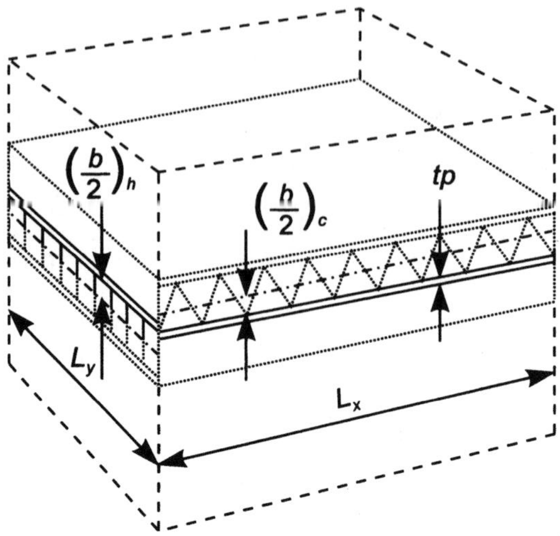

Fig.4.2 Design geometry for crossflow plate–fin exchanger

For the *contraflow* design, thermal sizing is eased by mentally reconfiguring slices of the original exchanger as an equivalent single-plate heat exchanger (Fig. 4.4) with total plate surface $S = EL$. Only the 'E' length is shown in Fig. 4.4.

Flow-friction and heat-transfer correlations are based on the cell geometries of full-height surfaces. For correct pressure loss the task is to calculate the loss for an individual flow channel. For correct heat flow the task is to associate heat-transfer

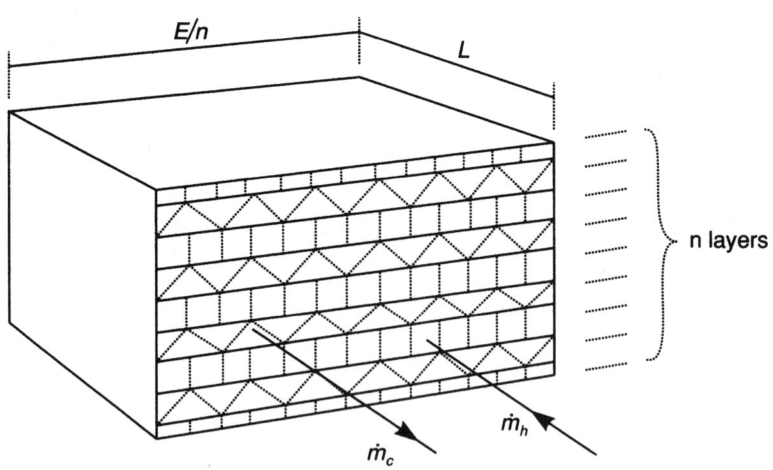

Fig.4.3 Contraflow plate–fin heat exchanger

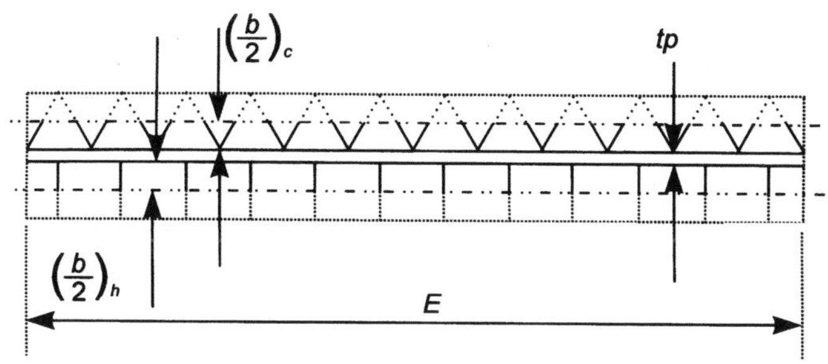

Fig.4.4 Design geometry for contraflow exchanger

coefficients with half-height fin surfaces, and relate these to the total plate surface ($S = EL$).

In *two-stream* designs, direct-sizing using 'slices' of plate–fin heat exchangers implies that transverse temperature distributions in all channels are everywhere symmetric, with zero temperature gradient at the centre of each channel. This should always be the case, unless flow maldistribution exists due to other causes, e.g. variation in channel geometry due to poor manufacturing techniques.

In *multi-stream contraflow* designs, transverse temperature symmetry is equally desirable, but this symmetry can be lost through other causes, even with perfect channel geometry, unless care is taken at plant configuration stage to match longitudinal temperature profiles throughout the multi-stream exchanger.

It is not always possible to secure absolute matching, particularly when specific heat (C) for one or more fluids may vary along a multi-stream exchanger. When channel transverse temperature symmetry has been lost, then special 'rating' procedures due to Haseler (1983) and Prasad (1993) become necessary in order to accommodate the 'cross-conduction' effect, see Section 11.6. However, in general, the multi-stream exchanger must first be 'sized' before 'rating' methods can be applied.

Several major reviews of heat exchanger design methodology have been published, e.g. London (1982), Shah (1983, 1988), Baclic & Heggs (1985), Manglik & Bergles (1990). This chapter will outline a new 'direct-sizing' approach.

4.2 Plate–fin surface geometries

In the pioneering design approach by Kays & London (1964), heat transfer is referred to total surface on one side of the exchanger, this possibly being an arrangement resulting from the single-blow experimental test procedure adopted when obtaining test data for the correlations.

For design, it is desirable to refer heat transfer to the base-plate surface. This can be done without changing the original correlations by Kays & London (1964) and London & Shah (1968), and is accomplished using the approach described in Sections 4.5 and 4.7.

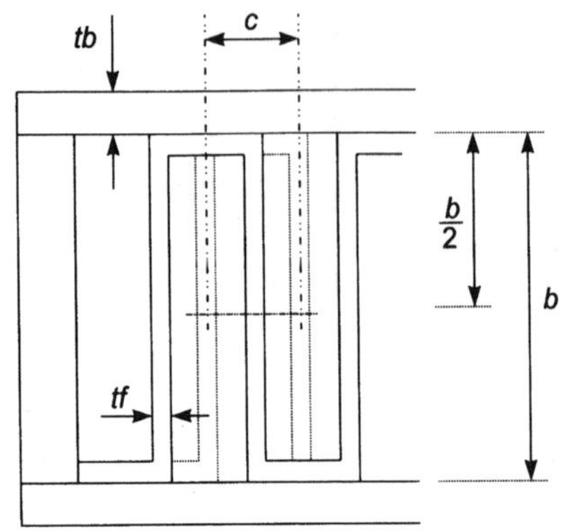

Fig.4.5 ROSF single-cell surface with strip length *x*

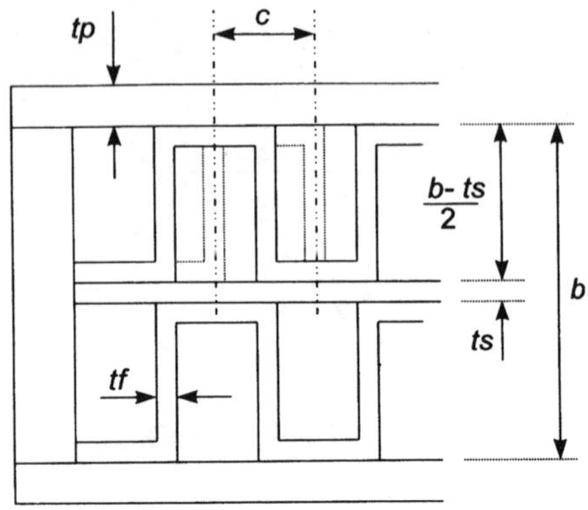

Fig.4.6 ROSF double-cell surface with strip length *x*

For variable rectangular offset strip-fin (ROSF) surfaces, the universal flow-friction and heat-transfer correlations found by Manglik & Bergles (1990) may be used providing geometrical parameters can be generated. Two cell configurations exist: the first with no splitter, the second with one central splitter (Figs 4.5 and 4.6).

Providing that at least one geometry is of the double-cell type, it is possible at lay-up time to fit a half-height layer at the block ends to provide the same pressure drop and a reasonable approximation to correct heat transfer with half the mass flowrate.

From references scanned, information did not seem to be available as to how the geometrical parameters of surfaces listed by Kays & London (1964) and by London & Shah (1968) were determined. For the double-cell geometry a problem exists as to whether the central surface areas should be considered as fin area contributing to fluid heat transfer. For symmetric transverse temperature fields the splitter (with attached fin surface) is isothermal, and its contribution to heat transfer might be neglected. However, Kays & London include this area, which would be proper for the

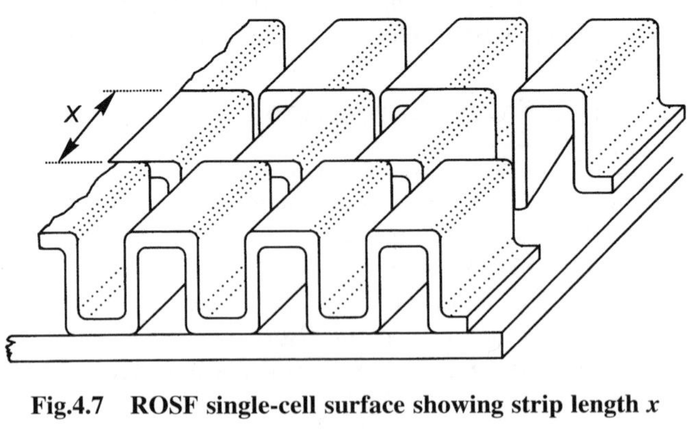

Fig.4.7 ROSF single-cell surface showing strip length *x*

case of non-symmetric transverse temperature fields (Haseler, 1983). The splitter and attached surfaces do, however, contribute to longitudinal conduction in all cases.

Geometrical parameters were calculated from scratch, and Table 4.12 lists the results. Table 4.12 confirms that a good match exists with Kays & London data for the single-cell geometries, but that values for double-cell geometries show some deviation.

4.3 Flow-friction and heat-transfer correlations

Presentation of correlations

Plate–fin heat exchangers are conventional, largely due to the extensive research of Kays & London (1964), London & Shah (1968), Webb (1994), and many other investigators who have presented flow-friction and heat-transfer correlations for plate–fin surfaces. A recent development is the vortex generator surface (Brockmeier *et al.*, 1993).

In rating methods and in direct-sizing methods it is most convenient to refer each extended surface to its base plate. This was not the approach adopted by early workers, however, the original experimental data developed by Kays & London (1964), London & Shah (1968), and Shah & London (1978) are independent of referral surface.

The preferred approach for using heat-transfer and flow-friction correlations in computer software is to fit the original data tables using an interpolating cubic spline-fit. The fit is frequently better than those to be seen in figures in the 1964 text by Kays & London.

For ROSF surfaces an alternative approach is to use the universal *f*- and *j*-correlations developed by Manglik & Bergles (1990). However, these only have an accuracy around ± 10 per cent, and are thus best used in exploring trends in surface selection. The reader should also note that there is no natural upper or lower limit on their application, and it is swiftly possible to find the design well out of range of the geometries on which the universal correlations are based. Further details of the Manglik & Bergles correlations will be found towards the end of this chapter.

4.4 Rating and direct-sizing design software

It is essential to have confidence in the design methods selected. To this end the author wrote four 'direct-sizing' programs and checked these against each other and against two 'rating' programs. The rating approach requires the designer to start with an exchanger block of known dimensions (Table 4.1).

Two direct-sizing programs were written for contraflow. The first employed tabulated data for *individual* correlations given by Kays & London (1964) and by London & Shah (1968) to create individual interpolating spline-fits, and used geometrical parameters for surfaces given by these authors.

Table 4.1 Design methods for contraflow

Data	*Rating method*	*Direct-sizing*
Given	*block size*	*thermal duty*
	$L \times W \times H$	$\dot{Q}, dP_h, dP_c$
Inlet values	$\dot{m}_h, C_h, T_{h1}$	$\dot{m}_h, C_h, T_{h1}$
	$\dot{m}_c, C_c, T_{c2}$	$\dot{m}_c, C_c, T_{c2}$
Find	*thermal duty*	*block size*

For crossflow, cold inlet and outlet temperatures are reversed.

The second pair of programs employed the *generalized* correlations for ROSF given by Manglik & Bergles (1990). This approach used basic dimensions of the geometries to generate the geometrical parameters. Two further direct-sizing programs were written for crossflow in a similar manner. The four *direct-sizing* programs are set out in Table 4.2.

In support of the above direct-sizing programs, the two *rating* programs summarized in Table 4.3 were also written. These were prepared so as to be able to recalculate existing published designs and remove any round-off that might be present in earlier hand calculations.

In the work which follows, the direct-sizing approach is emphasized.

Computational considerations

In running direct-sizing programs for steady-state crossflow, both outlet fluid temperatures are evaluated very near the end of computation. Only when the outlet temperatures are known can best estimates for mean bulk temperature, and thus more accurate mean physical properties, be obtained for inclusion in the original input data. Thus programs KAYSFIN or CROSSFIN should be run twice in succession, the second time with adjustments in physical properties.

In running direct-sizing programs for steady-state contraflow, viz. EDGEFIN or BERGFIN, all terminal temperatures are known in advance of computation. Thus

Table 4.2 Computer programs for *direct-sizing* design

Exchanger type	*Surface correlation*	*Surface geometry*	*Correlation accuracy*	*Software program*
Contraflow	K&L/L&S	Any listed	Spline-fit	EDGEFIN
	M&B	ROSF	±10%	BERGFIN
Crossflow	K&L/L&S	Any listed	Spline-fit	KAYSFIN
	M&B	ROSF	±10%	CROSSFIN

K&L, Kays & London (1964); L&S, London & Shah (1968); M&B, Manglik & Bergles (1990).

Table 4.3 Computer programs for *rating* method

Exchanger type	*Surface correlation*	*Surface geometry*	*Correlation accuracy*	*Software program*
Contraflow	K&L/L&S	Any listed	Spline-fit	CONTRRATE
Crossflow	K&L/L&S	Any listed	Spline-fit	CROSSRATE

K&L, Kays & London (1964); L&S, London & Shah (1968).

good estimates for mean physical properties can be included in the input data. However, one potential inaccuracy in contraflow design occurs when the fluid flow length is short, for this can lead to significant longitudinal conduction effects, and reduction in effective LMTD.

In Chapter 3, Section 3.4, a numerical method is developed for calculating reduction in LMTD due to longitudinal conduction, the matching finite-difference algorithm KROEGER2 being presented in Appendix B.5. Input data for LOGMEAN is output during a first run of either EDGEFIN or BERGFIN. These programs should then be run again applying a factor for reduction of LMTD.

It is not appropriate to use output data from steady-state direct-sizing designs as input for computation of transient response in contraflow, because transient algorithms require accurate physical properties at close computational stations along the length of the exchanger. Interpolating cubic spline-fits of physical properties against temperature also need to be created, and then incorporated in the computational program. Either transient programs need to be run from an initial isothermal state, or Crank–Nicholson software needs to be written for the steady state to match that for the transient solution. The Crank–Nicholson solution of three energy equations outlined in the supplement to Appendix B.8 would require simplification for a steady-state solution.

In Chapter 9, Appendix A, and Appendix B possible methods for calculating transients are discussed. In transient work, the Crank–Nicholson approach is restricted to time intervals smaller than the Courant–Friedrichs–Lewy conditions described later. An alternative approach might be to use the method of lines with Runge–Kutta algorithms at least for the mass and momentum equations.

A more detailed steady-state solution would raise further questions on assumptions made about inlet and outlet conditions. It is commonly assumed that a discontinuity of slope in the temperature occurs at both entry and exit. Nature does not behave in such a way, neither of course do heat exchangers. Just how fast the bulk fluid temperature changes at entry and exit may deserve exploration, but the effect may be small enough for one dimensional flow – see Fig. 3.29.

Total pressure loss

Only core pressure loss is used in direct-sizing and rating methods of design.

It is essential to return to the expression given by Kays & London (1964, Chapter 3 Section 3.11) for addition of inlet loss, acceleration loss, and outlet loss.

4.5 Direct-sizing of an unmixed crossflow exchanger

The best way of showing that 'direct-sizing' works is to use the data from a 'rating' example provided by Kays & London (Appendix B, Example 2 of their 1964 text) and calculate dimensions of the exchanger from scratch. Prior to comparison it was found necessary to rework Kays & London's original rating design using the CROSSRATE program to eliminate some 'round-off' errors.

For direct-sizing by KAYSFIN, values of mean bulk temperature required for evaluation of physical properties are usually first estimated from the known inlet temperatures, and the mean values $(T_{h1} + T_{c1})/2$ taken after a first design pass. However, here the same values were used as quoted in the published example. Data are first converted to SI units. Four numerical values for mean values of $\mathrm{Pr} = C\eta/\lambda$ and $p = \rho RT$ are included as input data, so that software can check that the assumed values are consistent.

Exchanger duty	$\dot{Q} = 4.854\ 82$ MW
Plate material	
Plate thickness, mm	$tp = 0.3048$
Plate thermal conductivity, J/(m s K)	$\lambda_w = 20.77$
Plate and fin density, kg/m^3	$\rho_w = 7030.0$
Surface geometry	
Warm gas	K–L plain 1/4-11.10 single-cell
Cool air	K–L louvre 3/8-06.06 single-cell

The crossflow exchanger has a unique design which satisfies all constraints simultaneously. This is because the pressure loss on one side determines channel length, and thus constrains edge length for the other side.

Only core pressure loss is used in the 'direct-sizing' treatment. It is about 98 per cent of the total loss in the example presented. Once the exchanger core is sized, then allowance for entrance effect, flow acceleration, and exit effects can be made.

The one-pass unmixed/unmixed crossflow exchanger is considered to be built up as a stack of single plates with half-height surfaces on each side. The concept is to size one plate in the stack using allowable pressure losses, and from that determine heat transfer for the whole stack. If the desired exchanger duty does not match calculated heat transfer then the solution is iterated until it does.

Given parameters are as set out in Table 4.4, with additional information given below. Physical properties for fluids and core material may be evaluated at the mean of the inter temperatures initially, and corrected later after the first design pass has been completed. Here identical values as used by Kays & London are adopted.

Crossflow direct-sizing – KAYSFIN program

The concept is based on designing a single plate, with half-height surfaces on each side.

Table 4.4 Input data for direct-sizing of crossflow exchanger

Flow stream parameters	*Hot LP gas*	*Cold HP air*
Specified parameters	Side-1	Side-2
Mass flowrate, kg/s	$\dot{m}_h = 24.683$	$\dot{m}_c = 24.318$
Inlet absolute pressure, bar	$p_{h1} = 1.027\,35$	$p_{c2} = 9.1014$
Allowable (core) pressure loss, N/m^2	$\Delta p_h = 2659.63$	$\Delta p_c = 3562.93$
Inlet temperatures, K	$T_{h1} = 702.59$	$T_{c1} = 448.15$
Mean values	Side-1	Side-2
Prandtl number	$\mathrm{Pr}_h = 0.670$	$\mathrm{Pr}_c = 0.670$
Specific heat at constant pressure, J/(kg K)	$C_h = 1084.8$	$C_c = 1051.90$
Absolute viscosity, kg/(m s)	$\eta_h = 3.0145 \times 10^{-5}$	$\eta_c = 2.850 \times 10^{-5}$
Thermal conductivity, J/(m s K)	$\lambda_h = 0.048\,817$	$\lambda_c = 0.044\,744$
Density by gas laws	Side-1	Side-2
Gas constant, J/(kg K)	$R_h = 287.07$	$R_c = 287.07$
Density, kg/m^3	$\rho_h = 0.576\,79$	$\rho_c = 5.8268$

Geometrical parameters at *cell-level* are taken from Kays & London (1964).

Side-1 (HOT)	***Side-2 (COLD)***
Plain 1/4-11.10 single-cell fins	Louvre 3/8-06.06 single-cell fins
alpha1 = b1*beta1/(b1 + 2*tp + b2)	alpha2 = b2*beta2/(b1 + 2*tp + b2)
beta1 = GIVEN data	beta2 = GIVEN data
gamma1 = GIVEN data	gamma2 = GIVEN data
kappa1 = b1*beta1/2	kappa2 = b2*beta2/2
lambda1 = gamma1*kappa1	lambda2 = gamma2*kappa2
sigma1 = beta1*D1/4	sigma2 = beta2*D2/4
omega1 = alpha1/kappa1	omega2 = alpha2/kappa2

Now estimate a value for Re1 within validity range of correlations on Side-1 (HOT), and use the friction factor correlation to obtain friction factor (f1), then:

Side-1 (HOT)

Mass velocity, kg/(m^2 s)	G1 = Re1*mu1/D1
Flow length for Side-1, m	Lp1 = dP1*2*rho1*D1/(4*f1*G1*G1)
Edge length for Side-2, m	E2 = Lp1

Repeat the process for Side-2.

We now have the single plate size

single plate, m^2 Splate = E1*E2

Next use geometric parameters to find mass flows for single plate.

Side-1 (HOT), single plate

Frontal area for Side-1, m^2	Afront1 = E1*(b1/2) (half-height fins)
Flow area for Side-1, m^2	Aflow1 = Afront1*sigma1
Mass flowrate for Side-1, kg/s	mp1 = G1*Aflow1 (single-plate)

Repeat the process for Side-2.

Number of plates required in stack
N1 = TRUNC(m1/mp1) + 1
N2 = TRUNC(m2/mp2) + 1
Test if (N1 = N2), if not then choose smaller value, call this (N)

Height of stacked plates (m)	H = N*(b1/2 + tp + b2/2)
Volume of exchanger (m^3)	Vol = E1*E2*H
Total plate surface (m^2)	Sexchr = N*Splate

With parameters Re1, Pr1, mu1, D1 evaluate heat-transfer coefficients from correlation (h1). Repeat for Side-2.

Now relate heat-transfer coefficients to plate surface.

Side-1 (HOT) parameters

Fin height (m)	Y1 = b1/2
Parameter for fin efficiency	mY1 = Y1*SQRT(2*h1/(kf1*tf1))
Hyperbolic tangent	tanh1 = (EXP(2*mY1) − 1)/(EXP(2*mY1) + 1)
Fin performance ratio	phi1 = tanh1/mY1
Parameter	eta1 = 1 − gamma1*(1 − phi1)
Heat-transfer coefficient, J/(m^2 s K)	h1t = eta1*h1 (referred to Stotal)
Heat-transfer coefficient, J/(m^2 s K)	u1 = h1t*kappa1 (referred to Sexchr)

Side-2 (COLD) parameters

Same as for Side-1

Overall heat-transfer coefficient and Ntu

Wall coefficient	uw = kp/tp
Overall coefficient	U = 1/(1/u1 + 1/uw + 1/u2)
Number of transfer units	Ntu1 = U*Sexchr/(m1*Cp1)

To obtain dimensionless mean temperature difference in crossflow, use Ntu values in algorithm CROSSTD (see Chapter 3, Section 3.5). Multiply by difference in inlet temperatures (Tspan) to obtain TDmean.

Mean temperature difference, K	TDmean = 44.64

Exchanger duty

Performance, MW	$\dot{Q}$ = U*Sexchr*TDmean/1000000

If the value of Q does not correspond to the specified duty (Q = 4.854 82 MW), then iterate Re1 and repeat the procedure until a match is obtained. Only then can the hot and cold fluid outlet temperatures be calculated from the energy balances

$$\dot{Q} = \dot{m}_1 C_1 (T_{h1} - T_{h2})$$
$$\dot{Q} = \dot{m}_2 C_2 (T_{c1} - T_{c2})$$

In practice it is best first to explore the range of validity of Reynolds number in spaced equal intervals so as to construct a graph, and then return to seek the iterated final solution. In this way it becomes possible to assess possible directions of improvement (Fig. 4.8).

For the example chosen, computed thermal duty in Fig. 4.8 is a sloping almost straight line, and the design point is where the horizontal desired duty line intersects this sloping line. The vertical line from this design point can be drawn to intersect the curve for design core volume.

The graph shows where exchangers sit within the permissible operational envelope. Table 4.5 presents the results of computation.

Final adjustment on each side is to add inlet, acceleration/deceleration, and exit losses to core pressure losses (see Chapter 3, Section 3.11).

Table 4.5 Design results for crossflow exchanger

Flow stream parameters	*Hot LP gas*	*Cold HP air*
Dimensionless numbers	Side-1	Side-2
Reynolds numbers	$\text{Re}_h = 1366.07$	$\text{Re}_c = 4079.98$
Stanton number	$\text{St}_h = 0.005\,894$	$\text{St}_c = 0.009\,581$
Ntu values	$Ntu_h = 4.0371$	$Ntu_c = 4.2261$
Heat transfer	Side-1	Side-2
Correlation coeffs, J/(m^2 s K)	$\alpha_h = 85.381$	$\alpha_c = 263.343$
Plate-referred coeffs, J/(m^2 s K)	$u_h = 289.652$	$u_c = 551.421$
Outlet temperatures, K	$T_{h2} = 521.283$	$T_{c1} = 637.944$
Flow friction	Side-1	Side-2
Correlation coefficients	$f_h = 0.015\,673$	$f_c = 0.037\,180$
Velocity, m/s	$Vel_h = 23.1496$	$Vel_c = 4.4844$
Design results		
Overall heat transfer coeff., J/(m^2 s K)	U	189.373
Number of separating plates	N	345
Plate surface area, m^2	*Splate*	1.6565
Total surface area, m^2	*Sexchr*	571.50
Performance comparison		
Exchanger duty, MW	$\dot{Q}$	4.854 82
Mean temp. difference, K	$\Delta\theta_m$	44.858
Specific performance, kW/(m^3 K)	$\dot{Q}/(V\Delta\theta_m)$	28.456 58

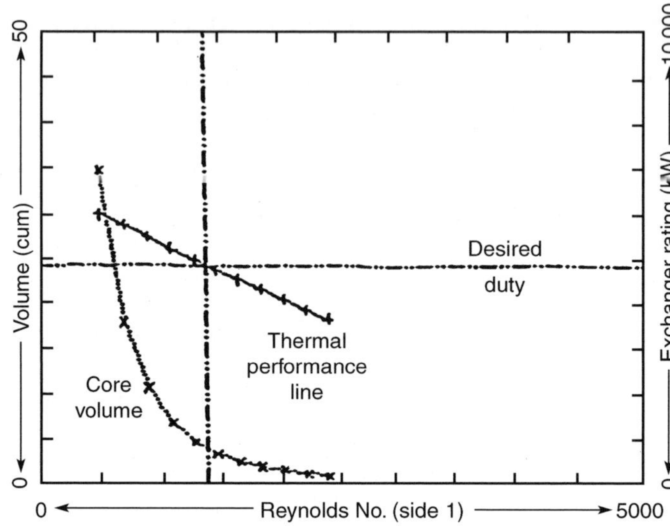

Fig.4.8 Direct-sizing design plot for one-pass (unmixed/unmixed) crossflow exchanger

Table 4.6 Comparison of rating and sizing methods

Design parameters	*Kays & London (rating method)*		*Direct sizing*
Block dimensions:	(ft)	(m)	(m)
Gas-flow path length	3.0	0.9144	0.9103
Air-flow path length	6.0	1.8288	1.8197
Block height	7.511*	2.2893	2.2959

*Original height adjusted to fit specified surfaces.

Computed block dimensions obtained in 'direct-sizing' cross-flow by EDGEFIT are compared in Table 4.6 with input data used in the CROSSRATE 'rating' method. The match is found to be good.

4.6 Concept of direct-sizing in contraflow

The concept of explicit analytical direct-sizing design will be discussed first, followed by comparison of rating and direct-sizing designs. The concept is based on treating the whole exchanger as a single plate, with half-height surfaces on both sides (Fig. 4.4). Once plate surface ($S = EL$) has been determined from flow length L and edge length E, then E may be divided into n equal strips to create a suitable aspect ratio for the finished block exchanger.

A triangular fin-cell geometry will be employed here for the simplicity of its heat-transfer and flow-friction correlations.

Explicit heat-transfer and flow-friction correlations for equilateral triangle cells (special case of isosceles cells) over a small Reynolds number range are given by Kays & London (1964) as

$$\mathrm{St(Pr)}^{2/3} = 2.7/\mathrm{Re} \tag{4.1}$$

$$f = 14/\mathrm{Re} \tag{4.2}$$

Edge length E, and flow path length L (exchanger block length), are the unknowns to be determined.

The number of cells multiplied by cell pitch equals edge length on both sides

$$E = z_1 c_1 = z_2 c_2 \tag{4.3}$$

Heat transfer

Mass velocity, kg/(m^2 s)

$$G = \frac{\dot{m}}{az}$$

Reynolds number

$$\mathrm{Re} = \frac{DG}{\eta}$$

Heat-transfer coefficient from equation (4.1)

$$\alpha = \frac{2.7\eta C}{(\mathrm{Pr})^{2/3} D}$$

Fin performance

$$\hat{m} = \sqrt{\frac{2\alpha}{\lambda_{fin} tf}} \qquad \phi = \frac{\tanh(\hat{m}Y)}{\hat{m}Y}$$

The above correlations may be evaluated for both stream 1 and stream 2, allowing for differing cell geometries. Referring fin surface (S_f) and exposed base surface (S_{xb}) to plate surface, the heat-transfer coefficients become

$$\bar{\alpha}_1 = \alpha_1\left[1 + \phi_1\left(\frac{S_f}{S_{xb}}\right)_1\right]\left(\frac{base_1}{c_1}\right) \qquad \alpha_w = \frac{\lambda_w}{tp}$$

$$\bar{\alpha}_2 = \alpha_2\left[1 + \phi_2\left(\frac{S_f}{S_{xb}}\right)_2\right]\left(\frac{base_2}{c_2}\right)$$

giving

$$\frac{1}{U} = \frac{1}{\bar{\alpha}_1} + \frac{1}{\alpha_w} + \frac{1}{\bar{\alpha}_2}$$

$$EL = \frac{\dot{Q}}{U\Delta\theta_m} = const. \tag{4.4}$$

Design is valid *above* this rectangular hyperbola.

Pressure loss

Core loss greatly exceeds other losses, and is the loss considered in direct-sizing. Using equation (4.2) in the expression for friction factor

$$f = \frac{14\eta}{DG} = \frac{14\eta az}{D\dot{m}}$$

and inserting f in the pressure loss equation

$$\Delta p = \frac{4fG^2}{2\rho}\left(\frac{L}{D}\right)$$

then, using the chosen triangular cell geometry to evaluate (a, D, G)

$$L = const. \times E, \qquad \text{say} \tag{4.5}$$

and design is valid *to the right* of this straight line.

For streams 1 and 2,

$$L_1 = Konst_1 \times E \tag{4.6}$$

$$L_2 = Konst_2 \times E \tag{4.7}$$

Curves for equations (4.4), (4.6), and (4.7) may be plotted together (Fig. 4.9). The greater of the two values $E = (const./Konst_1)^{1/2}$ and $E = (const./Konst_2)^{1/2}$ is taken as the solution from which L may be determined. This identifies the 'controlling' pressure loss. The non-controlling pressure loss for the other side has to be evaluated anew, as this is no longer the maximum permitted value.

Full use of the two allowable pressure losses is achieved when these two equations become equal at the design point. Returning to equations (4.4), (4.6), and (4.7) it can be seen that there are three simultaneous equations with only two unknowns. For a solution which is not over-determined it is desirable that equations (4.6) and (4.7) provide the same solution at the design point.

This implies that $Konst_1 = Konst_2$ and it quickly follows that

$$\frac{\Delta p_1}{\Delta p_2} = \left[\frac{(a_2/c_2)D_2^2}{(a_1/c_1)D_1^2}\right] \times const. \tag{4.8}$$

Thus the ratio of usable pressure losses may be fully determined for any given pair of compact surfaces. When this heat exchanger constraint can be combined with a system pressure loss constraint such as equations (2.34) or (2.36) then pressure

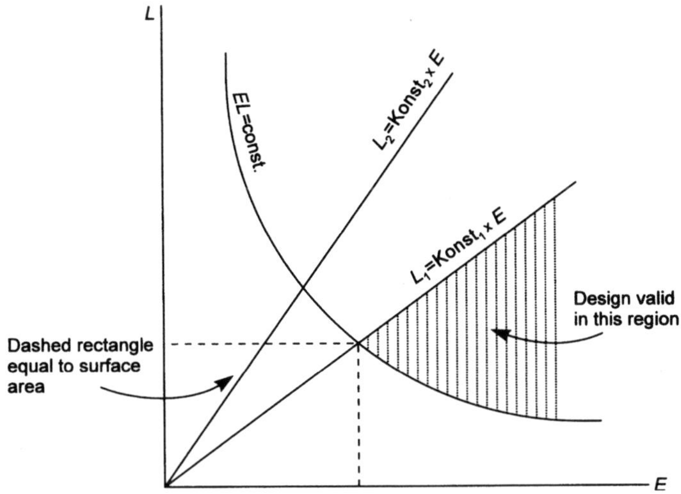

Fig.4.9 Design solution plot

losses need not be specified as input data. For direct-sizing however, the design message is: make pressure loss curve solutions coincide on the E–L plot.

In general, if both pressure losses are specified then it will be necessary to adjust cell geometries progressively to achieve the design. If, however, only one pressure loss is specified and the other 'floats' then less work will be required in design, but some 'fine-tuning' may subsequently be required.

4.7 Direct-sizing of a contraflow exchanger

Design concept

Direct-sizing is a method of arriving at exchanger dimensions without any prior assumption as to their values. We choose the same data values as for the previous crossflow design in Section 4.5 to see what kind of block size will emerge. Input data are given in Table 4.7.

Input data

In good design an overall exergy analysis of the whole plant should be made first, as this will provide the best specification of exchanger duty ($\dot{Q}$), two mass flowrates ($\dot{m}_1, \dot{m}_2$), all four terminal temperatures ($T_{h1}, T_{h2}, T_{c1}, T_{c2}$), and two inlet pressure levels (p_{h1}, p_{c2}). Pressure losses are usually dealt with later in the design.

Mean fluid temperatures can be calculated, and all physical property values obtained from interpolating cubic spline-fits of tabulated data as appropriate. Here, however, the values used were kept exactly the same as for the rating design example.

Table 4.7 Input data for contraflow exchanger

Flow stream parameters	*Hot LP gas*	*Cold HP air*
Specified parameters	Side-1	Side-2
Mass flowrate, kg/s	$\dot{m}_h = 24.683$	$\dot{m}_c = 24.318$
Inlet absolute pressure, bar	$p_{h1} = 1.027\,35$	$p_{c2} = 9.1014$
Inlet temperature, K	$T_{h1} = 702.59$	$T_{c2} = 448.15$
Outlet temperature, K	$T_{h2} = 521.68$	$T_{c1} = 637.94$
Mean values	Side-1	Side-2
Specific heat at const. press., J/(kg K)	$C_h = 1084.5$	$C_c = 1051.90$
Absolute viscosity, kg/(m s)	$\eta_h = 3.015 \times 10^{-5}$	$\eta_c = 2.850 \times 10^{-5}$
Thermal conductivity, J/(m s K)	$\lambda_h = 0.048\,817$	$\lambda_c = 0.044\,744$
Density by gas law	Side-1	Side-2
Gas constant, J/(kg K)	$R_h = 287.07$	$R_c = 287.07$
Mean density, kg/m^3	$\rho_h = 0.596\,18$	$\rho_c = 5.709\,94$

The surface geometries used in the previous Kays & London crossflow example may not be best suited to a contraflow design in which the surface plate edge length at flow entry will be the same for both surfaces. In changing from crossflow to contraflow, the main difference is that where previously the hot gas and cold air flows had independent plate edge lengths at flow entry, in contraflow the plate edge length will be the same for both fluids, which will not be optimum.

We try the same surfaces first to see what kind of improved performance can be obtained using existing geometries. Further development would then involve more appropriate choices of compact surfaces.

Exchanger duty $\dot{Q} = 4.854\,82$ MW

Plate material

Plate thickness, mm	$tp = 0.3048$
Plate thermal conductivity, J/(m s K)	$\lambda_w = 20.77$
Plate and fin density, kg/m^3	$\rho_w = 7030.0$

Surface geometry

Warm gas	K–L plain 1/4-11.10 single-cell
Cold air	K–L louvre 3/8-06.06 single-cell

Pressure losses

The exergy analysis of Chapter 2, Section 2.13, develops the following expression as an aid to selecting appropriate pressure losses

$$\left[\frac{\dot{m}_h R_h}{\dot{m}_c R_c}\right] \times \left(\frac{\Delta p_h}{p_{h1}}\right) + \left(\frac{\Delta p_c}{p_{c2}}\right) = \left[\frac{\dot{S}_{press}}{\dot{m}_c R_c}\right] \tag{4.9}$$

Usually the pressure loss is limited for the side having the lowest absolute pressure level. When the right-hand side of equation (4.9) is fixed by exergy analysis, then this equation provides means for estimating the second pressure loss. Adjustment of pressure loss pairs can then be made without much change in the exergy analysis.

In the present case, no exergy analysis from a complete plant was available, therefore both input pressure losses (dp_h, dp_c) were input, and the constant on the right-hand side of equation (4.9) was first evaluated. Future pressure loss pairs could then be compared with the original right-hand side of equation (4.9) even if surface geometries were changed.

When no prior information is available concerning allowable pressure losses then it is appropriate to fix the pressure loss on the low-pressure side so as to provide an acceptable outlet pressure. Pressure loss on the high-pressure side may then be set as high as the direct-sizing plot will allow, as the direct-sizing design process automatically provides the correct pressure loss to pair with that specified for the low-pressure side.

Cautionary remark. One matter of potential concern is high values of flow velocity which arise from relatively low values of Reynolds number. One constraint is for the Mach number not to exceed 0.1, to approach incompressible flow.

Contraflow direct-sizing – EDGEFIN program

The concept is based on treating the whole exchanger as a single plate, with half-height surfaces on each side (Fig. 4.4).

When the terminal temperatures are known, log mean temperature may be found (LMTD), and the product ($U\ S$) determined from $\dot{Q}$/LMTD.

Geometrical parameters at *cell level* required for evaluation of flow-friction and heat transfer correlations are given below:

Side-1 (HOT)		***Side-2 (COLD)***	
Plain 1/4-11.10 single-cell fins		louvre 3/8-06.06 single-cell fins	
alpha1	= b1*beta1/(b1 + 2*tp + b2)	alpha2	= b2*beta2/(b1 + 2*tp + b2)
beta1	= GIVEN data	beta2	= GIVEN data
gamma1	= GIVEN data	gamma2	= GIVEN data
kappa1	= b1*beta1/2	kappa2	= b2*beta2/2
lambda1	= gamma1*kappa1	lambda2	= gamma2*kappa2
sigma1	= beta1*D1/4	sigma2	= beta2*D2/4
omega1	= alpha1/kappa1	omega2	= alpha2/kappa2

For reduced plate edge length with hot gas flow, a taller fin height might be preferable. For increased plate edge length with cold air flow, a smaller fin height might be preferable. When the first comparative design is complete (or found unsatisfactory) then improved choices of finned surfaces can be made.

Several area parameters can now be evaluated.

For Re1 specified in scan of Reynolds number range

Side-1 (HOT) parameters

Mass velocity, kg/(m^2 s)	G1 = Re1*mu1/D1
Total flow area, m^2	Aflow1 = m1/G1
Total frontal area, m^2	Afront1 = Aflow1/sigma1
Edge length, m	E = Afront1/(b1/2)
Number of cells (rounded)	z1 = TRUNC(E/c1) + 1

Note: edge length in contraflow is (z1*c1 = E = z2*c2).

Side-2 (COLD) parameters

Number of cells (rounded)	z2 = TRUNC(E/c2) + 1
Total frontal area, m^2	Afront2 = E*(b2/2)
Total flow area, m^2	Afront2*sigma2
Mass velocity, kg/(m^2 s)	G2 = m2/Aflow2
Forced Reynolds number	Re2 = D2*G2/mu2

Mass velocities and Reynolds numbers for each side are now available.

Flow-friction and heat-transfer correlations are replaced by interpolating cubic spline-fits generated from the tabulated data in Kays & London (1964). This means:

(a) all the detail involved in the correlations is reproduced precisely
(b) there is no possibility of extrapolating beyond data limits

Geometrical parameters and corresponding flow-friction and heat-transfer interpolating spline-fits may be accessed from the pre-compiled software module SURFACE.

Spline-fitted correlations for flow-friction (f) and the Colburn j-factor (j) are now used with the Reynolds number pairs (Re_h, Re_c) to determine flow-friction and heat-transfer values on each side of the exchanger. Both Reynolds numbers must lie within the validity range of each spline-fitted correlation at all times.

It remains to refer the cell heat-transfer coefficients (h_h, h_c) to the common plate surface heat-transfer coefficients (u_h, u_c), allowing for half-height cell surfaces acting as fins.

Side-1 (HOT) parameters

Fin height, m	Y1 = b1/2
Parameter for fin efficiency	mY1 = Y1*SQRT(2*h1/(kf1*tf1))
Hyperbolic tangent	tanh1 = (EXP(2*mY1) − 1)/(EXP(2*mY1) + 1)
Fin performance ratio	phi1 = tanh1/mY1
Parameter	eta1 = 1-gamma1*(1-phi1)
Heat-transfer coeff., J/(m^2 s K)	h1t = eta1*h1 (referred to Stotal)
Heat-transfer coeff., J/(m^2 s K)	u1 = h1t*kappa1 (referred to Sexchr)

Side-2 (COLD) parameters

Same as for Side-1.

Overall heat-transfer coefficient and Ntu

Wall coefficient	uw = kp/tp
Overall coefficient	U = 1/(1/u1 + 1/uw + 1/u2)
Number of transfer units	Ntu1 = U*Sexchr/(m1*Cp1)

A check is now possible on *Ntu* values from terminal temperatures alone, by solving the simultaneous LMTD–*Ntu* rate and energy equations, (Chapter 2, Section 2.15).

Rate equation

$$\frac{T_{h1} - T_{c1}}{T_{h2} - T_{c2}} = \exp(N_h - N_c) \tag{4.10}$$

Energy equation

$$\frac{T_{c1} - T_{c2}}{T_{h1} - T_{h2}} = \frac{N_c}{N_h} \tag{4.11}$$

Appropriate tests are to check that ABS(termNtu1-heatNtu1) < 0.001 for Side-1, repeating the test for Side-2. If the conditions are not satisfied, then check that

$$\dot{Q} = \dot{m}_h C_h (T_{h1} - T_{h2}) = \dot{m}_c C_c (T_{c1} - T_{c2})$$

still holds within the input data.

The heat-transfer procedure can now evaluate the following parameters

Surface area of exchanger	Sexchr = (US)/U (plate referred)
Flow length for heat transfer	Lh = Sexchr/E

The flow-friction procedures give separate values for flow lengths.

Side-1 (HOT) parameters

Mass velocity, kg/(m^2 s)	Re1*mu1/D1
Flow length for Side-1 pressure loss	Lp1 = dP1*2*rho1*D1/(4*f1*G1*G1)

Similar values can be evaluated for Side-2.

For an initially estimated value of Re1, we now have three values for flow length (L_h, L_{p1}, L_{p2}). By varying values of Re1 three curves can be plotted on a graph for parameters (L_h, L_{p1}, L_{p2}) (Fig. 4.10).

When the correct pair of pressure losses are specified, the two pressure length curves will cut the heat-transfer curve at the same point. Usually this is not the case, then the design length is taken to be where the right-most intersection takes place. If this should be for the L_{p2} curve, then this is taken as flow length for the plate and the reduced pressure loss for Side-1 is calculated. The reverse procedure applies if the right-most intersection is with L_{p1}.

Design is now complete, as the edge length (E) and the flow length (L) are now available. Half-height surfaces exist on each side of the plate. The exchanger volume is found from the following expression

$$V = L * E * \left(\frac{b_1}{2} + tp + \frac{b_2}{2}\right) \tag{4.12}$$

The height of the finished block is open to choice by cutting the edge length (E) into a suitable number (N) of equal sections, and then stacking them.

If the pressure loss values obtained are not acceptable, or if there is no intersection of both pressure loss curves with the heat-transfer curve, the options are:

(a) select different allowable pressure loss values, following recommendations given with equation (4.9)
(b) change one or more surface geometries

Compared with crossflow, the contraflow exchanger volume has gone down by a factor of 0.697, while volumetric rating has gone up by a factor of 1.432.

The specific performance has gone down by a factor 0.928 due entirely to the larger mean temperature difference in the equivalent contraflow design. We can bring the temperature profiles closer in contraflow to reduce temperature exergy loss (with modest increase in pressure exergy loss), see Chapter 8. Keeping the original surface geometries and inlet face areas, the overall coefficient (U) would be unchanged and the same duty ($Q = US\Delta\theta_{lmtd}$) simply requires a longer flow length to leave the product $S\Delta\theta_{lmtd}$ unchanged.

In the examples presented the difference in *Ntu* values between crossflow and contraflow designs is quite remarkable in view of the improved volumetric rating of the contraflow design.

Contraflow	*Crossflow*
$Ntu_{\mathrm{h}} = 2.6134$	$Ntu_{\mathrm{h}} = 4.0371$
$Ntu_{\mathrm{c}} = 2.7347$	$Ntu_{\mathrm{c}} = 4.2261$

Inspection of $\varepsilon - Ntu$ curves for contraflow and crossflow (see e.g. Kays & London, 1964), shows that these values are necessary to give the same effectiveness, which was one intention of the comparison. Clearly *Ntu* and effectiveness values are not appropriate parameters for performance comparison of different types of exchanger.

Alternative contraflow design

The previous contraflow example kept the same thermal duty as for crossflow but allowed the mean temperature difference to grow by a factor of 1.543. The same Kays & London compact surfaces were used in each case. This produced a block volume for the contraflow case smaller by a factor of 0.698, but *Ntu* vales were only 0.647 the size of crossflow values. This confirmed that *Ntu* values are not suitable for comparing different designs (*Ntu* is just the ratio of two temperature differences).

An improved method of designing for minimum block volume in contraflow is developed in Appendix J – Compact surface selection for sizing optimization,

Table 4.8 Design results for contraflow exchanger

Flow stream parameters	*Hot LP gas*	*Cold HP air*
Dimensionless numbers	Side-1	Side-2
Prandtl numbers	$\mathrm{Pr}_h = 0.670$	$\mathrm{Pr}_c = 0.670$
Reynolds numbers	$\mathrm{Re}_h = 1555.85$	$\mathrm{Re}_c = 2324.57$
Ntu values	$Ntu_h = 2.6134$	$Ntu_c = 2.7347$
Heat transfer	Side-1	Side-2
Stanton numbers	$\mathrm{St}_h = 0.005\,765$	$\mathrm{St}_c = 0.011\,305$
Correlation coeff., $\mathrm{J/(m^2\ s\ K)}$	$\alpha_h = 95.065$	$\alpha_c = 177.066$
Plate-referred coeff., $\mathrm{J/(m^2\ s\ K)}$	$u_h = 318.762$	$u_c = 393.768$
Flow-friction	Side-1	Side-2
Correlation coefficients	$f_h = 0.014\,630$	$f_c = 0.041\,177$
Pressure losses, bar	$dp_h = 2659.63$	$dp_c = 519.25$
Velocity, m/s	$Vel_h = 25.504$	$Vel_c = 2.607$
Design results		
Overall heat-transfer coeff., $\mathrm{J/(m^2\ s\ K)}$	U	175.704
Flow length, m	L	0.7230
Edge length, m	E	550.652
Total surface area, $\mathrm{m^2}$	*Sexchr*	398.1308
Performance comparison		
Mean temp. difference, K	*LMTD*	69.224
Volumetric rating, $\mathrm{MW/m^3}$	Q/V	1.827 705
Specific performance, $\mathrm{kW/(m^3\ K)}$	$Q/(V \times LMTD)$	26.40265
Exchanger dimensions		
Length, m	L	0.7230
Width, m	W	open choice
Number of plates	N	E/W
Height, m	H	*

(Exchanger duty, $\dot{Q} = 4.854\,82$ MW)

*Height of block, $H = N \times \left(\frac{b_1}{2} + tp + \frac{b_2}{2}\right)$.

with the same mean temperature difference as the crossflow example, but using plain rectangular fins. The duty in MW increased by a factor of 1.127 over both previous examples in this Chapter, while the volume was 0.877 of that for the smaller contraflow example.

The optimized design procedure begins with inspection of performance trend curves presented by the author in Appendix C – Optimization of rectangular offset strip-fin surfaces, in the first edition of this book (Smith, 1997). Appendix C of this volume provides only a brief overview of the sets of figures given in the first edition.

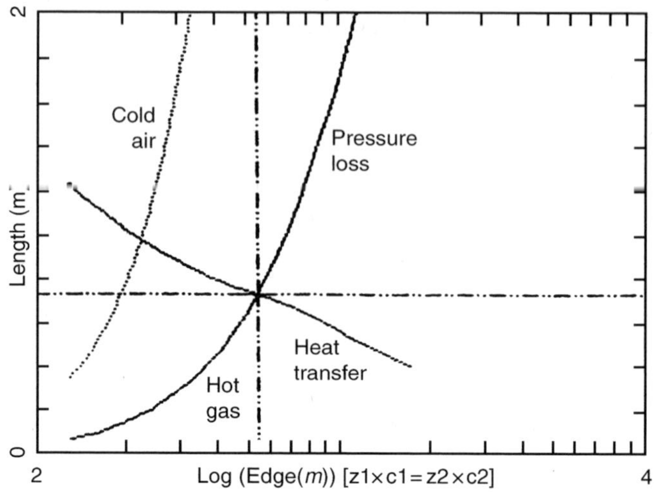

Fig.4.10 Direct-sizing design plot for contraflow, showing original pressure-loss curves

Performances are compared by using the specific performance parameter, and the values obtained are presented in Table J.7.

Once a contraflow exchanger is sized, allowance is made for longitudinal conduction effects. The LMTD reduction factor can be calculated as discussed in Chapter 3, allowing software to run again with a reduced value of LMTD. Final adjustment on each side is to add inlet, acceleration/deceleration, and exit losses, to core pressure losses (see Section 3.11).

The optimized contraflow design is preliminary as there was a lack of detailed heat-transfer and flow-friction correlations, but the results obtained point clearly to the reduced set of experimental tests necessary to obtain completely valid correlations.

4.8 Best of plain rectangular and triangular ducts

Duct constraints

We calculate performance of particular rectangular and triangular ducts in order to assess their relative merit. For mathematical simplicity both ducts shall be assumed to have a wall thickness that is very small in comparison to the duct dimensions, with the same height (b) between separating plates. Further, we impose the following constraints:

- the height of the rectangular duct (b) shall be twice the base (c) width.
- the base of the triangular duct shall be (a)

- for performance comparison both ducts shall be arranged to have the same hydraulic diameter (D) and the same mass velocity (G)
- flow in both ducts shall have the same pressure loss (Δp), and both half-height contraflow exchanger surfaces shall have the same edge length (E)

Local nomenclature

A	flow area of both ducts	P	duct wetted perimeter
f	friction factor	$\dot{Q}$	heat transferred
k	fin thermal conductivity	Re	Reynolds number
L	length of duct	s	sloping side of triangular duct
m	factor in fin performance	S	surface area for heat transfer
$\dot{m}$	mass flowrate	t	thickness of fin material
Nu	Nusselt number	Y	height of fin material
p	cell pitching		

Greek

α	heat-transfer coefficient	ρ	density of fluid flowing
η	absolute viscosity	ϕ	fin performance ratio
λ	thermal conductivity of fluid		

Duct geometries

Parameter	*Rectangular duct*	*Triangular duct*
Flow area	$A = bc$	$A = \frac{1}{2}ab$
Wetted perimeter	$P = 2(b + c)$	$P = a + \sqrt{4b^2 + a^2}$
Hydraulic diameter	$D = \dfrac{4A}{P} = \dfrac{2bc}{b + c}$	$D = \dfrac{4A}{P} = \dfrac{2ab}{a + \sqrt{4b^2 + a^2}}$

Hydraulic diameter

For equality

$$D = \frac{2bc}{b + c} = \frac{2ab}{a + \sqrt{4b^2 + a^2}} \qquad \text{whence} \qquad a = \frac{2bc}{\sqrt{b^2 - c^2}}$$

always subject to the constraint that ($b > c$). Imposing the constraint ($b = 2c$), then $a = (4/\sqrt{3})c$ and the same hydraulic diameter $D = (4/3)c$ exists for both ducts.

Sloping side of triangular duct

Sloping side $s = \sqrt{b^2 + (a^2/4)}$ from which $s = b^2/\sqrt{b^2 - c^2}$, and with ($b = 2c$) we get $s = (4/\sqrt{3})c$ hence the triangle is equilateral.

Simplified flow area and mass flowrate

Parameter	*Rectangular duct*	*Triangular duct*
Flow area	$A = bc = 2c^2$	$A = \frac{1}{2}ab = \left(\frac{4}{\sqrt{3}}\right)c^2$
Mass flowrate	$\dot{m} = AG = 2c^2G$	$\dot{m} = \left(\frac{4}{\sqrt{3}}\right)c^2G$

Heat-transfer and flow-friction correlations (laminar flow)

From the results of Shah & London (1978)

Parameter	*Rectangular duct*	*Triangular duct*
Nusselt number	$\mathrm{Nu}_H = 4.123$	$\mathrm{Nu}_H = 3.111$
Friction factor	$f\,\mathrm{Re} = 15.548$	$f\,\mathrm{Re} = 13.333$

Heat-transfer coefficient

Parameter	*Rectangular duct*	*Triangular duct*
Heat-transfer coefficient	$\alpha = \frac{\mathrm{Nu}\lambda}{D} = 4.123\left(\frac{\lambda}{D}\right)$	$\alpha = \frac{\mathrm{Nu}\lambda}{D} = 3.111\left(\frac{\lambda}{D}\right)$

remembering that hydraulic diameter is the same for both ducts.

Reynolds numbers

$\mathrm{Re} = \frac{DG}{\eta}$ is the same for both ducts.

Flow length

The Fanning core pressure loss is given by

$$\Delta p = \frac{4fG^2}{2\rho}\left(\frac{L}{D}\right)$$

from where

$$L = \frac{2\rho D}{4fG^2}\Delta p = \frac{1}{f}\left[\frac{\rho D \Delta p}{2G^2}\right] = \frac{1}{(f\,\mathrm{Re})}\left[\frac{\mathrm{Re}\rho D \Delta p}{2G^2}\right] \propto \frac{1}{(f\,\mathrm{Re})}$$

Parameter	*Rectangular duct*	*Triangular duct*
Flow length	$L = \frac{1}{15.548}\left[\frac{\mathrm{Re}\rho D \Delta p}{2G^2}\right]$	$L = \frac{1}{13.333}\left[\frac{\mathrm{Re}\rho D \Delta p}{2G^2}\right]$

Fin length

Parameter	*Rectangular duct*	*Triangular duct*
Fin length	$Y = \frac{b}{2} = c$	$Y = \frac{s}{2} = \left(\frac{2}{\sqrt{3}}\right)c$

Fin performance (rectangular cross-section)

Parameter	*Rectangular duct*	*Triangular duct*
Parameter m	$m = \sqrt{\frac{2\alpha}{kt}} = \sqrt{\frac{2 \times 4.123}{kt}\left(\frac{\lambda}{D}\right)}$	$m = \sqrt{\frac{2\alpha}{kt}} = \sqrt{\frac{2 \times 3.111}{kt}\left(\frac{\lambda}{D}\right)}$
	$= 2.8716\sqrt{\frac{\lambda}{ktD}}$	$= 2.4944\sqrt{\frac{\lambda}{ktD}}$
Parameter mY	$mY = 2.8716(c)\sqrt{\frac{\lambda}{ktD}}$	$mY = 2.4944\left(\frac{2}{\sqrt{3}}c\right)\sqrt{\frac{\lambda}{ktD}}$
	$= 2.8716\sqrt{\frac{\lambda c^2}{kt}\left(\frac{3}{4c}\right)}$	$= 2.8803\sqrt{\frac{\lambda c^2}{kt}\left(\frac{3}{4c}\right)}$
	$= 2.4869\sqrt{\frac{\lambda c}{kt}}$	$= 2.4944\sqrt{\frac{\lambda c}{kt}}$

The difference in the two values of mY is less than 0.302 per cent, thus fin performance ratio $\phi = \tanh(mY)/mY$ is almost the same for each duct.

Effective duct surface

The pitch of rectangular fins is c and one fin is associated with this base surface.

The pitch of triangular fins is

$$\left(\frac{a}{2} = \frac{2}{\sqrt{3}}c\right)$$

and one fin is associated with this base surface.

Parameter	*Rectangular duct*	*Triangular duct*
Surface area	$S = \left(c + \phi\frac{b}{2}\right)L = c(1+\phi)L$	$S = \left(\frac{a}{2} + \phi\frac{s}{2}\right)L = \frac{2c}{\sqrt{3}}(1+\phi)L$

Number of cells

Parameter	*Rectangular duct*	*Triangular duct*
Cell pitch	$p = c$	$p = \frac{a}{2} = \frac{2c}{\sqrt{3}} = 1.1547c$

For identical contraflow edge length (E) we need 11 547 rectangular fins for 10 000 triangular fins.

Check that the flow areas are the same:

Flow area for rectangular fins $= 11\,547 \times 2c^2 = 23\,094c^2$

Flow area for triangular fins $= 1000 \times \dfrac{4}{\sqrt{3}}c^2 = 23\,094c^2$

Thermal performance

Volume of each half-height surface is $V = EcL$, then from $\dot{Q} = \alpha S \Delta T_{mean}$, the specific thermal performance is:

Parameter	*Rectangular duct*	*Triangular duct*
$\dfrac{Q}{V\Delta T_{mean}} = \dfrac{\alpha S}{V}$	$= 4.123\left(\dfrac{\lambda}{D}\right) \times c(1+\phi)L$	$= 3.111\left(\dfrac{\lambda}{D}\right) \times \dfrac{2c}{\sqrt{3}}(1+\phi)L$
	$= 3.0923(1+\phi)\lambda L$	$= 2.6942(1+\phi)\lambda L$
	$= 3.0923(1+\phi)\dfrac{\lambda}{Ec}$	$= 2.6942(1+\phi)\dfrac{\lambda}{Ec}$

Finned area for longitudinal conduction

Parameter	*Rectangular duct*	*Triangular duct*
Cross-section for longitudinal conduction	$11\,547tc$	$10\,000t\left(\dfrac{4}{3}\right)c = 13\,333tc$

Should sinusoidal fins passing through corners of the triangular geometry be used, longitudinal conduction would be increased.

Performance ratios

Specific thermal performance

$$\frac{\text{rectangular}}{\text{triangular}} = \frac{3.0923}{2.6942} = 1.148$$

Comparative exchanger length

$$\frac{\text{rectangular}}{\text{triangular}} = \frac{13.333}{15.548} = 0.8575$$

LMTD reduction due to longitudinal conduction

$$\frac{\text{rectangular}}{\text{triangular}} = \frac{11\,547}{13\,333} = 0.8660$$

Comparison

For the above special cases, thermal performance of the rectangular duct is 14.8 per cent better, the exchanger is 14.25 per cent smaller, and the LMTD reduction factor is 13.4 per cent smaller. The results confirm the general conclusions of Webb (1994, pp. 115–116).

Longitudinal conduction losses

Rectangular offset strip fins will have even smaller longitudinal conduction losses. On the above considerations the software LOGMEAN for calculating LMTD reduction factors was prepared for rectangular fins only. The input data for LOGMEAN is produced directly from running either EDGEFIN or BERGFIN source listings.

4.9 Best small, plain rectangular duct

If the pressure levels on opposite sides of an exchanger are widely separated, and the respective mass flows are roughly the same, then it is likely that the cell flow areas will differ significantly.

One of the best performing compact surfaces is the rectangular offset strip-fin (ROSF) geometry. But before considering how this surface might be used it is important to consider what happens when the cell geometry is varied. This is most easily accomplished by assessing performance of the plain rectangular duct.

For laminar flow, theoretical studies of the performance of plain rectangular ducts in fully developed laminar flow were presented in book form by Shah & London (1978). A summary of some results pertaining to plain rectangular ducts subsequently appeared as a single page in the texts by Webb (1994, p. 43) and by Hewitt *et al.* (1994, p. 303). A relevant portion of these tables is reproduced below as Table 4.9 in which the base of the duct is designated c, the height as b, and the flow area is $b \times c$. Here we are not concerned with the niceties of proper allowance for fin thickness, or with half-height fins, as is required when dealing with direct-sizing surfaces.

Heat transfer (Nu_H) is the Nusselt number for constant heat flux, and flow friction is the product of friction factor and Reynolds number.

Comparison of the effects of geometric aspect ratio (b/c) on the performance of similar ducts begins with the requirement that all ducts shall have the same cross-sectional area. Notice also that results for the sequence given by Shah & London (1978) can be extended to aspect ratios 2/1, 4/1, 6/1, and 8/1.

What now is required is some information about mass flowrate, allowable pressure loss and duct cross-section, together with the physical properties of a suitable

Table 4.9 Extract from results of Shah & London (1978) (Fully developed laminar flow in plain rectangular ducts)

Duct aspect	Nu_H	$f \times Re$
1/8	6.490	20.585
1/6	6.049	19.702
1/4	5.331	18.233
1/2	4.123	15.548
1/1	3.608	14.227

fluid (e.g. nitrogen), such that the Reynolds number will remain in the laminar region for all cases investigated. Such information cannot be plucked from thin air, it can however be obtained by assessing duct performance.

Tall, thin fin surfaces perpendicular to the plate surface are well covered in published correlations. We explore the problem of dealing with plain ducts of low height which are much less dependent on fin performance.

Flow conditions

Mass velocity, kg/(m^2 s)	$G = 16$
Pressure loss, N/m^2	$\Delta p = 3000$

What matters is the duct footprint on the separating plates of a compact exchanger. Let us assume a fixed mass velocity, and a fixed allowable pressure loss, and compare the performance of eight square ducts of 2 × 2 mm sides against performance of a single rectangular duct of 2 × 16 mm sides, with the long side lying parallel to the separating plate.

There are eight square cells for one rectangular flow cell. If we take the surface area transferring heat from one side of a cell to be simply the product of cell width times flow length, then:

Eight square cells	*One rectangular cell*
$S = 8cL$	$S = cL = 0.016 \times 1.8888$
$= 8 \times 0.002 \times 0.8786$	$= 0.030\,331\ \text{m}^2$
$= 0.014\,058\ \text{m}^2$	

Table 4.10 Comparison of single duct performance

Cell parameters	*Aspect $b/c = 1/1$*	*Aspect $b/c = 1/8$*
Geometry		
Cell height, mm	$b = 2.0$	$b = 2.0$
Cell width, mm	$c = 2.0$	$c = 16.0$
Hydraulic diameter, m, D	$2bc/(b+c) = 0.002$	$2bc/(b+c) = 0.003\,55$
Flow area, m^2, A	$b \times c = 4 \times 10^{-6}$	$b \times c = 32 \times 10^{-6}$
Dimensionless numbers		
Nusselt number	$\text{Nu}_H = 3.608$	$\text{Nu}_H = 6.490$
Friction number	$f\text{Re} = 14.227$	$f\text{Re} = 20.585$
Friction factor		
Abs. viscosity, kg/(m s)	$\eta = 0.000\,03$	$\eta = 0.000\,03$
Reynolds number, Re	$DG/\eta = 1066.67$	$DG/\eta = 1866.67$
Friction factor, f	$(f\text{Re})/\text{Re} = 0.013\,3378$	$(f\text{Re})/\text{Re} = 0.011\,0277$
Flow length, m, L	$\Delta pD/(2fG^2) = 0.8786$	$\Delta pD/(2fG^2) = 1.8888$
Heat transfer		
Thermal cond., J/(m s K)	$\lambda = 0.05$	$\lambda = 0.05$
Heat-transfer coeff., J/(m^2 s K)	$\alpha = \text{Nu}\lambda/D = 90.20$	$\alpha = \text{Nu}\lambda/D = 91.28$

Comparative thermal performance per unit temperature difference is then:

Eight square cells

$$\left(\frac{\dot{Q}}{\Delta\theta}\right) = hS = 90.20 \times 0.014\,058$$
$$= 1.268\,03\ \text{watts/K}$$

One rectangular cell

$$\left(\frac{\dot{Q}}{\Delta\theta}\right) = hS = 91.28 \times 0.030\,331$$
$$= 2.7686\ \text{watts/K}$$

There will be fin effect on cell sides, and there are eight more sides on the square cells than on the rectangular cells. Assuming fin efficiency $\phi = 1.0$, and cell spacing to be 1.0 mm, we might redefine cell surface areas to be:

Eight square cells

$$\begin{aligned} S &= 8\left(c + 2\phi\frac{b}{2}\right)L \\ &= 8(c+b)L \\ &= 8 \times (0.004) \times 0.8786 \\ &= 0.028\,115\ \text{m}^2 \end{aligned}$$

One rectangular cell

$$\begin{aligned} S &= \left(c + 2\phi\frac{b}{2}\right)L \\ &= (0.016 + 0.002) \times 1.8888 \\ &= 0.0340\ \text{m}^2 \end{aligned}$$

There may be some uncertainty about fin efficiency at these small sizes when the fin is ‘chunky’, however, this is still not sufficient to make the square cell better than the rectangular cell because the new comparative performance is:

Eight square cells

$$\left(\frac{\dot{Q}}{\Delta\theta}\right) = 2.535\,97\ \text{watts/K}$$

One rectangular cell

$$\left(\frac{\dot{Q}}{\Delta\theta}\right) = 3.103\,50\ \text{watts/K}$$

However, for equal mass velocities, the rectangular cell will reach a Reynolds number of 2000 first.

Where winding of flow passages becomes necessary to combine small total flow area with large surface area for heat transfer, the square duct might be preferred so as to better approach the contraflow ideal, but numerical evaluation would reveal whether rectangular ducts showed better performance.

In choosing a winding geometry, the single, flattened, multi-start helical coil would be better than a multi-start serpentine platen. See Hausen (1950), Section 49, p. 213 onwards, for analysis of coils and platens having short lengths.

4.10 Fine-tuning of ROSF surfaces

Fine-tuning becomes possible when working with rectangular offset strip-fin (ROSF) geometries using BERGFIN or CROSSFIN software. This is because the universal flow-friction and heat-transfer correlations of Manglik & Bergles (1990) allow surface geometries to be adjusted at will (see Appendix C) so as to approach the desired optimum exchanger block (e.g. minimum block volume, minimum frontal area, etc.). The major restriction in designing with universal correlations is that local surface dimensions must fall within the dimensional envelope of the geometries used to produce the original correlations, the second restriction being that accuracy of the correlations will lie only within ± 10 per cent.

When more exact match of ROSF geometries is desired, then experimental testing is unavoidable. Single-blow testing (Chapter 10) can be used to provide original flow friction and heat transfer data for subsequent interpolative cubic spline-fitting (Appendix B.6).

4.11 Overview of surface performance

At this point it is useful to overview the situation to assess whether our choices so far are appropriate. We do this by examining the performance of an isolated plain rectangular duct, which is the basic building block for sub-compact heat exchangers. Subsequently we shall interpret our findings to match the performance of other surfaces as appropriate.

For laminar flow, theoretical studies of the performance of plain rectangular ducts in fully developed laminar flow were presented in book form by Shah & London (1978). The extract presented as Table 4.9 was discussed in Section 4.9.

What now is required is some information about mass flowrate, allowable pressure loss, and duct cross-section, together with the physical properties of a suitable fluid, e.g. nitrogen, such that the Reynolds number will remain in the laminar region for all cases investigated. Such information cannot be plucked from thin air, it can however be obtained for a single duct from an appropriate direct-sized design.

It will be assumed that the single duct is embedded in a compact heat exchanger such that the vertical side walls of the ducts form the inside surfaces of fins of thickness tf, and that the horizontal sides of the duct form the surfaces of separating plates.

Flow conditions

Mass flowrate, kg/s	$\dot{m}_g = 0.0001$
Pressure loss, N/m^2	$\Delta p = 3000.0$
Duct cross-sectional area, mm^2	$A = 8.00$
Fin thickness, mm	$tf = 0.10$

Physical properties

Fin material density, kg/m^3	$\rho_f = 8906.0$
Fin thermal conductivity, J/(m s K)	$\lambda_f = 20.0$
Gas density, kg/m^3	$\rho_g = 0.550$
Gas thermal conductivity, J/(m s K)	$\lambda_g = 0.05$
Gas absolute viscosity, kg/(m s)	$\eta_g = 0.000\,03$

The characteristic dimension for the duct is found

$$\left(d_{hyd} = \frac{4 \times \text{area for flow}}{\text{wetted perimeter}} = \frac{2bc}{b+c}\right)$$

allowing the heat-transfer coefficient (α) to be obtained from the corresponding value of Nu_H. The fin efficiency follows from ($\phi = \tanh(mY)/mY$), where

$$\left(m = \sqrt{\frac{2\alpha}{\lambda_f tf}}\right) \quad \text{and} \quad \left(Y = \frac{b}{2}\right)$$

and the effective heat-transfer surface for the plain duct to a single plate surface per unit length of duct is

$$\text{Effective surface of duct}, S_{eff} = 2(c + \phi b)L \tag{4.13}$$

From pressure loss, the Reynolds number is calculated as

$$\left(\mathrm{Re} = \frac{d_{hyd}\dot{m}_g}{\eta_f A}\right)$$

and the flow-friction factor is found from ($f = f\,\mathrm{Re}/\mathrm{Re}$). The length and volume of duct may then be obtained as

$$\text{Duct length}, \ L = \frac{\Delta p \times 2\rho\, d_{hyd}}{4fG^2} \tag{4.14}$$

$$\text{Duct volume}, V = b \times c \times L \tag{4.15}$$

From the simple expression for heat transferred ($Q = \alpha S_{eff}\Delta\theta$), the value for specific performance may be found as

$$\dot{Q}_{spec} = \frac{\dot{Q}}{V\Delta\theta} = \frac{\alpha S_{eff}}{V}\left(\frac{\text{Watts}}{\text{m}^3\text{K}}\right) \tag{4.16}$$

Results of the computation, plotted to a base of LOG(duct base/duct height) are shown in Fig. 4.11. The worst choice is the square duct, which is represented at the centre of the figure. Specific performance is best for flat, thin ducts on the right, being somewhat poorer for tall, thin ducts on the left.

This explains the success of the printed-circuit heat exchanger (PCHE) primary surfaces, however, an exchanger design based on the flat thin ducts would introduce many more separating plates. This leads to the lower porosity of PCHE blocks, making them more susceptible to parasitic longitudinal conduction losses. In practice this can be mitigated by using stainless steel instead of aluminium which reduces thermal conductivity by approximately one order of magnitude, however, the wall thicknesses through which heat is to be transferred would also be required to be reduced by the same order of magnitude.

On the left-hand side of Fig. 4.11, specific performance can be improved when ROSF surfaces are used as higher heat-transfer coefficients can be obtained due to continual restarting of the boundary layer. In laminar flow, simple flat plate theory predicts the mean heat-transfer coefficient to be twice that at the trailing edge of the plate. Thus we may anticipate that the $\dot{Q}_{spec}$ curve on the left would be much higher for ROSF surfaces.

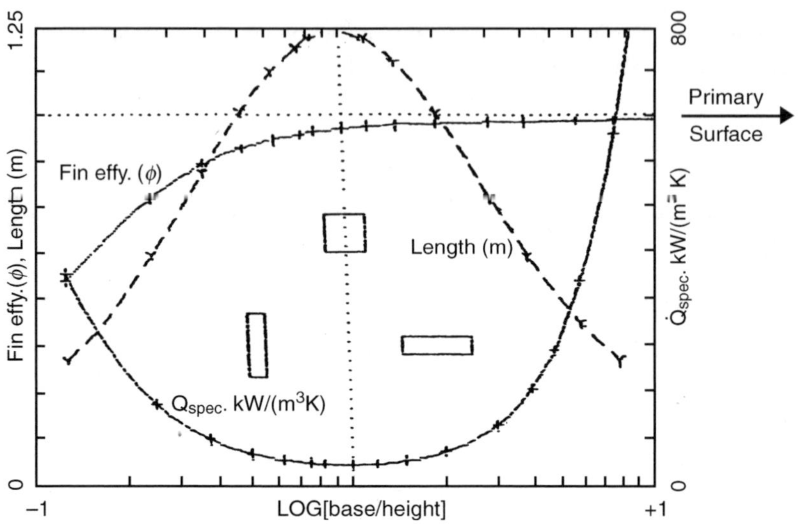

Fig.4.11 Specific performance comparison of plain rectangular ducts

Figure 4.11 also reveals that it is not desirable to go to extreme left or right limits of the diagram, as this leads to shorter flow lengths and correspondingly greater susceptibility to longitudinal conduction losses. For hot, low-pressure flows a good starting point would be to choose an effectiveness of 0.8 (see Fig. 4.11 and also Appendix J). The distance between separating plates is governed by flow area requirements. Hot, low-pressure flows need large flow areas and cold, high-pressure flows need small flow areas. A possible design philosopy would be to optimize the exchanger roughly, using plain ducts in laminar flow, before embarking on final design with ROSF or printed-circuit surfaces.

4.12 Headers and flow distributors

The subject of zero pressure loss in headers is dealt with in Chapter 8.

It may not be practicable to design a contraflow plate–fin heat exchanger without flow distributors. Problems created by introducing this extra surface include:

- allowing for additional pressure loss;
- allowing for additional heat transfer;
- allowing for variable 'phase-lag' in exchangers subject to transients.

Simple 'ribbing' of the distributor surface would create expanding and contracting flow channels at inlet and outlet. Included angles of less than 15 degrees would minimize separation losses. Pressure losses in the tapering rectangular ducts would have to be evaluated, both the mean width of ducts and the taper angle being reduced as the flow decreases to aim for equal pressure losses.

4.13 Multi-stream design (cryogenics)

It is possible to extend the contraflow design method to sizing of simple multi-stream exchangers. The case of three streams is straightforward, it only being necessary to ensure that the same pressure loss exists in both parts of the stream which is split, and that separate sections have the same length. When ROSF surfaces are used, length adjustment may be achieved by varying strip length (x).

Correction for longitudinal conduction is incorporated by adjustment of LMTD in the way described, but the problem of transverse conduction to non-adjacent streams will arise unless stream temperature profiles have already been matched in the earlier design process. This is a matter of careful layout of cryogenic plant at the system design stage, as problems can be reduced through proper attention to matching terminal temperatures and choice of streams. However, when stream temperature profiles do not match along the length of the exchanger, recourse to 'rating' design approaches like those of Haseler (1983), Prasad & Gurukul (1992), and Prasad (1993) become necessary (see Chapter 11).

4.14 Buffer zone or leakage plate 'sandwich'

Many aspects of hardware design have not been addressed in this volume. Taylor (1987) edited a guide to plate–fin heat exchangers which discusses mechanical construction including headering and pressure limitations. The Aluminium Plate–Fin Heat Exchanger Manufacturers' Association recently produced a set of standards (ALPEMA, 1994); Shah (1990) has discussed brazing methods; Haseler & Fox (1995) considered distributor models. Imperfections in construction lead to maldistribution and loss of performance, assessed by Weimer & Hartzog (1972).

One mechanical feature not previously discussed, and which directly affects thermal performance, is the leakage plate 'sandwich' used to prevent cross-contamination of two fluid streams. At lay-up each separating plate is replaced by two separating plates between which a shallow plain surface is placed. No end bars are fitted to the sandwich, so that any leakage may be to the external environment or to a leak detection system (McDonald, 1995). Both the plate spacing (b) and pitch (c) are small, while 'fin' thickness (tf) of the shallow plain surface is as large as practicable.

In thermal design it is a simple matter to treat the leakage plate 'sandwich' as a single plate, and proceed with direct-sizing as indicated earlier. The problem is to determine an equivalent thermal conductivity for the new barrier to heat flow. The following simple treatment provides an approach which may prove useful when more accurate data are not available.

Assume that the geometry of the shallow plain surface of thickness t is in the form of a sinusoid of pitch c and amplitude b. The staggered brazing better guarantees that no cross-leakage can occur. The surface may be represented by

$$y = \left(\frac{b-t}{2}\right)\sin\left(\frac{2\pi x}{c}\right) \tag{4.17}$$

By taking the derivative at $(x = 0,\ y = 0)$ and using Pythagoras, the horizontal distance across the shallow plain surface can be found. In any single pitch (c) there are two such horizontal distances. Mentally removing the metal surface, the air-gaps may be slid together horizontally giving an equivalent air-gap length which is easier to handle. The vertical heat flow length is $\ell_2 = b$ and the air-gap width is area per unit length of exchanger (A_2), given by

$$A_2 = c - 2t\sqrt{1 + \frac{c^2}{\pi^2(b-t)^2}} \tag{4.18}$$

For the metal surface the heat flow path is not at right angles to the separating plates. There are two heat flow paths of width t in any cell pitch c, hence the angled heat flow width per unit length of the exchanger is $A_1 = 2t$. Estimate the conduction length using gradient of the sinusoid at $(x = 0,\ y = 0)$ to obtain ℓ_1.

$$\ell_1 = b\sqrt{1 + \frac{c^2}{\pi^2(b-t)^2}} \tag{4.19}$$

To simplify notation replace the square-root expression by the single symbol χ in equations (4.18) and (4.19) and represent each heat flow path by a lumped form of Fourier's law $\dot{Q} = \lambda A(\Delta\theta/\ell)$, then

$$\dot{Q} = \lambda_1 A_1(\Delta\theta/\ell_1) + \lambda_2 A_2(\Delta\theta/\ell_2) = \left[\lambda_1\left(\frac{2t}{\chi c}\right) + \lambda_2\left(1 - \frac{2t\chi}{c}\right)\right]\left[\frac{\Delta\theta}{b}\right]$$

In practical cases $\chi \cong 1$, hence the equivalent conduction of the *gap* between the two 'leakage plates' becomes

$$\hat{\lambda} = \lambda_1\left(\frac{2t}{c}\right) + \lambda_2\left(1 - \frac{2t}{c}\right) \tag{4.20}$$

This is intuitively acceptable, and simple to incorporate in computer calculations. It follows that large values of t and small values of (b, c) are desirable, which is a manufacturing constraint.

Greater longitudinal conduction has now been built into the exchanger. For contraflow, direction of the sinusoids should be arranged at right-angles to the fluid flow directions. Cross-section for conduction in the single-plate design of Fig. 4.4 is then for two separating plates and one narrow plate, viz. $A = E(2 \times tp + t)$.

4.15 Consistency in design methods

Practical considerations

Plain fins are sometimes recommended for the gas-side of gas turbine recuperators, as plain fins can be cleaned effectively whereas rectangular offset strip fins cannot, (Webb, 1994). However, it is reported that when the cold air flow is by-passed then the hot-side fouling can be burnt off quite successfully.

Computational problems

When direct-sizing programs were run with the same input data, initially it was found that the predicted size of the exchanger might differ by about 1 per cent between programs. Differences were finally traced to slight discrepancies in the dimensions used for local surface geometry. One source of the problem was found to be the two values of hydraulic diameter quoted both in feet and in inches in Table 9.3 of Kays & London (1964), and these values do not quite correspond due to round-off.

Hydraulic diameter

Different definitions have been used for hydraulic diameter in generating the heat-transfer and flow-friction correlations. Earlier definitions used by different authors are to be found in the paper by Manglik & Bergles (1990), and correctness of heat-transfer coefficient and friction-factor values obtained depends on using the same definition as the original author(s). This of course is messy.

Manglik & Bergles developed an improved value for definition of hydraulic diameter given as

$$D = \frac{4shl}{2(sl + hl + th) + ts} \tag{4.21}$$

No explicit definition of hydraulic diameter was given by Kays & London in their 1964 text, but the writer provides means for defining this in Table 4.10. Matching the notation of Manglik & Bergles is done by re-defining dimensions thus

$$s = (c - tf), \qquad h = (b - tf), \qquad l = x, \qquad t = tf$$

when the hydraulic diameter obtained in this text is found to be

$$D = \frac{4shl}{2(sl + hl + th) + ts - t^2} \tag{4.22}$$

The numerical difference between alternative definitions is tiny. However, for consistency, the Manglik & Bergles expression should be used with their universal correlations.

Cautionary note. There is ± 10 per cent scatter in the Manglik & Bergles universal correlations, while for single surfaces a near exact match with experimental values can be obtained using interpolating cubic spline-fitting.

4.16 Geometry of rectangular offset strip fins

It is straightforward to generate ROSF surface parameters from basic fin dimensions. Surface specifications in Table 4.11 are based on cell dimensions only, and each side of the exchanger will normally produce different numerical values. The test of accuracy is to find that (omega1 = omega2).

Table 4.11 Geometries for rectangular offset strip-fin cells (cell surface valid over one strip length)

Parameter	*Single cell*	*Double cell*	*Notes*
Sbase/x	2(c − tf)	2(c − tf)	Exposed base
Splate/x	2c	2c	Plate surface
Vtotal/x	bc	bc	Total volume
Sfins/x	2(b − tf)	4[(b − ts)/2 − tf]	Fin sides
	+ 2(b − 2 tf) tf/x	+ 4[(b − ts)/2 − 2tf] tf/x	Fin ends
	+ (c)tf/x	+ 2(c) tf/x	Base ends
		+ 2(c − tf)	Splitter
Stotal/x	2(b − tf)	4[(b − ts)/2 − tf]	Fin sides
	+ 2(c − tf)	+ 2(c − tf)	Plates
	+ 2(b − tf) tf/x	+ 4[(b − ts)/2 − tf] tf/x	Fin ends
	+ (c) tf/x	+ 2(c) tf/x	Base ends
		+ 2(c − tf)	Splitter
Y	b/2	b/2	Fin height
Per (one cell)	2(b − tf)	2[(b − ts)/2 − tf]	Cell sides
	+ 2(c − tf)	+ 2(c − tf)	Cell ends
	+ 2(b − tf) tf/x	+ 2[(b − ts)/2 − tf] tf/x	Fin ends
	+ (c) tf/x	+ (c) tf/x	Base ends
Aflow (one cell)	(b − tf) (c − tf)	[(b − ts)/2 − tf] (c − tf)	Cell flow area
Afront	bc (one cell)	bc (two cells)	Cell frontal area

The following parameters can be evaluated from Table 4.11.

Side-1 cells	***Parameter***	***Side-2 cells***
Aflow1	Flow area on one side	Aflow2
Afront1	Frontal area on one side	Afront2
Per1	Effective perimeter of a cell	Per2
Sfins1	Fin surface on one side	Sfins2
Stotal1	Total surface on one side	Stotal2
Y1	Fin height	Y2
Vtotal1	Total flow volume on one side	Vtotal2
Splate1	Total surface area of separating plate	Splate2
Vexchr1	Volume of whole exchanger core	Vexchr2

Geometrical parameters (not all dimensionless):

Side-1 geometry	***Side-2 geometry***
alpha1 = (Stotal1/Vexchr1)	alpha2 = (Stotal2/Vexchr2)
beta1 = (Stotal1/Vtotal1)	beta2 = (Stotal2/Vtotal2)
gamma1 = (Sfins1/Stotal1)	gamma2 = (Sfin2/Stotal2)

Table 4.12 Comparison of Kays & London (1964) K&L values and calculated values for ROSF single-cell (S) and double-cell (D) surfaces

Geom. no.	*beta (1/mm)* K&L	*beta (1/mm)* Calc.	*gamma* K&L	*gamma* Calc.	*rh (mm)* K&L	*rh (mm)* Calc.	*Surface designation*	*L&S paper*
01	1.549	1.546	0.809	0.810	0.596	0.597	1/8-15.61 (S)	104 (S)
02	2.067	2.067	0.885	0.885	0.434	0.434	1/9-22.68 (S)	103 (S)
03	2.830	2.827	0.665*	0.664	0.302	0.303	1/9-24.12 (S)	105 (S)
04	2.359	2.359	0.850	0.850	0.373	0.374	1/9-25.01 (S)	101 (S)
05	2.490	2.486	0.611*	0.610	0.351	0.351	1/10-19.35 (S)	106 (S)
06	2.467	2.464	0.887	0.886	0.356	0.356	1/10-27.03 (S)	102 (S)
07	1.512	1.500	0.796	0.794	0.567	0.572	1/2-11.94 (D)	—
08	1.386	1.371	0.847	0.845	0.659	0.667	1/6-12.18 (D)	—
09	1.726	1.708	0.859	0.858	0.517	0.523	1/7-15.75 (D)	—
10	1.803	1.797	0.843	0.845	0.466	0.468	1/8-16.00 (D)	—
11	2.231	2.218	0.841	0.841	0.385	0.387	1/8-19.82 (D)	—
12	2.290	2.248	0.845	0.840	0.373	0.381	1/8-20.06 (D)	—

*Values quoted in Kays & London (3rd edn) are incorrect, and the above values are taken from the London & Shah (1968) paper.

kappa1 = (Stotal1/Splate1) kappa2 = (Stotal2/Splate2)
lambda1 = (Sfins1/Splate1) lambda2 = (Sfins2/Splate2)
sigma1 = (Aflow1/Afront1) sigma2 = (Aflow2/Afront2)
tau1 = (Sbase1/Splate1) tau2 = (Sbase2/Splate2)
omega1 = (Splate1/Vexchr1) omega2 = (Splate2/Vexchr2)

In Table 4.12 results of computation with these expressions compared with values quoted in the London & Shah paper of 1968 gave close agreement for single-cell surfaces, with some discrepancy for double-cell surfaces.

Manglik & Bergles universal correlations

For ROSF surfaces generalized explicit *f*- and *j*-correlations permit full optimization of heat exchanger cores. This allows continuous adjustment of basic cell geometry. The techniques used by Manglik & Bergles (1995) to obtain the correlations are also described by Webb (1994) and by Churchill & Usagi (1972), and seem to have been applied earlier to an entirely different problem by Clarke (1966), see Appendix I.

However, there are limits on the correlations. First, upper and lower limits must be observed for Reynolds number – and these may be different for different surface geometries. Second, upper and lower limits must be observed on the basic cell parameters, viz. cell height (b), cell pitch (c) and strip-length (x). Since experimental results for most ROSF geometries were obtained over a fairly limited range of cell aspect ratios (b/c), more experimental work on shorter, wider geometries seems desirable.

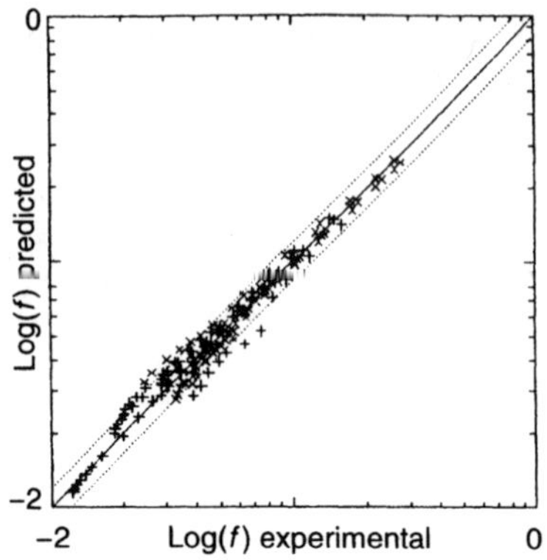

Fig.4.12 Manglik & Bergles flow-friction correlation for rectangular off-set strip fins

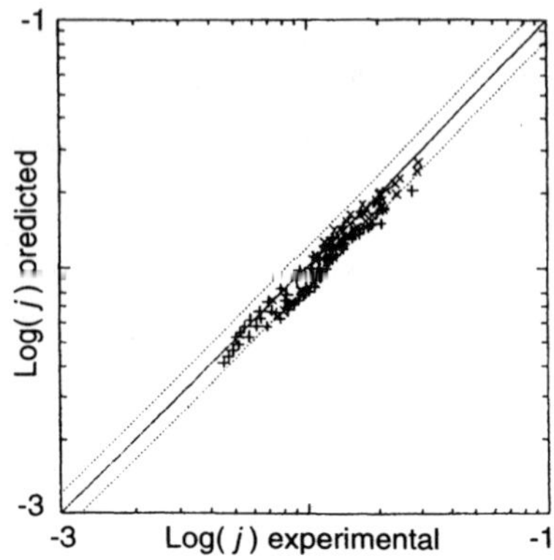

Fig.4.13 Manglik & Bergles heat-transfer correlation for rectangular offset strip fins

To confirm that the generalized Manglik & Bergles *f*- and *j*-correlations for rectangular offset strip-fin surfaces do provide a good representation of original data, six London & Shah single-cell and six Kays & London double-cell surfaces were re-assessed for fit, and the linear (log–log) fits presented in Figs 4.12 and 4.13 are very close to those originally given by Manglik & Bergles. Surfaces used are set out in Table 4.14 where (α, δ, γ) are geometrical factors used by Manglik & Bergles.

In the notation of this text the factor definitions of parameters are

$$\alpha = \frac{\text{cell pitch}}{\text{plate spacing}} = \left(\frac{c}{b}\right) \qquad \delta = \frac{\text{fin thickness}}{\text{strip length}} = \left(\frac{tf}{x}\right)$$

$$\gamma = \frac{\text{fin thickness}}{\text{cell pitch}} = \left(\frac{tf}{c}\right)$$

The correlation for flow friction is

$$\left.\begin{aligned} f &= 9.6243(\text{Re})^{-0.7422}(\alpha)^{-0.1856}(\delta)^{0.3053}(\gamma)^{-0.2659} \\ &\quad \times \left[1 + 7.669 \times 10^{-8}(\text{Re})^{4.429}(\alpha)^{0.920}(\delta)^{3.767}(\gamma)^{0.236}\right]^{0.1} \end{aligned}\right\} \tag{4.23}$$

The correlation for heat transfer is

$$\left.\begin{aligned} j &= 0.6522(\text{Re})^{-0.5403}(\alpha)^{-0.1541}(\delta)^{0.1499}(\gamma)^{-0.0678} \\ &\quad \times \left[1 + 5.269 \times 10^{-5}(\text{Re})^{1.340}(\alpha)^{0.504}(\delta)^{0.456}(\gamma)^{-1.055}\right]^{0.1} \end{aligned}\right\} \tag{4.24}$$

where the Colburn *j*-factor is $j = \text{St Pr}^{2/3}$

Table 4.13 Surfaces used to generate Manglik & Bergles *f*- and *j*-correlations

Geom. no.	*b (mm)*	*c (mm)*	*x (mm)*	*tf (mm)*	*ts (mm)*	*beta (1/mm)*	*gamma*	*rh (mm)*	*Surface designation*
01	6.350	1.627	3.170	0.102	—	1.549	0.809	0.596	1/8-15.61 (S)
02	7.645	1.120	2.820	0.102	—	2.067	0.885	0.434	1/9-22.68 (S)
03	1.905	1.053	2.822	0.102	—	2.831	0.665*	0.302	1/9-24.12 (S)
04	5.080	1.016	2.819	0.102	—	2.359	0.850	0.373	1/9-25.01 (S)
05	1.905	1.313	2.540	0.102	—	2.490	0.611*	0.351	1/10-19.35 (S)
06	6.350	0.940	2.540	0.102	—	2.467	0.887	0.356	1/10-27.03 (S)
07	6.020	2.127	12.70	0.152	0.152	1.512	0.796	0.567	1/2-11.94 (D)
08	8.966	2.085	4.521	0.102	0.152	1.386	0.847	0.659	1/6-12.18 (D)
09	7.722	1.613	3.629	0.102	0.152	1.726	0.895	0.517	1/7-15.75 (D)
10	6.477	1.588	3.175	0.152	0.152	1.803	0.843	0.466	1/8-16.00 (D)
11	5.207	1.282	3.175	0.102	0.152	2.231	0.841	0.385	1/8-19.82 (D)
12	5.015	1.266	3.175	0.102	0.152	2.290	0.845	0.373	1/8-20.06 (D)

*Values quoted in Kays & London (3rd edn) are incorrect, and the above values are taken from the London & Shah (1968) paper.

Table 4.14 General parameters for one side of an exchanger as first developed by Kays & London

Geometrical parameters	*Name*	*Kays & London all surfaces*	*This text (ROSF only)*
Stotal1/Vexchr1	alpha1	$b1 \times beta1/(b1 + 2tp + b2)$	$b1 \times beta1/(b1 + 2tp + b2)$
Stotal1/Vtotal1	beta1	GIVEN	Use Table 4.11
Sfins1/Stotal1	gamma1	GIVEN	Use Table 4.11
Stotal1/Splate1	kappa1	$b1 \times beta1/2$	Use Table 4.11
Sfins1/Splate1	lambda1	$kappa1 \times gamma1$	Use Table 4.11
Aflow1/Afront1	sigma1*	$beta1 \times D1/4$	Use Table 4.11
Sbase1/Splate1	tau1	—	Use Appendix B.4
Splate1/Vexchr1	omega1	$alpha1/kappa1$	$alpha1/kappa1$

*Note: the definition of parameter (sigma) differs from that used by Kays & London. The present form was found to be more convenient in programming.

It is not to be expected that exactly the same results will be obtained comparing design using interpolating cubic spline-fitted correlations against the universal Manglik & Bergles correlations which have to allow for ± 10 per cent scatter.

4.17 Compact fin surfaces generally

One of the best-performing surfaces for clean conditions is probably the ROSF surface. This is because the small strips continuously recreate the boundary layer and provide high heat-transfer coefficients, and because the discontinuous surface helps reduce effects of longitudinal conduction. The 'wavy' fin may show slightly better heat-transfer and flow-friction performance, but it lacks the ability to reduce the effects of longitudinal conduction (but see Fig. J.1).

Many alternative types of compact fin surface are possible candidates for use, including plain triangular, plain trapezoidal, plain sinusoidal, louvred triangular, louvred trapezoidal, plain wavy, etc. and new surface geometries continue to appear.

4.18 Conclusions

1. From nearness of approach of temperature profiles (Chapter 3), approximate maximum values of *Ntu* are likely to be as follows:
   Parallel flow, $Ntu = 2.0$ one-pass, unmixed crossflow, $Ntu = 4.0$
   Contraflow, $Ntu > 10.0$ two-pass, unmixed crossflow, $Ntu = 7.0$
2. Rectangular ducts offer one of the most convenient high-performance compact surface.
3. Trend curves for performance of single-cell and double-cell ROSF surfaces are presented in Appendix C, suggesting optimum directions for geometrical change.

4. A method of adjusting LMTD values to allow for longitudinal conduction in design of contraflow exchangers is available.
5. Assessment of small plain ducts indicates that a rectangular aspect ratio will give better performance than a square aspect ratio.
6. Low values of Reynolds number do not imply low values of flow velocity in compact heat exchanger designs. Check the Mach numbers.
7. Demonstration of 'direct-sizing' of a crossflow exchanger confirms the precision of the method. The similar approach to direct-sizing of contraflow exchangers is described. Brief discussion of extension of 'direct-sizing' to multi-stream exchangers is included.
8. Heat exchanger duty densities tending towards the following values appear practicable when surface geometries can be tuned, viz.

$$\text{specific performance}\ \left(\frac{\dot{Q}}{V\Delta\theta_m}\right) \Rightarrow 100\,\text{kW/m}^3\,\text{K}$$

9. The desirability of careful checking of published geometrical parameters of surfaces is emphasized. Any dimensional discrepancies found may influence the accuracy of heat-transfer and flow-friction correlations.
10. Pressure loss pairs can be adjusted for constant exergy generation in direct-sizing.

References

ALPEMA (1994) *The Standards of the Brazed Aluminium Plate-fin Heat Exchanger Manufacturer's Association*, ALPEMA, reprinted 1995.

Baclic, B.S. and **Heggs, P.J.** (1985) On the search for new solutions of the single-pass crossflow heat exchanger problem. *Int. J. Heat Mass Transfer*, **28**(10), 1965–1976.

Churchill, S.W. and **Usagi, R.** (1972) A general expression for the correlation of rates of transfer and other phenomena. *Am. Inst. Chem. Eng. J.*, **8**(6), 1121–1128.

Clarke, J.M. (1966) A convenient representation of creep strain data for problems involving time-varying stresses and temperatures. NGTE Report No. R284. National Gas Turbine Establishment, Pyestock, Hants.

Haseler, L.E. (1983) Performance calculation methods for multi-stream plate-fin heat exchangers. *Heat Exchangers: Theory and Practice*, (Eds. J. Taborek, G.F. Hewitt, and N. Afgan), Hemisphere/McGraw-Hill, New York, pp. 495–506.

Haseler, L.E. and **Fox, T.** (1995) Distributor models for plate-fin heat exchangers. In *4th UK National Heat Transfer Conference*, Manchester, 26–27 September 1995, pp. 449–456.

Hausen, H. (1950) *Wärmeübertragung im Gegenstrom, Gleichstrom und Kreuzstrom*, Springer, Berlin. (English edition: *Heat transfer in Counterflow, Parallel Flow and Cross-flow*, McGraw-Hill, 1976.)

Hewitt, G.F., **Shires, G.L.**, and **Bott, T.R.** (1994) *Process Heat Transfer*, CRC Press, Florida.

Kays, W.M. and **London, A.L.** (1964) *Textbook*: *Compact Heat Exchangers*, 2nd edn (1964), 3rd edn (1984), McGraw-Hill, New York.

London, A.L. (1982) Compact heat exchangers – design methodology. *Low Reynolds Number Flow Heat Exchangers.* (Eds. S. Kakaç, R.K. Shah, and A.E. Bergles), Hemisphere, New York, pp. 815–844.

London, A.L. and **Shah, R.K.** (1968) Offset rectangular plate-fin surfaces – heat transfer and flow-friction characteristics. *ASME J. Engng Power*, **90**, 218–228.

McDonald, C.F. (1995) Compact buffer zone plate-fin IHX – the key component for high temperature nuclear process heat realisation with advanced MHR. *Appl. Thermal Engng*, **16**(1), 3–32.

Manglik, R.M. and **Bergles, A.E.** (1990) The thermal hydraulic design of the rectangular offset strip-fin compact heat exchanger. *Compact Heat Exchangers – a festschrift for A.L. London*, (Eds. R.K. Shah, A.D. Kraus, and D. Metzger), Hemisphere, New York, pp. 123–149.

Manglik, R.M. and **Bergles, A.E.** (1995) Heat transfer and pressure drop correlations for the rectangular offset strip fin compact heat exchanger. *Exp. Thermal Fluid Sci.*, **10**, 171–180.

Prasad, B.S.V. (1993) The performance prediction of multistream plate-fin heat exchangers based on stacking pattern. *Heat Transfer Engng*, **12**(4), 58–70.

Prasad, B.S.V. and **Gurukul, S.M.K.A.** (1992) Differential methods for the performance prediction of multistream plate-fin heat exchangers. *ASME J. Heat Transfer*, **114**, 41–49.

Shah, R.K. (1982) Compact heat exchanger surface selection, optimisation and computer-aided thermal design. *Low Reynolds Number Flow Heat Exchangers* (Eds. S. Kakaç, R.K. Shah, and A.E. Bergles), Hemisphere, New York, pp. 845–876.

Shah, R.K. (1988) Plate-fin and tube-fin heat exchanger design procedures. *Heat Transfer Equipment Design* (Eds. R.K. Shah, E.C. Subbarao, and R.A. Mashelkar), Hemisphere, New York, pp. 256–266.

Shah, R.K. (1990) Brazing of compact heat exchangers. *Compact Heat Exchangers – a festschrift for A.L. London* (Eds. R.K. Shah, A.D. Kraus, and D. Metzger), Hemisphere, New York, pp. 491–529.

Shah, R.K. and **London, A.L.** (1978) *Laminar Forced Flow Convection in Ducts*, Supplement 1 to *Advances in Heat Transfer*, Academic Press, New York.

Smith, E.M. (1994) Direct thermal sizing of plate-fin heat exchangers. *The Industrial Sessions Papers, 10th International Heat Transfer Conference*, Brighton, UK, 14–18 August 1994, Institution of Chemical Engineers, UK.

Taylor, M.A. (Ed.) (1987) *Plate-Fin Heat Exchangers – Guide to their Specification and Use*, HTFS, Harwell (amended 1990).

Webb, R.L. (1994) *Principles of Enhanced Heat Transfer*, John Wiley.

Weimer, R.F. and **Hartzog, D.G.** (1972) Effects of maldistribution on the performance of multistream multipassage heat exchangers. In *Proceedings of the 12th Cryogenic Engineering Conference, Advances in Cryogenic Engineering*, vol. 18, Plenum Press, Paper B-2, pp. 52–64.

Bibliography

Barron, R.F. (1985) *Cryogenic Systems*, 2nd edn, Oxford University Press, p. 121.

Bassiouny, M.K. and **Martin, H.** (1984) Flow distribution and pressure drop in plate heat exchangers. Part 1 – U-type arrangements. Part 2 – Z-type arrangements. *Chem. Engng Sci.*, **39**(4), 693–700.

Bhatti, M.S. and **Shah, R.K.** (1987) Laminar convective heat transfer in ducts. *Handbook of Single-phase Heat Transfer* (Eds. S. Kakaç, R.K. Shah, and W. Aung), John Wiley.

Brockmeier, U., **Guentermann, Th.**, and **Fiebig, M.** (1993) Performance evaluation of a vortex generator heat transfer surface and comparison with different high performance surfaces. *Int. J. Heat Mass Transfer*, **36**(10), 2575–2587.

Campbell, J.F. and **Rohsenow, W.M.** (1992), Gas turbine regenerators: a method for selecting the optimum plate-finned surface pair for minimum core volume. *Int. J. Heat Mass Transfer*, **35**(12), 3441–3450.

Chapman, A.J. (1974) *Heat Transfer*, 3rd edn, Macmillan, p. 521.

Chiou, J.P. (1978) The effect of longitudinal heat conduction on crossflow heat exchanger. *ASME J. Heat Transfer*, **100**, 346–351.

Chiou, J.P. (1980) The advancement of compact heat exchanger theory considering the effects of longitudinal heat conduction and flow nonuniformity. *Compact Heat Exchangers – History, Technological Advancement and Mechanical Design Problems* (Eds. R.K. Shah, C.F. MacDonald, and C.P. Howard), ASME Winter Annual Meeting, Chicago, 16–21 November 1980, ASME HTD-Vol. 10, pp. 101–121.

Clark, S.H. and **Kays, W.M.** (1953) Laminar-flow forced convection in rectangular tubes. *Trans. ASME*, **75**, July, 859–866.

Cool, T., **Stevens, A.**, and **Adderley, C.I.** (1999) Heat exchanger optimisation using genetic algorithms. In *6th UK National Conference on Heat Transfer*, IMechE, London, pp. 27–32.

DeJong, N.C., **Zhang, L.W.**, **Jacobi, A.M.**, **Balachandar, S.**, and **Tafti, D.K.** (1998) A complementary experimental and numerical study of the flow and heat transfer in offset strip-fin heat exchangers. *Trans. ASME, J. Heat Transfer*, **120**, August, 690–698.

Dow, W.M. (1950) The uniform distribution of a fluid flowing through a perforated pipe. *Trans. ASME*, December 431–438.

Grassman, P. and **Kopp, J.** (1957) Zur gunstigen Wahl der Temperaturdifferenz und der Wärmeübergagszahl in Wärmeaustauschern. *Kaltetechnik*, 9 Jahregang, **10**, 306–308.

Herbein, D.S. and **Rohsenow, W.M.** (1988) Comparison of entropy generation and conventional method of optimising a gas turbine regenerator (actually recuperator). *Int. J. Heat Mass Transfer*, **31**(2), 241–244.

Hesselgreaves, J.E. (1993) Optimising size and weight of plate-fin heat exchangers. In *Proceedings of 1st International Conference on Heat Exchanger Technology*, Palo Alta, California, 15–17 February 1993 (Eds. R.K. Shah and A. Hashemi), Elsevier, pp. 391–399.

Incropera, F.P. and **DeWitt, D.P.** (1981) *Fundamentals of Heat Transfer*, John Wiley, p. 521.

Joshi, H.M. and **Webb, R.L.** (1987) Prediction of heat transfer and friction in the offset strip fin array. *Int. J. Heat Mass Transfer*, **30**, 69–84.

Kelkar, K.M. and **Patankar, S.V.** (1989) Numerical prediction of heat transfer and fluid flow in rectangular offset-fin arrays. *Int. J. Computer Methodology, Part A. Appl.: Numerical Heat Transfer*, **15**(2), March, 149–164. (Also, ASME Publication HTD-Vol. 52, pp. 21–28.)

Kreith, F. (1965) *Principles of Heat Transfer*, 2nd edn, International Textbook Co., Scranton, Pennsylvania, p. 497.

Kroeger, P.G. (1966) Performance deterioration in high effectiveness heat exchangers due to axial conduction effects. In *Proceedings of the 1966 Cryogenic Engineering Conference, Advances in Cryogenic Engineering*, vol. 12, Plenum Press, Paper E-5, pp. 363–372.

McDonald, C.F. and **Wilson, D.G.** (1996) The utilisation of recuperated and regenerated engine cycles for high-efficiency gas turbines in the 21st century. *Appl. Thermal Engng*, **16**(8/9), 635–653.

Najjar, Y.S.H. (1996) Relative effect of pressure losses and inefficiencies of turbomachines on the performance of the heat-exchange gas turbine cycle. *Appl. Thermal Engng*, **16** (8/9), 769–776.

Patankar, S.V. and **Prakash, C.** (1981) An analysis of the effect of plate thickness on laminar flow and heat transfer in interrupted plate passages. *Int. J. Heat Mass Transfer*, **24**(11), 1801–1810.

Reneaume, J.M., **Pingaud, H.**, and **Niclout, N.** (2000) Optimisation of plate fin heat exchangers – a continuous formulation. *Trans. IChemE*, **78**(A), September, 849–859.

Schmidt, F.W. and **Newell, M.E.** (1967) Heat transfer in fully developed laminar flow through rectangular and isoceles triangular ducts. *Int. J. Heat Mass Transfer*, **10**, 1121–1123.

Shah, R.K. (1981) Compact heat exchanger design procedures. *Heat Exchangers – Thermal-Hydraulic Fundamentals and Design* (Eds. S. Kakaç, A.E. Bergles, and F. Mayinger), Hemisphere, New York, pp. 495–536.

Soland, J.G., **Mack, W.M.**, and **Rohsenow, W.M.** (1978) Performance ranking of plate-fin heat exchanger surfaces. *ASME J. Heat Transfer*, **100**, August, 514–519.

Walsh, P.P. and **Fletcher, P.** (1998) *Gas Turbine Performance*, Blackwell Science, Oxford.

Weiting, A.R. (1975) Empirical correlations for heat transfer and flow friction characteristics of rectangular offset fin heat exchangers. *Trans. ASME. J. Heat Transfer*, **97**, 488–490.

CHAPTER 5
Direct-Sizing of Helical-Tube Exchangers

Practical design example

5.1 Design framework

Theoretical expressions are developed for the geometrical arrangement of the tube bundle in a simple helical-tube, multi-start coil heat exchanger, and in exchangers with central ducts. Consistent geometry provides uniform helix angles, uniform transverse and longitudinal tube pitches, and identical tube lengths throughout the bundle.

'Sizing' of a contraflow exchanger begins when both mean temperature difference $\Delta\theta_m$ and the product US of the overall heat-transfer coefficient and the surface area have been determined. Given tube geometry and both tube-side and shell-side pressure losses, a method is presented for arriving at an optimal tube-bundle configuration for the heat exchanger with single-phase fluids.

In developing the 'direct-sizing' method, simplified tube-side flow-friction and heat-transfer correlations for straight tubes are employed to permit a clean solution. This starts from knowledge of local tube and pitching geometry, and when the 'design window' is open (see Fig. 5.10) we arrive at an optimum tube-bundle configuration satisfying specified shell-side and tube-side heat-transfer and pressure-loss constraints.

However, tube curvature has an effect on heat transfer and pressure loss. For design-critical conditions, once the exchanger has been sized it is practicable to fine tune the design by tube coil length adjustment so that constant pressure loss occurs everywhere across the shell-side and also across the tube-side.

In this chapter a fully explicit design approach can be demonstrated because all correlations for heat transfer and pressure loss are available as algebraic expressions. A numerical design approach is also possible, and is probably to be preferred for practical design purposes. Setting up the numerical solution is left as an exercise.

The helical-tube, multi-start coil heat exchanger (Fig. 5.1), has no internal baffle leakage problems. It permits uninterrupted crossflow through the tube bank for high local heat-transfer coefficients, and provides advantageous counterflow terminal temperature distribution in the overall exchanger.

Some modification to LMTD is required when the number of tube turns is less than about ten, and this analysis has been presented by Hausen in both his German (1950) and his English (1983) texts.

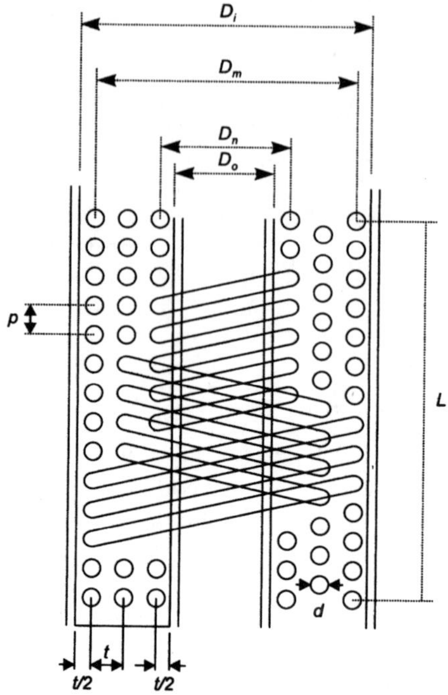

Fig.5.1 Helical tube bundle with start factor r = 1

The design is largely restricted to non-fouling fluids, and is particularly useful when exchange is required between high-pressure–low-volume flow and low-pressure–high-volume flow as often encountered in cryogenics. Flow areas on both sides may, however, be usefully varied. Thermal expansion can be accommodated by deflection of the ends of the coiled tube bundle.

This type of exchanger was patented by Hampson (1895), and subsequently repatented by L'Air Liquide (1934). However, formal geometry of the helical-tube, multi-start coil heat exchanger does not seem to have been given before 1960 when it was presented in an industrial report (Smith, 1960). A very brief note outlining the principal results was published (Smith, 1964).

Since that time, programmes of experimental work on heat transfer in helical-coil tube bundles have been published (Gilli, 1965; Smith & Coombs, 1972; Abadzic, 1974; Smith & King, 1978; Gill *et al.*, 1983). Further geometrical results have been derived, and a direct method of arriving at the design of the tube bundle has been obtained, both of which are included in this chapter.

A substantial amount of international work has been done on the helical-coil design. It has been applied in gas-cooled nuclear reactor plant, both marine and land-based pressurized water reactors (PWRs), and in cryogenic applications including LNG plant. Weimer & Hartzog (1972), have preferred the helical coil heat exchangers for LNG service, as the design is less sensitive to flow maldistribution.

In heat exchanger sizing, both LMTD–*Ntu* and ε–*Ntu* methods deliver the product of the overall heat-transfer coefficient, and the related surface area (*US*), leaving the design configuration to be determined by other methods. It is the purpose of this chapter to describe an approach to direct-sizing starting from the product *US* and the LMTD.

The method applies to tube arrangements in which the local geometry of the bundle is independent of the number of tubes in the exchanger, i.e. the shell-side area for flow for a single tube can be determined, and true counterflow is achieved without the use of redirecting baffles. A minimum value of $y = 10$, the number of times that shell-side fluid crosses a tube turn, is desirable, see (Hausen (1950, 1983). Before proceeding to thermal design, certain geometrical expressions for the helical-coil geometry have to be developed below.

5.2 Consistent geometry

Start factor (r)

If, as a simplification, the effect of tube curvature on heat transfer and pressure loss through the tube is neglected, then for the shell-side fluid, each parallel flow path should have the same axial configuration. For the tube-side fluid each tube should have the same length.

The simplest method of satisfying the above conditions is to give every tube the same helix angle, and to adopt an annular arrangement where the central coil has one tube, the second coil has two tubes, the third coil three tubes and so on. The mean coil diameters are selected so that the shell-side fluid everywhere passes over exactly the same number of tube turns in traversing the bundle. This layout will be especially satisfactory when a small area for flow in the tube bundle is required compared with the shell-side flow area.

It is possible to generalize the above case by multiplying the number of tubes in all coils by a constant factor r, which is an integer and which may take the values 1, 2, 3, etc. This increases the number of tubes in the exchanger and the area for flow on the tube-side r times. For the same heat-transfer surface it reduces the required length of individual tubes, and increases the helix angle of the tube coils.

In the expressions given below, the outermost coil is denoted as the m-th and contains rm tubes, whereas an intermediate coil is denoted as the z-th coil and contains rz tubes. For complete generality a central axial cylinder is introduced (Fig. 5.1), and this results in an innermost coil which is denoted as the n-th coil and contains rn tubes.

Mean diameter of the z-th coil (D_z)

This parameter is required for finding shell-side flow area. Noting that $p > d/\cos\phi$ always, and $t > d$ always, then for every tube in the exchanger (Fig. 5.2),

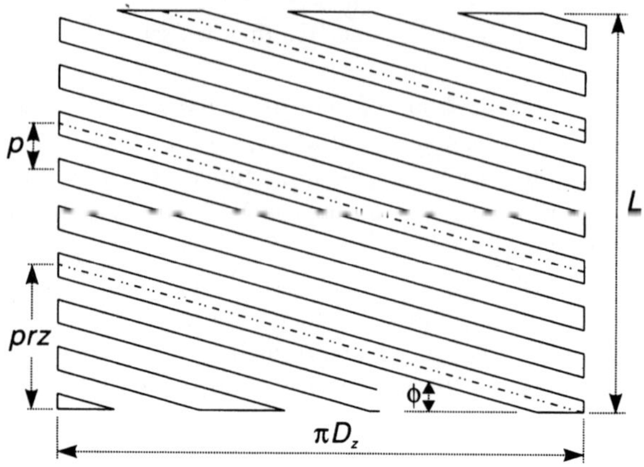

Fig.5.2 Developed z-th coil

$$\tan\phi = (prz)/(\pi D_z)$$
$$D_z - D_{z-1} = (2t) = (pr)/(\pi \tan\phi)$$

$$\boxed{D_z = 2zt} \quad (5.1)$$

Helix angle of coil (ϕ)

$$\boxed{\tan\phi = (rp)/(2\pi t)} \quad (5.2)$$

Length of the tube bundle (L)

For every tube in the exchanger (Fig. 5.2)

$$\sin\phi = L/\ell \quad (5.3)$$

then, using relationships (5.2) and (5.3)

$$\boxed{\left[\frac{\ell}{L}\right]^2 = 1 + \left[\frac{2\pi t}{rp}\right]^2} \quad (5.4)$$

from which L may be obtained.

Number of tubes in exchanger (N)

The z-th coil contains rz tubes, so that

$$N = r[n + (n+1) + (n+2) + \cdots + (m-1) + m]$$

hence

$$\boxed{N = r(m+n)(m-n+1)/2} \tag{5.5}$$

Number of times that shell-side fluid crosses a tube turn (y)

$$\boxed{y = L/p} \tag{5.6}$$

Length of tubing in one longitudinal tube pitch (ℓ_c)

Knowledge of the dimension ℓ_c is required in heat-transfer design for condensation.

$$y\ell_c = \text{total length of tubing} = N\ell$$

thus using equation (5.6)

$$\boxed{\ell_c = Np(\ell/L)} \tag{5.7}$$

Tubing in a projected transverse cross-section (ℓ_p)

Parameter required in evaluation of shell-side minimum area for flow. Clearly,

$$\ell_p = \pi \sum_n^m D_z$$

hence, from equation (5.1)

$$\ell_p = 2\pi t[n + (n+1) + \cdots + (m-1) + m] = \pi t(m+n)(m-n+1)$$

$$\boxed{\ell_p = (2\pi Nt)/r} \tag{5.8}$$

Shell-side minimum area for axial flow (A_{min})

This is required for axial crossflow through the tube bundle. From equation (5.1) and Fig. 5.1, the outside diameter of the central axial cylinder (core mandrel) is given by

$$\boxed{D_o = D_n - t = (2n-1)t} \tag{5.9}$$

Similarly the inside diameter of the exchanger shell (or bundle wrapper) is

$$\boxed{D_i = D_m + t = (2m+1)t} \tag{5.10}$$

Considering smooth tubes only, the shell-side projected face area for flow is

$$\ell_p(t-d) = 2\pi Nt(t-d)/r$$

hence the face area for axial flow, shell-side is

$$A_s = 2\pi Nt(t-d)/r$$

Using equations (5.5), (5.9), and (5.10), or proceeding directly from Fig. 5.3

$$A_s = \pi(m+n)(m-n+1)t(t-d) = \pi(D_i^2 - D_o^2)(1-d/t)/4$$

Denoting annular area between the central axial cylinder and the exchanger shell as

$$\boxed{A_a = \pi(D_i^2 - D_o^2)/4} \tag{5.11}$$

it follows that the correction for face area is

$$\boxed{A_s/A_a = 1 - d/t} \tag{5.12}$$

For flow-friction and heat-transfer correlations the fluid velocities in 'staggered' and 'in-line' tube-bundle arrangements are generally taken at the point of minimum gap between adjacent tubes. The use of alternate right- and left-hand coils in a multi-start helical-tube heat exchanger ensures a homogeneous mixture of all crossflow geometries between 'in-line' and 'staggered' in the tube bundle, independent of any axial displacement of individual coils (Fig. 5.1).

This will give an effective minimum shell-side flow area (A_{min}) which is greater than the minimum 'line-of-sight' flow area (A_s). The value of A_{min} for a multi-start coil helical-tube heat exchanger is found by considering Figs 5.4a and b. Figure 5.4a gives a three-dimensional view of a portion of the tube bundle that is developed to give straight tubes. 'AB' represents the distance between the centre-lines of adjacent rows of tubes when the tube bundle may be considered as in-line, and 'FG' represents the same distance when the tubes are staggered.

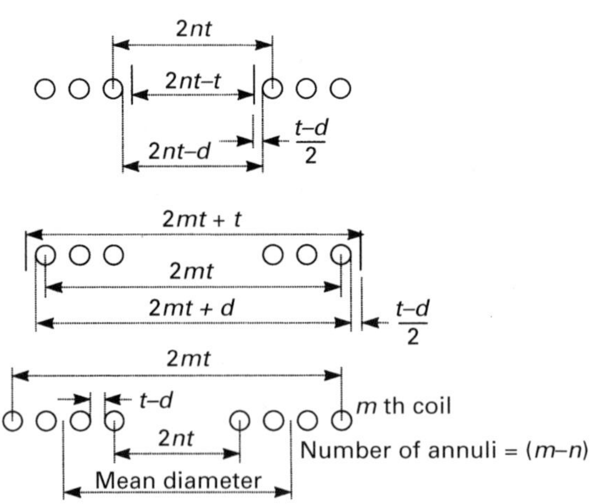

Fig.5.3 Shell-side area for flow area = $\sum$[π(mean diameter)(number of annuli)(width of annulus)]

Because of symmetry it is sufficient to calculate the effective minimum area for flow which lies between AB and FG (A_{min}), and to compare this with the corresponding face area for flow (A_s). At any distance from AB, the vertical distance between K and L is pbr. Thus

$$KL = \left[\frac{pr}{\pi t}\right]\sqrt{(b^2 + [\pi t^2/(pr)]^2)}$$

and assuming that the tube cross-section can be taken as circular, the gap between the tubes is

$$\left[\frac{pr}{\pi t}\right]\sqrt{(b^2 + [\pi t^2/(pr)]^2)} - d$$

There is a slight error in the above expression due to the assumption that tube cross-sections are circular and not elliptical.

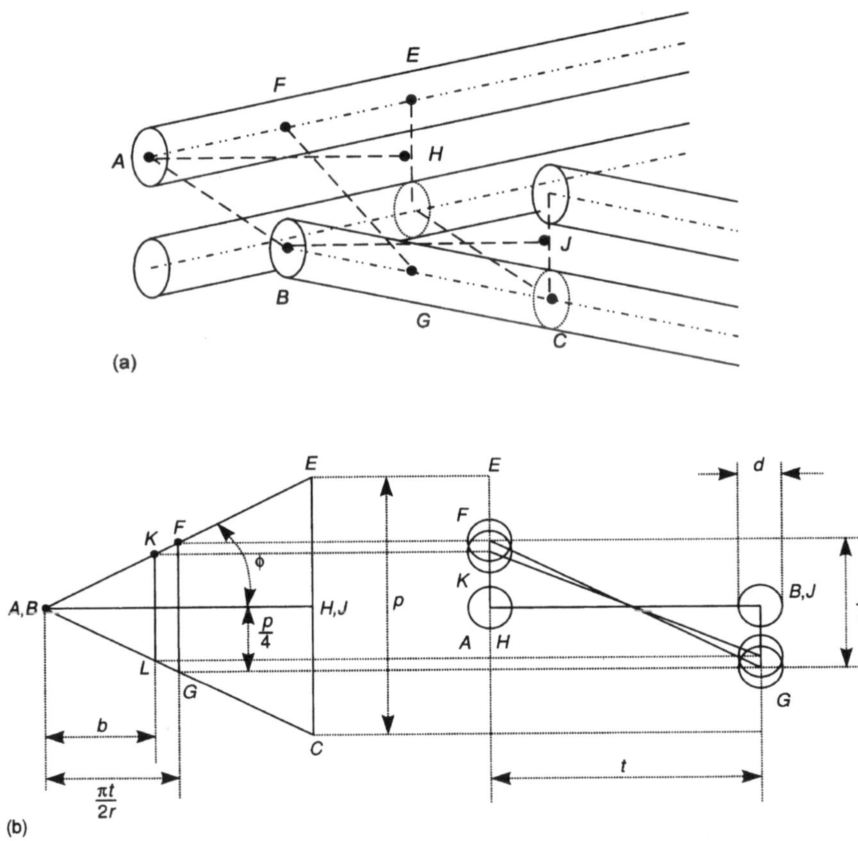

Fig.5.4 Minimum area for flow

The minimum area for flow between AB and FG (a_{min}) is found by integrating the above expression between the limits $b = 0$ and $b = \pi t/(2r)$. The area is

$$a_{min} = \int_0^{\pi t/(2r)} \left\{ \left[\frac{pr}{\pi t}\right] \sqrt{(b^2 + [\pi t^2/(pr)]^2)} - d \right\} db$$

This expression is of the form

$$\int \left\{ \sqrt{(x^2 + A^2)} - B \right\} dx$$

and has the solution

$$\left(\frac{1}{2}\right)\left[x\sqrt{(x^2 + A^2)} + A^2 \ln\left\{x + \sqrt{(x^2 + A^2)}\right\}\right] - Bx$$

hence

$$a_{min} = \left[\frac{\pi t^2}{4r}\right] ROOT + \left[\frac{\pi t^3}{2pr}\right] \ln\left[\left(\frac{p}{2t}\right) + ROOT\right] - \frac{d\pi t}{2r}$$

where

$$ROOT = \sqrt{\left[\left(\frac{p}{2t}\right)^2 + 1\right]}$$

The corresponding face area between AB and FG is

$$a_s = \left(\frac{\pi t}{2r}\right)(t - d)$$

thus the correction factor for helix angle in the bundle is

$$\boxed{A_{min}/A_s = a_{min}/a_s = \left(\frac{1}{t-d}\right)\left[\left(\frac{t}{2}\right) ROOT + \left(\frac{t^2}{p}\right) \ln\left(\frac{p}{2t} + ROOT\right) - d\right]} \tag{5.13}$$

The shell-side area for flow is determined from equations (5.11), (5.12), and (5.13) as

$$A_{min} = A_a \left(\frac{A_s}{A_a}\right)\left(\frac{A_{min}}{A_s}\right) \tag{5.14}$$

$$\boxed{A_{min} = \left[\frac{2\pi N t^2}{r}\right]\left[\frac{A_s}{A_a}\right]\left[\frac{A_{min}}{A_s}\right]} \tag{5.15}$$

This expression will not apply when there is no spacing, or very little spacing between the tubes, for then an entirely new shell-side geometry is created, and correlations may be better expressed in terms of tube-bundle porosity (see below).

Tube-side area for flow (A_t)

$$A_t = N(\pi/4)d^2 \tag{5.16}$$

Shell-side to tube-side flow area ratio (a_r)

This parameter is independent of the number of tubes, which permits direct-sizing of helical-tube, multi-start coil heat exchangers.

$$a_r = \left(\frac{A_{min}}{A_t}\right) = \left(\frac{8}{r}\right)\left(\frac{t}{d_i}\right)^2\left(\frac{A_s}{A_a}\right)\left(\frac{A_{min}}{A_s}\right) \tag{5.17}$$

independent of number of tubes N.

Shell-side porosity (P_y)

This parameter is used in correlating friction factor data for shell-side flow in the helical-coil tube bundle (Smith & Coombs, 1972; Smith & King, 1978).

The external volume taken up by the tubes is,

$$V_t = N(\pi/4)d^2\ell$$

The annular volume between exchanger shell and internal duct is

$$V_a = (\pi/4)\left(D_i^2 - D_o^2\right)L = \pi(m+n)(m-n+1)t^2L = \left(\frac{2\pi}{r}\right)Nt^2L$$

Thus shell-side porosity is $P_y = 1 - \text{(tube volume)}/\text{(annulus volume)}$

$$P_y = 1 - \left(\frac{r}{8}\right)\left(\frac{d}{t}\right)^2\left(\frac{\ell}{L}\right) \tag{5.18}$$

Exchanger with central duct

A central by-pass duct with a flow-control valve may be an advantage when close control of fluid temperature is necessary. It is convenient first to establish the leading dimensions of a simple exchanger to obtain the length of the by-pass duct required. The inside diameter of the central duct is then selected on the basis of pressure loss in the duct, and the outside diameter will then determine the value of n for the innermost coil using equation (5.9).

5.3 Simplified geometry

When a wire space of diameter $t - d$ is used to space tubes in both transverse and longitudinal directions, then several of the geometrical relationships are simplified as follows:

$$\cos\phi = t/p \tag{5.19a}$$

$$\sin\phi = r/(2\pi) \tag{5.19b}$$

Using equation (5.13)

$$\frac{L}{l} = \frac{r}{2\pi} \tag{5.20}$$

and in equation (5.18)

$$P_y = 1 - \left(\frac{\pi}{4}\right)\left(\frac{d}{t}\right)^2 \tag{5.21}$$

Substituting equation (5.20) in equation (5.4)

$$\left(\frac{t}{p}\right)^2 = 1 - \left(\frac{r}{2\pi}\right)^2 \tag{5.22}$$

We note that r may take only integer values (1 to 6) for real values of t/p.

For axial flow in the bundle it is best if the angle of inclination does not exceed 20° if heat-transfer correlations are to remain valid, thus $r = 1$ or 2.

For radial flow exchangers it may be preferable to have the tubes of successive coils set at right angles to each other, thus $r = 4$ or 5.

In obtaining the expression (5.15) for shell-side minimum area for axial flow, certain assumptions were made concerning the location of minimum flow area. It is necessary to show that at the locally 'staggered' section of the bundle the minimum flow area will always lie in the diagonal direction.

With the simplified geometry, when the intermediate coil of the other coiling hand is in the 'middle' position:

Transverse area for flow $= (2t - d)$

Diagonal area for flow $= 2[t\sqrt{(1 + 4t/d)} - d]$

If $p/d \geq \sqrt{(1 + 4t/d)}$ then minimum flow area is in transverse direction.

If $p/d \leq \sqrt{(1 + 4t/d)}$ then minimum flow area is in diagonal direction.

For values of r up to 4, Fig. 5.5 shows that the assumption made earlier is certainly valid for the simplified geometry. The assumption should probably be checked for values of r greater than 2 for more general tube-pitching arrangements.

Again, when either the transverse pitch ratio (t/d) or the longitudinal pitch ratio (p/d) is unity, then shell-side flow area will need to be reassessed. When both ratios are equal to unity, then porosity of the tube bundle will be the relevant parameter.

Table 5.1 Simplified geometry for tube bundle

r	1	2	3	4	5	6
p/t	1.013	1.055	1.138	1.297	1.651	3.369
ℓ/L	6.283	3.142	2.094	1.571	1.257	1.047
ϕ	9.158	18.56	28.52	39.54	52.73	72.74

5.4 Thermal design

Input data

To illustrate the design method, data for one of the OECD Dragon helium/steam heat exchangers (ENEA, 1960–1964), were modified to provide a single-phase problem. Constant (mean) fluid properties are employed, but the technique can be extended to piece-wise calculation of exchangers in which change in fluid properties is significant. Terminal temperatures, log mean temperature difference ($\Delta\theta_{lmtd}$), and exchanger duty ($\dot{Q}$) are known data, which will provide the product $U \times S$.

Exchanger performance

Exchanger duty, kW $\dot{Q} = 1500.0$

log mean temperature difference, K $\Delta\theta_{lmtd} = 39.152$

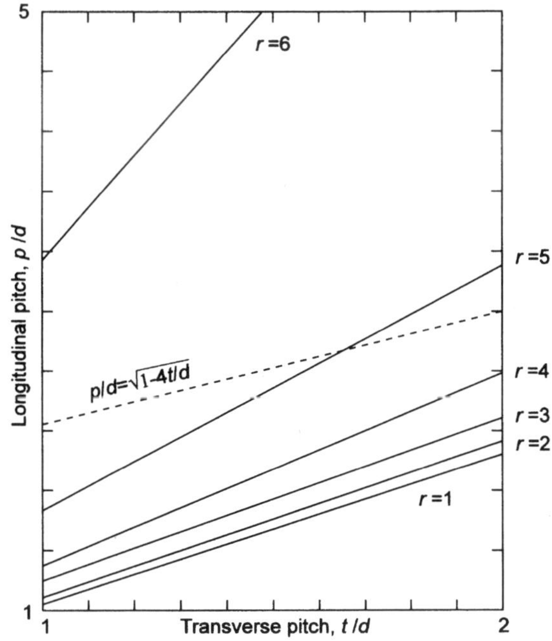

Fig.5.5 Location of shell-side minimum area for flow

Tube-side fluid (steam)

Mass flowrate, kg/s	$\dot{m}_t = 1.750$
Specific heat, J/(kg K)	$C_t = 6405.0$
Density, kg/m^3	$\rho_t = 88.00$
Thermal conductivity, J/(m s K)	$\lambda_t = 0.1040$
Absolute viscosity, kg/(m s)	$\eta_t = 0.000\,029\,78$
Prandtl number	$\Pr_t = 1.484$

Shell-side fluid (helium)

Mass flowrate, kg/s	$\dot{m}_s = 1.500$
Specific heat, J/(kg K)	$C_s = 5120.0$
Density, kg/m^3	$\rho_s = 1.200$
Thermal conductivity, J/(m s K)	$\lambda_s = 0.256$
Absolute viscosity, kg/(m s)	$\eta_s = 0.000\,038\,50$
Prandtl number	$\Pr_s = 0.770$

Local geometry

Tube external diameter, m	$d = 0.022$
Tube internal diameter, m	$d_i = 0.018$
Optimized tube spacing,[1] m	$t - d = 0.007\,61$
Tube minimum coiling diameter, m	$D_m = 0.200$
Tube thermal conductivity, J/(m s K)	$\lambda_w = 190.0$
Coiling start factor	$r = 1$

Correlations and constraints

Tube-side correlations

$$\text{Heat transfer, Nu} = 0.023(\text{Re})^{0.8}(\text{Pr})^{0.4} \tag{5.23}$$

$$\text{Flow friction,} \quad f = 0.046(\text{Re})^{-0.2} \tag{5.24}$$

Shell-side correlations

$$\text{Heat transfer, Nu} = 0.0559(\text{Re})^{0.794} \tag{5.25}$$

$$\text{Friction factor,} \ f = P_y \times 0.26(\text{Re})^{-0.117} \tag{5.26}$$

Equation (5.23) is the standard result for turbulent flow in a straight tube with the viscosity term omitted for simplicity. Equation (5.24) follows from (5.23) using Reynolds analogy. The Dean number correlation for flow in curved tubes is omitted as this correlation would introduce complications in the first optimization.

[1]The optimized tube spacing corresponds to $t/d = 1.346$. This can only be obtained after the computational runs required to construct Figs 5.8 and 5.9, and it corresponds to maximum utilization of available pressure losses. Its use at this point avoids extensive listing of data which do not correspond to the design point. The t/d ratio is also a constraint, in that it must lie 'within range' of values used in the test programme that established the shell-side correlations ($1.125 \leq t/d \leq 1.500$).

The design may be varied later using correlations selected from papers by Ito (1959), Mori & Nakayama (1965, 1967a,b), Ozisik & Topakoglu (1968), Jensen & Bergles (1981), Yao (1984), and Gnielinski (1986). These papers contain references to further publications.

Equations (5.25) and (5.26) were obtained during an experimental programme of work on investigation of shell-side flow in helical-coil heat exchangers (Smith & Coombs, 1972; Smith & King, 1978), and the constants can be modified to include the effects of Prandtl number, which Abadzic (1974) found to be $(Pr)^{0.36}$ with $Pr = 0.71$. More recent work is reported by Kanevets & Politykina (1989).

Each correlation (5.23) to (5.26) has a range of validity of Reynolds number. In addition, limiting velocities may exist for erosion (upper bound) and for fouling (lower bound). The problem of flow-induced tube vibration may also have to be considered (Chen, 1978). Maximum desired pressure losses are specified for both tube-side and shell-side flow.

The experimental Reynolds number ranges for equations (5.25) and (5.26) are not great. Abadzic (1974) examined heat transfer data from several sources and recommended three generalized equations for an extended range of Reynolds number, viz.

$$\mathrm{Nu} = 0.332(\mathrm{Re})^{0.6}\,(\mathrm{Pr})^{0.36} \qquad \text{for } (1 \times 10^{+3} < \mathrm{Re} < 2 \times 10^{+4})$$
$$\mathrm{Nu} = 0.123(\mathrm{Re})^{0.7}\,(\mathrm{Pr})^{0.36} \qquad \text{for } (2 \times 10^{+4} < \mathrm{Re} < 2 \times 10^{+5})$$
$$\mathrm{Nu} = 0.036(\mathrm{Re})^{0.8}\,(\mathrm{Pr})^{0.36} \qquad \text{for } (2 \times 10^{+5} < \mathrm{Re} < 9 \times 10^{+5})$$

These are straight-line (log–log) segments, and such multiple correlations can now be replaced by interpolating spline-fits of data with weighted errors.

Most of the data correlated by Abadzic corresponded to helix angles of around 9°, but the data are not for uniform helical-coil tube bundles. A comparison of the correlations presented by Smith & Coombs (1972) and Abadzic (1974) is presented in Fig. 5.6. Unfortunately Abadzic does not report friction-factor correlations. Shell-side correlations used in this paper are those which were developed specifically for helical-coil tube bundles, and of which the author has personal knowledge.

A comprehensive review of available heat-transfer and flow-friction correlations for helical-tube, multi-start coil heat exchanger bundles was given by Le Feuvre (1986), but not all of the winding geometries reported are consistent. When data outwith this extended range are required, the reader might turn to the extensive work on normal crossflow for in-line and staggered geometries by Zukauskas (1987), Zukauskas & Ulinskas (1988), or to Bejan's summary of these results (1993).

Tube-side constraints

Re for heat transfer and pressure loss	10^{+4} to 10^{+6}
Velocity, fouling/erosion, m/s	$1.5 < u_{fe} < 6.0$
Maximum design pressure loss, N/m^2	2000.0

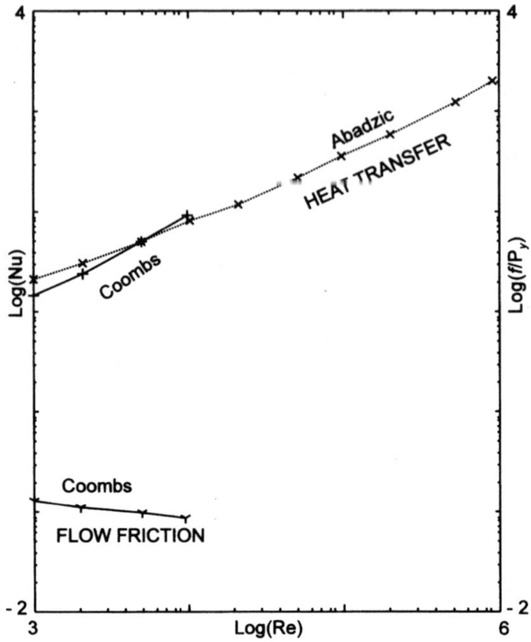

Fig.5.6 Shell-side correlations

Shell-side constraints

Re for heat transfer and pressure loss	10^{+3} to 10^{+4}
Velocity, tube vibration	Chen (1978)
Maximum design pressure loss, N/m^2	5000.0

The design approach has to work within the limits of the above envelope.

Local geometry

From relations (5.19) to (5.22) given earlier, p, ℓ/L, P_y, and t/p are evaluated numerically, then:

Number of tubes in exchanger, equation (5.5)

$$N = r(m+n)(m-n+1)/2 \text{ treat as unknown} \tag{5.27}$$

Tube-side area for flow, equation (5.16)

$$A_t = N(\pi/4)d_i^2 = 0.000\,254\,47\,(\text{m}^2) \tag{5.28}$$

Correction for face area, equation (5.12)

$$A_s/A_a = 0.257\,01 \tag{5.29}$$

Correction for helix angle, equation (5.13)

$$A_{min}/A_s = 1.160\,46 \tag{5.30}$$

Flow area ratio (shell-side/tube-side), equation (5.17)

$$a_r = \left(\frac{8}{r}\right)\left(\frac{t}{d_i}\right)^2\left(\frac{A_s}{A_a}\right)\left(\frac{A_{min}}{A_s}\right) = 6.456\,51 \tag{5.31}$$

which is independent of N, i.e. the shell-side area for flow corresponding to a single tube is determined. Hence shell-side area for flow is

$$A_s = a_r A_t = 0.001\,642\,99N\,(\mathrm{m}^2) \tag{5.32}$$

Velocity constraints

Because tube size is specified, the tube-side velocity bounds on Reynolds number and on fouling/erosion may be used to determine the number of tubes required. However, it is important to recognize that these are not the only considerations to be taken into account.

$$A_t = N(\pi/4)d_i^2 \qquad \dot{m}_t = \rho_t A_t u_t \qquad \mathrm{Re} = \rho_t u_t d_i/\eta_t$$

Re upper bound
- $V_{max} = 18.80$ (m/s)
- $A_{min} = 0.001\,0578$ (m^2)
- $N_{min} = 5$ tubes

Re lower bound
- $V_{min} = 0.188$ (m/s)
- $A_{max} = 0.105\,78$ (m^2)
- $N_{max} = 415$ tubes

Erosion upper bound
- $V_{max} = 6.0$ (m/s)
- $A_{min} = 0.003\,314$ (m^2)
- $N_{min} = 14$ tubes

Erosion lower bound
- $V_{min} = 0.1$ (m/s)
- $A_{max} = 0.1989$ (m^2)
- $N_{max} = 781$ tubes

Only restrictions on Reynolds number are known for shell-side flow, and these are handled in a slightly different manner.

$$A_s = a_r A_t, \qquad G_s = \dot{m}_s/A_s, \qquad \mathrm{Re}_s = dG_s/\eta_s$$

Re upper bound
- $N_{min} = 53$ tubes

Re lower bound
- $N_{max} = 521$ tubes

Overall heat transfer is now considered, carrying forward the unknown number of tubes (N) and the unknown tube length (ℓ), and referring all heat-transfer coefficients to the outside of the tube.

Heat-transfer constraints
The tube-side heat-transfer coefficient is determined from correlation (5.23) as follows:

$$G_t = \dot{m}_t/A_t = 6877.0/N$$
$$\mathrm{Re}_t = d_i G_t/\eta_t = 4156\,705.6/N$$

$$\mathrm{Nu}_t = 0.023(\mathrm{Re}_t)^{0.8}(\mathrm{Pr}_t)^{0.4} = 5312.52/(N)^{0.8}$$

$$\alpha_t = (\lambda_t/d_i)\mathrm{Nu}_t = 30\,694.6/(N)^{0.8}$$

and referring to tube outside diameter, multiplier (d_i/d)

$$\alpha_t = 25\,113.7/(N)^{0.8} \quad \mathrm{J/(m^2\,s\,K)} \tag{5.33}$$

The shell-side heat-transfer coefficient is determined from correlation (5.25) as follows:

$$G_s = \dot{m}_s/A_s = \dot{m}_s/(a_r A_t) = 912.971/N$$
$$\mathrm{Re}_s = dG_s/\eta_s = 521\ 697.5/N$$

$$\mathrm{Nu}_s = 0.0559(\mathrm{Re}_s)^{0.794} = 0.063\,235(\mathrm{Re}_s)^{0.794}(\mathrm{Pr}_s)^{0.36} = 1936.611/(N)^{0.794}$$

$$\alpha_s = (\lambda_s/d)\mathrm{Nu}_s = 22\,535.1/(N)^{0.794} \quad \mathrm{J/(m^2\,s\,K)} \tag{5.34}$$

The tube-wall heat-transfer coefficient is determined as

$$\alpha_w = 2\lambda_w/[d \cdot \ln(d/d_i)] = 86\,075.0 \quad \mathrm{J/(m^2\,s\,K)} \tag{5.35}$$

The expression for overall heat-transfer coefficient becomes

$$U = 22\,535.1/[0.8973(N)^{0.8} + 0.261808 + (N)^{0.794}] \quad \mathrm{J/(m^2\,s\,K)} \tag{5.36}$$

The given product $(U \times S) = (\dot{Q}/\Delta\theta_{lmtd}) = 38\ 311.93$ may now be employed,

$$38\,311.9 = [U \text{ from eqn (5.36)}]N\pi d \tag{5.37}$$

Thus

$$24.598 = (N\ell)/[0.89\ 732(N)^{0.8} + 0.26181 + (N)^{0.794}] \tag{5.38}$$

Equation (5.38) has to be satisfied to ensure correct heat transfer. Values of N already obtained under 'Velocity constraints' could now be substituted to obtain several values for tube length ℓ, but this would not necessarily guarantee performance as pressure losses have not yet been considered. All bounding values for N must be found before taking the design decision.

Pressure-loss constraints
Two further equations exist involving N and ℓ. These are related to tube-side and shell-side pressure losses, respectively.

For the tube-side

$$G_t = \dot{m}_t/A_t = 6877.065/N$$

$$\mathrm{Re}_t = d_i G_t/\eta_t = 4156\ 721.9/N$$

$$f_t = 0.046(\mathrm{Re})^{-0.2} = 0.002\ 1828(N)^{0.2} \qquad \text{from equation (5.24)}$$

$$\Delta p_t = \left(\frac{4f_t G_t^2}{2\rho_t}\right)\left(\frac{\ell}{d_i}\right) \tag{5.39}$$

and with the desired pressure drop of 2000.0 N/m^2

$$\ell = (N)^{1.8}/65.171\,64 \tag{5.40}$$

For the shell-side

$$G_s = \dot{m}_s/A_s = 912.971/N$$

$$\mathrm{Re}_s = dG_s/\eta_s = 521\,697.5/N$$

$$P_y = 1 - \left(\frac{\pi}{4}\right)\left(\frac{d}{t}\right)^2 = 0.5676$$

$$f_s = P_y \times 0.26(\mathrm{Re})^{-0.117} = 0.031\ 628(N)^{0.117} \qquad \text{from equation (5.26)}$$

$$\Delta p_s = \left(\frac{f_s G_s^2}{2\rho_s}\right)\left[\frac{S}{A_s}\right] = \left[\frac{4f_s G_s^2}{2\rho_s}\right]\left[\frac{d_o \ell}{a_r d_i^2}\right] \tag{5.41}$$

and with the desired pressure loss of 5000.0 N/m^2

$$\ell = (N)^{1.883}/92.416\,56 \tag{5.42}$$

Equations (5.41) and (5.42) each deliver the minimum number of tubes to satisfy respective pressure drops. Each is solved in turn with the heat-transfer equation (5.39) to obtain numerical values for N.

Tube-side	Shell-side
Eqns (5.38) and (5.40)	Eqns (5.38) and (5.42)
$N_{min} = 55$ tubes	$N_{min} = 55$ tubes

These values are identical because the design has already been optimized to satisfy tube-side and shell-side pressure losses simultaneously. If any other constraint had entered into the design consideration, then these two figures would be different.

Following solution of these equations, the design region satisfying thermal design may be identified, as in Fig. 5.7, in which the tube-side curve is shifted to the right for the purpose of illustration. In the present design the shell-side and tube-side curves are almost coincidental, and both cut the heat-transfer curve at the same point because the optimized tube spacing ($t - d$) has been used.

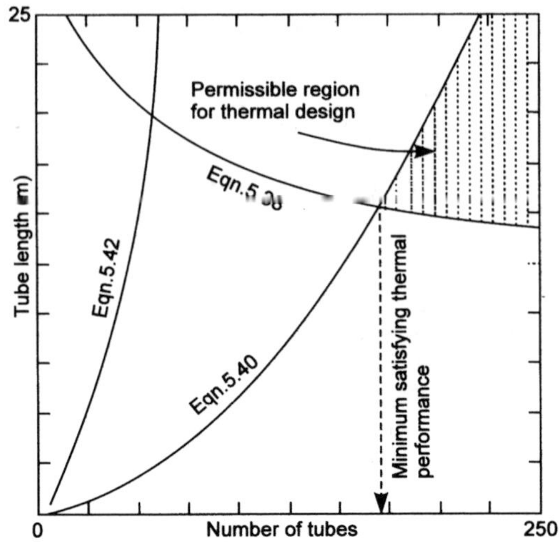

Fig.5.7 Direct-sizing of helical-coil exchanger (schematic)

5.5 Completion of the design

Values of N obtained in earlier sections are summarized in Table 5.2.

The minimum surface area for which the exchanger design is viable would occur with the largest value of N_{min}, namely $N = 55$ tubes. This will not necessarily prove to be the number of tubes in a helical-coil exchanger as further geometrical constraints have to be satisfied to ensure that every coiling station in the tube bundle is filled.

The size of the central duct is known from exchanger by-pass requirements, which determines the minimum coiling diameter. The smallest possible integer value of n is first obtained from minimum coiling diameter 0.200 m, by rounding up the value obtained from equation (5.43), and a first estimate of the integer m

Table 5.2 Design 'window' based on tubes

Tube-side		
Reynolds no.	5 (high Re)	415 (low Re)
Fouling/erosion	14 (erosion)	731 (fouling)
Heat transfer	55 (max Δp)	—
Shell-side		
Reynolds no.	53 (high Re)	521 (low Re)
Vortex shedding	No data	No data
Heat transfer	55 (max Δp)	—

obtained using equation (5.44).

$$0.200 = 2nt \tag{5.43}$$

$$N = r(m+n)(m-n+1)/2 \tag{5.44}$$

Equation (5.44) is employed again in reverse using these values of m and n, to obtain a first 'helical-coil' value for N.

If the new value of N is greater than 55 then the value of n is increased by 1 progressively until the closest match above 55 is obtained. In unusual circumstances it may be appropriate to decrease the value of m by 1 and adjust n to get a best match.

If the new value of N is below 55 then m is increased by 1, and n again increased progressively by 1 until the closest match above 55 is obtained. The nearest approach to the desired value of 55 is subsequently taken. Optimizing the bundle to the smallest number of tubes greater than 55, gives the following configuration:

$N = 56$	$(t/d) = 1.3459$
$m = 11$	$D_i = (2m+1)t = 0.6808\,\text{m}$
$n = 5$	$D_o = (2n-1)t = 0.2664\,\text{m}$

Using equation (5.38) individual tube length $\ell = 20.716$ m
Using equation (5.4), length of coiled bundle $L = 3.297$ m

The chosen value of $t/d = 1.3459$ makes near-maximum use of both allowable pressure losses and provides close approach to the smallest practicable tube-bundle volume. The design method might be extended by adding an iterative loop

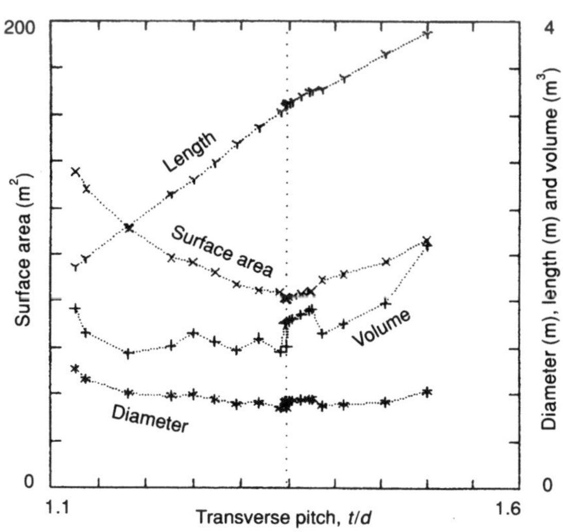

Fig.5.8 Optimized design of helical-coil exchanger

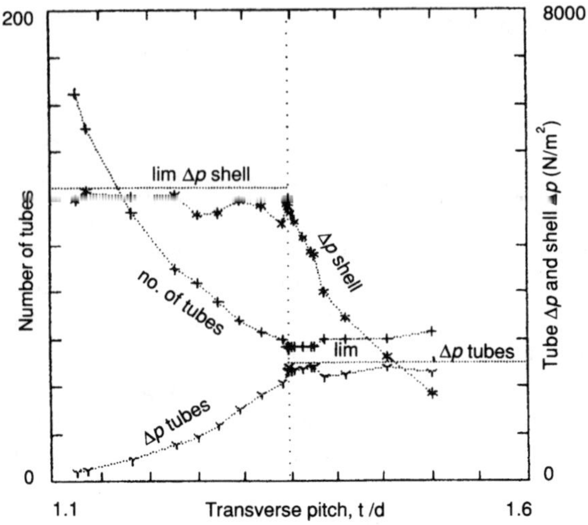

Fig.5.9 Optimized design of helical-coil exchanger

to vary tube spacing within the range of validity of the shell-side correlations. However, it was thought advisable to proceed by separate calculation for each value of $t - d$ as the several design constraints may invoke alternative limiting conditions, of which the designer should become aware.

Figure 5.8 illustrates the effect of varying transverse tube pitching. The jagged curves are a consequence of adjusting tube numbers to satisfy helical-coil tube geometry requirements. Limiting conditions for shell-side and tube-side pressure drop are shown in Fig. 5.9, but one or other, or both, could easily be replaced by maximum Re, or erosion/fouling velocity limitations for different input data.

5.6 Thermal design results for $(t/d) = 1.346$

The undernoted results were obtained by hand.

Surface area for heat transfer, m^2	$S = 80.180$
Tube length, m	$\ell = 20.716$
Bundle length, m	$L = 3.297$
Core mandrel outside diameter, m	$D_o = 0.266$
Bundle wrapper inside diameter, m	$D_i = 0.681$
Face area of bundle, m^2	$A_a = 0.3087$
Volume of bundle, m^3	$V_a = 1.0178$
Specific performance, $kW/(m^3K)$	$\dot{Q}/(V\Delta\theta_{lmtd}) = 37.64$

Tube heat-transfer coefficient, $J/(m^2\ s\ K)$	$\alpha_t = 1003.14$
Shell heat-transfer coefficient, $J/(m^2\ s\ K)$	$\alpha_s = 922.14$
Wall heat-transfer coefficient, $J/(m^2\ s\ K)$	$\alpha_w = 86\,075.0$
Overall heat-transfer coefficient, $J/(m^2\ s\ K)$	$U = 477.80$

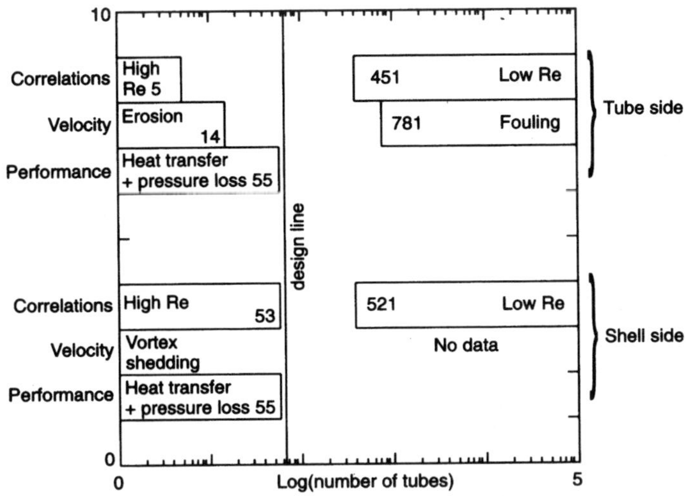

Fig.5.10 Design window for Dragon-type heat exchanger

Tube-side Reynolds number	$\mathrm{Re}_t = 74\,227.2$
Shell-side Reynolds number	$\mathrm{Re}_s = 9316.1$
Tube-side velocity, m/s	$u_t = 1.3906$
Shell-side velocity, m/s	$u_s = 13.586$
Tube-side friction factor	$f_t = 0.004\,883$
Shell-side friction factor	$f_s = 0.05065$
Tube-side pressure loss, N/m^2	$\Delta p_t = 1926.0$
Shell-side pressure loss, N/m^2	$\Delta p_s = 4078.7$
Tube-side area for flow, m^2	$A_t = 0.0143$
Shell-side area for flow, m^2	$A_s = 0.0920$
Shell/tube area ratio	$a_r = 6.456$
Exchanger duty ($U \times S \times \Delta\theta_{lmtd}$), kW	$\dot{Q} = 1500.0$

The optimized helical-tube design provides a useful standard against which competing designs may be assessed. This multi-start, helical-tube design is a preferred configuration when one fluid is evaporating, because the shell-side is fully interconnected for that fluid.

5.7 Fine tuning

Fine tuning of the design will now be developed to avoid thermodynamic mixing losses which degrade exchanger performance. Gnielinski's review (1986) of tube-side correlations omitted the significant earlier work of Ito and of Mori & Nakayama. The Gnielinski correlations are compared with the earlier work of Ito (1959) and of Mori & Nakayama (1965, 1967a,b), and a selection is made.

Résumé of correlations

All the undernoted correlations are consistent with the definition of friction factor for laminar flow in straight tubes leading to $f = 16/\text{Re}$.

Transition Reynolds number

The following expressions were examined by Gnielinski (1986)

$$\text{Ito: Re} = 2000(d/D)^{0.32} \tag{5.45}$$

$$\boxed{\text{Schmidt: Re} = 2300[1 + 8.6(d/D)^{0.45}]} \tag{5.46}$$

When plotted in the ranges $1000 < \text{Re} < 100\,000$ and $10 < D/d < 10\,000$ the Ito correlation is a straight line, but the Schmidt correlation is curved. There is, however, little to choose between them up to a value of $D/d = 100$ which is already a fairly large diameter heat exchanger. The Schmidt correlation is preferred as it correctly tends towards the straight tube transition at Re = 2300 for large values of Dean number, De = (D/d), see Figs 5.11a and b.

Laminar flow friction factor

Ito

$$\boxed{\frac{f_c}{f_s} = \frac{21.5\,\text{De}}{[1.56 + \log(\text{De})]^{5.73}} \quad \text{valid for} \quad (13.5 \le \text{De} \le 2000)} \tag{5.47}$$

Gnielinski

$$\frac{f_c}{f_s} = [1 + 0.33(\text{De})^4]\left(\frac{\eta_{wall}}{\eta}\right)^{0.27} \tag{5.48}$$

Mori

$$\frac{f_{c2}}{f_s} = \frac{0.1080\sqrt{\text{De}}}{(1 - 3.253/\sqrt{\text{De}})} \tag{5.49}$$

valid for $1 \le \text{Pr} \le \infty$.

When plotted for values of $D/d = 5$, 10, 20, 50, 100, and 200 the Gnielinski and Ito correlations were practically identical within the ranges $100 < \text{Re} < 100\,000$ and $0.001 < f_c < 1$. The Mori correlation is purely theoretical and was not compared.

The Ito correlation is preferred for the present analysis, as it does not require knowledge of bulk-to-wall properties.

Laminar flow heat transfer

Laminar heat transfer is the least well understood flow regime, and further experimental investigation into this aspect of performance is necessary.

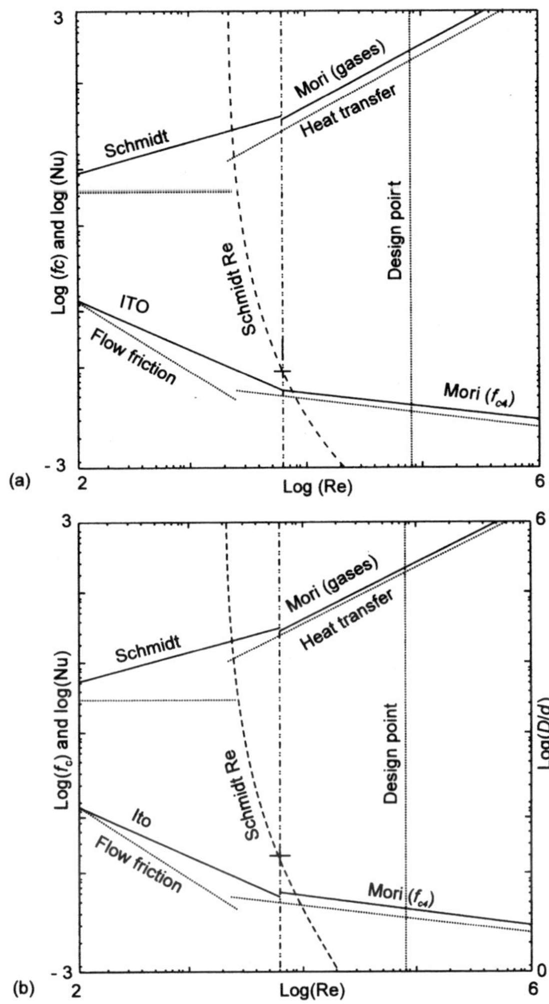

Fig.5.11 Correlations for flow in curved tubes with (a) D/d = 16.5, (b) D/d = 36.2

Schmidt

$$m = 0.5 + 0.2903(d/D)^{0.194}$$

$$\mathrm{Nu} = \left[3.65 + 0.08\left\{1 + 0.8\left(\frac{d}{D}\right)^{0.9}\right\}\right](\mathrm{Re})^{m}(\mathrm{Pr})^{0.33}\left(\frac{\mathrm{Pr}}{\mathrm{Pr}_{wall}}\right)^{0.14} \qquad (5.50)$$

Mori & Nakayama

For Pr $\leq$ 1.0 (gases) $\xi = \frac{1}{5}\left[2 + \sqrt{\{10/(\mathrm{Pr})^2 - 1\}}\right]$

For Pr $\geq$ 1.0 (liquids) $\xi = \frac{2}{11}\left[1 + \sqrt{\{1 + 77/(4 \times \mathrm{Pr}^2)\}}\right]$

$$\mathrm{Nu} = 0.864\frac{\sqrt{\mathrm{De}}}{\xi}\left[1 + \frac{2.35}{\sqrt{\mathrm{De}}}\right]$$

The last correlation was recommended in the third paper by Mori & Nakayama as being for practical use. Two other theoretical correlations developed in their earlier papers involved a great number of terms, and did not find favour.

When plotted for values of $D/d = 5$, 10, 20, 50, 100, and 200 large differences were found in predicted values for Nu within the ranges $100 < \mathrm{Re} < 100\,000$ and $1 < \mathrm{Nu} < 100$ between Schmidt and Mori & Nakayama. When very large values of D/d were tried in an effort to recover the straight tube correlation at low Reynolds numbers, the Schmidt correlation exhibited convergence towards the theoretical value, $(48/11) = 4.36$, for uniform heat flux.

The Schmidt correlation is based on experimental results and seems the better of the two, but requires knowledge of bulk-to-wall properties.

Turbulent flow friction factor

Ito

$$f_c = \frac{\sqrt{d/D}}{4}[0.029 + 0.304\{\mathrm{Re}(d/D)^2\}^{-0.25}]$$

valid for $[0.034 < \{\mathrm{Re}(d/D)^2\} < 300]$, i.e., moderate Re.

Gnielinski

$$f_c = \frac{[0.3164/(\mathrm{Re})^{0.25} + 0.03\sqrt{(d/D)}\{\eta_{wall}/\eta\}^{0.27}]}{4}$$

Mori & Nakayama

Moderate Re,

$$f_{c4} = \frac{0.305\left[1 + \frac{0.112}{\{\mathrm{Re}(d/D)^2\}}\right]\sqrt{(d/D)}}{4\{\mathrm{Re}(d/D)^2\}^{0.2}} \tag{5.51}$$

Large Re,

$$f_{c5} = \frac{0.192\left[1 + \dfrac{0.068}{\{\mathrm{Re}(d/D)^{2.5}\}^{0.167}}\right]\sqrt{(d/D)}}{4\{\mathrm{Re}(d/D)^{2.5}\}^{0.167}}$$

The Ito and Gnielinski correlations are both experimental and give identical results with $\eta_{wall}/\eta = 1$. The Mori correlations are theoretical, with the Mori f_{c4} correlation corresponding closely with Ito over the ranges $1000 < \mathrm{Re} < 100\,000$ and $0.001 < f_c < 0.1$. The integer 4 which appears in each of the above correlations adjusts the friction factor to the definition for straight tubes given earlier. The Mori f_{c4} correlation is employed.

Turbulent flow heat transfer

Gnielinski

The correlation makes use of the Gnielinski turbulent flow friction factor (f_c) from above,

$$\mathrm{Nu} = \frac{(f_c/2)\mathrm{Re}(\mathrm{Pr})^{0.4}(\mathrm{Pr}/\mathrm{Pr}_{wall})^{0.14}}{1 + 12.7[\mathrm{Pr}^{0.666} - 1]\sqrt{(f_c/2)}}$$

Mori & Nakayama

For Pr ≤ 1.0 (gases)

$$\mathrm{Nu} = \frac{\mathrm{Pr}(\mathrm{Re})^{0.8}(d/D)^{0.1}\left[1 + \dfrac{0.098}{\mathrm{Re}(d/D)^2}\right]^{0.2}}{26.2(\mathrm{Pr})^{0.666} - 0.074} \tag{5.52}$$

For Pr ≥ 1.0 (liquids)

$$\mathrm{Nu} = \frac{(Pr)^{0.4}(\mathrm{Re})^{5/6}(d/D)^{1/12}\left[1 + \dfrac{0.061}{\{\mathrm{Re}(d/D)^{2.5}\}^{0.167}}\right]}{41.0}$$

The Mori & Nakayama gas correlation gives virtually identical results to the Gnielinski correlation in the ranges $1000 < \mathrm{Re} < 100\,000$ and $1 < \mathrm{Nu} < 1000$ and is preferred as it does not require knowledge of bulk-to-wall properties.

Preferred correlations for curved tubes are shown as solid lines in Figs 5.11a and b for $D/d = 16.5$ and $D/d = 36.2$, respectively, for $\mathrm{Pr} = 1.484$, and matching of the transition between laminar and turbulent regions is remarkably good. The dashed line is the correlation for transition Reynolds number for flow in curved tubes. Flow-friction and heat-transfer correlations for straight tubes are also shown. Transition between laminar and turbulent flow for the curved tube is marked by a vertical chain-dotted line, and the exchanger tube-side design condition is marked by a vertical dotted line in the turbulent region.

5.8 Design for curved tubes

Straight-tube correlations

The reference design employed straight-tube turbulent heat-transfer and flow-friction correlations for tube-side flow, viz.

$$\mathrm{Nu} = 0.023(\mathrm{Re})^{0.8}(\mathrm{Pr})^{0.4} \tag{5.53a}$$

$$f = 0.046/(\mathrm{Re})^{0.2} \tag{5.53b}$$

The resulting tube bundle configuration satisfying all constraints has

Inner-coil tube count	$n = 5$
Outer-coil tube count	$m = 11$
Total number of tubes	$N = 56$
Inner-coil Dean number (D/d)	$\mathrm{De}_n = 16.45$
Outer-coil Dean number (D/d)	$\mathrm{De}_m = 36.19$

Severest conditions occur at the innermost coiling diameter. Equations (5.53a) and (5.53b) may therefore be replaced by the straight-tube equivalents of equations (5.51) and (5.52) using the value of $t/d = 1.346$ found in the earlier optimization. For the innermost coil ($n = 5$), gives $D/d = 13.46$ which provides

$$\mathrm{Nu} = 0.0326(\mathrm{Re})^{0.7922}(\mathrm{Pr})^{0.4} \tag{5.54a}$$

$$f = 0.0661/(\mathrm{Re})^{0.2089} \tag{5.54b}$$

This ensures that all constraints are properly taken into consideration. The design program is run again with the following results,

Inner-coil tube count	$n = 6$
Outer-coil tube count	$m = 12$
Total number of tubes	$N = 63$
Inner-coil Dean number (D/d)	$\mathrm{De}_n = 19.74$
Outer-coil Dean number (D/d)	$\mathrm{De}_m = 39.48$

Values of tube-side and shell-side pressure drops obtained are within the required limits

Tube-side pressure drop, $\mathrm{N/m^2}$	$\Delta p_t = 1797.1$
Shell-side pressure drop, $\mathrm{N/m^2}$	$\Delta p_s = 3455.8$

Fine tuning with curved-tube correlations

The approach to fine tuning is determined by examination of the equations governing thermal performance for a single tube, viz.

$$\dot{Q} = US\Delta\theta_{lmtd} = \dot{m}_t C_t(T_{t1} - T_{t2}) = \dot{m}_s C_s(T_{s1} - T_{s2}) \tag{5.55}$$

If a cryogenic exchanger is involved, thermodynamic mixing losses are to be avoided as temperature differences of a few Kelvins then become important. Applying the same philosophy of having constant exit temperatures everywhere in this high-temperature exchanger, design LMTD and design terminal temperatures have to be maintained at each coiling diameter of the helical-coil design, thus

$$R_m = (\dot{m}_t/\dot{m}_s) = const. \tag{5.56}$$

Tube-side pressure loss reduces with increase in coiling diameter. Adjustment of tube-side flow by orificing becomes necessary in order to maintain constant terminal temperatures. Since the ratio $R_m = (\dot{m}_t/\dot{m}_s)$ must also remain constant, the corollary is increase in the length of shell-side flow path with increase in coil diameter.

Individual coil design

The number of coils is known, and thermal performance of a single tube may be determined at each coiling diameter.

Solution involves simultaneous equations with two unknowns, tube length (ℓ), and shell-side mass flowrate ($\dot{m}_s$). Track is kept of these unknowns in the equations which follow by enclosing them in angle brackets $\langle\ \rangle$, and by assigning the parameter symbol K to that part which can be evaluated numerically.

Shell-side pressure loss

This is constant at each coiling diameter,

$$G_s = \left(\frac{\dot{m}_s}{A_s}\right) = \left[\frac{4}{\pi a_r d_i^2}\right]\langle\dot{m}_s\rangle$$

where a_r is dependent only on local geometry,

$$\mathrm{Re}_s = \left[\frac{dG_s}{\eta_s}\right]\langle\dot{m}_s\rangle$$

$$f_s = P_y \times 0.26(\mathrm{Re}_s)^{-0.117}\langle\dot{m}_s^{-0.117}\rangle$$

$$\Delta p_s = \frac{f_s G_s^2}{2\rho_s}\left(\frac{L}{r_h}\right) \qquad \text{with} \quad r_h = \frac{AL}{S}$$

where $S = \pi d\ell$ and $A = a_r \pi d_i^2/4$.

Then

$$\begin{aligned}\langle \ell \dot{m}_s^{(2-0.117)}\rangle &= \frac{2\rho_s a_r \Delta p_s}{4 f_s d}\left(\frac{d_i}{G_s}\right)^2 \\ &= K1, \text{ say}\end{aligned} \tag{5.57}$$

Shell-side heat transfer coefficient

$$\begin{aligned}\mathrm{Nu}_s &= 0.063\,235(\mathrm{Re}_s)^{0.794}(\mathrm{Pr}_s)^{0.36}\\ \alpha_s &= (\lambda_s/d)\times 0.063\,235(\mathrm{Re}_s)^{0.794}(\mathrm{Pr}_s)^{0.4}\langle \dot{m}_s^{0.794}\rangle\\ &= K2\langle \dot{m}_s^{0.794}\rangle,\ \text{say}\end{aligned} \tag{5.58}$$

Tube-side heat-transfer coefficient

$$\mathrm{Nu}_t = \frac{\mathrm{Pr}_t(\mathrm{Re}_t)^{0.8}(d_i/D)^{0.1}[1+0.098/\{\mathrm{Re}(d_i/D)^2\}^{0.2}]}{[26.2(\mathrm{Pr}_t)^{2/3}-0.074]} \tag{5.59}$$

where Pr_t and d_i/D are specified.

$$G_t = \frac{\dot{m}_t}{A_t} = \frac{4R_m}{\pi d_i^2}\langle \dot{m}_s\rangle$$

$$\mathrm{Re} = \frac{d_i G_t}{\eta_t}\langle \dot{m}_s\rangle$$

$$\alpha_t = K3\times(\mathrm{Re}_t)^{0.8}\langle \dot{m}_s^{0.8}\rangle\left[1+\frac{K4}{(\mathrm{Re}_t)^{0.2}\langle \dot{m}_s^{0.2}\rangle}\right],\ \text{say} \tag{5.60}$$

where

$$K3 = \left(\frac{\lambda_t}{d_i}\right)\left[\frac{\mathrm{Pr}_t}{26.2(\mathrm{Pr}_t)^{2/3}}-0.074\right]\left(\frac{d_i}{D}\right)^{0.1}$$

and

$$K4 = \frac{0.098}{(d_i/D)^{0.4}}$$

Referring α_t to outside diameter

$$\alpha_t = K5\times\langle \dot{m}_s^{0.8}\rangle + K6\times\langle \dot{m}_s^{0.6}\rangle \tag{5.61}$$

where $K5 = K3\times(\mathrm{Re}_t)^{0.8}(d_i/d)$ and $K6 = K3\times K4\times(\mathrm{Re}_t)^{0.6}$.

Overall heat-transfer coefficient

$$\frac{1}{U} = \left[\frac{1}{\alpha_t}+\frac{1}{\alpha_w}+\frac{1}{\alpha_s}\right] = \frac{1}{K5\langle \dot{m}_s^{0.8}\rangle + K6\langle \dot{m}_s^{0.6}\rangle}+\frac{1}{\alpha_w}+\frac{1}{K2\langle \dot{m}_s^{0.794}\rangle} \tag{5.62}$$

Heat transfer (referred to outside tube surface)

$$\begin{aligned}&U(\pi d\ell)\Delta\theta_{lmtd} = \dot{m}_s C_s(T_{s1}-T_{s2})\\ &\langle \ell/\dot{m}_s\rangle = K7/U,\ \text{say}\end{aligned} \tag{5.63}$$

where

$$K7 = \frac{C_s(T_{s1} - T_{s2})}{\pi d \Delta\theta_{lmtd}}$$

The tube length (ℓ) is eliminated from simultaneous equations (5.57) and (5.63), and $\dot{m}_s$ is found by binary search. A good first estimate for $\dot{m}_s$ is available from the reference design. Once $\dot{m}_s$ is determined, ℓ and $\dot{m}_t$ can be found, and the tube-side pressure loss may be evaluated.

Tube-side pressure loss (coiled)

Tube performance in each coiling diameter may now be evaluated

$$G_t = \frac{4\dot{m}_t}{\pi d_i^2}, \qquad \mathrm{Re}_t = \frac{d_i G_t}{\eta_t}, \qquad f_{c4} = \text{equation (5.51)}$$

$$\Delta P_{coil} = \frac{4 f_{c4} G_t^2}{2\rho}\left(\frac{\ell}{d_i}\right)$$

Required tube orifice pressure drops are obtained as $\Delta P_t - \Delta P_{coil}$.

Variations in mass flowrates

The results of fine tuning are given in Table 5.3. The variation in mass flowrates across the bundle can be seen. When the program was first run, all constraints were satisfied with the exception that cumulative mass flowrates obtained were 104.3 per cent of the desired values. This is a consequence of using curved-tube heat-transfer and pressure drop correlations for the innermost coil to ensure that all flow constraints were observed when calculating the reference configuration, and also because a greater number of tubes was subsequently obtained.

Table 5.3 Flowrates and tube-side pressure loss across tube bundle

Coil no.	$\dot{m}_s$ (kg/m^2)	$\dot{m}_t$ (kg/m^2)	ℓ (m)	L (m)	Δp_{coil} (N/m^2)	$\sum S$ (m^2)
6	0.023 995	0.027 994	18.052	2.873	1761.5	7.486
7	0.023 930	0.027 918	18.146	2.888	1739.3	16.265
8	0.023 873	0.027 852	18.226	2.901	1720.4	26.343
9	0.023 824	0.027 794	18.298	2.912	1704.1	37.725
10	0.023 779	0.027 743	18.362	2.922	1689.8	50.416
11	0.023 740	0.027 696	18.420	2.932	1677.1	64.419
12	0.023 704	0.027 654	18.472	2.940	1665.7	79.740

Constraints to be satisfied:

Cumulative tube-side flow, kg/s	$cum\ \dot{m}_t = 1.75$
Cumulative shell-side flow, kg/s	$cum\ \dot{m}_s = 1.50$
Tube-side pressure loss, N/m^2	$\Delta p_t = 1797$
Shell-side pressure loss, N/m^2	$\Delta p_s = 3456$

To accommodate this problem, a small reduction is applied to the Nusselt number in the reference design calculation only, by means of a multiplier. The multiplier is adjusted until correct values of cumulative mass flow appear in the fine-tuning calculation, which takes account of the changing thermal performance across the bundle. As the two programs are run in sequence, this takes only a few moments.

5.9 Discussion

Actual exchangers for the Dragon reactor have $n = 6$ and $m = 12$ with a start factor $r = 1$, all tubes having a right-handed helix angle of 16° (Gilli, 1965). This would lead to a ratio $p/t = 1.800$, which is much greater than in the design presented here.

For the present design, Fig. 5.12 shows the variation in tube-side flowrates across the bundle. Note that the right-hand scale has a suppressed zero. Shell-side mass flowrates follow exactly the same trend as tube-side mass flowrates. The extent of tube-side orificing required is apparent. Figure 5.13 presents the outline of the

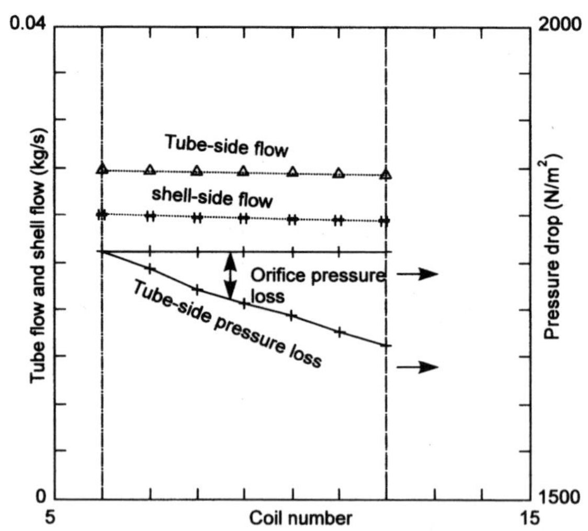

Fig.5.12 Fine tuning of Dragon-type exchanger

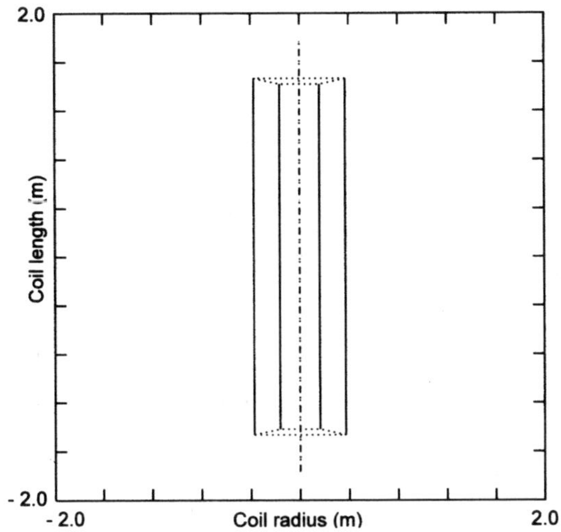

Fig.5.13 Aspect of fine-tuned heat exchanger tube bundle

fine-tuned bundle, illustrating the required variation in tube length against coiling diameter.

Inner-coil tube count	$n = 6$
Outer-coil tube count	$m = 12$
Total number of tubes	$N = 63$
Helix angle of tubes, deg.	$\phi = 9.158$
Inner mean coiling diameter ($2nt$), m	$D_n = 0.355$
Outer mean coiling diameter ($2mt$), m	$D_m = 0.711$
Inner bundle length, m	$L_n = 2.874$
Outer bundle length, m	$L_m = 2.940$
Core outer diameter $(2n - 1)t$, m	$D_o = 0.326$
Shell inner diameter $(2m + 1)t$, m	$D_i = 0.740$

The terminal temperatures used were as follows:

Shell-side inlet temperature (helium), °C	$T_{s1} = 600.0$
Shell-side outlet temperature (helium), °C	$T_{s2} = 404.7$
Tube-side outlet temperature (steam), °C	$T_{t1} = 522.4$
Tube-side inlet temperature (steam), °C	$T_{t2} = 388.5$
Thermal effectiveness	$(T_{s1} - T_{s2})/Tspan = 0.923$

The real Dragon primary heat exchangers were designed for boiling on the steam side (tube-side) and consequently the LMTD was also considerably different, ENEA (1960–1964), thus present results that cannot be directly compared although the final number of tubes in the present exchanger is exactly the same as for the Dragon exchangers.

5.10 Part-load operation with by-pass control

Each Dragon heat exchanger was provided with a central by-pass duct to control exit gas temperature on the shell-side of the exchanger during part-load operation. Under these conditions pressure loss in the central duct + control valve is equal to the pressure loss in the tube bundle. The two pressure-loss equations can be used, together with the mixing equation at exit, to solve for the mass flowrates and the exit temperature. Heat-transfer and flow-friction correlations for straight tubes are adequate for the purpose, as the control valve makes any necessary adjustment.

5.11 Conclusions

1. Geometry relevant to the design of helical-coil exchangers has been presented.
2. Because the flow area ratio (shell-side/tube-side) is independent of the number of tubes in the exchanger, direct-sizing of the tube bundle becomes possible.
3. A simple example illustrating the method of thermal design has been presented. This highlights the constraining factor which may then be scrutinized.
4. Design optimization is possible by varying tube spacing $(t - d)$. Full optimization to minimize any selected parameter (e.g. bundle volume, face area, total tube length) may be carried out by repeating the process for each commercially available tube size.
5. Correlations published by different authors for flow friction factor and heat transfer in curved tubes show consistency of prediction, except for the case of heat transfer in laminar flow.
6. Flow-friction and heat-transfer correlations for flow in curved tubes match well at the transition between laminar and turbulent regions compared with those for straight tubes (Figs 5.11a and b).
7. Curved-tube correlations for tube-side flow should be used for fine tuning of the design when thermodynamic mixing losses are to be avoided. For exacting applications, adjustment of tube length may be required across the tube bundle. Orificing pressure loss may be allowed for in extended tube 'tails'.
8. The number of tubes in the Dragon primary heat exchangers is confirmed, even though coiling directions and helix angles are different, and steam-side heat transfer and LMTD are different.
9. The final configuration is represented in the 'design window' of Fig. 5.10 as a solid line.

References

Abadzic, E.E. (1974) Heat transfer on coiled tubular matrix. ASME Winter Annual Meeting, New York, 1974, ASME Paper 74-WA/HT-64.

Bejan, A. (1993) *Heat Transfer*, Section 5.5.4, John Wiley, pp. 270–273.

Chen, Y.N. (1978) General behaviour of flow induced vibrations in helical tube bundle heat exchangers. *Sulzer Tech. Rev.*, Special Number 'NUCLEX 78', 59–68.

ENEA Paris (1960–1964) *OECD High temperature reactor project (Dragon)*, Annual Reports.

Gill, G.M., **Harrison, G.S.**, and **Walker, M.A.** (1983) Full scale modelling of a helical boiler tube. In *International Conference on Physical Modelling of Multi-Phase Flow*, BHRA Fluid Engineering Conference, April 1983, Paper K4, pp. 481–500.

Gilli, P.V. (1965) Heat transfer and pressure drop for crossflow through banks of multistart helical tubes with uniform inclinations and uniform longitudinal pitches. *Nucl. Sci. Engng*, **22**, 298–314.

Gnielinski, V. (1986) Heat transfer and pressure drop in helically coiled tubes. In *8th International Heat Transfer Conference*, San Francisco, 1986, vol. 6, pp. 2847–2854.

Hampson, W. (1895) Improvements relating to the progressive refrigeration of gases. British Patent 10165.

Hausen, H. (1950), *Wärmeübertragung im Gegenstrom, Gleichstrom und Kreuzstrom*, 1st edn, Springer-Verlag, Berlin, pp. 213–228.

Hausen, H. (1983) *Heat Transfer in Counterflow, Parallel Flow and Cross Flow*, 2nd edn, McGraw-Hill, New York, pp. 232–248.

Ito, H. (1959) Friction factors for turbulent flow in curved pipes. *ASME J. Basic Engng*, **81**, June, 123–129.

Jensen, M.K. and **Bergles, A.E.** (1981) Critical heat flux in helically coiled tubes. *ASME J. Heat Transfer*, **103**, November, 660–666.

Kanevets, G.Ye. and **Politykina, A.A.** (1989) Heat transfer in crossflow over bundles of coiled heat exchanger tubes. *Appl. Thermal Sci.*, **2**(1), Jan–Feb, 38–41.

L'Air Liquide (1934) Improvements relating to the progressive refrigeration of gases. British Patent 416,096.

Le Feuvre, R.F. (1986) A method of modelling the heat transfer and flow resistance characteristics of multi-start helically-coiled tube heat exchangers. In *8th International Heat Transfer Conference*, San Francisco, 1986, vol. 6, pp. 2799–2804.

Mori, Y. and **Nakayama, W.** (1965) Study on forced convective heat transfer in curved pipes (1st report, laminar region). *Int. J. Heat Mass Transfer*, **8**, 67–82.

Mori, Y. and **Nakayama, W.** (1967a) Study on forced convective heat transfer in curved pipes (2nd report, turbulent region). *Int. J. Heat Mass Transfer*, **10**, 37–59.

Mori, Y. and **Nakayama, W.** (1967b) Study on forced convective heat transfer in curved pipes (3rd report, theoretical analysis under the condition of uniform wall temperature and practical formulae). *Int. J. Heat Mass Transfer*, **10**, 681–695.

Ozisik, M.N. and **Topakoglu, H.** (1968) Heat transfer for laminar flow in a curved pipe. *Heat Transfer*, August, 313–318.

Smith, E.M. (1960) The geometry of multi-start helical coil heat exchangers. Unpublished report.

Smith, E.M. (1964) Helical-tube heat exchangers. *Engineering*, 7 February, 232.

Smith, E.M. (1986) Design of helical-tube multi-start coil heat exchangers. In *ASME Winter Annual Meeting*, Anaheim, California, 7–12 December 1986, ASME Publication HTD-Vol. 66, pp. 95–104.

Smith, E.M. and **Coombs, B.P.** (1972) Thermal performance of cross-inclined tube bundles measured by a transient technique. *J. Mech. Engng Sci.*, **14**(3), 205–220.

Smith, E.M. and **King, J.L.** (1978) Thermal performance of further cross-inclined in-line and staggered tube banks. In *6th International Heat Transfer Conference*, Toronto, 1978, Paper HX-14, pp. 267–272.

Weimer, R.F. and **Hartzog, D.G.** (1972) Effects of maldistribution on the performance of multistream heat exchangers. In *Proceedings of the 1972 Cryogenic Engineering Conference, Advances in Cryogenic Engineering*, Vol. 18, Plenum Press, Paper B-2, pp. 52–64.

Yao, L.S. (1984) Heat convection in a horizontal curved pipe. *ASME J. Heat Transfer*, **106**, 71–77.

Zukauskas, A.A. (1987) Convective heat transfer in cross flow. *Handbook of Single-Phase Convective Heat Transfer*, Chapter 6 (Eds. S. Kakaç, R.K. Shah, and W. Aung), John Wiley, New York.

Zukauskas, A.A. and **Ulinskas, R.** (1988) *Heat Transfer in Tube Banks in Crossflow*, Hemisphere/Springer Verlag, New York.

Bibliography

Gouge, M.J. (1995) Closed cycle gas turbine nuclear power plant for submarine propulsion. *Naval Engrs J.*, November, 35–41.

CHAPTER 6
Direct-Sizing of Bayonet-Tube Exchangers

Practical design example

6.1 Isothermal shell-side conditions

Explicit design of the bayonet-tube heat exchanger is practicable when the shell-side fluid is essentially isothermal, i.e. for some condensing and evaporating conditions, and for isothermal crossflow. Analytical expressions and dimensionless plots are presented for the four possible configurations, giving full temperature profiles, exchanger effectiveness, position of closest shell to tube-side temperature approach, and direct determination of exchanger length.

In designing bayonet-tube heat exchangers for the case when the shell-side fluid is essentially isothermal (e.g. condensing, evaporating, or isothermal crossflow) it was found that a modified theoretical approach to that used by Hurd (1946) was necessary, and that explicit design conditions existed. Four configurations – A, B, C and D – illustrated in Fig. 6.1 will be examined in turn, two for evaporation and two for the condensing condition.

Notation is awkward for the bayonet-tube exchanger, first because fluid in the bayonet tube enters and exits from the same end, second because each pass of that fluid requires separate identification. As only overall heat-transfer coefficients will be involved in the analysis which follows, the symbols (α, β) can be used for parameters in the solution. The concepts of LMTD and meanTD are not useful.

It was found convenient to introduce the concept of 'perimeter transfer units' $(P, \hat{P})$ equivalent to '*Ntu* per unit length' of the exchanger surface. These parameters arise quite naturally in the differential equations, and conversion to *Ntu* values $(N, \hat{N})$ is straightforward once the solutions have been obtained.

$$P = \frac{N}{L} = \frac{UZ}{\dot{m}C} \qquad \text{where } Z \text{ is the perimeter of outer tube}$$

$$\hat{P} = \frac{\hat{N}}{L} = \frac{\hat{U}\hat{Z}}{\dot{m}C} \qquad \text{where } \hat{Z} \text{ is the perimeter of inner tube}$$

In the solutions which follow, all physical parameters remain constant.

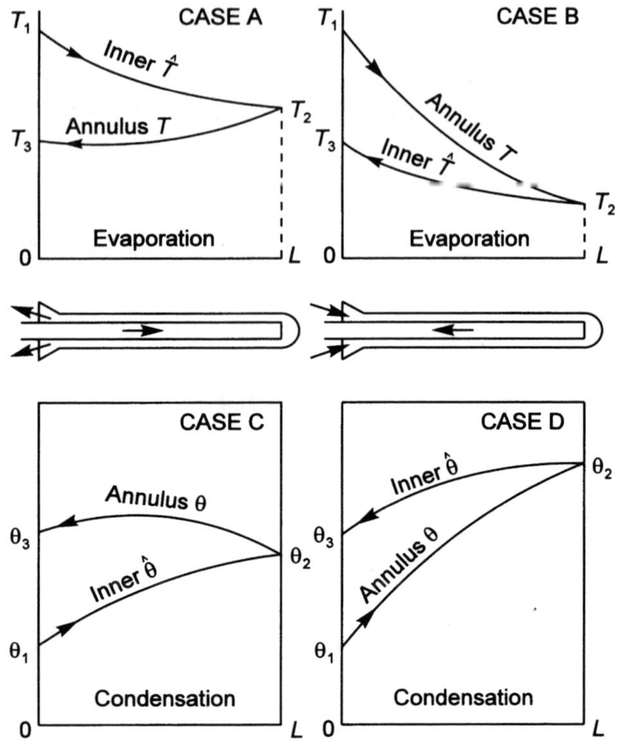

Fig.6.1 Alternative exchanger configurations. Condensation is reflected evaporation

6.2 Evaporation

Case A

An energy balance written for a differential length (dx) of the tube (Fig. 6.2) gives

Inner tube

$$\left\{\begin{matrix}\text{energy entering}\\ \text{with fluid}\end{matrix}\right\} - \left\{\begin{matrix}\text{energy leaving}\\ \text{with fluid}\end{matrix}\right\} - \left\{\begin{matrix}\text{heat transfered}\\ \text{to annulus}\end{matrix}\right\} = \left\{\begin{matrix}\text{energy stored}\\ \text{in fluid}\end{matrix}\right\}$$

$$\dot{m}C\hat{T} - \dot{m}C\left(\hat{T} + \frac{d\hat{T}}{dx}\delta x\right) - \hat{U}(\hat{Z}\delta x)(\hat{T} - T) = 0 \tag{6.1}$$

Annulus

$$\left[\dot{m}C\left(T + \frac{dT}{dx}\delta x\right) + U(Z\delta x)(\hat{T} - T)\right] - [\dot{m}CT + U(Z\delta x)(T - 0)] = 0 \tag{6.2}$$

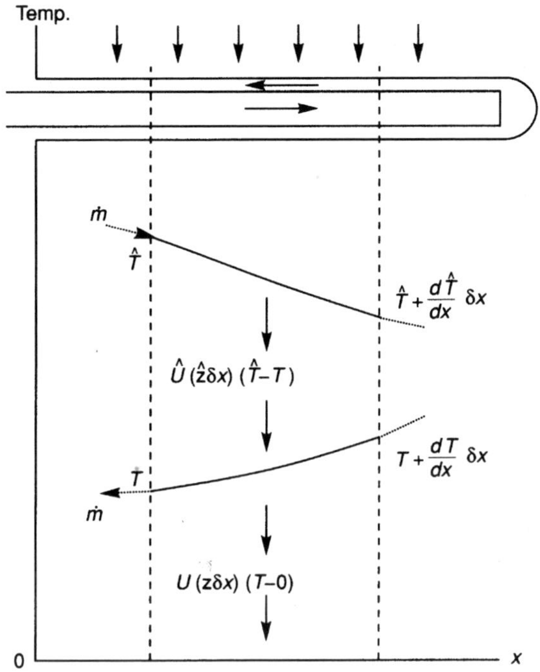

Fig.6.2 Differential energy balance for case A. Origin at flow entry and exit

giving, respectively,

$$\boxed{-\frac{d\hat{T}}{dx} = +\hat{P}(\hat{T} - T)} \tag{6.3}$$

$$\boxed{+\frac{dT}{dx} = -\hat{P}(\hat{T} - T) + PT} \tag{6.4}$$

Eliminating T from equations (6.3) and (6.4) produces

$$\frac{d^2\hat{T}}{dx^2} - P\frac{d\hat{T}}{dx} - P\hat{P}\hat{T} = 0 \tag{6.5}$$

which has the solution

$$\hat{T} = A_i e^{\alpha x} + B_i e^{\beta x} \tag{6.6}$$

with

$$\left.\begin{aligned} \alpha &= \frac{P}{2}\left[1+\sqrt{(1+4\hat{P}/P}\right] \\ \beta &= \frac{P}{2}\left[1-\sqrt{1+4\hat{P}/P}\right] \end{aligned}\right\} \tag{6.7}$$

$$P = (\alpha + \beta) \qquad \text{and} \qquad \frac{\hat{P}}{P} = \frac{-\alpha\beta}{(\alpha+\beta)^2} \tag{6.8}$$

An identical result exists for the other unknown temperature profile

$$T = A_o e^{\alpha x} + B_o e^{\beta x} \tag{6.9}$$

Annulus temperature profile (T)

Two boundary conditions are required, but only $T = T_3$ at $x = 0$ is immediately available. A second condition is obtained by noting that the overall energy balance must be satisfied, viz.

$$\dot{Q} = \dot{m}C(T_1 - T_3) = \int_{x=0}^{L} [U(Z\delta x)(T-0)]$$

$$(T_1 - T_3) = \left(\frac{UZ}{\dot{m}C}\right)\int_{x=0}^{L} T\delta x = P\int_{x=0}^{L} T\delta x = (\alpha+\beta)\int_{x=0}^{L} T\delta x \tag{6.10}$$

Inserting boundary conditions $T = T_3$ at $x = 0$ in equation (6.9)

$$T_3 = A_o + B_o$$

Substituting in equation (6.9)

$$T = (T_3 - B_o)e^{\alpha x} + B_o e^{\beta x} \tag{6.11}$$

and then in equation (6.10)

$$(T_1 - T_3) = (\alpha+\beta)\left[(T_3 - B_o)\left(\frac{e^{\alpha L} - 1}{\alpha}\right) + B_o\left(\frac{e^{\beta L}}{\beta}\right)\right]$$

from which B_o may be found for re-introduction in equation (6.11). Equation (6.11) is then solved for T_2 at $x = L$, and following some algebra too extensive to reproduce

$$T_2/T_3 = (termX + termY + termZ)/denom$$

where

$$\begin{aligned} termX &= +\alpha^2 e^{\alpha L}(e^{\beta L} - 1) \\ termY &= +\alpha\beta(e^{\beta L} - e^{\alpha L})(T_1 - T_3) \\ termZ &= -\beta^2 e^{\beta L}(e^{\alpha L} - 1) \\ denom &= (\alpha+\beta)\left[\alpha(e^{\beta L} - 1) - \beta(e^{\alpha L} - 1)\right] \end{aligned}$$

giving

$$T_2/T_3 = \phi[\alpha, \beta, L, (T_1/T_3)] \tag{6.12}$$

Inner temperature profile ($\hat{T}$)

Again two boundary conditions are required, but only $\hat{T} = T_1$ at $x = 0$ is immediately available. From equation (6.3) at $x = L$, $\hat{T} = T$, thus $d\hat{T}/dx = 0$. Three results from equation (6.6) are then obtained,

$$\left.\begin{aligned} T_1 &= A_i + B_i \\ T_2 &= A_i e^{\alpha L} + B_i e^{\beta L} \\ 0 &= \alpha A_i e^{\alpha L} + \beta B_i e^{\beta L} \end{aligned}\right\} \tag{6.13}$$

Solving the first two for A_i and B_i, and inserting in the third condition

$$\boxed{\frac{T_2}{T_1} = \frac{(\beta - \alpha)e^{(\alpha+\beta)L}}{\beta e^{\beta L} - \alpha e^{\alpha L}}} \tag{6.14}$$

Combining this result with equation (6.12)

$$\frac{T_1}{T_3} = \frac{T_1}{T_2} \cdot \frac{T_2}{T_3} = \frac{(\beta e^{\beta L} - \alpha e^{\alpha L})}{(\beta - \alpha)e^{(\alpha+\beta)L}} \phi[\alpha, \beta, L, (T_1/T_3)]$$

from which after substantial algebraic reduction there emerges

$$\boxed{\frac{T_1}{T_3} = \frac{\beta e^{\beta L} - \alpha e^{\alpha L}}{\beta e^{\alpha L} - \alpha e^{\beta L}}} \tag{6.15}$$

providing the explicit result

$$\boxed{L = \frac{1}{(\alpha + \beta)} \ln\left[\frac{\beta + \alpha(T_1/T_3)}{\alpha + \beta(T_1/T_3)}\right]} \tag{6.16}$$

Equation (6.15) delivers $\lim (T_1/T_3) = (-\alpha/\beta)$ as $L \to \infty$, thus the restriction $[1 < (T_1/T_3) < (-\alpha/\beta)]$ applies to equation (6.16).

From Fig. 6.1 it is evident that a minimum value may exist in the profile for T. This is readily found from equation (6.9) by solving

$$\left.\begin{aligned} T_3 &= A_o + B_o \\ T_2 &= A_o e^{\alpha L} + B_o e^{\beta L} \\ 0 &= \alpha A_o e^{\alpha X_{\min}} + \beta B_o e^{\beta X_{\min}} \end{aligned}\right\} \tag{6.17}$$

giving the position of the minimum as

$$X_{\min} = \frac{1}{(\beta - \alpha)} \ln\left[\frac{\alpha e^{\beta L} - (T_2/T_3)}{\beta e^{\alpha L} - (T_2/T_3)}\right] \tag{6.18}$$

(T_2/T_3) is found from equations (6.14) and (6.15) as

$$\boxed{\frac{T_2}{T_3} = \frac{(\beta - \alpha)e^{(\alpha+\beta)L}}{\beta e^{\alpha L} - \alpha e^{\beta L}}} \tag{6.19}$$

Inserting this result in equation (6.18), the locus of $X_{\min}$ is obtained as a straight line

$$\boxed{X_{\min} = L + \frac{1}{(\beta - \alpha)} \ln[(\alpha/\beta)^2]} \tag{6.20}$$

Minimum annulus temperature $T_{\min}$ is obtained by inserting $X_{\min}$ in the expression for the annulus temperature profile (T/T_3), viz.

$$\frac{T_{\min}}{T_3} = \frac{\beta e^{\alpha L} e^{\beta X_{\min}} - \alpha e^{\beta L} e^{\alpha X_{\min}}}{\beta e^{\alpha L} - \alpha e^{\beta L}} \tag{6.21}$$

This completes the analysis for Case A, but simplification is possible, noting from equation (6.7) that

$$\left.\begin{aligned} \alpha &= aP \qquad \text{where } a = \frac{1}{2}\left[1 + \sqrt{(1 + 4\hat{P}/P)}\right] \\ \beta &= bP \qquad \text{where } b = \frac{1}{2}\left[1 - \sqrt{(1 + 4\hat{P}/P)}\right] \end{aligned}\right\} \tag{6.22}$$

Table 6.1 offers a selection of values for (a, b) covering most applications.

The summary below recasts the above relationships in terms of (a, b) and $Ntu = N = PL$.

Summary of results for Case A

Exchanger length

$$\boxed{N = \frac{1}{(\beta - \alpha)} \ln\left[\frac{b + a(T_3/T_1)}{a + b(T_3/T_1)}\right]} \tag{6.16a}$$

Position of minimum ($T_{\min}$ if it exists)

$$\boxed{N_{\min} = N + \frac{1}{(b - a)} \ln\left[\left(\frac{b}{a}\right)^2\right]} \tag{6.18a}$$

Table 6.1 Useful values of ($\hat{P}/P$, a, b) for Case A

No.	($\hat{P}/P$)	$\sqrt{(root)}$	a	b	$1+a/b$
1	16/225 = 0.0711	17/15	16/15	−1/15	15/16
2	15/196 = 0.0765	16/14	15/14	−1/14	14/15
3	14/169 = 0.0828	15/13	14/13	−1/13	13/14
4	13/144 = 0.0903	14/12	13/12	−1/12	12/13
5	12/121 = 0.0992	13/11	12/11	−1/11	11/12
6	11/100 = 0.1100	12/10	11/10	−1/10	10/11
7	10/81 = 0.1235	11/9	10/9	−1/9	9/10
8	9/64 = 0.1406	10/8	9/8	−1/8	8/9
9	8/49 = 0.1633	9/7	8/7	−1/7	7/8
10	7/36 = 0.1944	8/6	7/6	−1/6	6/7
11	6/25 = 0.2400	7/5	6/5	−1/5	5/6
12	5/16 = 0.3125	6/4	5/4	−1/4	4/5
13	4/9 = 0.4444	5/3	4/3	−1/3	3/4
14	3/4 = 0.7500	4/2	3/2	−1/2	2/3
15	10/9 = 1.1111	7/3	5/3	−2/3	3/5
16	21/16 = 1.3125	10/4	7/4	−3/4	4/7
17	50/25 = 2.0000	15/5	10/5	−5/5	5/10
18	91/36 = 2.5278	20/6	13/6	−7/6	6/13
19*	45/16 = 2.8125	7/2	9/4	−5/4	4/9
20	144/49 = 2.9388	25/7	16/7	−9/7	7/16
21*	28/9 = 3.1111	11/3	7/3	−4/3	3/7
22	240/64 = 3.7500	32/8	20/8	−12/8	8/20
23	360/81 = 4.4444	39/9	24/9	−15/9	9/24
24*	77/16 = 4.8125	9/2	11/4	−7/4	4/11
25	600/100 = 6.0000	50/10	30/10	−20/10	10/30
26	900/121 = 7.4380	61/11	36/11	−25/11	11/36
27	1260/144 = 8.7500	72/12	42/12	−30/12	12/42
28	1764/169 = 10.4379	85/13	49/13	−36/13	13/49
29	2451/196 = 12.5051	100/14	57/14	−43/14	14/57
30	3484/225 = 15.4844	119/15	67/15	−52/15	15/67

$\sqrt{(root)} = \sqrt{(1+4\hat{P}/P)}$; $\quad a = [1+\sqrt{(root)}]/2$

$b = [1-\sqrt{(root)}]/2$, $\quad \varepsilon_{\lim} = 1 + \dfrac{a}{b}$

For heat-transfer design, possibly only results in the range 1–14 above may be of interest.
*Entries seem out of natural sequence.

Open-end temperature ratio

$$\boxed{\frac{T_3}{T_1} = \frac{be^{aN} - ae^{bN}}{be^{bN} + ae^{aN}} \qquad \text{with } \lim(T_3/T_1) = (-b/a) \text{ as } N \to \infty} \tag{6.15a}$$

Bayonet-end temperature ratio

$$\frac{T_2}{T_1} = \frac{(b-a)e^{aN}e^{bN}}{be^{bN} - ae^{aN}} \tag{6.14a}$$

Minimum annulus temperature (T_{min} if it exists)

$$\frac{T_{min}}{T_1} = \frac{be^{aN}e^{bN(X_{min}/L)} - ae^{bN}e^{aN(X_{min}/L)}}{be^{bN} - ae^{aN}} \tag{6.21a}$$

Inner temperature profile

$$\frac{\hat{T}}{T_1} = \frac{be^{bN}e^{aN(x/L)} - ae^{aN}e^{bN(x/L)}}{be^{bN} - ae^{aN}}$$

Annulus temperature profile

$$\frac{T}{T_1} = \frac{be^{aN}e^{bN(x/L)} - ae^{bN}e^{aN(x/L)}}{be^{bN} - ae^{aN}}$$

Limiting effectiveness

$$\varepsilon_{lim} = 1 + \frac{b}{a}$$

Figures 6.3a and b show different exchanger performance only for parameter Q/L, i.e. heat transfer per unit length. If exchanger performance can be controlled by keeping T_3 constant, then an optimum length of exchanger exists. If performance can be controlled by keeping T_1 constant then no useful optimum exists, and it may be best to limit exchanger length to some appropriate value, e.g. N_{min} from equation (6.18a).

Figure 6.4 illustrates actual temperature profiles referred to T_3, with a minimum at T_2. Figure 6.5 provides temperature profiles normalized to T_1, with the length of the exchanger now adjusted to N_{min}.

Case B

This case is simpler to analyse than Case A as the lower temperature profile does not include an intermediate minimum. Following a similar procedure the basic differential equations found are

$$\frac{d\hat{T}}{dx} = \hat{P}(\hat{T} - T) \tag{6.23}$$

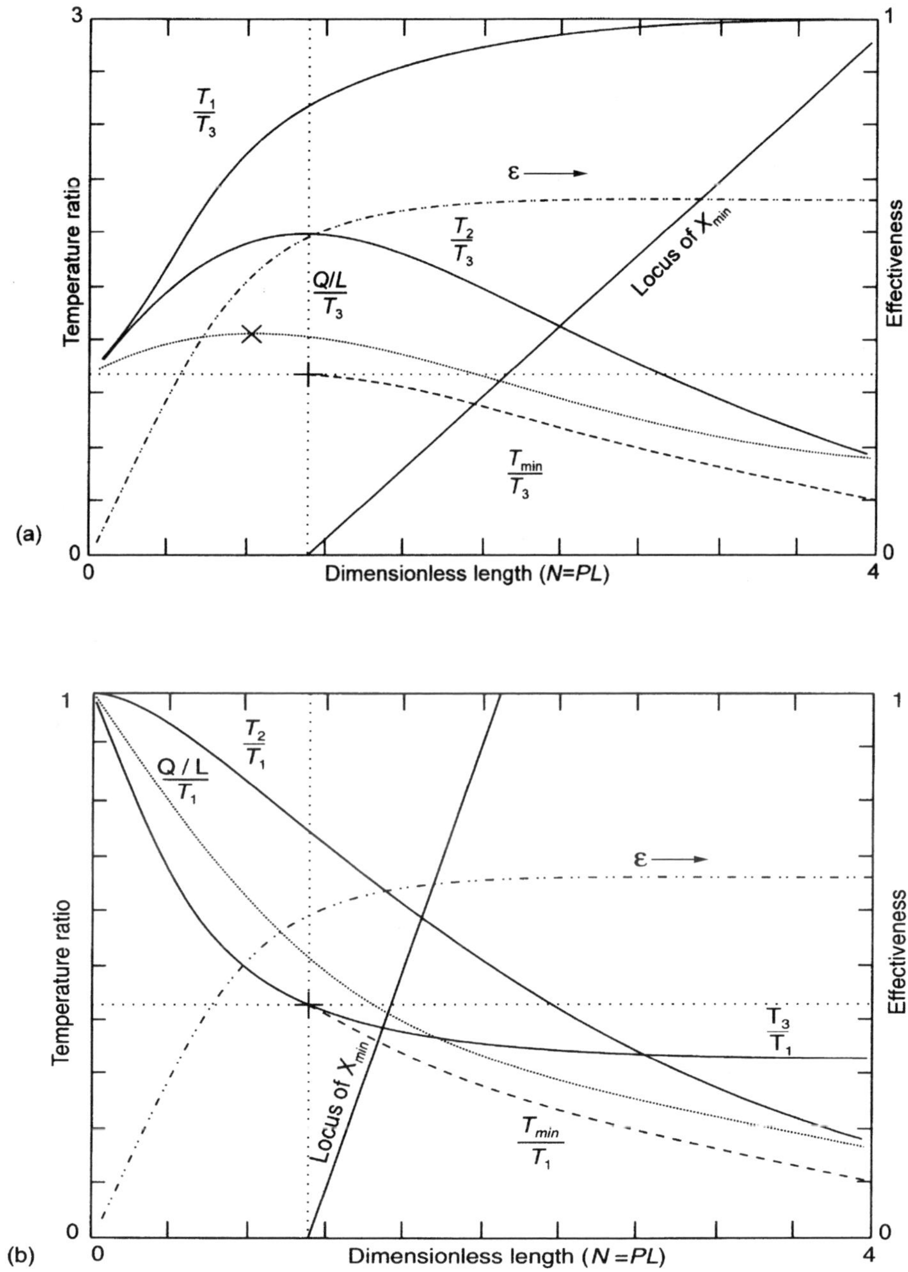

Fig.6.3 **(a) Variation of parameters for Case A with $a = +3/2$, $b = -1/2$ [(T_1/T_3), (T_2/T_3), (T_{min}/T_3), $(Q/L)/T_3$, X_{min}, and ε] versus ($N = PL$); (b) Variation of parameters for Case A with $a = +3/2$, $b = -1/2$ [(T_3/T_1), (T_2/T_1), (T_{min}/T_1), $(Q/L)/T_1$, X_{min}, and ε] versus ($N = PL$)**

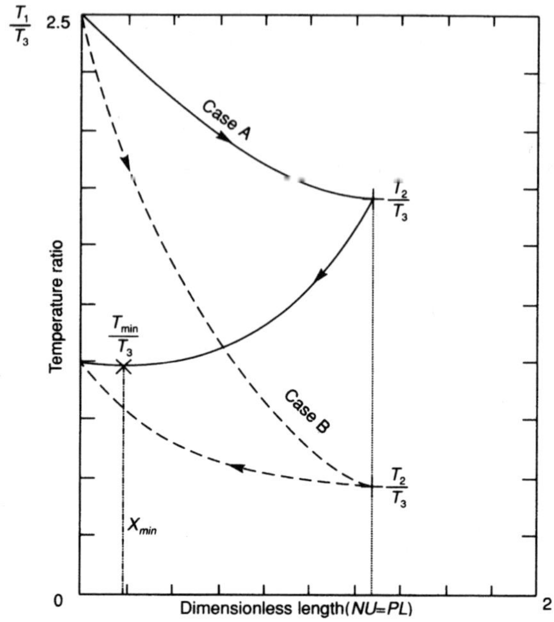

Fig.6.4 Actual temperature profiles for Case A ($a = +3/2$, $b = -1/2$) and Case B ($a = +1/2$, $b = -3/2$) showing existence of minimum for Case A

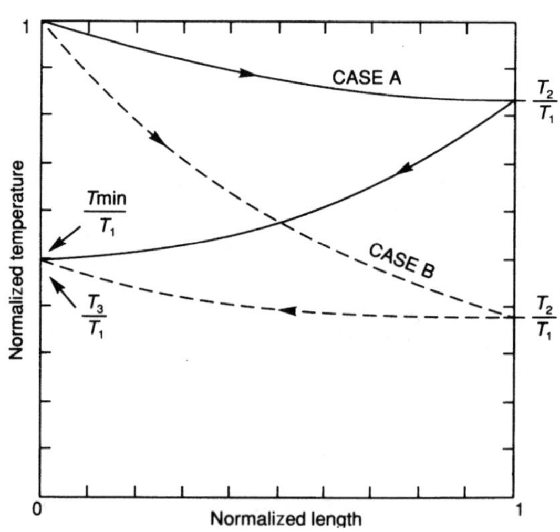

Fig.6.5 Normalized temperature profiles for Case A ($a = +3/2, b = -1/2$) and Case B ($a = +1/2, b = -3/2$) with minimum at fluid exit for Case A

$$\boxed{\frac{dT}{dx} = \hat{P}(\hat{T} - T) - PT} \tag{6.24}$$

Eliminating T as before

$$\frac{d^2\hat{T}}{dx^2} = P\frac{d\hat{T}}{dx} - P\hat{P}\hat{T} = 0 \tag{6.25}$$

which has the solution

$$\hat{T} = A_i e^{\alpha x} + B_i e^{\beta x} \tag{6.26}$$

with

$$\left.\begin{aligned} \alpha &= \frac{P}{2}\left[-1 + \sqrt{(1 + 4\hat{P}/P)}\right] \\ \beta &= \frac{P}{2}\left[-1 - \sqrt{(1 + 4\hat{P}/P)}\right] \end{aligned}\right\} \tag{6.27}$$

and

$$P = -(\alpha + \beta) \tag{6.28}$$

An identical result exists for the other unknown temperature profile

$$T = A_o e^{\alpha x} + B_o e^{\beta x} \tag{6.29}$$

From the annulus temperature profile and overall heat balance

$$T_2/T_1 = (termX + termY + termZ)/denom$$

where

$$\begin{aligned} termX &= +\alpha^2 e^{\alpha L}(e^{\beta L} - 1) \\ termY &= +\alpha\beta(e^{\beta L} - e^{\alpha L})(T_3/T_1) \\ termZ &= -\beta^2 e^{\beta L}(e^{\alpha L} - 1) \\ denom &= (\alpha + \beta)\big[\alpha(e^{\beta L} - 1) - \beta(e^{\alpha L} - 1)\big] \end{aligned}$$

giving

$$T_2/T_1 = \phi[\alpha, \beta, L, (T_3/T_1)] \tag{6.30}$$

Boundary conditions for the lower (inner) temperature profile, equation (6.24), again give $\hat{T} = T_3$ at $x = 0$ and $d\hat{T}/dx = 0$ at $x = L$.

$$\left.\begin{aligned} T_1 &= A_i + B_i \\ T_2 &= A_i e^{\alpha L} + B_i e^{\beta L} \\ 0 &= \alpha A_i e^{\alpha L} + \beta B_i e^{\beta L} \end{aligned}\right\} \tag{6.31}$$

which delivers

$$\boxed{\frac{T_2}{T_3} = \frac{(\beta - \alpha)e^{(\alpha+\beta)L}}{\beta e^{\beta L} - \alpha e^{\alpha L}}} \tag{6.32}$$

As before using equations (6.30) and (6.32)

$$\frac{T_1}{T_3} = \frac{T_1}{T_2}\frac{T_2}{T_3} = \frac{\beta e^{\alpha L} - \alpha e^{\beta L}}{\beta e^{\beta L} - \alpha e^{\alpha L}} \tag{6.33}$$

allowing the explicit result

$$\boxed{L = \frac{1}{(\alpha + \beta)} \ln\left[\frac{\alpha + \beta(T_1/T_3)}{\beta + \alpha(T_1/T_3)}\right]} \tag{6.34}$$

Here equation (6.34) delivers $\lim(T_1/T_3) = -\beta/\alpha$ as $L \to \infty$, thus the restriction $[1 < (T_1/T_3) < (-\beta/\alpha)]$ applies to equation (6.34).

This completes the analysis for Case B, which is a reversal of flow direction of the tube-side fluid for Case A. Again simplification is possible, recasting the above relationships in terms of (a, b) and $Ntu = N = PL$.

Summary of results for Case B

Exchanger length

$$\boxed{N = \frac{1}{(b - a)} \ln\left[\frac{a + b(T_3/T_1)}{b + a(T_3/T_1)}\right]} \tag{6.34a}$$

Open-end temperature ratio

$$\boxed{\frac{T_3}{T_1} = \frac{be^{bN} - ae^{aN}}{be^{aN} - ae^{bN}} \quad \text{with } \lim(T_3/T_1) = (-a/b) \text{ as } N \to \infty} \tag{6.33a}$$

Bayonet-end temperature ratio

$$\boxed{\frac{T_2}{T_1} = \frac{(b - a)e^{aN}e^{bN}}{be^{aN} - ae^{bN}}} \tag{6.30a}$$

Inner temperature profile

$$\boxed{\frac{\hat{T}}{T_1} = \frac{be^{bN}e^{aN(x/L)} - ae^{aN}e^{bN(x/L)}}{be^{aN} - ae^{bN}}}$$

Annulus temperature profile

$$\boxed{\frac{T}{T_1} = \frac{be^{aN}e^{bN(x/L)} - ae^{bN}e^{aN(x/L)}}{be^{aN} - ae^{bN}}}$$

Limiting effectiveness

$$\boxed{\varepsilon_{\mathrm{lim}} = 1 + \frac{a}{b}}$$

A design solution with the same $\hat{P}/P$ ratio as Case A is presented in Fig. 6.4. But note that the values for (a, b) are reversed, i.e. $a = +1/2$ and $b = -3/2$.

Evaluation of equations (6.33) and (6.34) produces $T_1/T_2 = 2.5$ at $x = 0$ and $N = 1.2825$, identical with Case A. Heat transfer and tube length are unaffected by direction of flow of the tube-side fluid, but temperature profiles are different.

6.3 Condensation

Case C

Symbols $(\theta, \hat{\theta})$ are adopted for temperatures so as to permit transformation to $(T, \hat{T})$ later in the analysis. Proceeding as for Case A, energy balances are again written for differential lengths of tube, viz.

Inner tube

$$\begin{Bmatrix}\text{energy entering}\\ \text{with fluid}\end{Bmatrix} - \begin{Bmatrix}\text{energy leaving}\\ \text{with fluid}\end{Bmatrix} + \begin{Bmatrix}\text{heat transferred}\\ \text{from annulus}\end{Bmatrix} = \begin{Bmatrix}\text{energy stored}\\ \text{in fluid}\end{Bmatrix}$$

$$\dot{m}C\hat{\theta} - \dot{m}C\left(\hat{\theta} + \frac{d\hat{\theta}}{dx}\delta x\right) + \hat{U}(\hat{Z}\delta x)(\theta - \hat{\theta}) = 0 \tag{6.35}$$

Annulus

$$\left[\dot{m}C\left(\theta + \frac{d\theta}{dx}\delta x\right) + \hat{U}(\hat{Z}\delta x)(\theta_c - \theta)\right] - [\dot{m}C\theta + U(Z\delta x)(\theta - \hat{\theta})] = 0 \tag{6.36}$$

giving, respectively,

$$\boxed{+\frac{d\hat{\theta}}{dx} = \qquad + \hat{P}(\theta - \hat{\theta})} \tag{6.37}$$

$$\boxed{-\frac{d\theta}{dx} = +P(\theta_c - \theta) - \hat{P}(\theta - \hat{\theta})} \tag{6.38}$$

Putting

$$\hat{T} = \theta_c - \hat{\theta}, \qquad d\hat{T} = -d\hat{\theta}$$
$$T = \theta_c - \theta, \qquad dT = -d\theta$$

equations (6.37) and (6.38) become identical with equations (6.3) and (6.4) of Case A.

Case D

In a manner similar to case B, it may be shown that the governing equations for case D are identical to equations (6.23) and (6.24).

Hurd number

Kayansayan (1996) presented a thermal design method for evaporators and condensers in which he defined the Hurd number (Hu) as the ratio of inner tube number of transfer units divided by outer tube number of transfer units, however, in the present notation

$$Hu = \frac{\hat{N}}{N} = \frac{\hat{P}}{P} \quad \left(= \frac{ntu}{NTU} \text{ in Kayansayan's notation} \right)$$

Kayansayan also presented a schematic illustration of a shell-and-tube exchanger fitted with bayonet tubes.

6.4 Design Illustration

In the absence of pressure-loss data for the bayonet end and the consequent impossibility of calculating total pressure loss in the bayonet tube at this time, it is presently only possible to design on the basis of heat transfer alone. Figure 6.6 below show variation of ε against Ntu with $(\hat{P}/P)$ as parameter. Inspection of these curves leads to formulation of a direct method of design, viz.

1. For a given cross-section of the exchanger calculate:
   - both overall heat-transfer coefficients $(U, \hat{U})$
   - both perimeter transfer units $(P, \hat{P})$
   - obtain (a, b), say $a = +3/2$, $b = (-1/2)$
2. Calculate limiting value of temperature ratio

   $$\lim(T_1/T_3) = (-a/b) = 3$$

3. Calculate limiting effectiveness

   $$\varepsilon_{\lim} = 1 - (T_3/T_1) = 2/3$$

4. Take a design fraction of this, say $f = 0.9$ to obtain actual effectiveness

   $$\varepsilon = 0.6$$

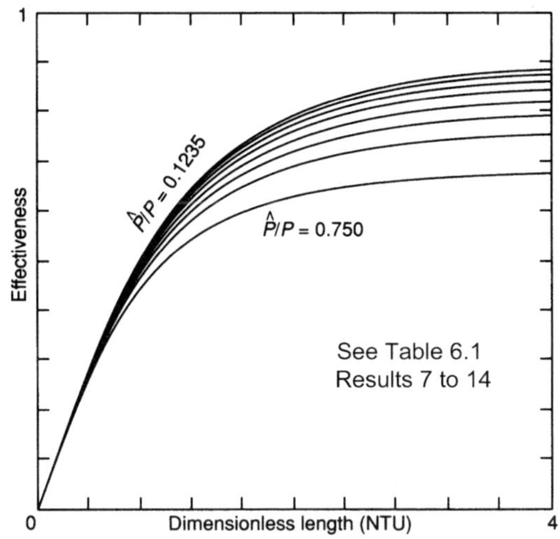

Fig.6.6 Effectiveness and limiting effectiveness

5. Solve for the actual temperature ratio (T_1/T_3), which is a measure of the heat transfer possible

   $(T_1/T_3) = 1/(1 - \varepsilon) = 5/2$

6. Solve equation (6.16) explicitly to obtain $N = 1.2825$. Resulting temperature profiles for the design selected are shown as solid lines in Fig. 6.4.

Helical annular flow between inner and outer tubes

For effective operation it is important that the heat-transfer coefficient in the annulus be significantly higher than the heat-transfer coefficient in the inner tube. With single-phase flow it becomes necessary to provide helical fins between the outside diameter of the inner tube so that the annulus fluid is forced to flow in a helical path. This provides a smaller characteristic dimension and a smaller flow area than for the simple concentric annulus. This same condition applies also to the design configurations which follow. For a narrow annulus a wire-wrap might be considered.

6.5 Non-isothermal shell-side conditions

An explicit design solution exists for the bayonet-tube heat exchanger with non-isothermal shell-side conditions only for the special case of equal water equivalents. Four possible flow configurations exist, each having four (reflected) temperature profiles. For the non-explicit solutions, selection of an appropriate configuration before numerical evaluation is eased when expected temperature profiles can be examined.

Table 6.2 Non-isothermal configurations

Exchanger class	*Case A*	*Case B*	*Case C*	*Case D*
Contra-flow	Type 1	Type 3	Type 2	Type 4
Parallel flow	Type 5	Type 7	Type 6	Type 8

Exchangers are classified as *contra-flow* if tube-side annulus flow and shell-side flow are in the opposite sense, and *parallel flow* if in the same sense. Each class has four possible cases (A, B and their reflections) corresponding to those for isothermal shell-side conditions, and as set out in Table 6.2.

For non-isothermal shell-side flow conditions, the origin is placed at the bayonet end, as this simplifies the analysis. In Figs 6.9 to 6.12 presented later, dashed lines represent flow in the inner tube of the bayonet. Narrow-band shading is used to denote heat transfer between shell-side fluid and annulus tube-side fluid.

Figure 6.7 illustrates the heat balances used to obtain the three coupled governing equations for contraflow – Case A (type 1). Note that all equations are for $x = 0$ at the bayonet end, selection of the origin being different from that for the isothermal solution.

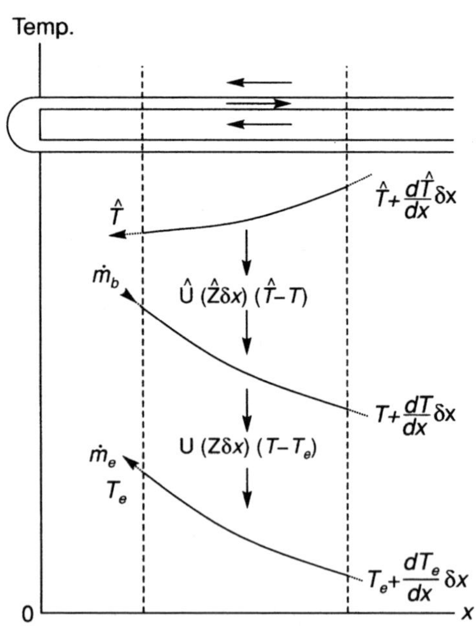

Fig.6.7 Differential energy balances for Case A (type 1). Origin at bayonet end

Temperatures (T, $\hat{T}$), mass flowrate ($\dot{m}_b$), and specific heat (C_b) will be used for the bayonet-tube fluid. Temperature (T_e), mass flowrate ($\dot{m}_e$), and specific heat (C_e) will be used for the external fluid.

Inner

$$\left\{\begin{matrix}\text{energy entering}\\ \text{with fluid}\end{matrix}\right\} - \left\{\begin{matrix}\text{energy leaving}\\ \text{with fluid}\end{matrix}\right\} - \left\{\begin{matrix}\text{heat transfered}\\ \text{to annulus}\end{matrix}\right\} = \left\{\begin{matrix}\text{energy stored}\\ \text{in fluid}\end{matrix}\right\}$$

$$\dot{m}_b C_b\left(\hat{T} + \frac{d\hat{T}}{dx}\delta x\right) - \dot{m}_b C_b \hat{T} - \hat{U}(\hat{Z}\delta x)(\hat{T} - T) = 0$$

giving

$$\boxed{+\frac{d\hat{T}}{dx} = +\hat{P}(\hat{T} - T) \quad \text{where } \hat{P} = (\hat{U}\hat{Z})/(\dot{m}_b C_b)} \tag{6.39}$$

Annulus (similarly)

$$\boxed{-\frac{dT}{dx} = -\hat{P}(\hat{T} - T) + P(T - T_e) \qquad \text{where } P = (UZ)/(\dot{m}_b C_b)} \tag{6.40}$$

External (similarly)

$$\boxed{-\frac{dT_e}{dx} = +P_e(T - T_e) \qquad \text{where } P_e = (UZ)/(\dot{m}_e C_e)} \tag{6.41}$$

The perimeter transfer units (P, $\hat{P}$, P_e) are positive, so for all configurations when temperature differences are kept positive, the sign of the gradient gives the slope of the temperature profile.

At the bayonet end $\hat{T} = T$, then

$$+\left.\frac{d\hat{T}}{dx}\right|_{x=0} = 0 \tag{6.39a}$$

$$-\left.\frac{dT}{dx}\right|_{x=0} = P(T - T_e) \tag{6.40a}$$

$$-\left.\frac{dT_e}{dx}\right|_{x=0} = P_e(T - T_e) \tag{6.41a}$$

When $\hat{T} = T$ and $x = 0$ are specified at the bayonet end then all gradients are known at $x = 0$, allowing generation of the temperature profile using finite differences from that end. See Section 6.8 on non-explicit solutions.

6.6 Special explicit case

The set of partial differential equations (6.39) to (6.41) are similar to those which Kroeger (1966) used in exploring the problem of longitudinal conduction in contra-flow heat exchangers. Kroeger found an explicit solution only for the special case of equal water equivalents, which is also the situation here.

Substitute equations (6.39) and (6.41) in equation (6.40)

$$\frac{dT}{dx} = \frac{d\hat{T}}{dx} + \left(\frac{P}{P_e}\right)\frac{dT_e}{dx}$$

$$\frac{d(\hat{T} - T)}{dx} = \left(\frac{-P}{P_e}\right)\frac{dT_e}{dx}$$

$$\hat{T} - T = \left(\frac{-P}{P_e}\right)T_e + const. \tag{6.42}$$

Boundary conditions

At $x = 0$, $T_e = T_{e2}$ and $\hat{T} - T = 0$

$$0 = \left(\frac{-P}{P_e}\right)T_{e2} + const. \qquad \text{thus } const. = \left(\frac{P}{P_e}\right)T_{e2}$$

At $x = L$, $T_e = T_{e1}$, and $\hat{T} - T = T_1 - T_3$

$$T_1 - T_3 = \left(\frac{P}{P_e}\right)(T_{e2} - T_{e1})$$

or

$$\dot{m}_b C_b(T_1 - T_3) = \dot{m}_e C_e(T_{e2} - T_{e1}) = \dot{Q} \tag{6.43}$$

which is the correct energy balance, and we do not gain another equation.

From equation (6.42) and the first boundary condition

$$\hat{T} - T = \left(\frac{P}{P_e}\right)(T_{e2} - T_e) \tag{6.44}$$

From equation (6.44)

$$T_e = T_{e2} - \left(\frac{P_e}{P}\right)(\hat{T} - T) \tag{6.45}$$

For inner temperature profile ($\hat{T}$), substitute equation (6.45) in equation (6.40)

$$-\frac{dT}{dx} = (P_e - \hat{P})T_e + (\hat{P} + P - P_e)T - PT_{e2} \tag{6.46}$$

From equation (6.39)

$$T = \hat{T} - \left(\frac{1}{\hat{P}}\right)\frac{d\hat{T}}{dx} \tag{6.47}$$

Differentiate

$$\frac{dT}{dx} = \frac{d\hat{T}}{dx} - \left(\frac{1}{\hat{P}}\right)\frac{d^2\hat{T}}{dx^2} \tag{6.48}$$

Substitute equations (6.47) and (6.48) in equation (6.46)

$$\frac{d^2\hat{T}}{dx^2} + (P - P_e)\frac{d\hat{T}}{dx} - \hat{P}P\hat{T} = -\hat{P}PT_{e2} \tag{6.49}$$

Solution of homogeneous equation is of the form $\hat{T} = e^{bx}$, thus

$$b = \frac{-(P - P_e) \pm \sqrt{(P - P_e)^2 + 4\hat{P}P}}{2} \tag{6.50}$$

roots are real, but one is negative

and internal temperature profile is

$$\hat{T} = e^{b_1 x} + e^{b_2 x} + const. \tag{6.51}$$

Boundary conditions

$$\hat{T} = e^{b_1 x} + e^{b_2 x} + const. \tag{6.52}$$

subject to $\hat{T} = T_1$ at $x = L$

$$\frac{d\hat{T}}{dx} = b_1 e^{b_1 x} + b_2 e^{b_2 x} \tag{6.53}$$

subject to $d\hat{T}/dx = 0$ at $x = 0$

From equation (6.53), $b_1 + b_2 = 0$. If the assumption ($b_1 = +b$, $b_2 = -b$) is made in order to get the special case analytical solution, then in equation (6.50) this implies

$$-(P - P_e) + \sqrt{(root)} = +(P - P_e) + \sqrt{(root)}$$

that is $P = P_e$, so that the constraint $\dot{m}_b C_b = \dot{m}_e C_e$ exists, implying

$$(T_{e2} - T_{e1}) = (T_1 - T_3)$$

Substituting in equation (6.50)

$$b = +\sqrt{\hat{P}P} \tag{6.54}$$

In equation (6.51), $const. = T_1 - (e^{+bL} + e^{-bL})$ which provides the explicit inner temperature profile.

6.7 Explicit solution

Bayonet tube

Inner profile ($\hat{T}$), Case A (type 1)

$$\boxed{\hat{T} = (e^{+bx} + e^{-bx}) + [T_1 - (e^{+bL} + e^{-bL})]} \tag{6.55}$$

Bayonet-end temperature (T_2), Case A (type 1)

At $x = 0$ put $\hat{T} = T = T_2$, say

$$\boxed{T_2 = 2 + [T_1 - (e^{+bL} + e^{-bL})]} \tag{6.56}$$

Inner profile gradient, Case A (type 1)

Differentiate equation (6.55)

$$\frac{d\hat{T}}{dx} = b(e^{+bx} - e^{-bx}) \tag{6.57}$$

which confirms that $d\hat{T}/dx = 0$ at $x = 0$.

Annulus

Annulus profile (T), Case A (type 1)

From equations (6.39) and (6.57)

$$T = \hat{T} - \left(\frac{b}{\hat{P}}\right)(e^{+bx} - e^{-bx}) \tag{6.58}$$

and substituting from equation (6.55)

$$\boxed{T = (e^{+bx} + e^{-bx}) + [T_1 - (e^{+bL} + e^{-bL})] - \left(\frac{b}{\hat{P}}\right)(e^{+bx} - e^{-bx})} \tag{6.59}$$

Check on bayonet-end temperature, Case A (type 1)

At $x = L$, put $T = T_3$

$$\boxed{T_3 = T_1 - \left(\frac{b}{\hat{P}}\right)(e^{+bL} - e^{-bL})} \tag{6.60}$$

At $x = 0$ put $T = \hat{T} = T_2$

$$T_2 = 2 + [T_1 - (e^{+bL} + e^{-bL})]$$

which checks with equation (6.56)

Exchanger duty, Case A (type 1)
From equation (6.60)

$$(T_1 - T_3) = \left(\frac{b}{\hat{P}}\right)(e^{+bL} - e^{-bL}) \tag{6.61}$$

$$\boxed{\dot{Q} = \dot{m}_b C_b (T_1 - T_3) = \dot{m}_b C_b \left(\frac{b}{\hat{P}}\right)(e^{+bL} - e^{-bL})} \tag{6.62}$$

Annulus profile gradient, Case A (type 1)

$$\frac{dT}{dx} = b(e^{+bx} - e^{-bx}) - \left(\frac{b^2}{\hat{P}}\right)(e^{+bx} + e^{-bx}) \tag{6.63}$$

At $x = 0$

$$\frac{dT}{dx} = -2\left(\frac{b^2}{\hat{P}}\right) \tag{6.64}$$

always negative

External profile

From equation (6.45),

$$T_e = T_{e2} - \left(\frac{P_e}{P}\right)(\hat{T} - T)$$

but $P = P_e$, thus $T_e = T_{e2} - (\hat{T} - T)$, and then from equation (6.58):

External profile, Case A (type 1)

$$\boxed{T_e = T_{e2} - \left(\frac{b}{\hat{P}}\right)(e^{+bx} - e^{-bx})} \tag{6.65}$$

At $x = L$ put $T_e = T_{e1}$

$$(T_{e2} - T_{e1}) = \left(\frac{b}{\hat{P}}\right)(e^{+bL} - e^{-bL}) \tag{6.66}$$

which corresponds with equation (6.61)

and obtain a check on the energy balance

$$(T_{e2} - T_{e1}) = \left(\frac{b}{\hat{P}}\right)(e^{+bL} - e^{-bL}) = (T_1 - T_2)$$

Using equation (6.66) in equation (6.65), an alternative expression for the external profile is:

Alternative external profile, Case A (type 1)

$$T_e = T_{e1} + \left(\frac{b}{\hat{P}}\right)[(e^{+bL} - e^{-bL}) - (e^{+bx} - e^{-bx})] \tag{6.65a}$$

Exchanger length

From equation (6.60),

$$e^{+bL} - e^{-bL} = \left(\frac{\hat{P}}{b}\right)(T_1 - T_3) = \left(\frac{\hat{P}}{b}\right)(T_{e2} - T_{e1}) = \frac{\hat{P}Q}{b\dot{m}_b C_b} = a, \text{ say} \tag{6.67}$$

Exchanger length, Case A (type 1)

$$L = \left(\frac{1}{b}\right)\ln\left[\frac{a + \sqrt{a^2 + 4}}{2}\right] \tag{6.68}$$

positive root only

Non-dimensional profiles

Inner profile, Case A (type 1)

$$\frac{\hat{T} - T_{e1}}{T_1 - T_{e1}} = 1 - \frac{(e^{+bL} + e^{-bL}) - (e^{+bx} + e^{-bx})}{(T_1 - T_{e1})} \tag{6.69}$$

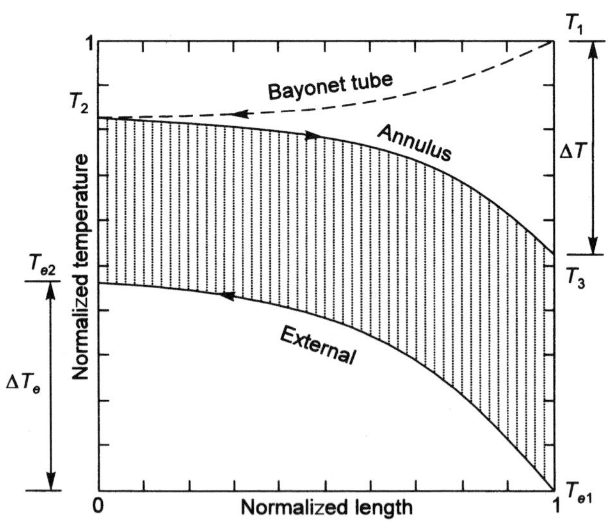

Fig.6.8 Temperature profiles for explicit solution of non-isothermal case with $P = P_e$, $\dot{m}_b C_b = \dot{m}_e C_e$ and $\Delta T = \Delta T_e$

Annulus profile, Case A (type 1)

$$\frac{T - T_{e1}}{T_1 - T_{e1}} = \left(\frac{b}{\hat{P}}\right) \frac{(e^{+bx} - e^{-bx})}{(T_1 - T_{e1})} \tag{6.70}$$

External profile, Case A (type 1)

$$\frac{T_e - T_{e1}}{T_1 - T_{e1}} = \left(\frac{b}{\hat{P}}\right) \frac{(e^{+bx} - e^{-bx}) - (e^{+bL} - e^{-bL})}{(T_1 - T_{e1})} \tag{6.71}$$

Typical temperature profiles for the only explicit solution with $P = P_e$ and $\dot{m}c = \dot{M}C$ are given in Fig. 6.8.

6.8 General numerical solutions

The analytical approach to solution of the same coupled partial differential equations has already been given by Kroeger (1966), but proceeding from equations (6.39a), (6.40a), and (6.41a) it is straightforward to set up a numerical method for generating temperature profiles. Starting from the bayonet end, it is necessary to assume a value for the bayonet-end temperature difference $(T_2 - T_{e2})$.

The resulting temperature profiles shown in Figs (6.9)–(6.12) were obtained by using the same arbitrary data for each configuration, given in Table 6.3. These solutions do not take longitudinal conduction in the tube walls into account. In laminar flow, a short-length bayonet-tube exchanger may not allow fully developed profiles to appear.

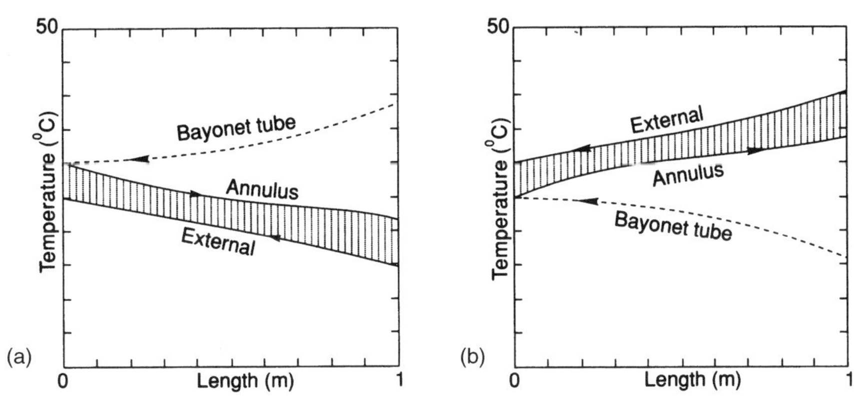

Fig.6.9 (a) Type 1, contraflow Case A; (b) Type 2, contraflow reflected A

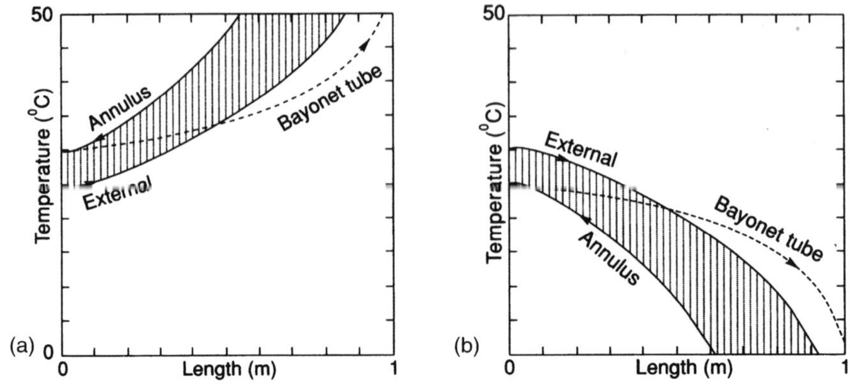

Fig.6.10 (a) Type 3, contraflow Case B; (b) Type 4, contraflow reflected B

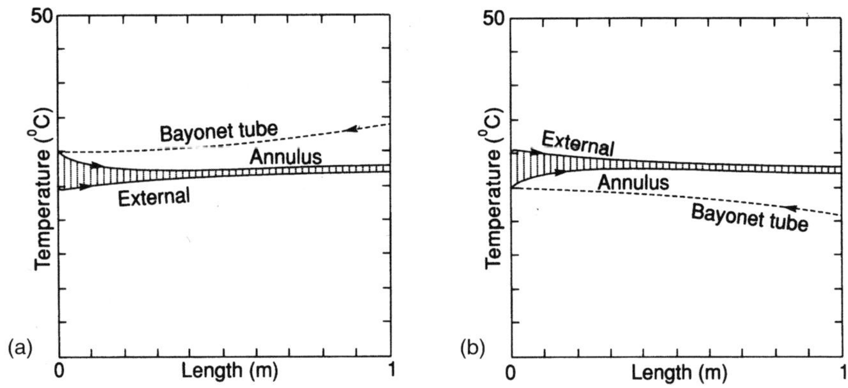

Fig.6.11 (a) Type 5, parallel flow Case C; (b) Type 6, parallel flow reflected C

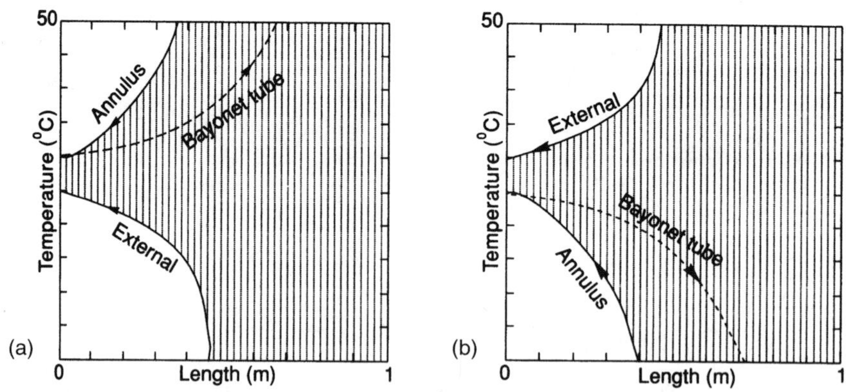

Fig.6.12 (a) Type 7, parallel flow Case D; (b) Type 8, parallel flow reflected D

Table 6.3 Numerical parameters used in solution

Parameter	*Internal*	*External*
Tube mean diameter, m	0.015	0.025
Mass flowrate, kg/s	0.010	0.070
Specific heat, J/(kg K)	4200.0	1000.0
Heat-transfer coefficient, J/(m^2 s K)	1000.0	2000.0

6.9 Pressure loss

To obtain higher heat-transfer coefficients in the annulus it becomes essential to reduce annular flow area by forming helical channels.

It is necessary to know pressure loss in the bayonet-tube heat exchanger before direct-sizing becomes possible. While sufficient information exists to calculate pressure loss in the central tube, at present there is sparse information about pressure loss for helical channel flow in the bayonet-tube annulus, and for the bayonet end which will become highly flow-direction dependent.

For a given outer-tube diameter (D), there may be one inner-tube diameter/thickness and one inner-tube/bayonet-end spacing which would provide the optimum (possibly least) pressure loss for a given reference Reynolds number. For the isothermal case only one pressure-loss curve will exist.

Some Russian data on pressure loss at the bayonet end for simple annular flow have been published by Idelchik & Ginzburg (1968) and are reported in the textbook by Miller (1990). This information was provided by D. Graham, a senior honours student at Heriot-Watt University, Edinburgh and further (unpublished) data for simple annular flow were obtained in an experimental programme under Dr B. Burnside.

While adequate bayonet-tube end and helical-channel pressure-loss data may not be available at this time, the question of pressure loss in the circular channel in which the 'non-isothermal' bayonet tube may be placed can be explored, and it seems worthwhile to obtain the correct analytical value for hydraulic diameter in the annulus for laminar flow conditions. There will then be consistency when attempting to fit heat-transfer and flow-friction data to turbulent correlations.

Simple annular flow

Steady laminar flow in a tube may be analysed using the cylindrical coordinate system to be found in most standard texts. An expression for flow velocity (u) along the tube is given by

$$u = \frac{-(p_1 - p_2)r}{4\eta\ell} + A\ln(r) + B$$

Inserting boundary conditions (u) *finite* at $r = 0$, and $u = 0$ at $r = a$ we may obtain an expression for velocity profile across the tube

$$u = \frac{(p_1 - p_2)}{4\eta\ell}(a^2 - r^2)$$

from which a mean velocity $\bar{u}$ say, may be found by integration, giving, for a circular tube,

$$\Delta p = \frac{8\eta\bar{u}\ell}{a^2} \tag{6.72}$$

An identical analysis for pressure loss in steady laminar flow within an annulus of outer radius a and inner radius b gives

$$\Delta p = \frac{8\eta\bar{u}\ell}{(a^2+b^2)-(a^2-b^2)/\ln(a/b)} \tag{6.73}$$

and the two equations (6.72) and (6.73) may be equated to give an expression for equivalent frictional diameter of an annulus

$$d_f = \sqrt{(D^2+d^2)-(D^2-d^2)/\ln(D/d)} \tag{6.74}$$

As $d \to 0$, in the limit $d_f \to D$, i.e. the plain circular tube is recovered. The ratio of d_f/D from equation (6.74) is plotted in Fig. 6.13 as a solid line.

When a similar analysis is made for flow in a very narrow annulus (in the limit, flow between two flat plates of spacing, s), then with Cartesian coordinates the expression for pressure loss becomes

$$\Delta p = \frac{12\eta\bar{u}\ell}{s^2} \tag{6.75}$$

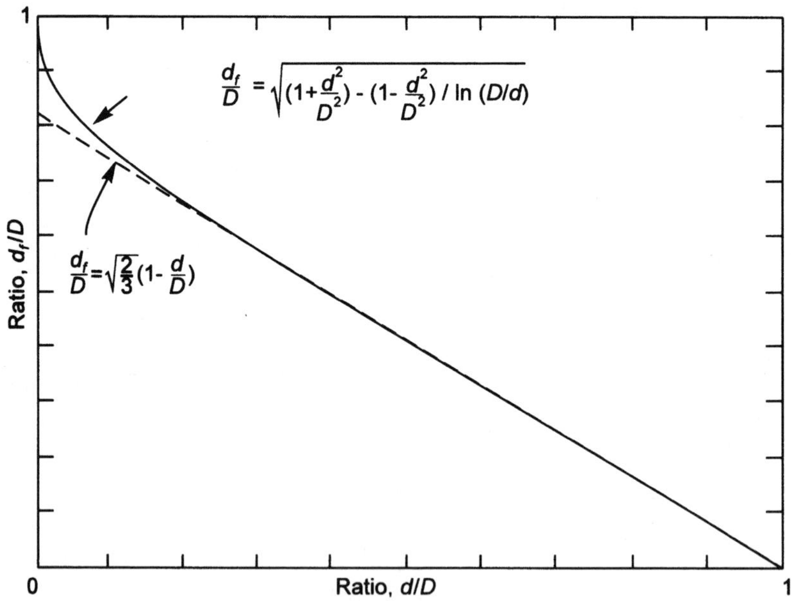

Fig.6.13 Laminar flow friction equivalent diameter for concentric annulus

Again by analogy with the solution for a circular tube

$$\left(\frac{12\eta\bar{u}\ell}{s^2} = \frac{8\eta\bar{u}\ell}{a^2}\right)$$

from which the equivalent frictional diameter for a very narrow annulus is obtained as

$$d_f = \sqrt{\frac{2}{3}}(D - d) \tag{6.76}$$

The ratio d_f/D from equation (6.76) is plotted in Fig. 6.13 as a dashed line, and it is remarkable how well it matches the value of d_f/D for an annulus over much of the diameter ratios. Indeed this may be seen as supporting the approximate equivalent diameter for flow friction in an annulus as

$$d_f = \frac{4 \times \text{area for flow}}{\text{wetted perimeter}} = (D - d)$$

because the constant $\sqrt{2/3}$ may be assimilated in the empirical constant of a correlation.

Helical annular flow

An abstract survey covering the last 10 years suggests that published data on helical annular flow in near-rectangular ducts are very sparse, and only the paper by Wang & Andrews (1995) provides the correct analysis for helical annular flow, plus references to the few papers of interest. With the additional effect of the 180° return at the bayonet-tube end, pressure loss becomes highly flow-direction dependent.

With the bayonet-tube fluid entering the central tube, flow at the bayonet-tube end should be mainly radial and longitudinal in character. With the bayonet-tube fluid entering the helical annulus, an additional tangential component is introduced to affect flow conditions.

The only experimental work on helical annular flow in rectangular ducts so far noted is that by Joye (1994) and by Joye & Côté (1995).

For heat transfer, it may also be that temperature profiles derived earlier would be affected to second-order of magnitude by the slight discontinuities now introduced by helical annular flow.

Review

An up-to-date review on bayonet-tube heat exchangers was published by Lock & Minhas (1997) shortly after the first edition of this text appeared. Applications and design features are discussed in depth, and several of the relevant papers are listed in the references below. Considerable attention is paid to flow patterns and pressure losses at the bayonet end.

6.10 Conclusions

1. The bayonet-tube exchanger transfers useful heat only from the outer tube, and the annulus should have helical channels for effective performance. This implies a substantial experimental programme to produce correlations.
2. Pressure losses in the bayonet-tube end will be flow-direction dependent, and a research programme to determine these is also needed.

Isothermal shell-side

3. Explicit temperature profiles are presented for the bayonet-tube exchanger having evaporation, condensation, or isothermal crossflow on the shell-side.
4. Overall heat exchange and optimum length of exchanger are unaffected by the direction of tube-side flow.
5. Temperature profiles are significantly affected by direction of tube-side flow, and this may be relevant in some design situations, e.g. Case A would be preferred to Case B when freezing of the tube-side fluid is to be avoided, and Case C preferred to Case D when boiling of the tube-side flow is to be avoided.
6. Possible applications include freezing of wet ground in order to stabilize conditions for excavation, and ice formation around sunken objects as a means of flotation.

Non-isothermal shell-side

7. The present derivation of temperature profiles for an individual bayonet-tube exchanger assumes that a constrained external longitudinal flow exists, which is a possible design situation – e.g. superheating secondary steam at the top of a PWR fuel element channel.
8. An explicit solution for temperature profiles has been obtained for the case of equal water equivalents ($\dot{m}c = \dot{M}C$). The explicit solution provides a check on numerical solutions.
9. For the more common case of unequal water equivalents, information helpful in selecting a suitable flow configuration has been provided, and sufficient information has been gathered to allow intelligent attacks on actual design problems.
10. One possible application is the use of a single, vertical bayonet tube at the centre of a large cryogenic storage tank, with external natural convection. Such an exchanger provides axi-symmetric cooling in the tank, and may encourage slow controlled circulation of the contents of the tank, thus helping to inhibit 'roll-over' incidents.
11. Bayonet-tube heat exchangers are suitable for heat recovery at high temperatures where metals are not strong enough. Silicon carbide bayonet tubes can be used.

References

Hurd, N.L. (1946) Mean temperature difference in the field or bayonet tube. *Ind. Engng Chemistry*, **38**(12), December, 1266–1271.

Idelchik, I.E. and **Ginzburg, Ya.L.** (1968) The hydraulic resistance of 180° annular bends. *Thermal Engng*, **15**(4), 109–114.

Joye, D.D. (1994) Optimum aspect ratio for heat transfer enhancement in curved rectangular channels. *Heat Transfer Engng*, **15**(2), 32–38.

Joye, D.D. and **Côté, A.S.** (1995) Heat transfer enhancement in annular channels with helical and longitudinal flow. *Heat Transfer Engng*, **16**(2), 29–34.

Kayansayan, N. (1996) Thermal design method of bayonet-tube evaporators and condensers. *Int. J. Refrigeration*, **19**(3), 197–207.

Kroeger, P.G. (1966) Performance deterioration in high effectiveness heat exchangers due to axial conduction effects. In *Proceedings of the 1966 Cryogenic Engineering Conference*, Boulder Colorado, 13–15 June 1966. (Also in *Cryogenic Engineering*, vol. 12, Plenum Press, 1967, pp. 363–372.)

Lock, G.S.H. and **Minhas, H.** (1997) Bayonet tube heat exchangers. *Appl. Mech. Rev.*, **50**(8), August, 415–472.

Miller, D.S. (1990) *Internal Flow Systems*, 2nd edn, BHRA (Information Services), pp. 218–225.

Wang, J.-W. and **Andrews, J.R.G.** (1995) Numerical simulation of flow in helical ducts. (helical co-ordinate system and equations for flow in helical ducts). *AIChE J.*, **41**(5), May, 1071–1080.

Bibliography

Chung, H.L. (1981) Analytical solution of the heat transfer equation for a bayonet tube exchanger. In *ASME Winter Annual Meeting*, Paper no 81-WA NE-3.

Guedes, R.O.C., **Cotta, R.M.**, and **Brum, N.C.L.** (1991) Heat transfer in laminar flow with wall axial conduction and external convection. *J. Thermophysics*, **5**(2), October–December, pp. 508–513.

Hernandez-Guerrero, A. and **Macias-Machin, A.** (1991) How to design bayonet heat-exchangers. *Chem. Engng*, **79**, April, 122–128.

Jolly, A.J., **O'Doherty, T.**, and **Bates, E.J.** (1998) COHEX a computer model for solving the thermal energy exchanger in an ultra high temperature heat exchanger (ceramic bayonet tube to 1600°C). *Appl. Thermal Engng*, **18**(12), December, 1263–1276.

Lock, G.S.H. and **Wu, M.** (1991) Laminar frictional behaviour of a bayonet tube (pp. 429–440). Turbulent frictional behaviour of a bayonet tube (pp. 405–416). In *Proceedings of the 3rd International Symposium on Cold Regions Heat Transfer*, Fairbanks, Canada, 1991.

Luu, M. and **Grant, K.W.** (1985) Heat transfer to a bayonet heat exchanger immersed in a gas-fluidised bed. In *Symposium on Industrial Heat Exchanger Technology*, pp. 159–173.

Minhas, H. and **Lock, G.S.H.** (1996) Laminar turbulent transition in a bayonet tube. *Int. J. Heat Fluid Flow*, **17**, 102–107.

Pagliarini, G. and **Barozzi, G.S.** (1991) Thermal coupling in laminar flow double-pipe heat exchangers. *ASME J. Heat Transfer*, **113**, August, 526–534.

Smith, E.M. (1981) Optimal design of bayonet tube exchangers for isothermal shell-side conditions. In *20th Joint ASME/AIChemE National Heat Transfer Conference*, Milwaukee, Winsconsin, 2–5 August 1981, ASME Paper 81-HT-34.

Todo, I. (1976) Dynamic response of bayonet-type heat exchangers. Part I: response to inlet temperature changes. *Bull. Japan. Soc. Mech. Engrs*, **19**(136), October, 1135–1140.

Todo, I. (1978) Dynamic response of bayonet-type heat exchangers. Part II: response to flow rate changes. *Bull. Japan. Soc. Mech. Engrs*, **21**(154), April, 644–651.

Ward, P.W. (1985) Ceramic tube heat recuperator – a user's experience. *Advances in Ceramics*, vol. 14. *Ceramics in Heat Exchangers* (Eds. B.D. Foster and J.B. Patton), American Ceramics Society.

Zaleski, T. (1984) A general mathematical model of parallel-flow, multichannel heat exchangers and analysis of its properties (includes bayonet tube exchangers). *Chem. Engng Sci.*, **39**(7/8), 1251–1260.

CHAPTER 7
Direct-Sizing of RODbaffle Exchangers

Practical design example

7.1 Design framework

The direct-sizing approach suggested in this chapter is provisional. It should be checked against the established rating method.

The RODbaffle exchanger can be a better performing shell-and-tube design than conventional tube-and-baffle designs. Design methods proposed by the originators of this exchanger type require prior knowledge of the diameter of the exchanger shell, thus these methods can be classed only as 'rating' methods. Direct 'sizing' of an exchanger becomes possible when the tube bundle can be designed with reference to 'local' geometry only, and this paper indicates an approach to such a method.

When the 'local geometry' in a heat exchanger is fully representative of the whole geometry, then direct methods of thermal sizing become possible (Smith, 1986, 1994). Both compact plate–fin and helical-tube heat exchangers are amenable to this approach, and the present chapter makes the case that the RODbaffle design may be handled in the same manner.

The paper by Gentry *et al.* (1982) presents a method for rating RODbaffle heat exchangers. This is based on test results obtained from experimental rigs on real heat exchangers. In setting out the Gentry *et al.* design approach, several decisions were taken which effectively prevents their method from being used for direct-sizing of RODbaffle heat exchangers, viz.:

- The exchanger inner-shell surface area is incorporated in the hydraulic diameter for pressure loss on the shell-side.
- Coefficients C_L and C_T in heat-transfer correlations for laminar and turbulent flow include expressions for A_ℓ/A_s and L/D_{bo}, each of which requires knowledge of exchanger shell diameter (see Notation).
- Coefficients C_1 and C_2 in the pressure loss correlation for baffle sections each require knowledge of exchanger shell diameter (see Notation).

It is the purpose of this chapter to set out an alternative approach to design to permit direct thermal sizing of RODbaffle heat exchangers. As no experimental work has been carried out to confirm the approach at this time, direct-sizing should be used only for preliminary design, and the method of Gentry *et al.* should be used to complete the final design.

Minor changes to the notation used by Gentry *et al.* will be used in the interests of clarity.

7.2 Configuration of the RODbaffle exchanger

The RODbaffle exchanger is essentially a shell-and-tube exchanger with conventional plate baffles (segmental or disc-and-doughnut) replaced by grids of rods (see Fig. 1.3). Unlike plate-baffles, RODbaffle sections extend over the full transverse cross-section of the exchanger.

Square pitching of the tube bundle is practicable with RODbaffles. To minimize blockage one set of vertical rods in a baffle section is placed between every second row of tubes. At the next baffle section the vertical rods are placed in the alternate gaps between tubes not previously filled at the first baffle section. The next two baffle sections have horizontal rod spacers, similarly arranged. Thus each tube in the bank receives support along its length.

7.3 Approach to direct-sizing

As the RODbaffle design is based on a set of four baffles, two with horizontal rods, and two with vertical rods, this may not seem consistent with having constant local geometry throughout the bundle. However, fluids with no memory do not recognize when a set of four baffles begins, thus length design to at least the nearest baffle pitch becomes practicable, neglecting flow distributions between the shell nozzles and the first and last baffles. Also, Hesselgreaves (1988) shows that RODbaffle flow creates von Kármán vortex streets, well distributed in the shell-side fluid. Hesselgreaves took street length as the pitch between adjacent RODbaffles; however, it may be that street length is longer.

Published correlations for heat-transfer and shell-side pressure loss were assessed for direct-sizing (see references) but in the end, data presented in Figs 6 and 8 of Gentry *et al.* (1982) for shell-side heat transfer and RODbaffle pressure loss were spline-fitted to obtain data for their ARA bundle configuration.

The RODbaffle pressure loss data of Gentry *et al.* claims to take into account both loss through the plane of the baffle, and friction on the inner shell surface. This seems an awkward concept, for it implies that baffle hydraulic diameter must change with shell diameter, which contravenes the basic concept of 'local action' in continuum mechanics.

An alternative concept of evaluating longitudinal leakage flow between the shell and the outside of the bundle might be employed. Because of the need to locate the baffle rods it is necessary to fit baffle rings between the tube bundle and the exchanger shell; this may permit leakage flow. Shell-side flow through the tube bundle could be evaluated first using local geometry concepts, and this same pressure loss then used to calculate the leakage flow between the baffles and the exchanger shell. The final outlet temperature would be the result of mixing of both streams.

With the above proposal, when the shell-side flow is being heated there will be some diffusion from the shell-side of the tube bundle into the leakage stream, and an opposite effect when the shell-side fluid is being cooled. However it is likely that the major contribution to leakage pressure loss would occur in the small clearance gaps around the baffle rings.

Experimental data for pressure loss due to leakage between baffle and shell is available in the thesis by Bell (1955) and in the papers by Bell & Bergelin (1957) and Bergelin *et al.* (1958). Dimensions for the baffle rings are provided in the paper by Gentry (1990). Further discussion of the development of this concept is presented in Appendix D.

The present design approach will simply assume that leakage flow losses can be included in the baffle loss coefficient (k_b). This permits the direct-sizing approach.

7.4 Flow areas

Flow areas per single tube

Tube-side

$$a_t = \frac{\pi}{4} d_i^2 \tag{7.1}$$

Shell-side (plain tubes)

$$a_s = p^2 - \frac{\pi}{4} d^2 \tag{7.2}$$

Shell-side (baffle section)

$$a_b = p^2 - \frac{\pi}{4} d^2 - pr \tag{7.3}$$

Total flow areas

Tube-side total flow area

$$A_t = Za_t \tag{7.4}$$

Shell-side total flow area (plain tubes)

$$A_s = Za_s \tag{7.5}$$

Shell-side total flow area (baffle section)

$$A_b = Za_b \tag{7.6}$$

7.5 Characteristic dimensions

For shell-side heat transfer in the interior of a tube bundle the Reynolds number can be based on local geometry only, allowing a definition of hydraulic diameter (D_s) for

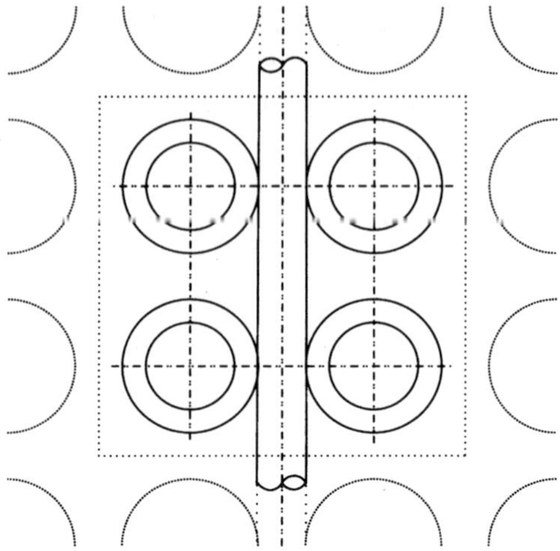

Fig.7.1 Local geometry of tube bundle at a RODbaffle section

plain-tubes, viz.

$$d_s = \frac{4 \times \text{area for flow}}{\text{heated perimeter}} = \frac{4p^2 - \pi d^2}{\pi d} \tag{7.7}$$

For shell-side pressure loss two characteristic dimensions are required, one for plain tubes only and one for the baffle section. The above expression for D_s can be used for plain-tubes, and an expression for the baffle ring section may be evaluated over a tube length equal to the thickness of the baffle (i.e. $d_r = 2r$), whence from Fig. 7.1,

$$d_b = \frac{4 \times \text{area for flow}}{\text{wetted perimeter}} = \frac{4p^2 - \pi d^2 - 4pr}{\pi(d + p/2)} \tag{7.8}$$

7.6 Design correlations

Whenever explicit algebraic correlations for heat-transfer and friction factor can be used throughout, it becomes possible to seek a direct algebraic solution for L and Z, although tracing missing numerical values through the analysis requires some care (see Chapter 5 on helical-tube, multi-start coil heat exchangers).

Here it is the intention to use the correlations provided by Gentry *et al.* in graphical form, and to spline-fit the correlations for heat-transfer, flow-friction, and baffle loss coefficient on the shell-side. This avoids the need to know exchanger shell diameter and baffle-ring diameters before design commences. A possible case for making this simplification can be seen by inspection of the graphs provided by Gentry *et al.* Scatter around each correlation is within usually acceptable limits, suggesting that it is possible to avoid detailed building of the main correlations from sub-correlations involving shell diameters and tube-bundle length.

It is to be expected that different correlations would be necessary for different tube-bundle arrangements. This is beyond the present task, which is to establish that direct-sizing is possible, but see Appendix D.

The procedure is first to evaluate Reynolds number constraints on both shell-side and tube-side correlations. Valid Reynolds number values on the shell-side can then be scanned, and corresponding Reynolds number values on the tube-side forced. Design within the valid envelope can then be completed.

7.7 Reynolds numbers

Shell-side (heat transfer)

With an assumed value for shell-side Re_s

$$G_s = \frac{\mathrm{Re}_s \eta_s}{d_s}, \qquad A_s = \frac{\dot{m}_s}{G_s}, \qquad Z = \frac{A_s}{a_s}$$

and the number of tubes is determined.

Tube-side (heat transfer and pressure loss)

The forced tube-side Reynolds number may now be obtained

$$a_t = \frac{\pi}{4} d_i^2, \qquad A_t = Z a_t, \qquad G_t = \frac{\dot{m}_t}{A_t}$$

$$\mathrm{Re}_t = \frac{d_i G_t}{\eta_t} \tag{7.9}$$

Shell-side (pressure loss)

Two Reynolds numbers are involved. The plain-tube value is identical with that assumed for heat transfer. The baffle-section Reynolds number is obtained as follows.

$$G_b = \frac{\dot{m}_s}{A_b}, \qquad \mathrm{Re}_b = \frac{D_b G_b}{\eta_s} \tag{7.10}$$

7.8 Heat transfer

Shell-side

The heat-transfer correlation shown in Fig. 6 of the paper by Gentry *et al.* is depicted as two straight-line segments, but in the text the curve is described as exhibiting a gradual change of slope. This feature is preserved in the spline-fit of Fig. 7.2.

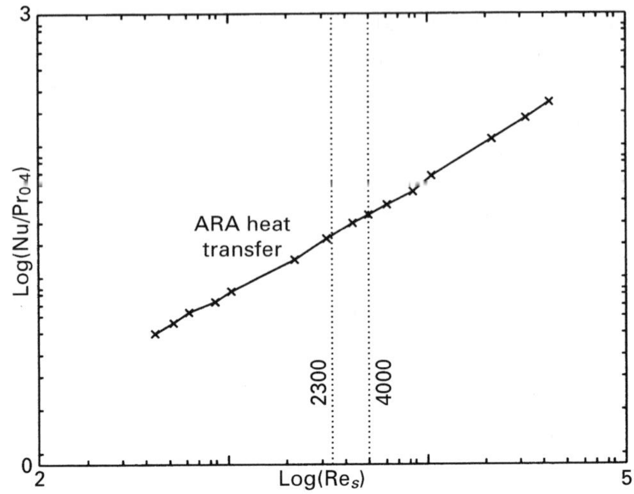

Fig.7.2 Heat-transfer correlation for configuration ARA (adapted from Gentry *et al.*, 1982)

Assuming that the viscosity ratio term is unity, then Nusselt numbers can be determined. The shell-side heat-transfer coefficient becomes

$$\alpha_s = \left(\frac{\lambda_s}{d}\right)\mathrm{Nu}_s \tag{7.11}$$

Tube wall

The tube-wall heat-transfer coefficient may be written as

$$\alpha_w = \frac{2\lambda_w}{d\ln(d/d_i)} \tag{7.12}$$

Tube-side

The conventional tube-side correlation (without viscosity correction) might be used, viz.

$$\mathrm{Nu} = 0.023(\mathrm{Re})^{0.8}(\mathrm{Pr})^{0.4}$$

or a more comprehensive correlation due to Churchill (1977, 1988, 1992).

With the 'forced' value for Re_t, and correcting to outside diameter we obtain

$$\alpha_t = \left(\frac{d_i}{d}\right)\left(\frac{\lambda_t}{d_i}\right)\mathrm{Nu}_t = \left(\frac{\lambda_t}{d}\right)\mathrm{Nu}_t \tag{7.13}$$

Overall coefficient

$$U = (1/\alpha_s + 1/\alpha_w + 1/\alpha_t)$$

Heat-transfer equation

$$LZ = \frac{\dot{Q}}{\pi d U \Delta\theta_{lmtd}} \tag{7.14}$$

7.9 Pressure loss tube-side

The total pressure loss is made up of three components, one due to friction, one due to flow acceleration/deceleration, and one due to entrance/exit effects. The largest of these is due to friction, sometimes reaching 98 per cent of the total pressure loss. In direct-sizing only the frictional loss is considered, but the other losses should be evaluated once dimensions of the exchanger are known.

Chen (1979) provides an explicit correlation for turbulent friction factor in a pipe over the Reynolds number range ($4000 < \mathrm{Re} < 4 \times 10^8$) taking into account roughness ε/d in the range ($5 \times 10^{-7} < (\varepsilon/d) < 0.05$), see Fig. 7.3.

Using the standard friction factor expression

$$\frac{1}{\sqrt{f}} = -2.0 \log\left[\frac{(\varepsilon/d)}{3.7065} - \frac{5.0452}{(\mathrm{Re})} \log\left\{\frac{(\varepsilon/d)^{1.1098}}{2.8257} + \frac{5.8506}{(\mathrm{Re})^{0.8981}}\right\}\right] \tag{7.15}$$

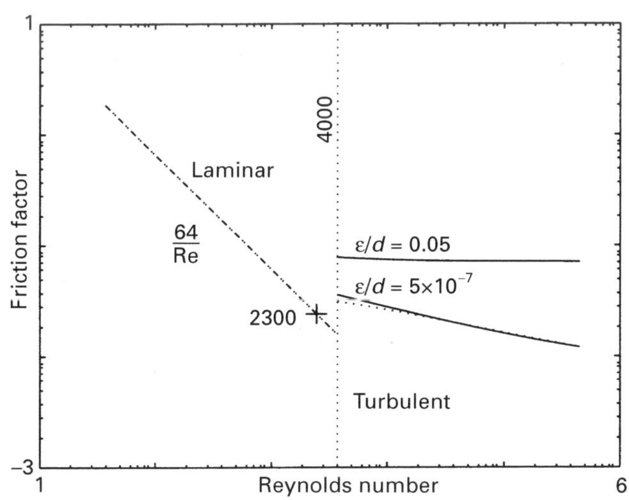

Fig.7.3 Plain-tube flow-friction correlation: laminar (Moody, 1944) and turbulent (Chen, 1979)

with the pressure-loss expression

$$\Delta p_t = 4f_t\left(\frac{G_t^2}{2\rho}\right)\left(\frac{L}{d_i}\right) = \left(\frac{4f_t}{2\rho}\right)\left(\frac{\dot{m}_t^2}{d_i a_t^2}\right)\frac{L}{Z^2}$$

the pressure loss equation for the tube-side may be written

$$L = Z^2\Delta p_t\left\{\left(\frac{2\rho}{4f_t}\right)\left(\frac{\dot{m}_t^2}{d_i a_t^2}\right)\right\}$$

$$L = Z^2\Delta p_t\{\text{bracket}\} \tag{7.16}$$

7.10 Pressure loss shell-side

The total pressure loss is made up of four components, one due to friction on plain-tubes, one due to baffle losses, one due to flow acceleration/deceleration, and one due to entrance/exit effects. Only flow-friction and baffle losses are considered in direct-sizing, but the other losses should be evaluated once dimensions of the exchanger are known.

The technical literature contains papers on longitudinal flow in tube bundles, but these are mainly concerned with reporting results of heat-transfer and pressure-loss tests on nuclear fuel element bundles. Often a near-triangular pitch is adopted which makes the results of less interest for the RODbaffle exchanger. Additionally, very close spacing of the fuel rods is employed (Rehme, 1992). When results for square pitching are to be found, these tend to be for high Reynolds numbers (Tong, 1968).

In shell-side flow pressure loss in the plain-tube section of a RODbaffle exchanger tends to be an order of magnitude less than for that in the rod baffles. In the absence of exact information plain-tube pressure loss is calculated from the correlation for flow inside smooth tubes, using an appropriate hydraulic diameter.

Plain tubes

The Chen friction factor correlation may be used again, then

$$G_s = \frac{\dot{m}_s}{A_s} = \frac{\dot{m}_s}{Za_s}$$

$$\Delta p_f = 4f_s\left(\frac{G_s^2}{2\rho}\right)\frac{L}{d_s} \tag{7.17}$$

Baffle section

Baffle loss coefficient (k_b) is obtained from a spline-fit of k_b versus Re_b (Fig. 7.4), then

$$G_b = \frac{\dot{m}_s}{A_b} = \frac{\dot{m}_s}{Za_b}$$

$$\Delta p_b = k_b\left(\frac{G_b^2}{2\rho}\right)\frac{L}{L_b} \tag{7.18}$$

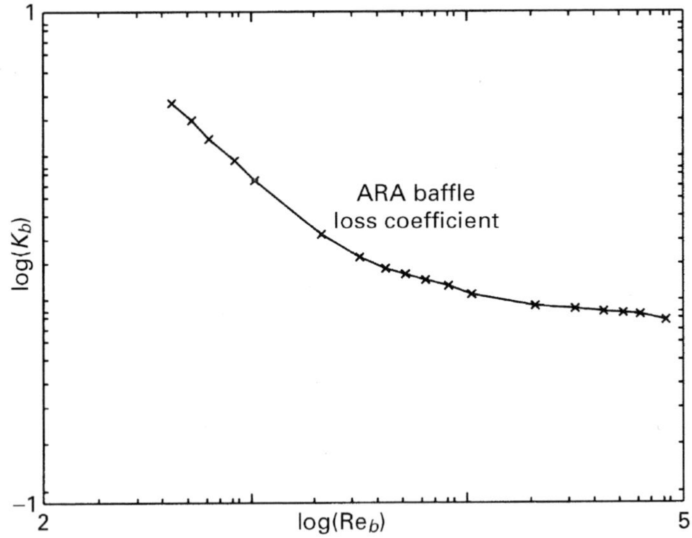

Fig.7.4 Baffle loss coefficient (k_b) for configuration ARA (adapted from Gentry *et al.*, 1982)

Additive pressure loss

$$
\begin{aligned}
\Delta p_s &= \Delta p_p + \Delta p_b \quad (\text{plain tube loss} + \text{baffle loss}) \\
&= 4 f_p \left(\frac{G_s^2}{2\rho} \right) \frac{L}{d_s} + k_b \left(\frac{G_b^2}{2\rho} \right) \frac{L}{L_b} \\
&= \left\{ \frac{4 f_p}{2\rho} \left(\frac{\dot{m}_s}{a_s} \right)^2 \frac{1}{d_s} + \frac{k_b}{2\rho} \left(\frac{\dot{m}_s}{a_b} \right)^2 \frac{1}{L_b} \right\} \frac{L}{Z^2} \\
&= \{\text{bracket}\} \frac{L}{Z^2}
\end{aligned}
$$

$$L = Z^2 \frac{\Delta p_s}{\{\text{bracket}\}} \tag{7.19}$$

7.11 Direct-sizing

To complete the design it is now appropriate to plot equations (7.14), (7.16), and (7.19) together with abscissa Z and ordinate L and construct a direct-sizing design solution plot (Fig. 7.5). The procedure is to assume a value of Z, and evaluate the three values of L corresponding to equations (7.14), (7.16), and (7.19). Repeating the process as far as is necessary produces Fig. 7.7, staying within the validity limits of the correlations.

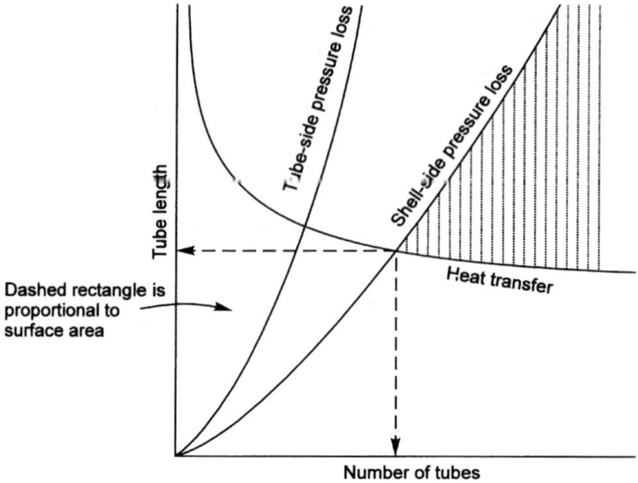

Fig.7.5 Schematic design solution plot. The real direct-sizing plot is Fig. 7.7

It is not known beforehand if shell-side pressure loss or tube-side pressure loss will lie to the right and be 'controlling'. Whichever curve provides the solution tube length may be chosen such that a whole number of shell-side baffles is obtained. This constraint is perhaps not absolute because shell-side flow at entry to and exit from the bundle may be transverse to the tubes.

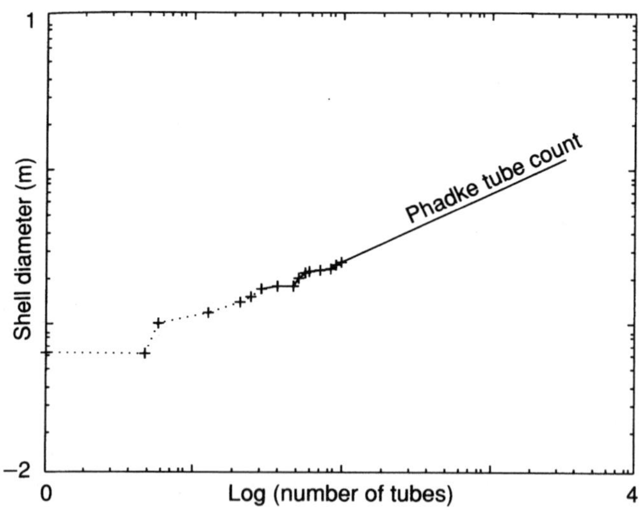

Fig.7.6 Plot of tube count (Z) versus inside shell diameter (D_i)

7.12 Tube-bundle diameter

The diameter of the tube bundle may be determined using the results provided by Phadke (1984). Using data generated from the Phadke expressions a log–log plot of tube count (Z) versus inside shell diameter (D_i) is very nearly a straight line for tube numbers greater than 100 (Fig. 7.6).

For accurate values it is necessary to consult the Phadke paper. In doing so, it is seen that Phadke always assumed one tube at the exact centre of the tube plate. There exists another square pitch tube layout in which there is no tube at the centre of the tube plate. This may give a slightly different tube count.

It is very likely that the calculated value of Z does not provide the correct number of tubes to completely fill the tube plate without leaving gaps. The number of tubes must now be adjusted to an appropriate value.

This may be done either by selecting appropriate values for Z and L which lie in the shaded area of the design solution plot, and then completing a 'rating' design, or alternatively by adjusting values of allowable pressure loss in the 'sizing' design until suitable values of Z and L appear.

7.13 Practical design

The paper by Gentry (1990) provides information on the design of a gas–gas heat exchanger used as a feed preheater in a catalytic incinerator process for a petrochemical plant.

While a substantial amount of information is available, there are some gaps which need filling. As no information was provided as to gas composition the physical properties of nitrogen are used. The specific heat of the cold shell-side fluid is found from a heat balance, allowing an estimation of specific heat for the hot gas. Terminal temperatures in the exchanger can then be found, plus *Ntu* values from the (LMTD–*Ntu*) relationships.

The undernoted data were used in direct-sizing of the heat exchanger surface:

Exchanger specification

Performance:

Exchanger duty, kW	$\dot{Q} = 5450.0$
Log mean temperature difference, K	$\Delta\theta_{lmtd} = 106.2$

Fluid properties

Shell-side fluid (gas)

Mass flowrate, kg/s	$\dot{m}_s = 14.946$
Bulk mean temperature, K	$T_s = 467.0$
Absolute viscosity, kg/(m s)	$\eta_s = 0.000\,0245$
Thermal conductivity, J/(m s K)	$\lambda_s = 0.0366$
Specific heat, J/(kg K)	$C_s = 1027.1$
Prandtl number	$\mathrm{Pr}_s = 0.686$

Tube-side fluid (gas)

Mass flowrate, kg/s	$\dot{m}_t = 15.674$
Bulk mean temperature, K	$T_t = 573.0$
Absolute viscosity, kg/(m s)	$\eta_t = 0.000\,0283$
Thermal conductivity, J/(m s K)	$\lambda_t = 0.0426$
Specific heat, J/(kg K)	$C_t = 1034.8$
Prandtl number	$\mathrm{Pr}_t = 0.685$

Local geometry

RODbaffle configuration ARA

Tube outside diameter, mm	$d = 38.10$
Tube inside diameter, guesstimate, mm	$d_i = 31.75$
Baffle rod diameter, mm	$d_r = 6.35$
Baffle-section spacing, mm	$L_b = 150$

Pressure losses quoted by Gentry were outside the design 'window' for direct-sizing. In direct-sizing pressure loss data are for friction in the heat exchange surface only, and it was then realized that the Gentry data were for total losses in the exchanger. On the tube-side, tube entry and exit effects, flow acceleration, and nozzle outlet losses will exist in addition to flow-friction in the tubes. On the shell-side, nozzle inlet and outlet losses will be present, plus flow distribution losses at the ends, plus flow-friction in the baffled shell-side.

It seemed as if a design comparison would not be possible, but by using maximum values of pressure loss, constrained by the heat-transfer curve in direct-sizing, a good approximation to the tube-bundle design described by Gentry is obtained. Since no data are used beyond the experimental limits given by Gentry, the pressure losses actually determined may be close to those of the exchanger described.

Absolute values of pressure are required to evaluate gas densities, and these are guesstimates.

Shell-side pressure, guesstimate, bar	$p_s = 5.5$
Tube-side pressure, guesstimate, bar	$p_t = 1.1$
Shell-side density, kg/m^3	$\rho_s = 4.1026$
Tube-side density, kg/m^3	$\rho_t = 0.6687$
Shell-side pressure loss, max. possible, N/m^2	$\Delta p_s = 11\,305.46$
Tube-side pressure loss, max. possible, N/m^2	$\Delta p_t = 8493.63$

Direct-sizing results

Figure 7.7 is the actual design solution plot obtained from the computer in which the pressure-loss curves are seen to be coincident.

Leading dimensions

Tube length, m	$L = 11.437$
Number of tubes	$Z = 519$
Number of RODbaffles	$B = 76$

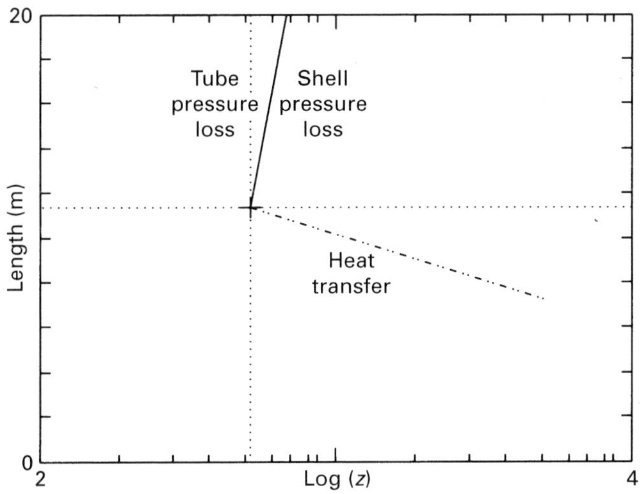

Fig.7.7 Actual direct-sizing solution plot

Reynolds numbers

Shell-side, plain-tubes	$\text{Re}_s = 40\,000.0$
Tube-side	$\text{Re}_t = 43\,601.9$

Heat-transfer coefficients (referred to o.d.)

Shell-side heat-transfer coefficient, $\text{J}/(\text{m}^2\ \text{s K})$	$\alpha_s = 205.39$
Tube-side heat-transfer coefficient, $\text{J}/(\text{m}^2\ \text{s K})$	$\alpha_t = 134.98$
Tube-wall heat-transfer coefficient, $\text{J}/(\text{m}^2\ \text{s K})$	$\alpha_w = 12\,956.3$
Overall heat-transfer coefficient, $\text{J}/(\text{m}^2\ \text{s K})$	$U = 72.27$

Number of transfer units

Ntu shell-side	$N_s = 3.34$
Ntu tube-side	$N_t = 3.16$
Ntu overall	$Ntu = 3.34$

Actual pressure losses

Shell-side, N/m^2	$\Delta p_s = 11\,305.46$
Tube-side, N/m^2	$\Delta p_t = 8493.63$

Phadke shell sizing

Shell/baffle-ring clearance, m	radial gap $= 0.0035$
Baffle-ring/tubes clearance, m	$d_r/2 = 0.0032$
Nearest number of tubes to fill shell	$Z = 529$
Corresponding inside shell diameter, m	$D_i = 1.217$

RODbaffle performance

Volume of exchanger, m^3	$V = 13.304$
Specific performance, $\text{kW}/(\text{m}^3\text{K})$	$\dot{Q}/(V\Delta\theta_m) = 3.857$

Table 7.1 Design comparison

Parameter	*Gentry design*	*Direct-sizing*
Number of tubes	516	519
Shell diameter, m	1.219	1.217
Number of baffles	78	76
Baffled tube length, m	11.582	11.437
Unsupported tube length, m	0.610	—
Total surface area, m^2	753.0	709.94#
Overall heat-transfer coeff., $J/(m^2\ s\ K)$	61.3	72.27#
Shell-side pressure loss, N/m^2	35 700.0*	11 305.46†

#No allowance for leakage flow.
*Total loss.
†Bundle only.

For the results given in Table 7.1 direct-sizing assumes that RODbaffles are spaced over the complete bundle, with 76 baffles compared with 78 in the Gentry design. Some 0.610 m of tube bundle in the Gentry design was said to be unsupported and adding this to the length found in direct-sizing gives 12.047 m, which is within 1.2 per cent of the tube length quoted by Gentry (1990). Leakage flow past the RODbaffle rings may be the reason for lower coefficients calculated by Gentry *et al.* (1982). The shell diameter found using Phadke (1984) would accommodate 529 tubes, but some of this space is taken up by support rings and tie rods.

Shell-side flow between the end RODbaffles and entry/exit nozzles does not necessarily relate to data for conventionally baffled shell-and-tube exchangers, and the writer could not find data for these end-effects. More information is required for shell-side distributional flow between the inlet/outlet nozzles and the end RODbaffles. This is particularly so when an annular vapour belt is provided, for then flow in the tube bundle will be radial.

Apart from the inlet/outlet end discrepancy and the difference in values of the overall heat-transfer coefficients there seems close enough correspondence between the Gentry design and direct-sizing to warrant further investigation of direct-sizing.

7.14 Generalized correlations

Shell-side heat transfer

The generalized correlations proposed by Hesselgreaves (1988) interpret the results for shell-side heat-transfer and pressure loss tests in terms of parameters derived from the detailed geometry of the RODbaffle exchanger. Such generalized correlations can be very useful in optimization, see the Manglik & Bergles (1993) correlations for plate–fin exchangers. For the shell-side, Hesselgreaves found that geometric values of the ratio

$$\frac{D_{hyd}}{(L_b + L')/2} = \frac{\text{hydraulic diameter of intertube space}}{\text{mean of(baffle spacing + vortex spacing*)}}$$

*spacing on one side of the von Kármán street only

were necessary parameters in his correlations. See Daugherty *et al.* (1985), Duncan *et al.* (1975), and White (1986) for discussion of the von Kármán vortex street.

Shell-side heat transfer was reasonably correlated over the whole range of Reynolds number by combining vortex street (laminar) Nusselt number ($\mathrm{Nu_L}$), and turbulent Nusselt number ($\mathrm{Nu_T}$), in the form

$$\boxed{\frac{\mathrm{Nu}}{\mathrm{Pr}^{0.4}} = \left[\left(\frac{\mathrm{Nu_L}}{\mathrm{Pr}^{0.4}}\right)^2 + \left(\frac{\mathrm{Nu_T}}{\mathrm{Pr}^{0.4}}\right)^2\right]^{1/2}}$$

where $\mathrm{Nu_L}$ and $\mathrm{Nu_T}$ are to be found from the undernoted expressions recommended by Gnielinski (1990)

$$\mathrm{Nu_L} = \frac{0.664}{P_r^{1/6}}\sqrt{\left(Pe\frac{d_{hyd}}{L'}\right)}$$

$$\mathrm{Nu_T} = \frac{(f/8)\,\mathrm{Pr}\,(\mathrm{Re} - 1000)}{1 + 12.7\sqrt{\{(f/8)(\mathrm{Pr}^{2/3} - 1)\}}}\left[1 + \left(\frac{d_{hyd}}{L'}\right)^{2/3}\right]$$

The turbulent friction factor was by Chen (1979), viz.

$$\frac{1}{\sqrt{f}} = -2.0\log_{10}\left(\frac{5.0452}{\mathrm{Re}}\log_{10}\frac{5.8506}{\mathrm{Re}}\right)$$

Shell-side baffle pressure loss

Hesselgreaves found that an approximate representation of the drag coefficient (C_D) could be represented by

$$C_D = 1.00 + \frac{3000}{\mathrm{Re}_b}$$

allowing the pressure loss per baffle to be written

$$\boxed{\Delta p = \frac{1}{2}\rho u_b^2 C_D\left(\frac{A_{total}}{A_{flow}}\right)}$$

where

$$\frac{A_{total}}{A_{flow}} = \frac{4p^2}{4p^2 - \pi d^2 - 4pr}$$

may be understood by reference to Fig. 7.1, and u_b is baffle velocity at the minimum flow area.

The attempt at a generalized correlation for baffle pressure loss was less successful, with relatively high scatter in the data as is evident in the figures of the paper by Hesselgreaves (1988). Further work was recommended, but there may now be sufficient information to establish the structure of an optimization procedure.

The generalized data fit for heat transfer appears as good as that found by Manglik & Bergles (1993) for rectangular offset strip fins, but for baffle pressure loss there is considerable scatter. Present designs should perhaps use experimental results for individual RODbaffle geometries which can be spline-fitted.

7.15 Recommendations

All experimental work to date appears to involve testing of RODbaffle geometries complete with baffle rings and RODbaffle ring/shell by-pass leakage. If the proposed concept for dealing with separate leakage flow (Section 7.3, Appendix D.2) is to be accurately assessed, then thermal performance of RODbaffle geometry should be carried out in a square duct with no by-pass leakage. This will provide the 'local' thermal performance, without the unknown contribution due to by-pass leakage.

Bell & Bergelin's (1957) method for calculating by-pass leakage flow may then be applied with more accuracy, and the degradation in performance due to leakage will then be clearly seen.

On the question of tube counts relating to inside shell diameter, it is relevant to enquire whether a second 'Phadke' analysis is possible, starting with four tubes placed symmetrically at the centre of the tube plate.

From heat-transfer coefficients calculated, enhancement of tube-side heat-transfer would produce significant reduction in surface area. The prospect of using twisted-tape inserts has recently been examined by Manglik and Bergles (1993) in two extensive papers and the possibility of using internally rippled tubes seems worth considering.

7.16 Other shell-and-tube designs

Conventionally baffled shell-and-tube

Direct-sizing is not immediately recommended for accurate design of conventional shell-and-tube heat exchangers because of the need to allow for leakage paths in the clearance between tubes and tube holes in the baffle plates, and because of the changing flow directions of shell-side flow traversing the tube bundle.

However, the paper by Polley & Dominique (1994) reports a study of conventionally baffled shell-and-tube exchangers which provides information on optimum tube-bundle layout including values of window–crossflow area ratio, tube pitch/tube o.d. ratio, and tube layout angle. This study was based on use of existing shell-and-tube computer design programs developed by the Heat Transfer and

Fluid Flow Service and D. Butterworth. The design approach generally proceeds through rating of a number of different exchanger configurations.

When unbalanced water equivalents exist, the temperature profiles along a contraflow exchanger will not be parallel, giving different values of overall mean temperature differences between each pair of baffles. A step-wise rating design approach may then become preferable, which also allows end-flow conditions at the shell inlet and outlet nozzles to be treated differently.

Direct-sizing methods could be applied, using the generalized shell-side correlation given by Kern (1950) which is reported in the text by Hewitt *et al.* (1994). The Bell-Delaware method is an extension of the Kern approach and introduces correction factors for shell-side leakage flow paths, by-passing flow and baffle shape. This may give a rapid way of arriving at a first estimate of exchanger size.

Flattened and helically twisted tubes

The helically twisted, flattened-tube heat exchanger discussed in Section 1.6 seems first to have been developed for aircraft applications in Russia as indicated in the texts by Danilov *et al.* (1986) and by Dzyubenko *et al.* (1990). See also Ievlev *et al.* (1982a,b). In parallel the Brown Fintube Company is developing industrial-scale shell-and-twisted-tube heat exchangers (Butterworth *et al.*, 1996). This geometry is equally suited to direct-sizing providing the tube pitching is given equal triangular or square pitching, and the tubes are oriented so that the tubes are aligned along their length.

A more compact design than the RODbaffle configuration is possible, but equally the same uncertainties exist regarding transverse mixed flow at the ends of the bundle near shell-side nozzles or annular inlet/outlet connections. The problem of filling a circular shell with exactly the right number of tubes is the same (Phadke, 1984), and the same problem exists of having extra shell-side flow space next to the shell.

Helically baffled heat exchangers

Norwegian development of the spiral shell-side flow, angle-baffled shell-and-tube heat exchanger was reported by Næss *et al.* (1990), and later by Austergard *et al.* (1993, 1994, 1996). This has now been superseded by more commercial work on the *Helixchanger* concept (Král & Nemcansky, 1996), which differs from the helically baffled exchanger in a number of significant ways. The Helixchanger employs a set of four quadrant segmental baffles, angled to direct flow along the exchanger, each set being repeated along the exchanger. The significant improvement was to arrange for flow leaving one baffle to pass on either side of the next baffle plate, thus minimizing dead space on the rear of a baffle plate.

The Helixchanger has seen service in power plants, chemical plants, oil refineries, and other industrial processes in Czechoslovakia and the former USSR since the 1970s. The design provides near plug flow conditions and reduces dead space within the shell space. Relatively high values of shell-side heat-transfer coefficient, low pressure drop, and low shell-side fouling are also obtained.

The Helixchanger is now under design research and commercial development by ABB Lummus, which should lead to enhanced understanding of the concept.

Stehlík *et al.* (1994) have compared correction factors for shell-and-tube heat exchangers with segmental and helical baffles. Using an extension of the Bell–Delaware method, for helical baffles they found a heat-transfer enhancement 1.39 times that for ideal crossflow, accompanied by a reduction of pressure loss varying from 0.26 to 0.60. Thus performance improvement can be expected.

One question which remains to be settled is whether heat-transfer (and flow friction) is consistent everywhere on the shell-side. If flow mixing is good, then one way to confirm 'goodness' of this design would be to measure the radial (and angular) exit temperature distribution on the tube-side.

7.17 Conclusions

1. A method of direct-sizing of RODbaffle exchangers has been proposed and results from *one* example of the approach have been compared with the design approach recommended by Gentry *et al.* (1982).
2. It would be of considerable assistance in design to have dimensionless plots of baffle loss coefficient and heat-transfer correlation for each RODbaffle configuration, and to have tabulated the recommended smoothed data used in constructing these curves, as is done in the text, *Compact Heat Exchangers*, by Kays & London (1964).
3. Existing design information on shell-side flow between the entry/exit nozzles and the end baffles needs to be assessed and incorporated into the design.
4. Shell-side leakage flow should be examined to assess the prospect of handling losses in a different method from that presently recommended by Gentry *et al.* (1982).
5. In its present form, direct-sizing seems to provide a reasonable method for arriving at the shell diameter, thus allowing the original method of design recommended by Gentry *et al.* to proceed.
6. The helically twisted, flattened-tube exchanger can probably be sized directly in exactly the same way.

References

Austergard, A., **Næss, E.**, and **Sønju, O.K.** (1993) Three dimensional flow modelling of shell-side flow in a novel helical flow shell-and-tube heat exchanger. In *New Developments in Heat Exchangers, ICHMT International Symposium*, Lisbon. Available from Norwegian Institute of Technology, Trondheim.

Austergard, A., **Næss, E.**, and **Sønju, O.K.** (1994) Cross flow pressure drop in tube bundles. Presented at *Open Poster Forum at 10th International Heat Transfer Conference*, Brighton, UK, August 1994. Available from Norwegian Institute of Technology, Trondheim.

Austergard, A., **Næss, E.**, and **Sønju, O.K.** (1996) Three dimensional flow modelling of shell-side flow in a novel helical flow shell-and-tube heat exchanger. *New Developments* *in Heat Exchangers* (Eds. N. Afgan, M. Carvalho, A Bar-Cohen, and D. Butterworth), Gordon & Breach, pp. 487–499.

Bell. K.J. (1955) Annular orifice coefficients with application to heat exchanger design, PhD Thesis, University of Delaware, Newark, USA.

Bell. K.J. and **Bergelin, O.P.** (1957) Flow through annular orifices. *Trans. ASME*, **79**, 593–601.

Bergelin, O.P., **Bell, K.J.**, and **Leighton, M.D.** (1958) Heat transfer and fluid friction during flow across banks of tubes. IV. The effect of internal leakages within segmentally baffled exchangers. *Trans. ASME*, **80**, 53–60.

Butterworth, D., **Guy, A.R.**, and **Welkey, J.J.** (1996) Design and application of twisted-tube heat exchangers. *Advances in Industrial Heat transfer*, IChemE, pp. 87–95.

Chen, N.H. (1979) An explicit equation for friction factor in pipe. *Ind. Engng Chemistry, Fundamentals*, **80**(19), 229–230.

Churchill, S.W. (1977) Comprehensive correlating equations for heat, mass and momentum transfer in fully developed flow in smooth tubes. *Ind. Engng Chemistry, Fundamentals*, **16**, 109–115.

Churchill, S.W. (1988) The role of mathematics in heat transfer. *AIChE Symp. Ser., Heat Transfer*, **84**(263), 1–13.

Churchill, S.W. (1992) The role of analysis in the rate process. *Ind. Engng Chemistry, Res.*, **31**(3), 643–658.

Danilov, Yu.I., **Dzyubenko, B.V.**, **Dreitser, G.A.**, *et al.* (1986) *Teploobmen i gidrodinamika v kanalakh slozhonoy formy* (Heat transfer and hydrodynamics in complex geometry channels), Mashinostroyeniye Moscow.

Daugherty, R.L., **Franzini, J.B.**, and **Finnemore, E.J.** (1985) *Fluid Mechanics with Engineering Applications*, 8th edn, McGraw-Hill, New York pp. 317–319.

Duncan, W.J., **Thom, A.S.**, and **Young, A.D.** (1975) *Mechanics of Fluids*, 2nd edn, Arnold, pp. 204–205.

Dzyubenko, B.V., **Dreitzer, G.A.**, and **Ashmantas, L.-V.A.** (1990) *Unsteady Heat and Mass Transfer in Helical Tube Bundles* (Title refers to helically twisted flattened-tube heat exchangers), Hemisphere, New York.

Gentry, C.C. (1990) RODbaffle heat exchanger technology. *Chem. Engng Prog.*, **86**(7), July, 48–57.

Gentry, C.C., **Young, R.K.**, and **Small, W.M.** (1982) RODbaffle heat exchanger thermal-hydraulic predictive methods. In *Proceedings of the 7th International Heat Transfer Conference*, Munich, 6–10 September 1982, vol. 6, pp. 197–202. (Data for ARA configuration.)

Gnielinski, V. (1990) Forced convection in ducts. *Hemisphere Handbook of Heat Exchanger Design*, Chapter 2.5.1 (Coordinating Ed. G.F. Hewitt), Hemisphere, Washington.

Hesselgreaves, J.E. (1988) A mechanistic model for heat transfer and pressure drop in rod-baffled heat exchangers. In *2nd UK National Conference on Heat Transfer*, University of Strathclyde, 14–16 September 1988, Paper C197/88, pp. 787–800.

Hewitt, G.F., **Shires, G.L.**, and **Bott, T.R.** (1994) *Process Heat Transfer*, Section 3, CRC Press, Florida, pp. 271–292.

Ievlev, V.M., **Dzyubenko, B.V.**, **Dreitser, G.A.**, and **Vilemas, Yu.V.** (1982a) In-line and crossflow helical tube heat exchangers. *Int. J. Heat Mass Transfer*, **25**(3), 317–323.

Ievlev, V.M., **Kalinin, E.K.**, **Danilov, Yu.I.**, **Dzyubenko, B.V.**, and **Dreitser, G.A.** (1982b) Heat transfer in the turbulent swirling flow in a channel of complex shape. In *Proceedings* *of the 7th International Heat Transfer Conference*, Munich, vol. 3, Hemisphere, New York, General papers, pp. 171–176.

Kays, W.M. and **London, A.L.** (1964) *Compact Heat Exchangers*, 2nd edn, McGraw-Hill, New York.

Kern, D.Q. (1986) *Process Heat Transfer*, McGraw-Hill.

Král, D. and **Nemcansky, J.** (1996) The Helixchanger – helically baffled heat exchanger. *New Developments in Heat Exchangers* (Eds. N. Afgan, M. Carvalho, A. Bar-Cohen, and D. Butterworth), Gordon & Breach, pp. 467–477.

Manglik, R.M. and **Bergles, A.E.** (1993) Heat transfer and pressure drop correlations for twisted-tape inserts in isothermal tubes. Part I – Laminar flows. Part II – Transition and turbulent flows. *ASME J. Heat Transfer*, **115**, November, 881–889, and 890–896.

Moody, L.F. (1944) Friction factors in pipe flow. *Trans. ASME*, **66**, 671–684.

Næss, E., **Knusden, H.**, and **Sønjy, O.K.** (1990) Technical note on a novel high performance shell-and-tube heat exchanger. Presented at *Open Poster Forum at the 9th International Heat Transfer Conference*, Jerusalem, 20–24 August 1990. Available from Norweigian Institute of Technology, Trondheim.

Phadke, P.S. (1984) Determining tube counts for shell-and-tube exchangers. *Chem. Engng*, **91**(18), September, 65–68.

Polley, G.T. and **Dominique, D.** (1994) Optimal tube bundle geometries. In *Proceedings of the 10th International Heat Transfer Conference*, Brighton, UK, 14–18 August 1994, *The Industrial Sessions Papers*, Paper 1/1-DES-4, IChemE, pp. 19–24.

Rehme, K. (1992) The structure of turbulence in rod bundles and the implications on natural mixing between the sub-channels. *Int. J. Heat Mass Transfer*, **35**(2), 567–581.

Smith, E.M. (1986) Design of helical-tube multi-start coil heat exchangers. In *Advances in Heat Exchanger Design, ASME Winter Annual Meeting*, Anaheim, California, 7-12 December 1986, ASME HTD-Vol. 66, pp. 95-104 (revised 3rd December 1994).

Smith, E.M. (1994) Direct thermal sizing of plate-fin heat exchangers. In *Proceedings of the 10th International Heat Transfer Conference*, Brighton, UK, 14–18 August 1994, The Industrial Sessions Papers, Institution of Chemical Engineers, Rugby.

Stehlík, P., **Nemcansky, J.**, **Král, D.**, and **Swanson, L.W.** (1994) Comparison of correction factors for shell-and-tube heat exchangers with segmental or helical baffles. *Heat Transfer Engng* **15**(1), January–March, 55–65.

Tong, L.S. (1968) Pressure drop performance of a rod bundle. In *Heat Transfer in Rod Bundles, Symposium at Winter Annual Meeting of ASME*, ASME Nucleonics Heat Transfer Committee, K-13, 1968.

White, F.M. (1986) *Fluid Mechanics*, 2nd edn, McGraw-Hill, New York pp. 262–267.

Bibliography

Barrington, E.A. (1973) Acoustic vibration in tubular exchangers. *Chem. Engng Prog.*, **69**(7), July, 62–68.

Eilers, J.F. and **Small, W.M.** (1973) Tube vibration in a thermosiphon reboiler. *Chem. Engng Prog.*, **69**(7), 57–61.

Gentry C.C. (1995) Experimental databases for RODbaffle geometries. *Private communication*, Phillips Petroleum Company, 3 April 1995.

Gentry, C.C. and **Small, W.M.** (1981) The RODbaffle heat exchanger. In *Proceedings of the 2nd Symposium on Shell and Tube Heat Exchangers*, Houston, Texas, 14–18 September 1981, pp. 389–409. (Data for ARG configuration.)

Gentry, C.C. and **Small, W.M.** (1985) RODbaffle exchanger thermal-hydraulic predictive models over expanded baffle spacing and Reynolds number ranges. In *23rd National Heat Transfer Conference*, Denver, AIChE Symposium Series No. 245, vol. 81.

Gentry, C.C., **Young, R.K.**, and **Small, W.M.** (1980) The RODbaffle heat exchanger thermal-hydraulic performance. In *20th National IMIQ/AIChE Conference*, Acapulco, October 1980.

Gentry, C.C., **Young, R.K.**, and **Small, W.M.** (1982) Conceptual RODbaffle nuclear steam generator design. *IEEE/ASME/ASCE 1982 Joint Power Generation Conference*, 17–21 October 1982, EI Conference no. 02416.

Gentry, C.C., **Young, R.K.**, and **Small, W.M.** (1984) RODbaffled heat exchanger thermal-hydraulic predictive methods for bare and low-finned tubes. In *AIChE Symposium Series, No. 236, Heat Transfer*, Niagara Falls, 1984, vol. 80, 104–109.

Jegede, F.O. and **Polley, G.T.** (1992) Optimum heat exchanger design. *Trans. IChemE*, **70**(A), March, 133–141.

Miller, P., **Byrnes, J.J.**, and **Benforado, D.M.** (1956) Heat transfer to water flowing parallel to a rod bundle. *J. AIChE*, **2**(2), 226–234. (Triangular pitch only – nuclear fuel elements.)

Peters, M.S. and **Timmerhaus, K.D.** (1958) *Plant Design and Economics for Chemical Engineers*, 3rd edn, McGraw-Hill, New York, chapter on Heat Transfer Equipment Design and Costs, pp. 678–710. (Chapter on shell-and-tube optimization.)

Polley, G.T., **Panjeh Shahi, M.H.**, and **Picon Nunez, M.** (1991) Rapid design algorithms for shell-and-tube and compact heat exchangers. *Trans. IChemE*, **69**(A), November, 435–444.

Roetzel, W. (1994) Effect of baffle/shell leakage flow on heat transfer in shell-and-tube heat exchangers. *Exp. Thermal Fluid Sci.*, **8**(1), January, 10-20.

Roetzel, W. and **Lee, D.** (1993) Experimental investigation of leakage in shell-and-tube heat exchangers with segmental baffles. *Int. J. Heat Mass Transfer*, **36**(15), 3765–3771.

Sangster, W.A. (1968) Calculation of rod-bundle pressure loss. *ASME Paper no. 68-WA/HT-35.*

Small, W.M. and **Young, R.K.** (1979) The RODbaffle heat exchanger (design and operations). *Heat Transfer Engng*, **1**(2), October–December, 21–27.

Smyth, R. (1981) Comparative assessment of RODbaffle shell-and-tube heat exchangers. *Heat Transfer Engng*, **2**(3–4), January–June, 90–94.

Sun, S.-Y., **Lu, Y.-D.**, and **Yan, C.-Q.** (1993) Optimisation in calculation of shell and tube exchanger. *Int. Comm. Heat Mass Transfer*, **20**(5), September–October, 675–686.

CHAPTER 8
Exergy Loss and Pressure Loss

Exergy loss may define optimum temperature profiles. Pressure loss control ensures good flow distribution

EXERGY LOSS

8.1 Objective

There is a need for unambiguous standards for measuring the performance of heat exchangers. Best temperature recovery is easy to understand, minimization of avoidable work requires more thought.

In contraflow, minimization of *entropy generation* produces a temperature pinch point at the hot end, while minimization of *exergy loss* produces a temperature pinch point at the cold end. Reaching particular temperature levels is important for best chemical reaction rates in chemical process plants. Reduction of avoidable work loss is important in cryogenic liquefaction plant and in power production plant.

Exergy is a measure of the potential for work production at any stage of a thermodynamic process. Exergy loss will be seen to depend on thermodynamic inlet and outlet conditions for any process, and in this sense it applies to all contraflow, parallel flow, and crossflow heat exchangers. Internal or terminal fluid mixing situations existing in crossflow are considered part of exchanger design. Where conceptual difficulties may arise, these are due to any heat or fluid leakage.

For contraflow, minimization of the exergy loss number corresponds exactly to achieving the Grassmann & Kopp optimum dimensionless local temperature ratio everywhere in the exchanger

$$\frac{\text{local temperature difference}}{\text{local hot (or cold) absolute temperature}} = \frac{\Delta\theta}{T_h} = a = const, \tag{8.1}$$

where a may be made as small as may be convenient.

A case is made for adopting the exergy loss number [equation 8.6] as parameter for characterizing the performance of heat exchangers, especially in plant involving work production or work consumption.

8.2 Historical development

A number of authors have developed expressions describing heat exchanger performance. Some have studied minimum entropy creation, others minimum exergy loss. These two parameters provide appropriate headings under which to assess previous work.

Minimum entropy generation

- Bejan (1993) defined a dimensionless performance ratio $N_s = (\dot{S}_{gen}/\dot{m}C_p)$, and developed a closed-form solution for the case of a balanced contraflow exchanger involving temperatures and pressures, viz.

$$N_s = (1-\varepsilon)\frac{(T_2 - T_1)^2}{T_1 T_2} + \frac{R}{C_p}\left[\left(\frac{\Delta p}{p_1}\right) + \left(\frac{\Delta p}{p_2}\right)\right] \tag{8.2}$$

  He was only able to do this because balanced heat capacities reduced by one the number of unknowns in the analysis (effectiveness ε replaced two outlet temperatures). The presence of rate of entropy loss evident in his starting expression allows placement of Bejan's analysis in this class.

- Witte & Shamsunder (1983) chose a dimensionless efficiency expression

$$\eta = 1 - \left(\frac{\dot{S}_{gen}}{\dot{Q}}\right) \times T_0 \tag{8.3}$$

  in which the minimum entropy rate is present. They concluded that the highest efficiency was obtained when the heat capacity rates were equal (see Bejan above). When the capacity rates were not equal, better efficiency was obtained when the hot fluid had the largest capacity, namely when a hot-end pinch point existed.

- Hesselgreaves (2001) selected as his dimensionless parameter

$$N_{s1} = \left(\frac{\dot{S}_{gen}}{\dot{Q}}\right) \times T_1 \tag{8.4}$$

  where the cold inlet temperature (T_1) was used to replace the dead-state temperature T_0 to ease subsequent analysis. Hesselgreaves' findings about minimizing the rate of entropy creation $\dot{S}_{gen}$ required the hot fluid to have the maximum heat capacity. Hesselgreaves also noted that only in the case of equal heat capacities could the value of N_{s1} approach zero.

A good way of assessing the above findings is to consider the combined plot of effectiveness (ε) versus parameters involving number of transfer units ($N_c - N_h$) for contraflow, ($N_c + N_h$) for parallel flow (Fig. 2.16). The limiting effectiveness for parallel flow is $N_c/(N_c + N_h)$. Only in the case of contraflow does effectiveness approach unity for every ratio of heat capacity rates (N_h/N_c), at which point the temperature difference must approach zero, i.e. in the limit *the best exchanger is a fictitious exchanger.*

Minimum exergy loss

- London & Shah (1983) considered a number of different system applications in their paper, but the expression obtained for exchanger heat transfer and fluid friction irreversibilities was

$$\left.\frac{Ir\dot{r}ev_E}{\dot{q}}\right|_{\dot{q},\Delta p} = \left[\left(\frac{s_{out} - s_{in}}{h_{out} - h_{in}}\right)_C - \left(\frac{s_{in} - s_{out}}{h_{in} - h_{out}}\right)_H\right] \times T_{WF} \tag{8.5}$$

where T_{WF} is a temperature-weighting factor to be chosen by the analyst. They noted that when T_{WF} was put equal to the dead-state temperature T_0 the irreversibility expression reduced to the exergy loss expression (8.13) of the present chapter. London & Shah also mention heat and fluid leaks. Later in this chapter heat and fluid leaks are assessed, some types of leak could be accommodated analytically while others introduced conceptual difficulties.

- The author's approach was to minimize the dimensionless exergy loss number

$$N_x = \left(\frac{\Delta\dot{B}}{\dot{Q}}\right) = \left[-\left(\frac{b_{out} - b_{in}}{h_{out} - h_{in}}\right)_{hot} + \left(\frac{b_{out} - b_{in}}{h_{out} - h_{in}}\right)_{cold}\right] \tag{8.6}$$

which expression includes the dead-state temperature T_0 quite naturally. The Grassmann & Kopp optimum temperature profile was found compatible with the expression for minimum exergy loss, *giving non-fictitious design in all cases.*

Reference temperature

London & Shah claimed that T_0 was a special case of T_{WF} where the latter is a temperature weighting-factor to be specified by the analyst. There is good reason for using T_0, first because it arrives naturally in the exergy expression, and second because it should be specified by the analyst.

In design of plant required to operate under both Arctic and Equatorial conditions, different values for T_0 can be specified for each region, the plant being required to come up to a different specification for both these duties. In outer space, the value of T_0 may be 3 or 4 K, and it would be inappropriate to apply such a value to most Earth-bound plant.

In cryogenic plant design it is useful to keep to a single value of T_0 for all components. The relative importance of each heat exchanger down towards liquefaction then becomes apparent. Additionally, simply multiplying the dimensionless exergy loss number by thermal duty then gives the actual exergy loss of the component under investigation.

8.3 Exergy change for any flow process

Consider a finite fluid source with temperature T and pressure p. Heat (δ_q) is negative when taken from the fluid, while ($T - T_0$) is positive. Work ($p\delta_v$) is positive when taken from the fluid, while $(p - p_0)$ is positive. The opposite is true when heat is delivered to the finite fluid source.

For unit mass of fluid, reversible abstraction of dq units of heat from the fluid will give $(T - T_0/T)(-d\dot{q})$ units of work from a Carnot engine operating between T and T_0.

The *specific available work* is then

$$\begin{aligned} da &= -\left(\frac{T - T_0}{T}\right) d\dot{q} + (p - p_0)dv \\ &= -d\dot{q} + T_0\left(\frac{d\dot{q}}{T}\right) + p\,dv - p_0\,dv \\ &= (-d\dot{q} + p\,dv) + T_0\,ds - p_0\,dv \end{aligned}$$

By the First law $(dq = de + p\,dv)$, thus

$$da = -de + T_0\,ds - p_0\,dv$$

Integrating between state (1) and the reference (dead) state (0) the *specific availability* for state (1) becomes

$$\begin{aligned} a_1 - a_0 &= -\int_1^0 de + T_0 \int_1^0 ds - \int_1^0 p_0\,dv \\ &= -(e_0 - e_1) + T_0(s_0 - s_1) - p_0(v_0 - v_1) \\ &= (e_1 - T_0 s_1 + p_0 v_1) - (e_0 - T_0 s_0 + p_0 v_0) \end{aligned}$$

The *specific exergy difference* between an initial state (1) and a final state (2) is the difference between specific availabilities of the two states

$$\begin{aligned} b &= a_2 - a_1 \\ &= (e_2 - T_0 s_2 + p_0 v_2) - (e_1 - T_0 s_1 + p_0 v_1) \\ &= (e_2 - T_0 s_2) - (e_1 - T_0 s_1) + p_0(v_2 - v_1) \end{aligned}$$

For a flow process, displacement work terms have to be added. For fluid (1) this is $p_1 v_1$ less the work expended on the surroundings $p_0 v_1$. The exergy difference for a flow process thus eliminating dead-state pressure terms, viz.

$$\begin{aligned} \Delta b &= b_2 - b_1 \\ &= [(e_2 - T_0 s_2) - (e_1 - T_0 s_1) + p_0(v_2 - v_1)] + [(p_2 - p_0)v_2 - (p_1 - p_0)v_1] \\ &= (e_2 - T_0 s_2) - (e_1 - T_0 s_1) + (p_2 v_2 - p_1 v_1) \\ &= (h_2 - T_0 s_2) - (h_1 - T_0 s_1) \end{aligned}$$

Kinetic energy (pure work) terms involving inlet and exit fluid velocities may be added

$$\Delta b = (h_2 - T_0 s_2) - (h_1 - T_0 s_1) + \left[\frac{(u_2)^2}{2} - \frac{(u_1)^2}{2}\right] \tag{8.7}$$

In many engineering applications velocities at inlet and exit are nearly equal. The numerical magnitude of these terms is small in any case, and they are usually neglected.

8.4 Exergy loss for any heat exchanger

For any two-stream exchanger

$$\Delta b_{hot} = [(h_{out} - T_0 s_{out}) - (h_{in} - T_0 s_{in})]_{hot} \tag{8.8}$$

$$\Delta b_{cold} = [(h_{out} - T_0 s_{out}) - (h_{in} - T_0 s_{in})]_{cold} \tag{8.9}$$

The hot stream suffers a loss of exergy, while the cold stream shows a gain in exergy. The loss is always greater than the gain.

Total exergy change rate for exchanger is

$$\Delta \dot{B} = \dot{m}_{hot} \Delta b_{hot} + \dot{m}_{cold} \Delta b_{cold} \tag{8.10}$$

$$\Delta \dot{B} = [-\dot{m}_{hot}(s_{out} - s_{in})_{hot} + \dot{m}_{cold}(s_{out} - s_{in})_{cold}] \times T_0$$

Total heat transfer rate for exchangers without heat leak

$$\dot{Q} = \dot{m}_{hot}(h_{in} - h_{out})_{hot} = \dot{m}_{cold}(h_{out} - h_{in})_{cold} \tag{8.11}$$

Heat leakage usually cannot be evaluated until the exchanger has been sized, and discussion of this aspect is covered in a later section. The first part of the paper is restricted to discussion of the case for zero heat leakage.

For a two-stream exchanger the dimensionless measure of performance (including a sign change to make the exergy loss positive)

$$N_x = \frac{\Delta \dot{B}}{\dot{Q}} = \left[-\left(\frac{b_{out} - b_{in}}{h_{out} - h_{in}}\right)_{hot} + \left(\frac{b_{out} - b_{in}}{h_{out} - h_{in}}\right)_{cold}\right] \tag{8.12}$$

which simplifies to

$$N_x = \frac{\Delta \dot{B}}{\dot{Q}} = \left[-\left(\frac{s_{out} - s_{in}}{h_{out} - h_{in}}\right)_{hot} + \left(\frac{s_{out} - s_{in}}{h_{out} - h_{in}}\right)_{cold}\right] \times T_0 \tag{8.13}$$

The exergy loss number is a function of the terminal conditions, and in this respect is independent of temperature profiles and exchanger type. It is not quite independent of heat capacity rates as these affect temperature differences and thus enthalpy and entropy values. If evaluation of N_x is required it would need interpolation of thermodynamic properties in two dimensions, unless ideal gases are involved.

The expressions in plain brackets are independent of the order in which they are taken, providing numerator and denominator receive the same treatment. This makes equation (8.12) immediately applicable to all two-stream contraflow, parallel flow, and crossflow exchangers, and allows further extension to multi-stream exchangers.

Instantaneous exergy loss

The differential form of equation (8.13) applies to differential length (i.e. zero pressure loss)

$$N_x = \left[-\left(\frac{\partial s}{\partial h}\right)_{hot} + \left(\frac{\partial s}{\partial h}\right)_{cold} \right] \times T_0 \tag{8.14}$$

Maxwell's relationships (see Kestin 1978) allow the instantaneous exergy loss number to be expressed as $T = (\partial h/\partial s)_p$, then

$$N_x = \left[-\left(\frac{1}{T_h}\right) + \left(\frac{1}{T_c}\right) \right] \times T_0 = \left[\frac{T_h - T_c}{T_h} \right] \times \left(\frac{T_0}{T_c}\right)$$

$$\text{Carnot} \times \text{Adjustment} \tag{8.15}$$

The adjustment ratio reveals the dependence of exergy loss number on absolute temperature level. Pressure terms are absent from this instantaneous expression.

Multi-stream exchangers

The exergy loss number for multi-stream exchangers can be constructed by extending the fundamental equations (8.10) and (8.11), thus

$$N_x = \left[-\left\{ \frac{\sum_1^n (\dot{m}\Delta b)}{\sum_1^n (\dot{m}\Delta h)} \right\}_{hot} + \left\{ \frac{\sum_1^m (\dot{m}\Delta b)}{\sum_1^m (\dot{m}\Delta h)} \right\}_{cold} \right] \tag{8.16}$$

where the number of hot fluid flows (n) need not equal the number of cold fluid flows (m), always providing the following heat balance is correct.

$$\dot{Q} = \left[\sum_1^n (\dot{m}\Delta h) \right]_{hot} = \left[\sum_1^m (\dot{m}\Delta h) \right]_{cold} \tag{8.17}$$

The equations for multi-stream exchangers should remain valid for fluids that have inlets or outlets somewhere along the length of the exchanger, but this situation has not been tested.

8.5 Contraflow exchangers

Consider the heat exchanger of Table 8.1. First evaluate exergy loss for the hot fluid, and exergy gain for the cold fluid. Using equations (8.8) and (8.9) rewritten

Table 8.1 Terminal conditions for gas turbine recuperator

Parameter	*Inlet*	*Outlet*
Hot mass flowrate, kg/s	$\dot{m}_h = 6.9976$	
Hot fluid pressure, bar	$p_{h1} = 1.100\,00$	$p_{h2} = 1.056\,00$
Hot fluid temperature, K	$T_{h1} = 825.0$	$T_{h2} = 564.4$
Hot mean specific heat, J/(kg K)	$C_h = 1096.74$	
Hot gas constant, J/(kg K)	$R_h = 287.07$	
Cold mass flowrate, kg/s	$\dot{m}_c = 6.1282$	
Cold fluid pressure, bar	$p_{c2} = 14.000\,00$	$p_{c1} = 13.972\,84$
Cold fluid temperature, K	$T_{c2} = 475.0$	$T_{c1} = 776.0$
Cold mean specific heat, J/(kg K)	$C_c = 1084.25$	
Cold gas constant, J/(kg K)	$R_c = 287.07$	
Dead-state temperature, K	$T_0 = 300.0$	

for ideal gases

$$\begin{aligned}\Delta\dot{b} &= [(h_{out} - T_0 s_{out}) - (h_{in} - T_0 s_{in})]\\ &= [C\{(T_{out} - T_{in}) - T_0 \ln(T_{out}/T_{in})\}_{temp} + \{RT_0 \ln(p_{out}/p_{in})\}_{press}]\end{aligned}$$

then

$$\begin{cases} \Delta\dot{B}_{hot} = \dot{m}_{hot}\Delta\dot{b}_{hot} = \{-1\,999\,987.2 + 874\,023.9\}_{temp} + \{-24\,601.0\}_{press} \\ \Delta\dot{B}_{cold} = \dot{m}_{cold}\Delta\dot{b}_{cold} = \{+1\,999\,994.8 - 978\,411.4\}_{temp} + \{-1\,024.9\}_{press} \end{cases}$$

As expected, the only exergy gain involves the cold fluid temperature bracket.

The canonical exergy loss equation (8.13) for heat exchangers for ideal gases could be written

$$N_x = \left[-\left\{ \frac{+\ln(T_{out}/T_{in}) - \dfrac{R}{C}\ln(p_{out}/p_{in})}{T_{out} - T_{in}} \right\}_{hot} + \left\{ \frac{+\ln(T_{out}/T_{in}) - \dfrac{R}{C}\ln(p_{out}/p_{in})}{T_{out} - T_{in}} \right\}_{cold} \right] \times T_0 \tag{8.18}$$

Evaluating the undernoted expression shows a negative value of exergy change, i.e. a loss of exergy

$$N_x = \frac{\Delta\dot{B}_{hot} + \Delta\dot{B}_{cold}}{\dot{Q}_{mean}} = \frac{-1\,150\,564.3 + 1\,020\,558.5}{1\,999\,990.8} = -0.065\,03 \text{ (6.50\% loss)}$$

Optimum temperature profiles

We explore the effect of imposing optimum temperature profiles on the exchanger of Table 8.1. The Grassmann & Kopp (1957) condition for optimum temperature profiles (Chapter 2, Section 2.11 and Appendix F) leads to the following consistent expressions

$$\frac{\Delta\theta}{T_h} = \frac{a}{1+a}, \qquad \frac{\Delta\theta}{T_c} = \frac{a}{1} \tag{8.19}$$

where a is a constant.

We maintain the cold fluid terminal temperatures at $T_{c1} = 776\,\text{K}$ and $T_{c2} = 475\,\text{K}$, and the design log mean temperature difference at $\Delta\theta_{LM} = 67.188\,\text{K}$ to keep the thermal duty constant. Valid hot fluid terminal temperature pairs are found by inserting values of T_{h2} in the expression for log mean temperature difference to find T_{h1}.

The exergy loss number for temperatures alone (zero pressure losses) is calculated from

$$N_x = \left[-\frac{\ln(T_{h1}/T_{h2})}{T_{h1} - T_{h2}} + \frac{\ln(T_{c1}/T_{c2})}{T_{c1} - T_{c2}} \right] \times T_0 \tag{8.20}$$

and by iteration the minimum exergy loss number for temperature alone was found to be $N_x = 0.048\,370$, occurring at the Grassmann & Kopp ratio

$$\left(\frac{\Delta\theta_2}{T_{c2}} = \frac{52.065}{475} \right) = \left(\frac{\Delta\theta_1}{T_{c1}} = \frac{84.989}{776} \right) = a = 0.1095$$

Thermal duty remained the same, but the ratio of heat capacities for the original exchanger was changed from 1.155 02 to

$$\frac{\dot{m}_h C_h}{\dot{m}_c C_c} = \frac{T_{c1} - T_{c2}}{T_{h1} - T_{h2}} = 0.901\,106 \qquad \text{at minimum exergy loss}$$

Results would be slightly different if we had chosen to fix the hot fluid terminal temperatures, while varying cold fluid terminal temperatures, but similar conclusions would be reached.

For temperatures alone, the original exergy loss number was $N_x = 0.052\,192$, to be compared with a minimum exergy loss number of $N_x = 0.048\,370$. Improvement is 7.32 per cent for a cold inlet temperature of 475 K, but the operational temperature level can make a significant difference, as will be shown in the next section.

8.6 Dependence of exergy loss number on absolute temperature level

For simplicity, the temperature of operation of an exchanger will here be taken as the temperature of the cold fluid entering the exchanger. To obtain a conceptual picture of what the absolute temperature level of operation does to the exergy loss number we shall use equation (8.20) in which pressure losses are neglected.

For an optimum contraflow exchanger governed by the results of the Grassmann & Kopp equation, the following arbitrary choices are made:

- fix a value for the cold fluid temperature rise (100 K)
- fix a value for the constant ($a = 0.1$)
- choose a value for the cold fluid temperature inlet, say $T_{c2} = 20\,\text{K}$.

Then

$$T_{c2} = 20\,\text{K}, \qquad T_{c1} = 120\,\text{K}, \qquad T_{h1} = 132\,\text{K}, \qquad T_{h2} = 22\,\text{K}, \qquad T_0 = 300\,\text{K}$$

$$\frac{\dot{m}_h C_h}{\dot{m}_c C_c} = \frac{T_{c1} - T_{c2}}{T_{h1} - T_{h2}} = 0.909\,09, \qquad N_x = 0.488\,66, \qquad \Delta\theta_{LM} = 5.5811\,\text{K}$$

Keeping the cold fluid terminal temperatures and log mean temperature difference fixed, we allow the values of T_{h2} to vary, and calculate the corresponding values of T_{h1}. For each value of T_{h2} chosen we evaluate the exergy loss number (N_x), and the heat capacity ratio $(\dot{m}_h C_h)/(\dot{m}_c C_c)$ required. The data thus found are used to plot a single curve of heat capacity ratio against exergy loss number.

The process is then repeated choosing a different value of the cold inlet temperature (T_{c2}) for each new curve. The results of following this process are shown in Fig. 8.1.

Interpretation of diagram

Each curve on the plot of Fig. 8.1 is labelled with the cold fluid inlet temperature as parameter. With the dead state defined at 300 K, the cryogenic region lies below 150 K, and the heat pump/refrigerator region lies between 300 and 150 K (for these choices of temperature ranges see Fig. 2.21).

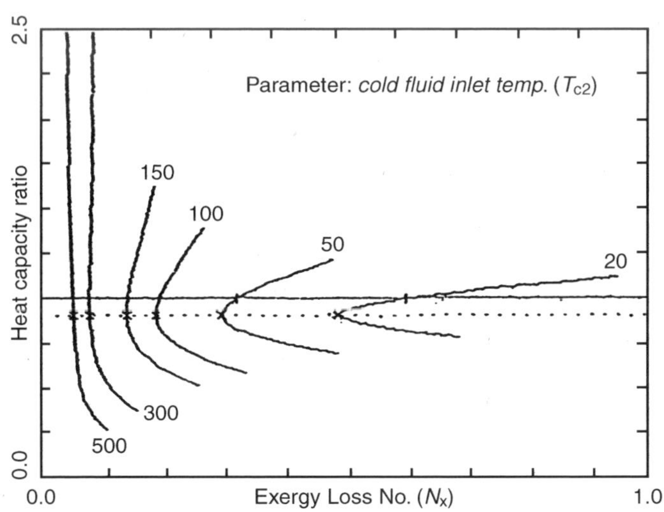

Fig.8.1 Plot of heat capacity ratio $(\dot{m}_h C_h)/(\dot{m}_c C_c)$ versus exergy loss number (N_x) for contraflow exchangers with cold inlet temperature (T_{c2}) as parameter

The balanced contraflow exchanger condition $(\dot{m}_h C_h)/(\dot{m}_c C_c) = 1.000$ is denoted by the horizontal solid line. The optimum contraflow exchanger condition is denoted by the horizontal dotted line $(\dot{m}_h C_h)/(\dot{m}_c C_c) = 0.909\,09$.

The constant value of $a = 0.1$ used in equations (8.19) ensured that the log mean temperature difference $(\Delta\theta_{LM})$ increased with the absolute temperature of operation.

Above 300 K the choice of heat capacity ratio did not affect the value of N_x over a wide range. In the heat pump range 300–150 K the validity of Bejan's analysis becomes marginal. Below 150 K it becomes essential to design for minimum N_x. However, the exergy loss number is fundamental over the complete temperature range.

The cold fluid temperature range $(T_{c1} - T_{c2})$ was kept fixed at 100 K in Fig. 8.1. Suppose we now half this to 50 K, keeping the lowest value of $T_{c2} = 20$ K and the value of $a = 0.1$, then

$$T_{c2} = 20\,\text{K}, \quad T_{c1} = 70\,\text{K}, \quad T_{h2} = 22\,\text{K}, \quad T_{h1} = 77\,\text{K}, \quad T_0 = 300\,\text{K}$$

$$\frac{\dot{m}_h C_h}{\dot{m}_c C_c} = \frac{T_{c1} - T_{c2}}{T_{h1} - T_{h2}} = 0.909\,09, \quad N_x = 0.683\,33, \quad \Delta\theta_{LM} = 3.9912\,\text{K}$$

Note the increase in exergy loss number.

Suppose we now calculate the performance of the second half of the original exchanger, keeping the original value of $a = 0.1$, then

$$T_{c2} = 70\,\text{K}, \quad T_{c1} = 120\,\text{K}, \quad T_{h2} = 77\,\text{K}, \quad T_{h1} = 132\,\text{K}, \quad T_0 = 300\,\text{K}$$

$$\frac{\dot{m}_h C_h}{\dot{m}_c C_c} = \frac{T_{c1} - T_{c2}}{T_{h1} - T_{h2}} = 0.909\,09, \quad N_x = 0.294\,00 \quad \Delta\theta_{LM} = 9.2765\,\text{K}$$

Note the reduction in exergy loss number, but the associated increase in log mean temperature difference.

This result suggests that the value of a can be reduced as absolute temperatures rise.

8.7 Performance of cryogenic plant

In liquefaction plant, the use of cryo-expanders allows optimum temperature distributions to be achieved in contraflow heat exchangers. Three prior constraints in choice of temperature span have to be considered, viz.:

- reduction of the temperature rise of the heat exchanger to the point where approximately linear temperature profiles exist (step-wise design preferred),
- the expansion limit imposed on cryo-expanders, which is set out for typical monatomic and diatomic gases in Table 11.2 of Chapter 11,
- the need to match the compression ratio of the final multi-stage compressor.

Balancing the sequence of multi-stage compressors is not for optimal compression alone. The final compression ratio has to match the allowable expansion ratio of the cryo-expanders. The choice of preceding compression ratios is directly related to the choice of Joule–Thompson expansion stages further down the liquefaction system – see Fig. 11.3, minimization of total compression work being the major target.

Work produced by cryo-expanders is small relative to the work required by the main compressors, and it is frequently dissipated using a small brake wheel [see Fig. 8.5.7. of Gosney (1982)]. Braking work output is carefully insulated from the refrigeration and product streams of the liquefaction plant.

Now let us consider relationships between the Grassmann & Kopp parameters, the log mean temperature difference, and the exergy loss number. We have from Grassmann & Kopp (1957) the local temperature relationships

$$\Delta\theta = T_h - T_c, \qquad \frac{\Delta\theta}{T_h} = \frac{a}{1+a}, \qquad \frac{\Delta\theta}{T_c} = \frac{a}{1}, \qquad \frac{T_h}{T_c} = 1 + a \tag{8.21}$$

From log mean temperature difference

$$\begin{aligned} \Delta\theta_{LM} &= \frac{\Delta\theta_1 - \Delta\theta_2}{\ln(\Delta\theta_1/\Delta\theta_2)} \\ &= \left(\frac{a}{1+a}\right)\left[\frac{T_{h1} - T_{h2}}{\ln(T_{h1}/T_{h2})}\right] = \left(\frac{a}{1}\right)\left[\frac{T_{c1} - T_{c2}}{\ln(T_{c1} - T_{c2})}\right] \end{aligned} \tag{8.22}$$

From exergy loss number

$$N_x = \left[-\frac{\ln(T_{h1}/T_{h2})}{T_{h1} - T_{h2}} + \frac{\ln(T_{c1}/T_{c2})}{T_{c1} - T_{c2}}\right] \times T_0 = \left(\frac{a^2}{1+a}\right)\frac{T_0}{\Delta\theta_{LM}} \tag{8.23}$$

noting that there remains a functional connection between the constant a and the log mean temperature difference $\Delta\theta_{LM}$.

Substituting from equation (8.22) into equation (8.23), then

$$N_x = \left(\frac{a^2}{1+a}\right)\frac{T_0}{\Delta\theta_{LM}} = \left(\frac{a^2}{1+a}\right)\left(\frac{1+a}{a}\right)\left[\frac{\ln(T_{h1}/T_{h2})}{T_{h1} - T_{h2}}\right] \times T_0 = a\left[\frac{\ln(r)}{r-1}\right]\frac{T_0}{T_{h2}}$$

where $r = T_{h1}/T_{h2}$. Differentiating N_x with respect to r

$$\frac{dN_x}{dr} \propto \frac{1}{(r-1)^2}\left[1 - \frac{1}{r} - \ln(r)\right] \tag{8.24}$$

The minimum value of $r = T_{h1}/T_{h2}$ is unity, and slightly larger values reveal that the ratio of T_{h1}/T_{h2} (measure of duty) should be kept as small as practicable consistent with the three requirements for design of cryogenic plant given above. Also the log mean temperature difference (or the value of a) for each exchanger should be kept to as small a value as is practicable to minimize exergy loss.

Thermal design of individual cryogenic exchangers needs be carried out in step-wise fashion to accommodate such temperature departures from the optimum profiles as may still remain, see Chapter 11.

8.8 Allowing for leakage

Fluid leakage

There exist at least four basic leakage modes for two-stream exchangers, viz.:

- leakage of hot fluid to surroundings
- leakage of cold fluid to surroundings
- leakage of hot fluid to cold fluid
- leakage of cold fluid to hot fluid

The last two possibilities depend on which fluid has the greater pressure. The situation gets more complicated if leakage is both to surroundings and to the other fluid. Fluid leakage in multi-stream exchangers leads to especially difficult assessment problems, and frequently the block exchanger has to be scrapped.

For two-stream recuperators with fluid leakage, providing all inlet and outlet mass flowrates are known, and that any mixing is complete, it should be possible to define exergy loss from first principles using only terminal values of thermodynamic properties. It would also seem appropriate to compare such a value with that for the design non-leakage situation.

Heat leakage

For two-stream contraflow, allowance for longitudinal conduction has been treated in Section 3.4. For one-pass crossflow a similar treatment is presented in Section 3.8. For transient contraflow, appropriate terms could be included in the full set of seven simultaneous equations for transient performance which are set out in Appendix A.2 (such solution not so far tested). For contraflow multi-stream exchangers, cross-conduction effects can arise, and the treatment of these is discussed in Section 11.5.

For heat leakage from the outside shell to the environment, and heat leakage at the two ends of the separating wall (longitudinal conduction) for steady-state heat transfer, Jakob (1957, Vol.II, Section 34-4) provides analytical treatments.

Heat leakage in simple contraflow

For a two-stream contraflow exchanger:

- above the dead state, if heat leakage exists from the hot fluid to the surroundings, then it would have to be calculated separately before it could appear in the exergy loss equation,
- below the dead state, if heat leakage exists from the cold fluid to the surroundings then it would have to be calculated separately before it could appear in the exergy loss equation.

Excepting the above constraints, two heat balance equations could be written

$$\left.\begin{aligned} \dot{Q}_{hot} &= \dot{m}_{hot}(h_{in} - h_{out})_{hot} \\ \dot{Q}_{cold} &= \dot{m}_{cold}(h_{out} - h_{in})_{cold} \end{aligned}\right\} \tag{8.25}$$

If heat leakage is present, terms will have different signs above and below the dead state, and the heat balance expressions including a heat leak term might be written

$$\left.\begin{aligned} \dot{Q}_{hot} &= \dot{Q}_{cold} + \dot{Q}_{leak} \quad \text{for exchangers above the dead state} \\ \dot{Q}_{cold} &= \dot{Q}_{hot} + \dot{Q}_{leak} \quad \text{for exchangers below the dead state} \end{aligned}\right\} \tag{8.26}$$

Once the heat leak is identified, its commercial cost can be evaluated separately from assessment of exergy loss number, and will take the difference between the heat leak temperature and the dead-state temperature into account.

The problem of placing a magnitude on heat leak can be straightforward with traditional shell-and-tube exchangers when above the dead state the colder fluid is the shell-side fluid, and below the dead state when the hotter fluid is the shell-side fluid. In compact plate–fin exchangers this distinction is absent, and reliance may have to be placed on measurement of inlet and outlet temperature and mass flowrate conditions from the exchanger.

In the case of cryogenic exchangers the cost of heat leak is very high. The cold end of the exchanger column is often placed inside an evacuated 'cold-box' to minimize radiation losses. Radiation shields and super-insulation may become appropriate.

The relative importance of any heat leak may be gauged by comparing the modulus of Carnot efficiency versus absolute temperature level in the range 0–1000 K (see Fig. 2.21).

When there is imbalance in hot and cold thermal duties, the effect of heat leakage can be accommodated in the exergy loss number, and it seems appropriate to use the heat flow 'on offer' as the appropriate value for $\dot{Q}$ in the exergy loss number.

Above the dead state, the hot fluid is the fluid with heat 'on offer', and $\dot{Q} = \dot{Q}_{hot}$. Below the dead state, the cold fluid is the fluid with coolth 'on offer', and $\dot{Q} = \dot{Q}_{cold}$.

Consider the first case, then

$$\begin{aligned} N_x = \frac{\Delta \dot{B}}{\dot{Q}_{hot}} &= \left[\frac{+\dot{m}_{hot}\Delta b_{hot} + \dot{m}_{cold}\Delta b_{cold}}{\dot{m}_{hot}(h_{in} - h_{out})_{hot}}\right] \\ &= \left[-\left\{\frac{\Delta b_{hot}}{(h_{out} - h_{in})_{hot}}\right\} + \left(\frac{\dot{m}_{cold}}{\dot{m}_{hot}}\right)\left\{\frac{\Delta b_{cold}}{(h_{in} - h_{out})_{hot}}\right\}\right] \\ &= \left[-\left\{1 - T_0\left(\frac{s_{out} - s_{in}}{h_{out} - h_{in}}\right)_{hot}\right\} + \left(\frac{\dot{m}_{cold}}{\dot{m}_{hot}}\right)\right. \\ &\quad \left. \times \left\{\frac{(h_{out} - h_{in})_{cold}}{(h_{in} - h_{out})_{hot}}\right\}\left\{1 - T_0\left(\frac{s_{out} - s_{in}}{h_{out} - h_{in}}\right)\right\}_{cold}\right] \\ &= \left[-\left\{1 - T_0\left(\frac{s_{out} - s_{in}}{h_{out} - h_{in}}\right)_{hot}\right\} + \left\{\frac{\dot{Q}_{cold}}{\dot{Q}_{hot}}\right\}\left\{1 - T_0\left(\frac{s_{out} - s_{in}}{h_{out} - h_{in}}\right)\right\}_{cold}\right] \end{aligned} \tag{8.27}$$

When $\dot{Q}_{hot} = \dot{Q}_{cold}$ the expression reduces to the simple form of equation (8.13). This correspondence indicates that equation (8.27) should also be applicable to parallel flow and crossflow exchangers. The second case may be handled in a similar fashion.

8.9 Commercial considerations

The wider problem of exergy minimization for plant components operating under thermo-physical and thermo-physical-chemical change conditions involves cost analysis, and this is discussed more fully in the texts by Kotas (1985), Bejan *et al.* (1996), Bejan (1997), and Bejan & Mamut (1999).

For good design, Sama (1995) remarks '... to use excessively large or excessively small values of $\Delta\theta_{LM}$ in design is a mistake. There is a range of optimal temperature differences over which the annual cost is essentially the same, but within which the distribution between cost of capital and the cost of energy is significantly different'. Sama's remark directly affects the choice of the Grassmann & Kopp constant (a).

8.10 Conclusions

1. For exchangers not subject to fluid leakage the dimensionless exergy loss number (N_x) gives unambiguous indication of the specific performance of an individual exchanger, and need only be multiplied by the thermal duty ($\dot{Q}$) to obtain the actual exergy loss required in complete plant analysis.
2. Use of N_x in situations involving real fluids may require interpolation of thermodynamic properties in two or more dimensions.
3. The absolute temperature level of operation has a strong effect on the value of exergy loss number obtained. It is only appropriate to compare the exergy loss number of exchangers operating over the same duty range.
4. Some heat leaks can be accommodated in the exergy loss number.
5. The process of evaluating exergy loss reveals that thermal optimization can only be pursued so far. Full optimization of a single component of a complex energy system can only be achieved by including capital and running costs, and this will in every case involve analysis of the complete system.

Exergy loss references

Pressure loss references are given at the end of this chapter.

Bejan, A. (1993) *Advanced Engineering Thermodynamics*, John Wiley, p. 597.

Bejan, A. (1997) *Advanced Engineering Thermodynamics*, John Wiley.

Bejan, A. and **Mamut, E.** (1999) *Thermodynamic Optimisation of Complex Energy Systems*, NATO Science Series 3, High Technology, vol. 69, Kluwer Academic Publishers.

Bejan, A., Tsatsaronis, G., and **Moran, M.** (1996) *Thermal Design and Optimisation*, John Wiley.

Gosney, W.B. (1982) *Principles of Refrigeration*, Cambridge University Press, Cambridge.

Grassmann, P. and **Kopp, J.** (1957) Zur günstigen Wahl der temperaturdifferenz und der Wärmeübergangzahl in Wärmeaustauchern. *Kaltetechnik*, **9**(10), 306–308.

Hesselgreaves, J.E. (2001) *Compact Heat Exchangers – Selection, Design and Optimisation*, Pergamon Press.

Jakob, M. (1957) *Heat Transfer*, vol. II, John Wiley.

Kestin, J. (1978) *A Course in Thermodynamics*, vols I and II, Hemisphere/McGraw-Hill, New York.

Kotas, T.J. (1985) *The Exergy Method of Thermal Plant Analysis*, Butterworths, London.

London, A.L. and **Shah, R.K.** (1983) Cost of irreversbilities in heat exchanger design. *Heat Transfer Engng*, **4**(2), April–June, 59–73.

Sama, D.A. (1995) The use of second law of thermodynamics in process design. *Trans. ASME, J. Energy Resources Tech.*, **117**, 179–185.

Witte, L.C. and **Shamsunder, N.** (1983) A thermodynamic efficiency concept for heat exchange devices. *Trans. ASME, J. Engng Power*, **105**, 199–203.

Bibliography

Bruges, E.A. (1959) *Available Energy and the Second Law Analysis*, Butterworths, London.

Hewitt, G.F., **Shires, G.L.**, and **Bott, T.R.** (1994) *Process Heat Transfer*, CRC Press, Florida.

Linhoff, B., **Townsend, D.W.**, **Boland, D.**, **Hewitt, G.F.**, **Thomas, B.E.A.**, **Guy, A.R.**, and **Marsland, R.H.** (1982) *A User Guide on Process Integration for the Efficient use of Energy*, Institution of Chemical Engineers, Rugby, UK.

Smith, E.M. (1997) *Thermal Design of Heat Exchangers*, John Wiley, Chichester.

Vargaftik, N.B. (1983) *Handbook of Physical Properties of Liquids and Gases*, Hemisphere–Springer, New York–Berlin.

Walsh, P.P. and **Fletcher, P.** (1998) *Gas Turbine Performance*, Blackwell Science, Oxford.

PRESSURE LOSS

8.11 Control of flow distribution

Objective

Proper attention to the design of headers avoids poor flow distribution and hence less than anticipated design performance from the exchanger core.

Temperature-dependent fluid properties

The steady-state temperature fields for crossflow were first solved analytically by Nusselt (1911). Since that time many other analytical solutions have appeared, and it was left to Baclic & Heggs (1985) to show that these different approaches were mathematically equivalent. The author solved the same steady-state problem numerically, and the results of the computation are given in Figs 3.16 and 3.17.

The prior assumptions in all these approaches was that mass flowrate across each inlet face of the exchanger is constant across the width of that face. However, it only takes a glance at the temperature distributions to realize that once temperature-dependent physical properties are involved, the pressure loss will not be the same for every parallel flow path in the exchanger. Should poor manufacturing techniques also produce significant variations in flow channel geometry, then this too will affect performance.

If the flow channels are not interconnected, the mass flowrates will differ across the exchanger, thus questioning some of the initial assumptions made in solving the temperature fields. Worse, if the flow channels are interconnected there may be some migration of flow across the parallel flow paths. While these effects may not be large, they illustrate that it is pressure loss which controls the flow pattern in every heat exchanger.

8.12 Header design

Dow (1950) studied flow along a perforated pipe. His problem was that of establishing the same mass flow through each perforation for a gas burner, and he analysed the solution as follows:

> '... The pressure drop due to friction losses in the flow though the pipe has to be exactly balanced by the pressure increase due to the deceleration of the flow in the pipe which necessarily occurs when part of the flow escapes through the ports ... the pressure loss due to friction is related to the surface area of the tube while the pressure gain due to deceleration is related to the cross-sectional area of the pipe'.

(One referee noted that the above quote was not exact in modern terms, as *total* pressure reduces due to friction, while *static* pressure increases due to diffusion.)

Dow noted that it was possible to control the pressure by adjusting the ratio of these two areas, and he suggested three means of doing this, viz.:

- inserting a tapered plug in the pipe
- changing the diameter of the pipe
- constructing the burner manifold according to a geometric design

In developing a sound analytical basis for design he examined four cases:

- *Case 1* – laminar flow in the pipe
- *Case 2* – turbulent flow in the pipe
- *Case 3* – negligible friction in the pipe ($f = 0$)
- *Case 4* – manifold serving as infinite reservoir

observing that only some commercial designs corresponded to Case 4.

To illustrate Dow's approach, the balance of linear momentum for unsteady flow gives

$$\frac{\partial u}{\partial t} + u\frac{\partial u}{\partial x} = -\frac{1}{\rho}\frac{\partial p}{\partial x} - \frac{fu^2}{2r_{hyd}} \tag{8.28}$$

thus steady-state flow along a header duct from the open end can be written

$$u\frac{du}{dx} = -\frac{1}{\rho}\frac{dp}{dx} - \frac{fu^2}{2r_{hyd}} \tag{8.29}$$

Fluid velocities through the exchanger flow channels must be equal to give uniform distribution, thus (dp/dx) must be zero to give a level discharge across and into the core of the exchanger, giving

$$u\frac{du}{dx} = -\frac{fu^2}{2r_{hyd}} \tag{8.30}$$

If we now measure from the duct closed end making $y = 0$ at $x = L$

$$\frac{du}{dy} = +\frac{fu}{2r_{hyd}} \tag{8.31}$$

Balance of mass at any cross-section of the header requires

$$\dot{m} = \rho A u \tag{8.32}$$

As rate of flow through the header must vary linearly from the inlet maximum $(\dot{m}_0)$ to zero at the closed end, then also

$$\dot{m} = \dot{m}_0\left(\frac{y}{L}\right) \tag{8.33}$$

giving

$$\rho u = \left(\frac{\dot{m}_0}{L}\right)\frac{y}{A} \tag{8.34}$$

where $u = \phi(y, A)$. Differentiating equation (8.34) with respect to y, remembering that ρ is constant

$$\frac{du}{dy} = \left(\frac{\dot{m}_0}{\rho AL}\right)\left[1 - \frac{y}{A}\frac{dA}{dy}\right] \tag{8.35}$$

Substituting in equation (8.31)

$$\frac{A}{y} - \frac{dA}{dy} = \frac{A}{y}\left(\frac{\rho AL}{\dot{m}_0}\right)\frac{fu}{2r_{hyd}}$$

$$\frac{dA}{dy} = \frac{A}{y} - \frac{L}{y}\left(\frac{\rho Au}{\dot{m}_0}\right)\frac{Af}{2r_{hyd}}$$

$$\frac{dA}{dy} = \frac{A}{y} - \frac{L}{y}\left(\frac{\dot{m}}{\dot{m}_0}\right)\frac{Af}{2r_{hyd}}$$

Separating variables

$$\frac{dA}{A} = \frac{dy}{y}\left[1 - \left(\frac{\dot{m}}{\dot{m}_0}\right)\left(\frac{L}{2r_{hyd}}\right)f\right] \tag{8.36}$$

Equation (8.36) holds for any header cross-section.

Values for the frictional coefficient may be written

- laminar flow, $f = b/\mathrm{Re}$ (b constant)
- turbulent flow, $f = b/\mathrm{Re}^a$ (a, b constants)

An explicit solution is available for a circular cross-section and laminar flow, as equation (8.36) simplifies further to

$$\frac{dA}{A} = \frac{dy}{y}\left[1 - \left(\frac{\rho A u}{\dot{m}_0}\right)\left(\frac{L}{2r_{hyd}}\right)\frac{b}{\text{Re}}\right] \tag{8.37}$$

Expanding the Reynolds number

$$\frac{dA}{A} = \frac{dy}{y}\left[1 - \left(\frac{A}{\dot{m}_0}\right)\left(\frac{L}{2D}\right)\frac{b\eta}{D}\right]$$

the final expression within square brackets becomes independent of the variable D

$$\frac{dD}{D} = \frac{dy}{y}\left[\frac{1}{2} - \left(\frac{b}{16}\right)\left(\frac{\pi L\eta}{\dot{m}_0}\right)\right] \tag{8.38}$$

Writing equation (8.38) in simplified form

$$\frac{dD}{D} = a\frac{dy}{y}$$

subject to boundary conditions $y = L$ at $D = D_0$. This can be integrated immediately to give

$$\frac{D}{D_0} = \left(\frac{y}{L}\right)^a \qquad \text{where } a = \frac{1}{2} - \left(\frac{b}{16}\right)\left(\frac{\pi L\eta}{\dot{m}_0}\right) \tag{8.39}$$

For a tapering tube the changing diameter with length can now be determined, which provides a preliminary indication of the area changes required.

Dow tested a single-pipe gas burner, and showed that the flame height could vary along the length of the pipe, being high at inlet if the flowrate was small, and high at the closed end if the flowrate was large. Only the first of these situations is shown in Fig. 8.2. Readers are referred to the complete set of photographs in Dow's 1950 paper.

The solution he adopted was to fit a tapered insert into the pipe, and a schematic representation of the effect is shown in Fig. 8.2b. It seems that Dow took the expression (8.36) to be his final result, but where an insert is used the calculation of pressure loss should also take into account the surface area of the tapered insert.

Further work (with experimental confirmation) is desirable to discover whether the changes in surface area due to the presence of many orifices along the duct length will change the empirical expressions for friction factor along the duct length.

Headers of varying rectangular section

For heat exchangers, header design normally requires one flat face of constant width of length L and breadth B, requiring a change in header height z to satisfy equation (8.30). Treatment has to be numerical.

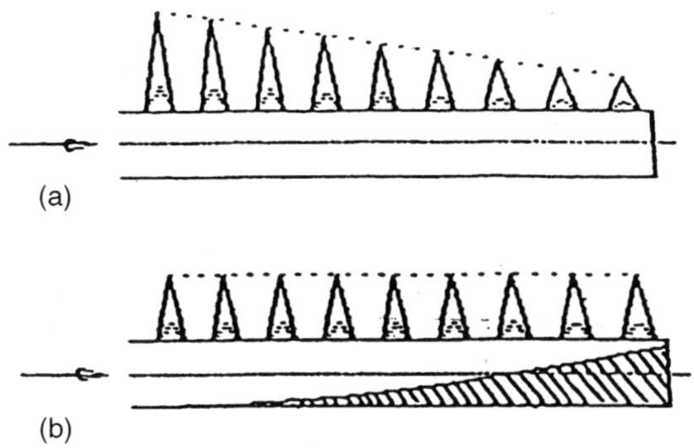

Fig.8.2 Gas burner using perforated pipe showing flow distribution: (a) without insert, low mass flowrate; (b) with insert, all mass flowrates

Combining Dow's equations (8.33) and (8.36)

$$\frac{\Delta A}{A} = \frac{\Delta y}{y}\left[1 - \left(\frac{yf}{2r_h}\right)\right] \tag{8.40}$$

At any section:

Flow area

$A = Bz$ where $z = \phi(y)$

(... A does not vary linearly with y)

Hydraulic radius

$$r_h = \frac{\text{area for flow}}{\text{wetted perimeter}} = \frac{A}{P}$$

Reynolds number

$$\text{Re} = \frac{4r_h\dot{m}}{\eta A} = \frac{4\dot{m}}{\eta P}$$

Friction factors*

$f = \dfrac{b}{\text{Re}}$ for laminar flow (b constant*)

$f = \dfrac{b}{\text{Re}^a}$ for turbulent flow (a, b constants*)

(*see under Discussion below)

Inlet section

At header inlet $y = L$ choose a limiting Re number for laminar flow, say 2000.

Knowing $(\dot{m}, \eta)$ the perimeter (P) of the duct can be found.

Knowing duct breadth (B) the required inlet area of the duct (A) can be found.

Hence the height of the duct (z) can be found.

Assuming that known duct length (L) has already been divided into a constant number of increments (Δy) then equation (8.40) can be solved for ΔA.

Next duct section

New duct length is $y = y - \Delta y$, new flow area is $A = A - \Delta A$.

New duct height is $(z = A/B)$, hence new perimeter is $P = 2(B + z)$.

New mass flowrate is $(\dot{m} = \dot{m}_0 y/L)$, hence new Reynolds number and friction factor can be calculated. Calculation proceeds as outlined earlier until $y = 0$.

The header is now optimum for one inlet flowrate only. In experiments Dow was able to vary gas flow at will and still maintain equal gas distribution. Further theoretical and experimental investigation is indicated.

Discussion

Friction factors will change as the duct aspect ratio changes. For laminar flow, Shah & London (1978) calculated values of $f \times \text{Re}$ for plain rectangular ducts of different aspect ratios. This work was aimed at developing understanding of flow in the channels of compact heat exchangers. Experimentally proven friction factors for ducts with porous surfaces are required. Until such results become available, it is suggested that the Shah & London data for laminar flow in plain ducts be used.

Cautionary notes

Design uncertainties lie primarily in determining the wetted perimeter (P) for a duct with one perforated face, and developing corresponding laminar and turbulent friction factor correlations. The header designs developed by London *et al.* (1968) are less than ideal, as it is assumed that pressure losses exist in the headers.

When no special attention is paid to header design, headers have to be made large enough to satisfy Case 4. As exchangers become smaller and smaller, or include many multi-stream paths, it may become necessary to consider Dow's approach.

A recent treatment of flow distribution and pressure loss in plate heat exchangers has been provided by Bassiouny & Martin (1984). The analysis considers both U-type and Z-type headering arrangements (see Fig. 8.3), and is not restricted to headering arrangements only. This work is comprehensive and should be read.

Exchanger aspect ratios

As chemical plant moves towards process intensification, miniaturization (without loss of throughput) will change the appearance of individual components such as heat exchangers. In particular we may expect to see ever smaller and shorter flow channels – which implies increased longitudinal conduction. This parasitic effect

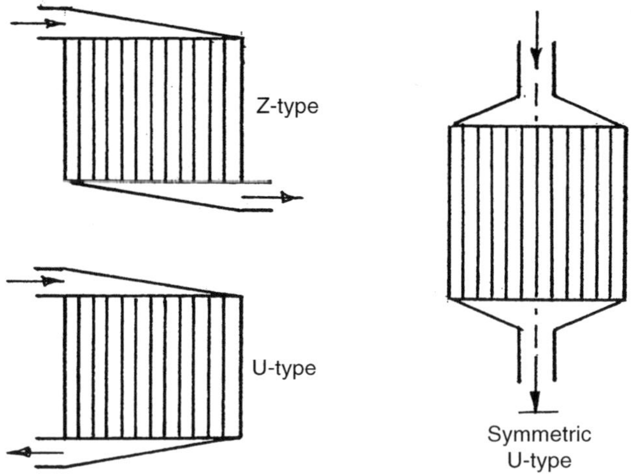

Fig.8.3 Z-type and U-type headering

can be kept under control, as outlined in Chapter 4. Care to achieve even flow distribution may also require a rethink of header design.

Directional headers

Das & Roetzel (1995) analysed the merits of Z-type and U-type header patterns for a contraflow exchanger. U-type headering was preferred when transients were involved as faster temperature response was then obtained. Headers of finite and constant cross-section gave better results in both steady and unsteady flow conditions with the U-type configuration. With tapering headers even better results can be expected when the U-type is used in symmetrical fashion (Fig. 8.3).

Compact flow distributors

Haseler & Fox (1965) have examined the problem of flow distribution in multistream block heat exchangers used in cryogenics, including design of distributor sections within the core block. This is primarily a method of directing injected streams to an appropriate part of the main exchanger, and it does not deal with design of the header itself.

Haseler & Fox (1965) provide information on methods of evaluating pressure losses for seven basic types of distributor. With careful design, proper allowance can be made for additional pressure losses. In certain two-stream contraflow heat exchangers, it is possible that distribution headers for one stream may lie completely in contact within the other flow stream. Additional heat-transfer surface then exists in the distribution headers, which will affect assumed entry and exit conditions for the true contraflow 'core'. Solutions to this problem do not seem to have reached the open literature yet.

8.13 Minimizing effects of flow maldistribution

Cowans (1978) studied maldistribution of flow in parallel channels in a counterflow exchanger, and found a simple method for compensating automatically for the maldistribution. His solution was to shape the passages differently in the two sides of the exchanger so that the shift in temperature in the middle of the exchanger strongly affected the hot flow only. Hot channels were narrowed mid-way along their length and cold channels were made correspondingly wider. His approach worked well in practice, and suggests significant potential gains through the use of printed, etched, and diffusion bonded methods of building heat exchanger cores; however, his theoretical analysis seems incomplete.

There is more satisfaction in Cowan's development of the primary surface exchanger shown in Fig. 8.4. Flat plates may be progressively deformed to create a simple regular heat exchanger pattern in which each hot flow channel is surrounded by four cold flow channels, and vice-versa. This is a stabilizing heat exchanger arrangement when maldistribution of flow is anticipated. It also suggests how individual plates may be configured at inlet and outlet of the exchanger so as to ease the headering problem, going progressively from end configuration [Fig. 8.4(a)] through the intermediate stage [Fig. 8.4(b)] to the main heat exchanger surface [Fig. 8.4(c)]. There is no difficulty in fitting end bars between plates, or in accommodating headers. However, a better channel aspect ratio may be rectangular rather than square, see right hand-side of Fig. 4.11 in Chapter 4.

To evaluate transient thermal performance of such cores, recourse to testing model heat exchangers may be desirable, using test cores of different lengths to

(a)
(b)
(c)

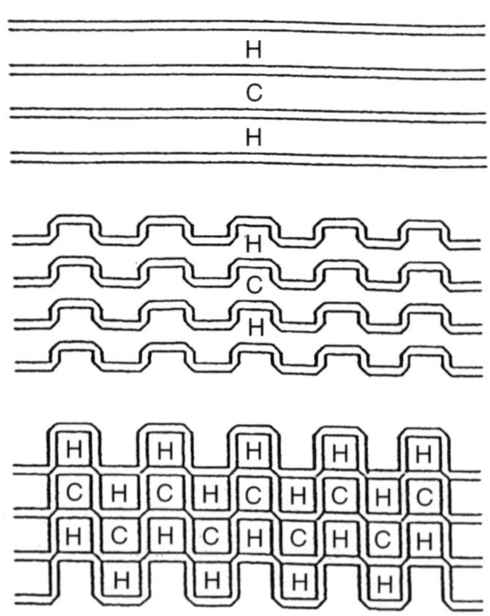

Fig.8.4 Primary surface heat exchanger development

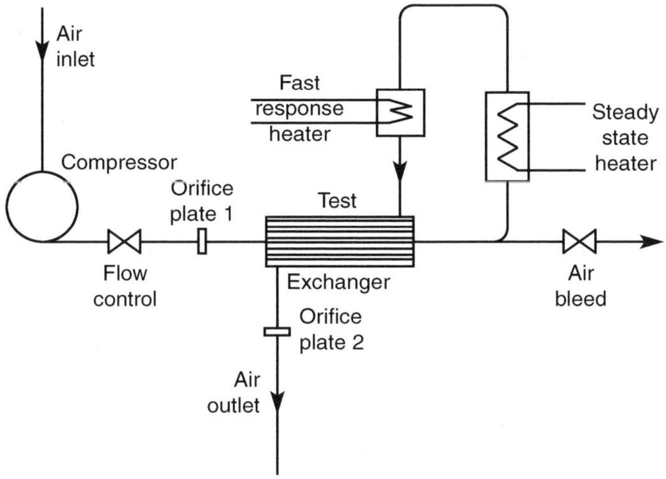

Fig.8.5 Test-rig for transients in model heat exchanger

separate out end-effects from core performance. This would have to be supported by analytical solution of the laminar flow performance of individual cells. A possible transient test-rig is shown in Fig. 8.5.

8.14 Embedded heat exchangers

In Chapter 2 attention was paid to optimized temperature differences and optimization of pressure losses. Chapter 4 described the direct-sizing of compact plate–fin heat exchangers in which the temperature profiles were not optimum but, through varying local surface geometry, a method of sizing the exchanger for minimum volume was found. Chapter 5 described an approach to direct-sizing of helical-tube, multi-start coil exchangers, which are often used in cryogenic plant, and Chapter 7 showed that direct-sizing could also be applied to the RODbaffle shell-and-tube design.

All of the above considered the exchanger in isolation, and from the above it would clearly be possible to design an exchanger with optimum temperature profiles, with optimized pressure losses and optimized for minimum core volume, etc.

Practical applications are not like this, however. Only in a few special applications does it become possible to employ the near-perfect exchanger – one example being that of cryogenic plant discussed in Chapter 11, where the arrangement of expansion turbines would allow adjustment of the cooling mass flowrate to obtain the optimized temperature profiles for a two-stream exchanger. But even here the commercial attraction of building multi-stream block exchangers can introduce a further complication – that of cross-conduction (Haseler, 1983 and Chapter 11).

Table 8.2 Contraflow exchanger parameters

$\dot{m}_h = 101.1\,\text{kg/s}$	$T_{h1} = 825.0\,\text{K}$	$p_{h1} = 1.0662\,\text{bar}$
$C_h = 1.150\,\text{kJ/(kg K)}$	$T_{h2} = 554.5\,\text{K}$	$\Delta p_h = 4260.0\,\text{N/m}^2$
$\dot{m}_c = 100.0\,\text{kg/s}$	$T_{c1} = 782.3\,\text{K}$	$p_{c2} = 10.765\,\text{bar}$
$C_c = 1.005\,\text{kJ/(kg K)}$	$T_{c2} = 469.4\,\text{K}$	$\Delta p_c = 32\,300.0\,\text{N/m}^2$

Allowing for streams to be 'by-passed' inside the multi-stream exchanger, or even introduced or extracted from the block exchanger may be sufficient to disturb the design away from optimum conditions.

Temperature profiles across the exchanger may have to be constrained to achieve least exergy loss for the whole plant. Kotas (1985) provides examples of exergy analysis for four different plants, viz.:

- Linde air liquefaction plant
- sulphuric acid plant
- gas turbine plant
- refrigeration plant

where such compromises can be seen. The extent of the compromise required may be seen from recuperator design for a gas turbine plant outlined by Walsh & Fletcher (1998), the terminal temperatures that follow from assuming full load effectiveness of 0.88 are given in Table 8.2.

Using the data in Table 8.2 the following parameters may be calculated

$$\varepsilon = \frac{\Delta T_{\max}}{Tspan} = \frac{782.3 - 469.4}{825.0 - 469.4} = 0.88$$

$$\dot{Q} = \dot{m}_h C_h (T_{h1} - T_{h2}) = \dot{m}_c C_c (T_{c1} - T_{c2}) = 31.4481\,\text{MW}$$

$$\Delta\theta_{LM} = \frac{\Delta\theta_1 - \Delta\theta_2}{\ln(\Delta\theta_1/\Delta\theta_2)} = 61.48\,\text{K}$$

$$US = \frac{\dot{Q}}{\Delta\theta_{LM}} = 511.517\,\text{kW/K}$$

$$N_h = \frac{US}{\dot{m}_h C_h} = 4.3994, \qquad N_c = \frac{US}{\dot{m}_c C_c} = 5.0895$$

Expression (3.8) derived in Chapter 3 may now be used to obtain the temperature profiles, viz.

$$\frac{T_{h1} - T_h}{T_{h1} - T_{h2}} = \frac{1 - \exp\left[-(N_h - N_c)\frac{x}{L}\right]}{1 - \exp[-(N_h - N_c)]} = \frac{T_{c1} - T_c}{T_{c1} - T_{c2}}$$

The resulting normalized profiles are shown in Fig. 8.6. The reader may compare these profiles against those of the thermally optimized exchanger given in Fig. 2.22

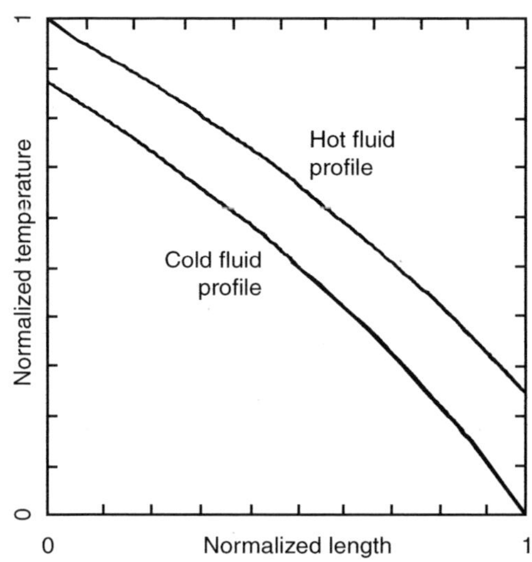

Fig.8.6 Normalized temperature profiles for embedded exchanger

to see how much temperatures depart from the optimum profile given by $(\Delta\theta/T_h) = const$.

Apart from the wider temperature difference in Fig. 8.6, it is also evident that the temperature difference at the hot end of the exchanger is smaller than the temperature difference at the cold end. This results in a different profile curvature from the optimum seen in Fig. 2.3.

As all heat exchangers have to be embedded in a complete plant, the complete thermodynamic process should be optimized first before specifying design parameters for the heat exchanger. If this is not done, and procurement of the heat exchanger (or indeed any other single component of the plant) proceeds in isolation, then there is danger of serious mis-match and poor overall plant performance.

8.15 Pumping power

The steady flow energy equation for each fluid may be written as

$$\dot{Q} - \dot{W} = \dot{m} \times \left\{ \left(e + \frac{p}{\rho} + \frac{u^2}{2} \right)_{outlet} - \left(e + \frac{p}{\rho} + \frac{u^2}{2} \right)_{inlet} \right\} \tag{8.41}$$

The heat flow (Q) is positive when inwards to the fluid, and the shaft work (W) is positive when outwards from the fluid. No shafts exist for a heat exchanger, thus W is zero in the thermodynamics of heat exchanger design. We cannot find pumping power by analysing components of the steady-flow equation for the heat exchanger alone.

Pumping power is to be calculated for a *pump* which produces a pressure rise equal to the pressure loss found for each fluid in the heat exchanger. Pumping power for compressible gas may be found from the single-stage compressor equation

$$\dot{W}_g = \dot{m}RT\left(\frac{\gamma}{\gamma-1}\right)\left\{\left(\frac{p_{inlet}}{p_{outlet}}\right)^{\gamma-1/\gamma} - 1\right\} \tag{8.42}$$

where the inlet and outlet pressure values are for the heat exchanger.

The compressor inlet temperature (T) may be taken as the heat exchanger outlet temperature as the pressure rise across the compressor is small. Note that the isentropic efficiency of the compressor $\eta_s = (W_s/W_{real})$ may also need to be known where component mechanical power is required.

Where pressure loss, $\Delta p = p_{inlet} - p_{outlet}$, is for incompressible flow only, the following simple expression holds

$$\dot{W} = \frac{\dot{m}\Delta p}{\rho} \tag{8.43}$$

Pressure loss references

Exergy loss references are given after Section 8.10.

Baclic, B.S. and **Heggs, P.J.** (1985) On the search for new solutions of the single-pass cross-flow heat exchanger problem. *Int. J. Heat Mass Transfer*, **28**(10), 1965–1976.

Bassiouny, M.K. and **Martin, H.** (1984) Flow distribution and pressure drop in plate heat exchangers. Part 1 – U-type arrangement. Part 2 – Z-type arrangement. *Chem. Engng Sci.*, **39**(4), 693–700 and 701–704.

Cowans, K.W. (1978) A countercurrent heat exchanger that compensates automatically for maldistribution of flow in parallel channels. *Adv. Cryogenic Engng*, **19**, 437–444.

Das, S.K. and **Roetzel, W.** (1995) Dynamic analysis of plate exchangers with dispersion in both fluids. *Int. J. Heat Mass Transfer*, **36**(6), 1127–1140.

Dow, M.W. (1950) The uniform distribution of a fluid flowing through a perforated pipe. *Trans. ASME, J. Appl. Mech.*, December, 431–438.

Haseler, L.E. (1983) Performance calculation methods for multi-stream plate-fin heat exchangers. *Heat Exchangers: Theory and Practice* (Eds. J. Taborek, G.F. Hewitt, and N. Afgan), Hemisphere, McGraw-Hill, New York, pp. 495–506.

Haseler, L.E. and **Fox, T.** (1965) Distributor models for plate-fin heat exchangers. In *4th UK National Heat Transfer Conference*, Manchester, 26–27 September 1995, pp. 449–456.

Kotas, T.J. (1985) *The Exergy Method of Thermal Plant Analysis*, Butterworths, London, p. 296.

London, A.L., **Klopfer, G.**, and **Wolf, S.** (1968) Oblique flow headers for heat exchangers. *Trans. ASME, J. Engng Power*, July, 271–286.

Nusselt, W. (1911) Der Warmeubergang im Kreuzstrom. *Zeitschrift des vereines deutscher Ingenieur*, **55**, 2021–2024.

Shah, R.K. and **London, A.L.** (1978) *Laminar Flow Convection in Ducts*, Supplement 1 to *Advances in Heat Transfer*, Academic Press, New York.

Walsh, P.P. and **Fletcher, P.** (1998) *Gas Turbine Performance*, Blackwell Science, Oxford, pp. 323–325.

Bibliography

Hewitt, G.F., **Shires, G.L.**, and **Bott, T.R.** (1994) *Process Heat Transfer*, Chapter 26, CRC Press, Florida, pp. 841–855.

Lalot, S., **Florent, P.**, **Lang, S.K.**, and **Bergles, A.E.** (1999) Flow maldistribution in heat exchangers. *Appl. Thermal Engng*, **19**, 847–863.

Linnhoff, B. (1983) New concepts in thermodynamics for better chemical process design. *Proc. R. Soc., Ser. A.*, **386** (1790), 8 March 1983, 1–33. (Also in *Chem. Engng Res. Design*, **61**(4), July, 207–223.)

Linnhoff, B. and **Flower, J.R.** (1978) Synthesis of heat exchanger networks. Part 1 – systematic generation of energy optimal networks. Part 2 – evolutionary generation of networks with various criteria of optimality. *AIChE J.*, **24**, 633–642 and 642–654.

Linnhoff, B. and **Hindmarsh, E.** (1983) The pinch design method for heat exchanger networks. *Chem. Engng Sci.*, **38**(5), 745–763.

Linnhoff, B., **Townsend, D.W.**, **Boland, D.**, **Hewitt, G.F.**, **Thomas, B.E.A.**, **Guy, A.R.**, and **Marsland, R.H.** (1982) *A User Guide on Process Integration for the Efficient Use of Energy*, Institution of Chemical Engineers, Rugby, UK.

Paterson, W.R. (1984) A replacement for the logarithmic mean. *Chem. Engng Sci.*, **39**(11), 1635–1636.

Underwood, A.J.V. (1933) Graphical computation of logarithmic mean temperature difference. *The Ind. Chem.*, May, 167–170.

CHAPTER 9
Transients in Heat Exchangers

Partial differential equations

9.1 Review of solution methods – contraflow

Three main approaches to solution of the simultaneous equations for transients exist, viz.:

- finite differences
- method of characteristics
- Laplace transformation

No method is without problems, but the finite-difference approach probably allows best understanding of the physical situation.

Finite differences

This approach is to work with the original set of partial differential equations and the original initial and boundary conditions. Partial differential equations are converted into finite-difference form, employing a variety of solution methods. Ames (1965) comments that '... there is some justification for operating with finite differences on the original equations without reducing them to characteristic form'.

Direct solution

Solution by direct finite-difference methods has been tackled by several workers. Sharifi *et al.* (1995) have presented results for solution by the fully implicit but unconditionally stable Crank–Nicholson scheme. Their partial differential equations did not include the second-order terms for longitudinal conduction. A better approach may be to use the method of lines with Runge–Kutta, at the very least for the mass and momentum equations.

MacCormack's method

Ontko (1989) and Ontko & Harris (1990) have presented a numerical method using a modified MacCormack predictor–corrector solution algorithm, providing results for both imposed temperatures and imposed mass flowrate steps, but with constant physical properties. Thermal storage in the wall is included, but longitudinal conduction was neglected. A description of the method by Ontko (1989) includes computer source listings which use the modified MacCormack algorithm.

Other approaches

Sharifi *et al.* (1995) solved the governing partial differential equations using a number of numerical techniques, using backward differencing for the convection term, and evaluating different time marching schemes. Lakshmann & Potter (1984, 1990) claimed that their 'cinematic' model requires the least amount of computation to provide fast and accurate results. They assumed constant physical properties and that the wall equation could be neglected, but as the wall equation plays a major role in provision of thermal storage, the solution is less than satisfactory.

Method of characteristics

This involves finding the characteristics of the problem, which are the 'natural coordinates' of the system, lines representing the (x, t) trajectories of fluid particles as they pass through the exchanger. The method provides good physical insight into the problem, and transient discontinuities pass through the exchanger travelling along the characteristics – only providing the characteristics can be found. Appropriate mathematical schemes can be developed for numerical solution, finding the unknown temperatures corresponds to integrating along the characteristics, but with some difficulties.

When account is taken of significant temperature variations in physical properties at points throughout the exchanger, then heat-transfer coefficients change, the density of the fluids change, transit times change, and mass flowrates change. All these factors lead to a change in direction of the characteristics which then become curved.

With variable mass flowrates, the effect is more pronounced, as these affect directly the transit times through the exchanger. The problem then grows to finding the curved characteristics, and of establishing a numerical scheme for calculating points along the exchanger, which is not a simple task. Much effort has gone into solving problems in fluid dynamics using the method of characteristics, particularly for handling shock waves but the method has not so far been developed satisfactorily for dealing with transients in heat exchangers.

Dzyubenko *et al.* (1990) completed a thorough study on the flattened and helically twisted tube design of exchanger, in which the full set of energy and momentum equations were first set out. The transient solution involved solving a one-dimensional energy equation by the method of characteristics and two Cauchy problems. Some simplifications were necessary to obtain a solution.

Laplace transforms with numerical inversion

Laplace transformation follows a hybrid approach. The method is to transform the partial differential equations plus initial conditions and boundary conditions, into ordinary differential equations containing the initial conditions plus transformed boundary conditions. Because the boundary conditions (i.e. transients) may not be recognizable mathematical functions, this can require Fourier representation of the disturbances. The ordinary differential equations are then solved by finite differences, and the results are then inverted numerically back into real time using fast-Fourier transforms.

The mathematical approach accommodates longitudinal conduction effects, but requires a significant amount of computing power. A number of workers in heat transfer have gone down this route, as can be seen from papers listed in the References, and some important solutions have been obtained by Roetzel & Xuan (1991, 1992a,b,c, 1993a,b,c,d), Roetzel & Lee (1993), Rotzel *et al.* (1994). In all cases the mathematical effort involved in producing a solution to a real problem is considerable. The fundamental paper is that of Roetzel & Xuan (1992a) which examined transient response of simple exchangers excluding axial dispersion terms. Since that time the main thrust has included plate-and-frame exchangers, serpentine tube panels, and conventional shell-and-tube exchangers, all of which types included consideration of *axial dispersion* in the momentum equation for design-critical applications; see Taylor (1954) and the several papers by Roetzel and others which include axial dispersion terms for the fluids as a means of representing fine detail in the flow. Roetzel & Lee (1993) confirmed that for values of Pe > 40 computed results are virtually identical with those for plug flow which is usually the assumption.

In examining the continuum equations governing transient flow the author felt that the *Rayleigh dissipation function* in the energy equations for the fluids was possibly a better expression to use when thermal transients were involved; however, both effects are small, and other more important considerations need attention.

For a solution in real time, inversion of the Laplace transform solution requires either the Gaver–Stehfest algorithm (Jacquot *et al.*, 1983), or the Fourier series approximation (Ichikawa & Kishima, 1972). Roetzel and coworkers found that the Gaver–Stehfest inversion took very little computational time, but was not suitable for handling disturbances containing rapid oscillatory components. The Fourier series approximation was preferred in handling oscillations, but convergence was slow. This may be speeded-up by using the fast-Fourier transform (Crump, 1976; Press *et al.*, 1992).

Boundary conditions are required as functions of dimensionless time. They can be expressed in terms of combinations of functions that can be transformed, e.g. step, ramp, exponential, sine. Care in selecting the appropriate solution method may be necessary, e.g. summation of infinite Fourier series does not represent square waveforms accurately, the overshoot remaining finite at 18 per cent at each change of amplitude, viz. Gibb's phenomenon (Mathews & Walker, 1970). However, real transients in heat exchangers tend to be mathematically smooth.

The Laplace transform method works with linear differential equations. Temperature dependence of physical properties seems most difficult to incorporate in a solution.

9.2 Contraflow with finite differences

Preliminary considerations

Transient equations for compressible flow with temperature-dependent physical properties are presented as equations (A.1) in Appendix A.1. These are derived from the fundamental equations of continuum mechanics. Some manipulation is necessary to bring the equations into computable form, and the four stages of development are presented in Appendix A.2.

Along the way, the Rayleigh dissipation function (Φ) is neglected as its contribution was small.

In the final set of equations the 'pressure-field' equations were dropped to allow stability of the numerical solution to be assessed as a first step. Results of the computation shown later (Figs 9.1–9.3) are for this first stage only. Once computational stability is confirmed, equations for a particular fluid can be incorporated. The cause of any new instability in computation can then be more closely identified.

In the supplement to Appendix B, the pressure-field equations for a perfect gas are developed and their straightforward incorporation in the finite-difference algorithms is explained. The introduction of pressure-field equations generates additional coupling of transients in density, velocity, and temperature parameters.

Selection of time intervals

For transient computation, selection of time intervals is constrained by the Courant–Friedrichs–Lewy (CFL) stability condition, see e.g. Fletcher (1991). The CFL condition depends on the local speed of sound in the fluid, and is given as

$$\Delta t \leq \left(\frac{\Delta x}{c}\right) \tag{9.1}$$

where for a perfect gas $c = \sqrt{\gamma RT}$.

Mechanical (pressure) disturbances travel at the speed of sound in a fluid. Thermal disturbances travel much slower. The idea is to keep disturbances in one space interval from reaching the next space interval. With the CFL condition in

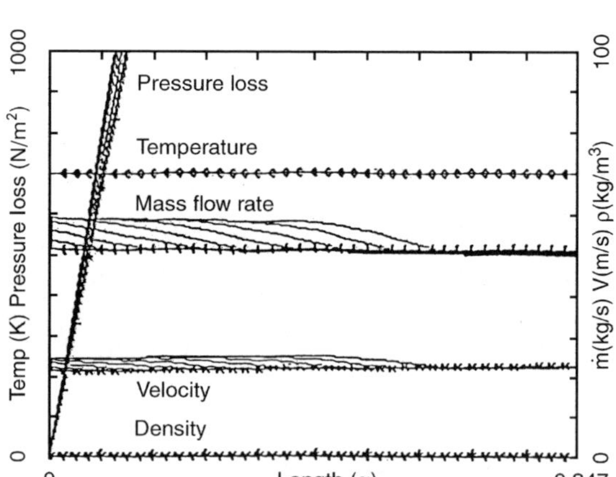

Fig.9.1 Response from disturbance of 15 per cent increase in inlet mass flowrate with heat transfer to duct wall (wall mass/100 and wall thermal conductivity × 100). Symbols used on the first curve for each parameter are as follows: ◇, temperature (K); #, pressure loss (N/m²); +, mass flowrate (kg/s); X, velocity (m/s); Y, density (kg/m³)

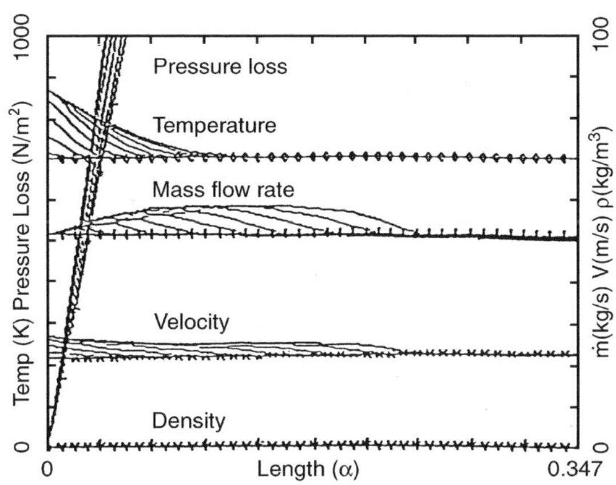

Fig.9.2 Response from disturbance of 25 per cent increase in inlet fluid temperature with heat transfer to duct wall (wall mass/100 and wall thermal conductivity × 100). For symbols see Fig.9.1 caption

mind the author used

$$\Delta t \leq \left(\frac{\Delta x}{|u| + c} \right) \tag{9.2}$$

where u was the local velocity of the fluid.

For every time interval, the CFL condition has to be evaluated for every space interval in the computation, and the smallest value of Δt is taken for the next time

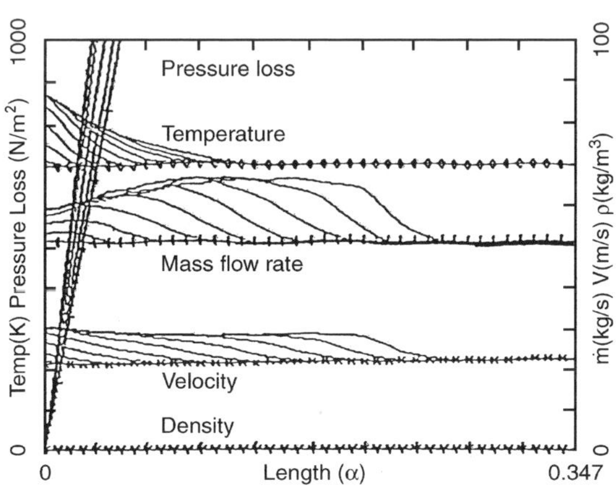

Fig.9.3 Response from combined disturbance of 15 per cent increase in inlet mass flow-rate and 25 per cent increase in inlet fluid temperature, with heat transfer to duct wall (wall mass/100 and wall thermal conductivity × 100). For symbols see Fig.9.1 caption

interval. Further, it may be desirable to multiply the time interval by, say, 0.95 as velocity values for the next interval are not yet known.

No instabilities were observed in computation, however pressure transients can travel both forwards and backwards in one space dimension, and in future computations the author would use

$$\Delta t \leq \left(\frac{\Delta x/2}{|u| + c}\right) \tag{9.3}$$

Events during the next time step are not yet known, but the definition of equation (9.3) should ensure that a gap exists between the end of the pre-selected value ($\Delta x/2$) and the x-value at the end of vectors $(+u + c)$ and $(+u - c)$.

Allowance for convective mesh drift

All convected transient equations contain a convected term on the left-hand side, consider

$$\frac{D}{Dt} = \frac{\partial}{\partial t} + u\frac{\partial}{\partial x}$$

Algorithms do exist for correcting convective mesh drift, e.g. the MacCormack predictor–corrector algorithm, or the method of lines with Runge–Kutta. The interesting question is whether there may exist reasonable means for adjusting for mesh drift when using the stable Crank–Nicholson algorithm. Appendices A.3 and B.8 contain further discussion of this concept which has never been applied by the author due to computational restrictions, and seems to require investigation plus development, followed by validation or rejection.

Pressure terms

Pressure terms involve both pressure gradient and pressure loss due to flow friction, viz.

$$\frac{\partial p}{\partial x} + \left(\frac{f\rho u^2}{2r_{hyd}}\right)$$

When the equation of state for pressure, plus the friction factor versus Reynolds number correlation for the channel under consideration, are inserted into the above expression, the solution becomes unique for a particular case.

The necessary procedures for incorporating pressure terms into the Crank–Nicholson algorithms for a perfect gas are explained in Appendix A.3 and Appendix B.8. Pressure gradients at flow entry and exit should be made zero, and be replaced by numerical values of losses due to entrance and exit effects. Any flow acceleration/deceleration will be computed automatically.

Shell heat leakage

An analysis of losses from the exchanger shell surface has been made by Nesselman (1928), with further treatment by Hausen (1950). These effects were not considered in steady-state treatments because modern insulating materials can minimize the effect. In transients, the study of this effect is not usually a first priority.

Longitudinal conduction

A term for longitudinal conduction in the separating walls is present in the full set of transient energy equations. When an additional pressure shell is used the effect of longitudinal conduction in the shell may also need consideration if the pressure shell is thick. For liquid metals terms for longitudinal conduction in the fluids themselves may become necessary.

Approximations remaining

In generating the simultaneous partial differential equations for transients in contraflow it is assumed that there was no temperature difference across the solid wall of the exchanger. In steady-state analysis it is straightforward to incorporate the thermal resistance of the wall, but in transient analysis thermal capacity of the wall itself may be more significant.

One-dimensional plug flow is assumed in both fluids, i.e. there was no attempt to distinguish between boundary layer and bulk flow. Allowance for any transverse flow would involve the Rayleigh dissipation function, reduced to suit the number of dimensions involved.

Physical properties

Temperature dependence of physical properties is most conveniently represented by interpolating cubic spline-fit. This is also true for heat-transfer and flow-friction correlations where high accuracy is required. Some physical constants acting as coefficients in the differential equations may need to be evaluated at each time interval and for each space interval during the computation.

General remarks

Transient solutions that do not include the solid wall equation are of little practical value. This is because energy storage in the wall is always significant. Three basic forms of transient inlet disturbances exist for each fluid, viz. temperature transients, mass flow transients, and pressure transients:

- in heat exchangers, Mach numbers are usually less than 0.05, however transient temperatures, pressures, and mass flowrates will be felt by a compressible fluid. Zero flow ('choking') or reversal of flow direction might be encountered even though the steady-state Mach number is a long way from sonic value
- temperature transients change some fluid densities (e.g. gases) and also affect physical properties in each fluid and in the wall. Where such properties are not primary unknowns in the differential equations they may need to be determined using interpolating cubic spline-fits for each finite-difference station along the exchanger. Fluid parameters involved may be specific heat at constant pressure (C), absolute viscosity (η), and thermal conductivity (λ). Reynolds and Prandtl numbers (Re, Pr) may be required to evaluate

heat-transfer coefficients (α) and flow-friction coefficients (f) locally. Solid parameters such as thermal conductivity (λ) and density (ρ) are required
- mass flow transients change fluid velocities
- pressure transients change densities, and thus velocities.

The above considerations indicate that the study of transients should either involve mass flow transients under isothermal conditions, or involve both mass flow transients and temperature transients together, as the two effects then cannot be separated.

Solution of the transient problem separates into three distinct problems, viz:

- solution of the velocity field for the hot fluid
- solution of the velocity field for the cold fluid
- solution of the coupled temperature fields for both fluids and the solid wall

Fluid (density) and density × velocity equations are solved independently, but require knowledge of the imposed temperature field. The momentum equation additionally involves pressure terms, which particularizes a given design solution.

Development of numerical algorithms for mass flow and momentum equations is only necessary for the hot fluid. The cold fluid can use algorithms for the hot fluid provided only that care is taken to renumber the finite-difference equations at input, and to reverse re-number the solutions at output.

Shape of disturbances

The disturbances used were in the form of a modified sine curve, viz.

$$\left.\begin{array}{ll}(t < t_{start}) & \text{form} = 0 \\ (t_{start} < t < t_{end}) & \text{form} = \dfrac{1}{2}\left[1 + \sin\left\{\dfrac{\pi}{2}\left(\dfrac{2t - B - A}{B - A}\right)\right\}\right] \\ (t > t_{end}) & \text{form} = 1\end{array}\right\} \tag{9.4}$$

where A is the time at start of disturbance and B is the time at the end of disturbance. The normalized disturbance is in the range $0 \leq$ form ≤ 1, and this multiplier is applied to any increase or decrease in mass flowrate or temperature, over any desired duration of disturbance. The sine curve was chosen to give a smooth approach to, and departure from, the chosen disturbance. A step change (and especially a shock wave) would introduce additional computational problems.

The disturbances, although brief, were not initiated at time zero. This was to provide time for the steady solution to settle down.

Order of solution

The solution must proceed from some isothermal steady state, so that all necessary physical constants are available to get the solution started.

Step 1. Search for the maximum value of fluid velocity in both hot and cold fluids and calculate the magnitude of the next time interval.

Step 2. Use temperatures to update physical properties along the exchanger using interpolating spline-fits.

Step 3. Obtain flow-friction coefficients from correlations.

Step 4. Solve the balance of mass and balance of linear momentum equations for the hot fluid sequentially. From new values of ρ and ρu obtain new values of velocity (u) required in solution of the simultaneous energy equations.

Step 5. For the cold fluid, reverse number stations along the exchanger and use the same algorithms developed for the hot fluid to obtain new values of velocity (u). Reverse re-number the solution.

Step 6. Obtain heat-transfer coefficients from correlations.

Step 7. Solve the energy equations simultaneously for new temperatures for each fluid and for the wall. The energy equations provide the only link between the two fluids.

Step 8. Return to Step 1 and repeat the process.

Results of computation (without pressure-field equations)

When pressure terms are omitted from the momentum equations, symmetry of the mass and momentum equations (Appendix A.2) suggested a first computation with the simplified equations to test stability of computation (flow-friction pressure losses were estimated later). Should instability appear with the full set of equations it is suggested that the advection terms be replaced with upwind differences for time and downwind differences for space (Fletcher, 1991, Section 9.1.2). Alternatively – or additionally – the addition of a diffusion term

$$\left(D\frac{\partial^2 T}{\partial x^2}\right) \Longrightarrow D_n^t\left(\frac{T_{n+1}^t - 2T_n^t + T_{n-1}^t}{(\Delta x)^2}\right)$$

may cure the problem.

Probably the most interesting feature to be seen is the mass flowrate disturbance caused by temperature alone (Fig. 9.2). In all computations, the mass of the exchanger was reduced by two orders of magnitude to obtain response curves that show trends. This is because the calculations were done on an 8 MHz computer with 1 MByte of RAM and without a hard disk. Thermal storage terms retained in the solid wall equation are still sufficient to delay outlet response to an input disturbance to the order of minutes (or tens of minutes) in the case of the contraflow example given by Campbell & Rohsenow (1992).

9.3 Further considerations

Phase lag, cross-conduction, and boundary conditions

Das & Roetzel (1995) solved the equations for a plate heat exchanger with phase lag in the fluid entry temperatures (Z-type headering system). This problem arises because of the time taken to transit the headering system. However the optimized header design discussed in Chapter 8 was not in use, and investigation of phase-lag effects for headers with inserts of tapering cross-section is a further area for research. At fluid exits from the exchanger core, Das & Roetzel *et al.* assumed zero temperature gradients.

With phase lag it becomes necessary to treat each flow channel separately because no two channels have the same entry condition, and will not behave in the same way.

Phase lag leads to cross-conduction effects in plate–fin exchangers. The solution scheme adopted by Das & Roetzel (1995) did not allow for the important cross-conduction effect analyzed by Haseler (1983) which is discussed in Chapter 11. Apart from this, the approach seems well developed and powerful and it has been applied to several exchanger configurations. With plate–fin exchangers, extra design flexibility is available by fitting half-height channels at each end of the block.

The fundamental differential equations for plate-and-frame exchangers can be rewritten to distinguish between flow in the end channels, which only transfer heat to or receive heat from a single surface, and the internal flow channels which transfer heat to or receive heat from two surfaces. Typical equations are presented by Sharifi *et al.* (1995).

For fast transients, Das & Roetzel (1995) found U-type headering gave a quicker response than Z-type headering, however phase-lag effects seem likely to be exacerbated by the use of headers with constant cross-sectional area. See header design discussed in Chapter 8, p. 249.

Shell-and-tube exchangers with small tube inclinations

Serpentine tube bundles and helical-coil tube bundles fall into the class of exchangers that may be directly sized as parallel or contraflow exchangers having crossflow heat-transfer coefficients, providing that a sufficient number of tube passes exist.

For a small number of tube passes (less than ten say) Hausen (1950, 1983) derived expressions for steady-state temperature distributions. For such serpentine tube bundles, Roetzel & Xuan (1991) found the dispersion model to be effective at modelling flow maldistributions in the steady-state. Dispersion may be more appropriate for isothermal flows, while the Rayleigh dissipation function may be preferable for non-isothermal flows. Extension of the dispersion model to transient conditions in serpentine exchangers was also reported by Roetzel & Xuan (1992b,c).

9.4 Engineering applications – contraflow

A recuperative gas-turbine propulsion system applied to a ship can produce very significant fuel savings of the order of 30 per cent, leading to cost savings of $1.5 million per annum (Cownie, 1993; Crisalli & Parker, 1993). Under high-speed manoeuvring conditions control of the gas turbine must take into account thermal storage effects in the heat exchanger. Due to manufacturing constraints, for maintenance, and for damage control, multiple units are desirable. A plate-and-frame design derivative operating in contraflow has been developed (Crisalli & Parker, 1993; Valenti, 1995). These units operate with an effectiveness of over 0.88 at full power, and over 0.95 at low power. U-type headering arrangements have been adopted for fast response, see analysis of plate-and-frame exchangers by Das & Roetzel (1995). Whether Dow's tapered inserts (Chapter 8) would further improve the gas–gas situation of Figs 9.1 to 9.3 is a matter for future investigation. Water-cooled compact plate–fin heat exchangers have been developed as intercoolers (Crisalli & Parker, 1993; Bannister *et al.*, 1994); these units are manufactured in segments.

9.5 Review of solution methods – crossflow

Summary of past work

A number of published solutions of the crossflow problem exist, but fewer, different, and less satisfactory solutions are available for crossflow than for contraflow.

In 1963 Evans & Smith published a classic but heavily analytical solution for transients in crossflow. The wall equation was not included, and the solutions obtained are mostly in the form of infinite series.

Myers *et al.* (1967) developed an approximate integral solution for the associated problems of crossflow, evaporation, and condensing in which one fluid is mixed. In a further paper, Myers *et al.* (1970) provide a solution for evaporation or condensation based on the method of characteristics, and compare this with Rizika's (1956) exact solution.

Gvozdenac (1986) provides the characteristics which are developed later, but he then simplified the equations to obtain an analytical solution.

Romie (1983) solved the transient response problem for gas–gas crossflow exchangers, with neither gas mixed, using a single Laplace transform; however, it was necessary to neglect fluid thermal capacity terms. Chen & Chen (1991) also neglect fluid thermal capacity terms, restricting their solution to gas–gas (unmixed–unmixed) heat exchangers.

Spiga & Spiga have published two solutions to the unmixed–unmixed problem for arbitrary inlet temperature disturbances only. The first paper (1987) is on gas–gas crossflow, and fluid thermal capacity terms are neglected. The second paper (1992) includes fluid thermal capacity terms. In both cases the solution approach required three-fold Laplace transformation.

A finite-difference solution is due to Yamashita *et al.* (1978) for one-pass crossflow with both fluids unmixed. Only a step change in temperature is considered, and fluid velocities are assumed to remain constant.

Kou & Yuan (1994) have studied the effects of conduction in the separator sheet during transient response with neither gas mixed.

There are similarities and some important differences between solution of transients for cross-flow compared with contraflow. In both cases it is not possible to separate temperature transients from mass flow (velocity) transients because temperature-induced changes in physical properties alone will result in changes in mass flowrate. Even in the steady state, it can be seen that individual channels have differing temperature profiles, Fig. 3.16.

The most closely mathematically defined configuration for crossflow is that of one-pass unmixed–unmixed flow. Even with this simple configuration, the constraint of having equal pressure loss through each channel on each side requires each channel to have a different mass flowrate, and each channel will thus require its own treatment to determine the mass flowrate fields. Such constraints are not

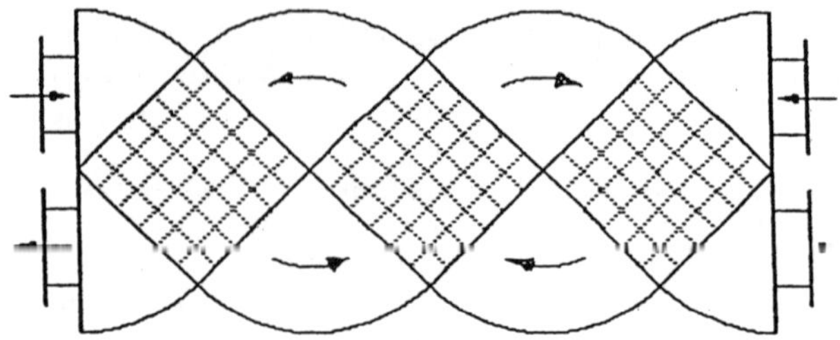

Fig.9.4 Schematic symmetric three-pass crossflow

usually considered in steady-state analysis, and may become more severe under transient conditions.

For two-pass unmixed–unmixed crossflow, mass flowrate differences would also require that no mixing takes place in the end headers. Such separation will introduce extra pressure loss. Without header separators one solution would appear to be to adopt a symmetric three-pass arrangement as shown in Fig. 9.4. There probably is little point in going to a greater number of passes as steady-state performance of a four-pass arrangement closely approaches that of simple contraflow.

Solution methods

A finite-difference solution would use equations similar to those outlined for contraflow in Appendix A.1. Solutions of the mass and momentum equations would be obtained first, followed by solution of the coupled temperature equations. However, as each channel of a crossflow exchanger is likely to have a different temperature profile under steady-state conditions, any transient solution might require channel-by-channel solutions.

The Lax–Wendroff method in two space dimensions may offer an alternative approach to straightforward finite differences. However, Mitchell & Griffiths (1980) indicate that while the Lax–Wendroff method attains the Courant–Friedrichs–Lewy limit for stability in one space dimension, the limit for stability in two space dimensions is much more severe. The MacCormack algorithm is a development of Lax–Wendroff.

9.6 Engineering applications – crossflow

The problem of transients in two-pass unmixed–unmixed crossflow is more difficult. Each half of the exchanger has to be iterated in turn at each time interval until matching intermediate temperatures are obtained. This problem is, however, of some importance in designing two-pass compact crossflow heat exchangers for vehicle propulsion, as the paper by Ward & Holman (1992) shows.

The problem of solving transients in single-pass crossflow involving changing mass flowrates with changing temperatures is more complex than solving comparable transients in contraflow. A first step might be to solve the problem of steady unmixed–unmixed crossflow with temperature-dependent fluid properties to determine how mass flowrates may vary across the width of the exchanger section for each fluid.

The above difficulties might make choice of crossflow less attractive for critical applications, but there may be less reflection of disturbances in crossflow.

References and further reading

Collected under: contraflow, crossflow, and textbooks

Contraflow

Acrivos, A. (1956) Method of characteristics technique, application to heat and mass transfer problems. *Ind. Engng Chem.*, **48**(4), April, 703–710.

Al-Nimr, M.A. (1998) Transient response of finite-wall capacitance heat exchanger with phase change. *Heat Transfer Engng*, **19**(1), 42–52.

Bannister, R.L., **Cheruvu, N.S.**, **Little, D.A.**, and **McQuiggan, G.** (1994) Turbines for the turn of the century. *Mech. Engng, ASME*, **116**(6), June, 68–75.

Beckman, L., **Law, V.J.**, **Bailey, R.V.**, and **von Rosenberg, D.U.** (1990) Axial dispersion for turbulent flow with a large radial heat flux. *AIChE J.*, **36**(4), April, 598–604.

Campbell, J.F. and **Rohsenow, W.M.** (1992) Gas turbine regenerators: a method for selecting the optimum plate-finned surface pair for minimum core volume. *Int. J. Heat Mass Transfer*, **35**(12), 3441–3450.

Carver, M.B. (1980a) Pseudo characteristic method of lines solution of the conservation equations. *J. Comp. Physics*, **35**(1), March, 57–76.

Carver, M.B. (1980b) Pseudo characteristic method of lines solution of first order hyperbolic equation systems. *Math. Comp. Simulation*, **22**(1), March, 30–35.

Chester, M. (1963) Second sound in solids. *Phys. Rev.*, **131**, 2013–2015.

Churchill, S.W. (1988) The role of mathematics in heat transfer. In *AIChE Symposium Series No. 263, Heat Transfer*, Houston, 1988, vol. 84, pp. 1–13.

Cownie, J. (1993) Aerospace R & D 90 years on. *Prof. Engng*, **6**(11), December, 17–19.

Crisalli, A.J. and **Parker, M.L.** (1993) Overview of the WR-21 intercooled recuperated gas turbine engine system. A modern engine for a modern fleet. In *International Gas Turbine and Aeroengine Congress and Exposition*, Cincinnati, Ohio, 24–27 May 1993, ASME Paper 93-GT-231.

Crump, K.S. (1976) Numerical inversion of Laplace transforms using a Fourier series approximation. *J. Ass. Comp. Mach.*, **23**(1), 89–96.

Das, S.K. and **Roetzel, W.** (1995) Dynamic analysis of plate exchangers with dispersion in both fluids. *Int. J. Heat Mass Transfer*, **38**(6), 1127–1140.

Gvozdenac, D.D. (1990) Transient response of the parallel flow heat exchanger with finite wall capacitance. *Ingenieur-Archive (Archive of Applied Mechanics)*, **60**(7), 481–490.

Haseler, L.E. (1983) Performance calculation methods for multi-stream plate-fin heat exchangers. *Heat Exchangers: Theory and Practice* (Eds. J. Taborek, G.F. Hewitt, and N. Afgan), Hemisphere/McGraw-Hill, New York, pp. 495–506.

Hausen, H. (1970) Berechnung der Wärmeübertragung in Regeneratoren bei zeitlische veranderlichem Mengenstrom. *Int. J. Heat Mass Transfer*, **13**, 1753–1766.

Herron, D.H. and **von Rosenberg, D.U.** (1966) An efficient numerical method for the solution of pure convective transport problems with split boundary conditions. *Chem. Engng Sci.*, **21**, 337–342.

Huang, Y.M. (1989) Study of unsteady flow in the heat exchanger by the method of characteristics. In *ASME Pressure Vessels and Piping Division Publication PVP*, vol. 156, *ASME Pressure Vessels and Piping Conference*, Honolulu, 23–27 July 1989, pp. 55–62.

Ichikawa, S. and **Kishima, A.** (1972) Application of Fourier Series techniques to Inverse Laplace Transforms. *Kyoto Univ. Memoires*, **34**, 53–67.

Jacquot, R.G., **Steadman, J.W.**, and **Rhodine, C.N.** (1983) The Gaver–Stehfest algorithm for approximate inversion of Laplace transforms. *IEEE Circuits and Systems Mag.*, **5**(1), March, 4–8.

Jaswon, M.A. and **Smith, W.** (1954) Countercurrent transfer processes in the non-steady state. *Proc. R. Soc. Ser. A*, **225**, 226–244.

Kabelac, S. (1988) Zur Berechnung des Zeitverhaltens von Wärmeübertragern (calculation of transients in heat exchangers). *Wärme und Stoffübertragung (Heat and Mass Transfer)*, **23**(6), 365–370.

Kahn, A.R., **Baker, N.S.**, and **Wardle, A.P.** (1988) The dynamic characteristics of a countercurrent plate heat exchanger. *Int. J. Heat Mass Transfer*, **31**(6), 1269–1278.

Lakshmann, C.C. (1994) Dynamic simulation of a countercurrent heat exchanger modelling start-up and frequency response ('cinematic' method). *Int. Comm. Heat Mass Transfer*, **21**(3), May–June, 421–434.

Lakshmann, C.C. and **Potter, O.E.** (1984) Dynamic simulation of heat exchangers and fluidised beds. In *Proceedings of the 12th Australian Chemical Engineering Conference CHEMECA 84*, vol. 2, 871–878.

Lakshmann, C.C. and **Potter, O.E.** (1990) Dynamic simulation of plate heat exchangers ('cinematic' method). *Int. J. Heat Mass Transfer*, **35**(5), 995–1002.

Longley, H.J. (1960) Methods of differencing in Eulerian hydrodynamics. Los Alamos Scientific Laboratory Report, LAMS-2379.

Luo, X. and **Roetzel, W.** (1995) Extended axial dispersion model for transient analysis of heat exchangers. In *Proceedings of the 4th UK National Heat Transfer Conference*, IMechE, Paper C510/141/95, pp. 411–416.

Massoud, M. (1990) Numerical analysis of transients in parallel and counter flow heat exchangers. *Thermal Hydraulics of Advanced Heat Exchangers, ASME Winter Annual Meeting* 1990, ASME Nuclear Engineering Division Publication NE, vol. 5, pp. 99–104.

Mercier, P. and **Touchon, P.** (1997) Analysis of turbulent flow and heat transfer in compact heat exchangers by pseudo-direct numerical simulation. In *Conference on Compact Heat Exchangers in the Process Industry*, Snowbird, Utah, 22–27 June 1997, Begell House, pp. 223–230.

Myers, G.E., **Mitchell, J.W.**, and **Lindeman, C.F. Jr** (1970) The transient response of heat exchangers having an infinite capacitance rate fluid (condensers, evaporators). *ASME J. Heat Transfer*, **92**, May, 269–275.

Nesselman, K. (1928) Der Einfluss der Wärmeverluste auf Doppelrohrwärmeaustauscher. *Zeitschrift gestalte Kalte-Industrie*, **35**, 62.

Ontko, J.S. (1989) A parametric study of counterflow heat exchanger transients. Institute for Aviation Research, Wichita State University, Wichita, Kansas, Report IAR 89–10.

Ontko, J.S. and **Harris, J.A.** (1990) Transients in the counterflow heat exchanger. *Compact Heat Exchangers – a festschrift for A.L. London* (Eds. R.K. Shah, A.D. Kraus, and D. Metzger), Hemisphere, New York, pp. 531–548.

Peyret, R. and **Viviand, H.** (1975) Computation of viscous compressible flows based on the Navier-Stokes equations. Advisory Group for Aerospace Research and Development, Report AGARD-AG-212.

Roetzel, W. and **Lee, D.** (1993) Experimental investigation of leakage in shell-and-tube heat exchangers with segmental baffles. *Int. J. Heat Mass Transfer*, **36**(15), 3765–3771.

Roetzel, W. and **Xuan, Y.** (1991) Dispersion model for divided-flow heat exchanger (serpentine tubes). *Design and Operation of Heat Exchangers* (Eds. W Roetzel, P.J. Heggs, and D. Butterworth), *Proceedings of the EUROTHERM Seminar no. 18*, Hamburg, 27 February–1 March 1991, Springer Verlag, Berlin, pp. 98–110.

Roetzel, W. and **Xuan, Y.** (1992a) Transient response of parallel and counterflow heat exchangers. *ASME J. Heat Transfer*, **114**(2), May, 510–511.

Roetzel, W. and **Xuan, Y.** (1992b) Transient behaviour of multipass shell-and-tube heat exchangers (serpentine tubes). *Int. J. Heat Mass Transfer*, **35**(3), 703–710.

Roetzel, W. and **Xuan, Y.** (1992c) Analysis of transient behaviour of multipass shell and tube heat exchangers with the dispersion model (serpentine tubes). *Int. J. Heat Mass Transfer*, **35**(11), 2953–2962.

Roetzel, W. and **Xuan, Y.** (1993a) The effect of core longitudinal heat conduction on the transient behaviour of multipass shell-and-tube heat exchangers. *Heat Transfer Engng*, **14**(1), January–March, 52–61.

Roetzel, W. and **Xuan, Y.** (1993b) Transient behaviour of shell and tube exchangers. *Chem. Engng Technol.*, **16**(5), October, 296–302.

Roetzel, W. and **Xuan, Y.** (1993c) The effect of heat conduction resistances of tubes and shells on transient behaviour of heat exchangers. *Int. J. Heat Mass Transfer*, **36**(16), 3967–3973.

Roetzel, W. and **Xuan, Y.** (1993d) Dynamics of shell-and-tube heat exchangers to arbitrary temperature and step flow variation. *AIChE J.*, **39**(3), 413–421.

Roetzel, W., **Das, S.K.**, and **Luo, X.** (1994) Measurement of the heat transfer coefficient in plate heat exchangers using a temperature oscillation technique. *Int. J. Heat Mass Transfer*, **37**(Suppl. 1), 325–331.

Romie, F.E. (1984) Transient response of the counterflow heat exchanger. *ASME J. Heat Transfer*, **106**, August, 620–626.

Romie, F.E. (1985) Transient response of the parallel flow heat exchanger. *ASME J. Heat Transfer*, **107**, August, 727–730.

Shah, R.K. (1981) The transient response of heat exchangers. *Heat Exchangers Thermal-Hydraulic Fundamentals and Design* (Eds. S. Kakac, A.E. Bergles, and F. Mayinger), Hemisphere, Washington, pp. 915–953.

Sharifi, F., **Golkar Narandji, M.R.**, and **Mehravaran, K.** (1995) Dynamic simulation of plate heat exchangers. *Int. Comm. Heat Mass Transfer*, **22**(2), March April, pp. 213 225.

Spang, B., **Xuan, Y.**, and **Roetzel, W.** (1991) Thermal performance of split-flow heat exchangers (serpentine tubes). *Int. J. Heat Mass Transfer*, **34**(3), 863–874.

Stehfest, H. (1970) Numerical inversion of Laplace Transforms. *Comm. Assoc. Comp. Mach.*, **13**, 47–49.

Tan, K.S. and **Spinner, I.H.** (1978) Dynamics of a shell-and-tube heat exchanger with finite tube-wall heat capacity and finite shell-side resistance. *Ind. Engng Chemistry Fundamentals*, **17**(4), 353–358.

Tan, K.S. and **Spinner, I.H.** (1984) Numerical methods of solution for continuous countercurrent processes in the nonsteady state. Part I – Model equations and development of numerical methods and algorithms. Part II – Application of numerical methods. *AIChE J.*, **30**(5), September, 770–779 and 780–786.

Tan, K.S. and **Spinner, I.H.** (1991) Approximate solutions for transient response of a shell and tube heat exchanger. *Ind. Engng Chemistry Res.*, **30**, 1639–1646.

Taylor, G.I. (1954) The dispersion of matter in turbulent flow through a pipe. *Proc. R. Soc. Ser. A*, **223**, 446–468.

Valenti, M. (1995) A turbine for tomorrow's Navy. *Mech. Engng, ASME*, **117**(9), September, 70–73.

Wang, C.C. and **Liao, N.S.** (1989) Transient response of a double-pipe condenser to change in coolant flow rate. *Int. Comm. Heat Mass Transfer*, **16**(3), May–June, 325–334.

Willmott, A.J. (1968a) Operation of Cowper stoves under conditions of variable flow. *J. Iron Steel Inst.*, **206**, 33–38.

Willmott, A.J. (1968b) Simulation of a thermal regenerator under conditions of variable mass flow. *Int. J. Heat Mass Transfer*, **11**, 1105–1116.

Willmott, A.J. (1969) The regenerative heat exchanger computer representation. *Int. J. Heat Mass Transfer*, **12**, 997–1014.

Wolf, J. (1964) General solution of the equations for parallel flow multichannel heat exchangers. *Int. J. Heat Mass Transfer*, **7**, 901–919.

Xuan, Y. and **Roetzel, W.** (1993a) Dynamics of shell-and-tube heat exchangers to arbitrary temperature and step flow variations. *AIChE J.*, **39**(3), March, 413–421.

Xuan, Y. and **Roetzel, W.** (1993b) Stationary and dynamic simulation of multipass shell and tube heat exchangers with the dispersion model for both fluids (serpentine tubes). *Int. J. Heat Mass Transfer*, **36**(7), 4221–4231.

Xuan, Y., Spang, B., and **Roetzel, W.** (1991) Thermal analysis of shell and tube exchangers with divided flow pattern. *Int. J. Heat Mass Transfer*, **34**(3), 853–861.

Crossflow (with reservations – see Section 9.5)

Abdallah, S. and **Rooke, S.** (1997) Transient response of a serpentine finned-tube crossflow heat exchanger to a step change in temperature. *Heat Transfer Engng*, **18**(3), 51–60.

Chen, H.-T. and **Chen, K.-C.** (1991) Simple method for transient response of gas-to-gas crossflow heat exchangers with neither gas mixed. *Int. J. Heat Mass Transfer*, **24**(11), 2891–2898.

Chen, H.-T. and **Chen, K.-C.** (1992) Transient response of crossflow heat exchangers with finite wall capacitance – technical note. *ASME J. Heat Transfer*, **114**(3), August, 752–755.

Evans, F. and **Smith, W.** (1963) Cross-current transfer processes in the non-steady state. *Proc. R. Soc. Ser. A*, **272**, 241–269.

Gvodzenac, D.D. (1986) Analytical solution of the transient response of gas-to-gas crossflow heat exchanger with both fluids unmixed. *ASME J. Heat Transfer*, **108**, November, 722–727

Gvozdenac, D.D. (1991) Dynamic response of the crossflow heat exchanger with finite wall capacitance. *Warme-und Stoffubertragung*, **26**, 207–212.

Jacquot, R.G., **Steadman, J.W.**, and **Rhodine, C.N.** (1983) The Gavir–Stehfest algorithm for approximate inversion of Laplace transforms. *IEEE Circuits System Mgmt*, **5**(1), 4–8.

Kou, H.S. and **Yuan, P.** (1994) Effect of longitudinal separator sheet conduction on the transient thermal response of crossflow heat exchangers with neither gas mixed. *Numerical Heat Trans., Part A*, **25**, 223–236.

Myers, G.E., **Mitchell, J.W.**, and **Norman, R.F.** (1967) The transient response of crossflow heat exchangers, evaporators, and condensers. *ASME J. Heat Transfer*, **89**, February, ASME Paper 66-WA/HT-34, pp. 75–80.

Myers, G.E., **Mitchell, J.W.**, and **Lindeman, C.F. Jr** (1970) The transient response of heat exchangers having an infinite capacitance rate fluid (condensers, evaporators). *ASME J. Heat Transfer*, **92**, May, 269–275.

Press, W.H., **Flannery, B.P.**, **Teukolsky, S.A.**, and **Vetterling, W.T.** (1989) *Numerical Recipies in Pascal*, Chapter 15.1, Cambridge University Press, Cambridge.

Rizika, J.W. (1956) Thermal lags in flowing incompressible fluid systems containing heat capacitors. *Trans. ASME*, **78**, 1407–1413.

Romie, F.E. (1983) Transient response of gas-to-gas crossflow heat exchangers with neither gas mixed. *ASME J. Heat Transfer*, **105**, August, 563–570.

Romie, F.E. (1994) Transient response of crossflow heat exchangers with zero core thermal capacitance. *ASME J. Heat Transfer*, **116**, August, 775–777.

San, J.-Y. (1993) Heat and mass transfer in a two-dimensional crossflow regenerator with a solid conduction effect. *Int. J. Heat Mass Transfer*, **36**(3), 633–643.

Shah, R.K. and **London, A.L.** (1978) Laminar flow forced convection in ducts. Supplement 1 to *Advances in Heat Transfer*, (Eds T.F. Irvine and J.P. Hartnett) Chapter XV11, Academic Press, New York, pp. 406–407.

Spiga, G. and **Spiga, M.** (1987) Two dimensional transient solutions for crossflow heat exchangers with neither gas mixed. *ASME J. Heat Transfer*, **109**, May, 281–286.

Spiga, M. and **Spiga, G.** (1988) Transient temperature fields in crossflow heat exchangers with finite wall capacitance. *ASME J. Heat Transfer*, **110**, February, 49–53.

Spiga, M. and **Spiga, G.** (1992) Step response of the crossflow heat exchanger with finite wall capacitance. *Int. J. Heat Mass Transfer*, **35**(2), 559–565.

Ward, M.E. and **Holman, L.** (1992) Primary surface recuperator for high performance prime movers. *Society of Automotive Engineers International Congress and Exposition*, Detroit, Michigan, 24–28 February 1992, SAE Paper 920150.

Yamashita, H., **Izumi, R.**, and **Yamaguchi, S.** (1978) Analysis of the dynamic characteristics of crossflow heat exchangers with both fluids unmixed. (On the transient response to a step change in the inlet temperature.) *Bulletin of the JSME*, **21**(153), March, Paper No. 153-16, pp. 479–485.

Textbooks

Ames, W.F. (1965) *Nonlinear Partial Differential Equations in Engineering*, Section 7.25 onwards, Academic Press, New York.

Ames, W.F. (1969) *Numerical Methods for Partial Differential Equations*, Section 4-14, Nelson.

Anderson, D.A., **Tannehill, J.C.**, and **Pletcher, R.H.** (1984) *Computational Fluid Mechanics and Heat Transfer*, Hemisphere, New York, p. 599.

Bellman, R., **Kalaba, R.E.**, and **Lockett, J.A.** (1966) *Numerical Inversion of the Laplace Transform*, American Elsevier, New York.

Collatz, L. *The Numerical Treatment of Differential Equations*, Springer, Berlin.

Courant, R. and **Hilbert, D.** (1953, 1962) *Methods of Mathematical Physics*, vol. I, *Ordinary Differential Equations*, Vol. II, *Partial Differential Equations*, John Wiley.

Dzyubenko, B.V., **Dreitser, G.A.**, and **Ashmantas, L.-V.A.** (1990) *Unsteady Heat and Mass Transfer in Helical Tube Bundles* (helically twisted tubes in a shell-and-tube exchanger), Hemisphere, New York.

Ferziger, J. and **Perić, M.** (1996) *Computational Methods for Fluid Dynamics*, Springer, Berlin.

Fletcher, C.A.J. (1991) *Computational Techniques for Fluid Dynamics, vols 1 and 2*, 2nd edn, Springer, Berlin.

Fox, L. (1962) *Numerical Solution of Ordinary and Partial Differential Equations*, Chapter 17 – Hyperbolic equations and characteristics, Chapter 18 – Finite-difference methods for hyperbolic equations, Chapter 28 – Characteristics in three independent variables, Pergamon Press, pp. 205–217, 218–229, 366–377.

Garabedian, P. (1964) *Partial Differential Equations*, John Wiley.

Hausen, H. (1950) *Wärmeübertragung im Gegenstrom, Gleichstrom und Kreuzstrom*, Springer, Berlin, pp. 213–229, (2nd edn, 1976).

Hausen, H. (1983) *Heat Transfer in Counterflow, Parallel Flow and Crossflow*, 2nd English edn, McGraw-Hill, New York.

Jeffrey, A. and **Taniuti, T.** (1964) *Non-linear Wave Propagation with Applications to Physics and Magnetohydrodynamics*, Academic Press, New York.

Mathews, J. and **Walker, R.L.** (1970) *Mathematical Methods of Physics*, 2nd edn, Addison-Wesley, pp. 98–99.

McAdams, W.H. (1954) *Heat Transmission*, 3rd edn, McGraw-Hill, pp. 75–80.

Mitchell, A.R. and **Griffiths, D.F.** (1980) *The Finite Difference Method in Partial Differential Equations*, John Wiley, Chichester.

Peyret, R. and **Taylor, T.D.** (1982) *Computational Methods for Fluid Flow*, Springer, Berlin.

Press, W.H., **Teukolsky, S.A.**, **Vetterling, W.T.**, and **Flannery, B.P.** (1992) *Numerical Recipies in FORTRAN*, 2nd edn (similar volumes available for C, Pascal, and BASIC), Cambridge University Press, Cambridge, p. 963.

Roetzel, B. and **Xuan, Y.** (1999) *Dynamic Behaviour of Heat Exchangers, Developments in Heat Transfer*, vol. 3, WIT Press, Southampton.

Chapter 10
Single-Blow Test Methods

Single-blow testing is for correlations on one side of an exchanger at a time

10.1 Features of the test method

This test method is used to obtain heat-transfer and flow-friction correlations for one side of a heat exchanger at a time. It is not suitable for testing real exchangers – a possible test-rig for this purpose is illustrated in Fig. 8.5.

The two essentials in single-blow testing are:

(a) a state-of-the-art experimental test-rig with measuring equipment of the highest accuracy
(b) theoretical modelling of the transient process, to match the real process

Experimental test-rig

The test equipment developed in the engineering laboratories at the University of Newcastle upon Tyne is fully described in the thesis by Coombs (1970). The open cycle rig comprised the following items starting at the flow inlet:

- flared inlet section 457 mm (18 in) square to eliminate flow separation
- honeycomb flow straightener to minimize swirl
- contraction section designed to National Physical Laboratory pattern to further decrease longitudinal component of turbulence
- wire-mesh screen at contraction outlet to flatten the velocity profile
- auxiliary in-plane heater to correct for variation in ambient temperature
- rapid-response in-plane heater controlled by thyristors (giving around 6 K temperature rise)
- inlet pressure tappings in duct wall
- inlet platinum resistance thermometer of 0.025 mm diameter wire
- test section (150 mm square × 240 mm long) in Tufnol sheet (6 mm thick) to receive model surfaces
- outlet platinum resistance thermometer of 0.025 mm diameter wire
- use of Rosemount d.c. resistance bridges to measure temperature
- pressure recovery section with a series of outlet pressure tappings to allow search for the point of maximum pressure recovery

- use of a Betz manometer to measure pressure loss across the test section
- transition section from 150 mm square to 100 mm round
- long circular tube 100 mm (4 in) diameter carrying a British Standard orifice plate with pressure tappings to BS 1042
- flow control valve, leading to compressor suction

Theoretical modelling

The requirement is that theoretical modelling should match the experimental technique as closely as possible. The experimental set-up and the theoretical model need to be considered together to achieve best results.

Ambient air is usually the test medium. The 6 K temperature rise is so limited to allow the assumption of constant physical properties. This permits assumption of steady-flow conditions, making it unnecessary to include solution of the balance of mass (density field) and balance of linear momentum (density $\times$ velocity field) equations, leaving only the balance of energy (temperature field) equations for solution. When the change in physical properties becomes significant, or the assumption of steady mass flowrate no longer holds, then a more comprehensive transient model is required.

The test method is to create a known temperature disturbance in the steady inlet flow to the test section and measure the resulting temperature response at outlet from the test section. The theoretical method is to model the inlet temperature disturbance and look at the predicted outlet temperature response.

The shape of the experimental time/temperature inlet temperature disturbance needs to be measured accurately, and when this can be made to correspond to an exponential temperature rise, theoretical modelling by Laplace transforms is greatly simplified. When an arbitrary inlet temperature disturbance is used then the theoretical model must be capable of matching the disturbance.

Once an inlet experimental disturbance curve has been matched, the task is to use the theoretical model to predict the outlet experimental response curve. This is done by progressively changing the theoretical value of *Ntu* until a best match between response curves is obtained. Best results are obtained by matching complete experimental curves with the theoretical model using a 'least-squares' method.

One test run produces a single point for the heat-transfer correlation, and a single point for the pressure-loss correlation.

10.2 Choice of theoretical model

Two approaches to modelling the transient may be used, e.g. the Laplace transform method produces the necessary temperature/time curves at the desired stations, or a full numerical method can be used to produce temperature/length plots with time as parameter (Fig. 9.2) from which the necessary temperature/time curves at desired stations can be obtained.

Table 10.1 Experimental papers

Input disturbance	*Representative experimental papers*	*Year*
Sine wave	Bell & Katz	1949
Sine wave	Meek	1962
Sine wave	Hart & Szomanski	1965
Step	Mondt	1961
Step	Pucci, Howard & Piersall	1967
Square wave	Close	1965
Exponential	Coombs	1970
Exponential	Liang & Yang	1975
Exponential	Elliott & Rapley	1988

Inlet disturbances

In testing it is essential that initial conditions in the rig are isothermal, as this is generally the assumption in the theoretical model. Ambient temperatures must therefore be measured during the tests.

While it is certainly possible to generate a good exponential inlet temperature/time disturbances by careful control of the power input to the fast response heater, it is to be remembered that each data point pair for heat transfer and flow friction implies a change in Reynolds number, and hence a new setting of mass flowrate. This requires adjustment of thyristor settings for the fast-response heater, and some time to find the correct settings.

For the fast-response heater, Coombs (1970) used tightly wound coils of nichrome wire supported on elliptical alumina ceramic rods with the longest axis of the ellipse set parallel with the flow stream. The resistance wires were then almost totally in the free air stream, touching the ceramic rods only at two points. This allowed the electrical resistance to be made high enough to allow shaping of the temperature disturbance through input power control by thyristors. Other researchers listed in Table 10.1 have used a variety of methods to produce the inlet disturbance, the latest being Adams *et al.* (2003) who employed a heater mesh arrangement.

The experimental procedure becomes much simpler if the theoretical model can accept any reasonable arbitrary shape of inlet disturbance.

10.3 Analytical and physical assumptions

Since the pioneer work of Anzelius (1926), Nusselt (1927), Schumann (1929), and Hausen (1929), a variety of theoretical solutions for thermal response of an initially isothermal matrix subject to specific forms of inlet fluid temperature disturbances have been formulated.

With zero longitudinal conduction, the single-blow transient test method may be used to determine heat transfer in crushed rock beds, plate–fin heat exchanger cores, tube banks, sphere beds, etc. When longitudinal conduction is present, e.g. compact-fin surfaces and RODbaffle tubes, a theoretical model including this condition is required.

Some desirable physical requirements are listed below:

1. Thermal and physical properties of the gas and matrix are independent of temperature (implying that the temperature change of inlet disturbance is small compared with the absolute temperature of the gas). This means that gas velocity (u_g) can be assumed constant, and solution of both balance of mass and balance of linear equations is not required.
2. Thermal conductivity of the matrix material is infinitely large in the direction normal to the flow, and infinitely small in the direction parallel to the flow (implying negligible heat loss from the test matrix casing, implying near-ambient test conditions, with a small axial conduction path within the matrix itself). Corrections can be applied for heat loss from the test matrix casings.
3. Thermal capacity of the gas in the matrix at any instant ($\tilde{m}_g$) is small compared with the thermal capacity of the matrix itself (M_b) (e.g. implying air as test fluid, copper for test matrix).
4. Surface temperatures and bulk temperatures for the solid matrix during thermal transients are indistinguishable (implying Biot number, Bi $\rightarrow$ 0, i.e. solid must be thin, and/or have high thermal conductivity).
5. Initial test conditions should be isothermal. Circumstances may require departure from the above conditions, e.g. the requirement to test at much higher temperatures may introduce heat loss from the matrix surface and therefore transverse temperature gradients within the test matrix and the gas. In this case the bulk temperature within the solid may have to be related to surface temperatures and longitudinal diffusion within the gas may become significant. Additional terms in the equations will then be required.

For the physical assumptions specified, a variety of mathematical attacks on the transient test technique have been published for different input disturbances. It seems useful to bring these together in a single general solution capable of accepting the range of input disturbances listed in Table 10.1. The analysis given is for initially isothermal conditions in the absence of longitudinal conduction. Theoretical and experimental aspects are discussed further in Appendix E.

There is no a priori reason why a finite-difference approach cannot be used for single-blow testing to accommodate arbitrary inlet temperature disturbances. This could simplify the experimental side of single-blow testing.

10.4 Simple theory

Coupled fluid and solid equations

Although the single-blow technique is for obtaining the heat-transfer coefficient between fluid and solid, internal conduction in the solid also exists. Thus two subscripts are involved in describing the solid: b for bulk properties and s for surface properties. The fluid is best chosen to be a perfect gas, and the subscript g is used for the fluid.

Energy balance
Equations for one fluid only

$$\begin{cases} \dfrac{\partial T_g}{\partial t} + u_g \dfrac{\partial T_g}{\partial x} = -\dfrac{u_g}{L}\left(\dfrac{\alpha S}{\dot{m}_g C_g}\right)(T_g - T_s) & \text{hot gas} \\ \dfrac{\partial T_b}{\partial t} - \kappa \dfrac{\partial^2 T_b}{\partial x^2} = +\left(\dfrac{\alpha S}{M_b C_b}\right)(T_g - T_s) & \text{solid matrix} \end{cases}$$

For transient solutions, $\theta = T - T_{ref}$ is used for temperature where the reference temperature is measured at the time of testing.

Fluid – perfect gas

$$\boxed{\frac{\partial \theta_g}{\partial t} + u_g \frac{\partial \theta_g}{\partial x} = -n_g\left(\frac{u_g}{L}\right)(\theta_g - \theta_b)} \tag{10.1}$$

Solid
Without longitudinal conduction the solution is further simplified.

$$\frac{\partial \theta_g}{\partial t} = +\left(\frac{\alpha S}{\dot{m}_g C_g}\right)\left(\frac{\dot{m}_g C_g}{M_b C_b}\right)(\theta_g - \theta_b) = n_g\left(\frac{\dot{m}_g C_g}{M_b C_b}\right)(\theta_g - \theta_b)$$

Defining residence mass $\tilde{m}_g = \dot{m}_g(L/u_g)$ and parameter $R_{bg} = M_b C_b/(\tilde{m}_g C_g)$

$$\boxed{\frac{\partial \theta_s}{\partial t} = +\frac{n_g}{R_{bg}}\left(\frac{u_g}{L}\right)(\theta_g - \theta_s)} \tag{10.2}$$

When we can assume thin sections with high thermal conductivity the surface temperature (θ_s) can be taken as equal to bulk temperature (θ_b), which simplifies the solution considerably.

The next step is non-dimensionalization and scaling. In the overall notation scheme X and T would normally be used, but this would take the notation away from that normally favoured by workers in Laplace transforms and it was considered preferable to use ξ and τ. The fluid residence time ($\tau_g = L/u_g$) will not be used in this analysis, we shall instead work with the right-hand expressions of these equations. The local value of $Ntu = n_g$ is the only value of Ntu in this solution, from which the heat-transfer correlation would eventually be constructed.

Non-dimensional scaling of length, $\xi = Ntu(x/L)$ and non-dimensional modification and scaling of time,

$$\tau = \left(\frac{Ntu}{R_{bg}}\right)\left(\frac{ut - x}{L}\right)$$

then as Bi $\to$ 0 we may put $\theta_s \to \theta_b$. With temperature excesses $B = \theta_b - \theta_i$ and $G = \theta_g - \theta_i$ over some initial value θ_i, equations (10.1) and (10.2) become

$$\frac{\partial G}{\partial \xi} = (B - G) \qquad \text{(Fluid)} \tag{10.3}$$

$$\frac{\partial B}{\partial \tau} = (G - B) \qquad \text{(Solid)} \tag{10.4}$$

Solution of basic equations

Analytical solutions by Laplace transforms or by fast-Fourier transforms are available, but when temperature-dependent physical properties are encountered numerical methods may be easier to implement. Taking Laplace transforms

$$L\left\{\frac{\partial G}{\partial \xi}\right\} = \frac{d\hat{G}}{d\xi}$$

$$L\left\{\frac{\partial B}{\partial \xi}\right\} = s\hat{B} - B(\xi, 0)$$

Term $B(\xi, 0)$ is the initial temperature distribution in the matrix. For isothermal conditions at the start of blow $B(\xi, 0) = 0$, which keeps the solution simple, see e.g. Kohlmayr (1968a), then

Fluid

$$\frac{d\hat{G}}{d\xi} = \hat{B} - \hat{G} \tag{10.5}$$

Solid

$$s\hat{B} = \hat{G} - \hat{B} \tag{10.6}$$

Combining equations (10.5) and (10.6) to obtain fluid temperatures

$$\frac{d\hat{G}}{d\xi} + \left(1 - \frac{1}{s+1}\right)\hat{G} = 0$$

which has the solution

$$\hat{G} = A \exp\left[\left(\frac{1}{s+1} - 1\right)\xi\right]$$

where A is to be determined from the boundary conditions.

Boundary conditions

At inlet

$$\xi = 0$$

$$\hat{G}_1(0, s) = A = \hat{g}(s)$$

where $\hat{g}(s)$ is defined as the Laplace transform of the inlet fluid disturbance. Thus

$$\hat{G} = \hat{g}(s)\exp\left[\left(\frac{1}{s+1} - 1\right)\xi\right] \tag{10.7}$$

At outlet

$$\xi = Ntu$$

$$\hat{G}_2(Ntu,\, s) = \hat{g}(s)\exp\left[\left(\frac{1}{s+1} - 1\right)Ntu\right]$$

Inverse transforms

Applying inverse Laplace transforms to outlet fluid temperature response

$$G_2 = \exp(-Ntu)L^{-1}\left\{\exp\left(\frac{Ntu}{s+1}\right)\hat{g}(s)\right\}$$

$$= \exp(-Ntu)\int_0^{\tau}\left[\delta(\sigma) + \exp(-\sigma)\frac{Ntu\, I_1(2\sqrt{Ntu\,\sigma})}{\sqrt{Ntu\,\sigma}}\right]G_1(\tau-\sigma)d\sigma$$

$$= \exp(-Ntu)\left[\int_0^{\tau} G_1(\tau-\sigma)\delta(\sigma)d\sigma + \int_0^{\tau} G_1(\tau-\sigma)\frac{e^{-\sigma}Ntu\, I_1(2\sqrt{Ntu\,\sigma})}{\sqrt{Ntu\,\sigma}}d\sigma\right]$$

where the Dirac δ-function has the property of 'sifting out' the value of another integrand at time zero, then

$$G_2 = \exp(-Ntu)\left[G_1(\tau) + \int_0^{\tau} G_1(\tau-\sigma)R(\sigma)d\sigma\right]$$

where

$$R(\sigma) = e^{-\sigma}\frac{Ntu \cdot I_1(2\sqrt{Ntu \cdot \sigma})}{\sqrt{Ntu \cdot \sigma}}$$

With non-dimensional inlet disturbances (D) given in Table 10.2 the general solution for outlet fluid temperature response becomes

$$\boxed{G^{\#} = \exp(-Ntu)\left[D(\tau) + \int_0^{\tau} D(\tau-\sigma)R(\sigma)d\sigma\right]} \tag{10.8}$$

When solid temperatures are required, combining equations (10.6) and (10.7)

$$\hat{B} = \hat{g}(s)\frac{1}{s+1}\exp\left[\left(\frac{1}{s+1} - 1\right)\xi\right] \tag{10.9}$$

Table 10.2 Inlet disturbance

Inlet disturbance	*Non-dimensional* $D(\tau)$	*At* $x = 0$
Step	1	$0 \rightarrow 1$
Exponential	$1 - k\exp(-\tau/\tau^*)$	$\tau/\tau^* = t/t^*$
First harmonic	$a_0 + a_1\cos(\omega\tau) + b_1\sin(\omega\tau)$	$(\omega\tau) = \omega t$

At outlet

$$\xi = Ntu$$

$$\hat{B}_2(Ntu, s) = \hat{g}(s)\,\frac{1}{s+1}\exp\left[\left(\frac{1}{s+1} - 1\right)Ntu\right]$$

Tables E.1, E.2 and E.3 of Laplace transforms given in Appendix E include inversions which were not to be found in the mathematical literature. Applying inverse Laplace transforms to the outlet matrix temperature response

$$\begin{aligned} B_2 &= \exp(-Ntu)\,L^{-1}\left\{\frac{1}{s+1}\exp\left(\frac{Ntu}{s+1}\right)\hat{g}(s)\right\} \\ &= \exp(-Ntu)\int_0^t e^{-\sigma} I_0(2\sqrt{Ntu\,\sigma})\,G_1(\tau-\sigma)d\sigma \\ &= \exp(-Ntu)\int_0^t G_1(\tau-\sigma)\,P(\sigma)d\sigma \end{aligned}$$

where $P(\sigma) = e^{-\sigma} I_0(2\sqrt{Ntu\,\sigma})$

With non-dimensional inlet disturbances D given in Table 10.2, the general solution for outlet matrix temperature response becomes

$$\boxed{B^{\#} = \exp(-Ntu)\int_0^t D(\tau-\sigma)P(\sigma)d\sigma} \qquad (10.10)$$

Temperatures elsewhere in the matrix may be found by inserting other values for ξ in equations (10.8) and (10.10) or by using fictitious values for L.

For the step input disturbance it is easily shown that the temperature difference (gas–solid) at outlet is

$$(G^{\#} - B^{\#})_{step} = \exp(Ntu\,\tau)I_0(2\sqrt{Ntu\,\tau})$$

and that the slope of the outlet response at any point is

$$\frac{dG^{\#}}{d\tau} = \exp(-Ntu)\,R(\tau)$$

and that in terms of an independent parameter (a), locus of maximum slope

$$Ntu\,P(\tau) = (1+\tau)R(\tau)$$

is given by

$$\tau = \frac{a}{2}\frac{I_0(a)}{I_1(a)} - 1$$

subject to $2 \le Ntu \le \infty$ with $Ntu = (a^2/4t)$ and $\beta = (\tau/Ntu)$. This last relationship was obtained in more complicated form by Kohlmayer (1968b), both expressions giving the same shaded curve in Fig. 10.1.

Attempts to obtain a similar expression for the locus of maximum slope of the outlet response for exponential input disturbance leads to the condition

$$\int_0^t \exp\left(\frac{\sigma}{\tau^*}\right) R(\sigma) d\sigma = \frac{\tau^* R(\tau)}{\exp(-\tau/\tau^*)}$$

Evaluation of this expression was not carried out. Instead the position of maximum slope was determined numerically during evaluation of $G^{\#}_{exp}$ and the results are plotted in Fig. 10.1.

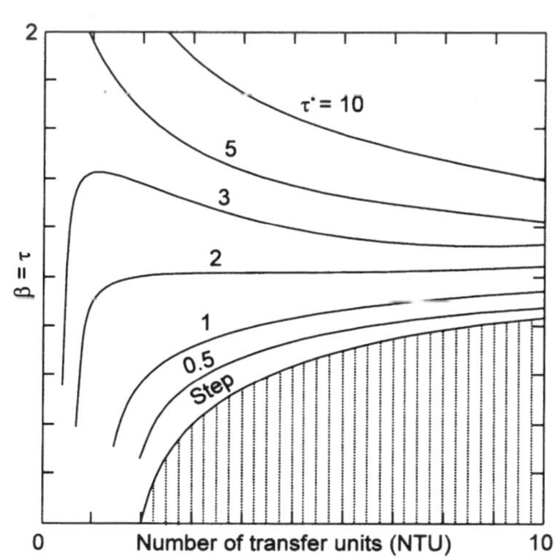

Fig.10.1 Locus of maximum slope for exponential input

10.5 Relative accuracy of outlet response curves in experimentation

Comparison of methods

The *Ntu* value corresponding to a given experimental outlet response curve is determined through seeking the mathematical outlet response prediction which has the identical shape. Four techniques of comparison have been proposed, namely:

- complete curve matching
- maximum slope
- initial rise
- phase angle and amplitude

Complete curve matching for both step and exponential inlet disturbances may be by least-squares fit or by using a direct optimization simplex method (Spendley *et al.*, 1962; Nelder & Meade, 1965; Coombs, 1970; Kohlmayer, 1971; Parkingson & Hutchinson, 1972). *Maximum slope* has been used by Locke (1950) and later by Howard (1964) with step inputs for the case of longitudinal conduction in the matrix.

For exponential input with zero longitudinal conduction, new *Ntu* versus locus of maximum slope curves are shown in Fig. 10.1. Practicable fast-response heaters have exponential time constants around $\tau^* = 0.2$ giving a locus of maximum slope curve close to that for a step response. By choosing an inlet disturbance constant ($\tau^* = 2.0$) an almost linear relationship between *Ntu* and $\tau_{max\ slope}$ may be obtained, and additionally it becomes possible to evaluate experimentally (with some resolution) values of *Ntu* down to about 1.0.

Figures 10.2 and 10.3 illustrate both step and exponential ($\tau^* = 2.0$) dimensionless response curves for zero longitudinal conduction calculated using equation (10.8). The *initial rise* technique proposed by Mondt & Siegla (1972) makes use of the fact that the intercept of the response to a step input at $\tau = 0$ has the value $\exp(-Ntu)$ from analytical solutions for both zero and infinite longitudinal conduction in the matrix (Mondt, 1961), it being postulated that the same result will hold for intermediate values.

No heater has been devised that will produce a perfect step input (Kramers & Alberda, 1953), and although output response curves for step and exponential input ($\tau^* = 0.2$) are virtually identical down to *Ntu* values of about 5.0 it is clear that the initial rise method should be avoided completely, and the maximum slope method used only with knowledge of τ^*. The complete curve matching technique is the safest.

Figure 10.4 for first harmonic responses (with $a_0 = 1$, $a_1 = -1$, $b_1 = 0$, and $\bar{\omega} = 1.0$) illustrates initial stages of steady-cyclic methods of testing, in which values of *Ntu* may be calculated either from the measurements of the ratio of amplitude of the varying fluid temperature at outlet to that of inlet, or alternatively from measurement of the phase lag between inlet and outlet fluid temperature variations.

A separate theoretical analysis may be used when steady-cyclic conditions have been attained, e.g. Bell & Katz (1949), Meek (1961), and Shearer (1962) who considered finite radial conductivity within the solid, and Stang & Bush (1972) who examined the case of longitudinal conduction within the matrix.

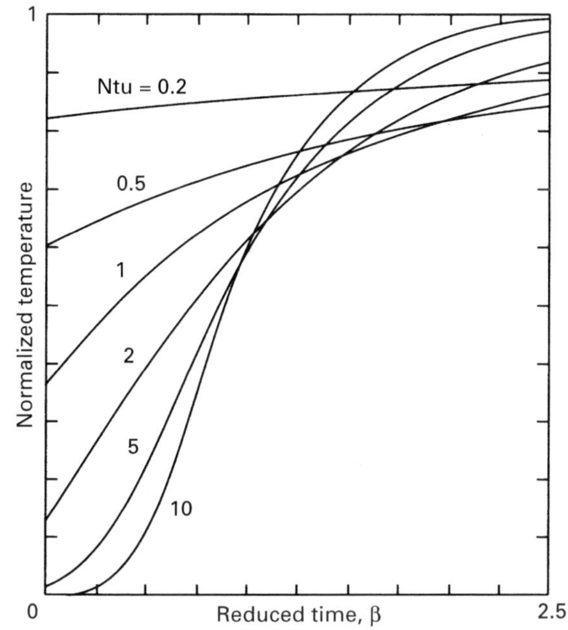

Fig.10.2 Response from step input disturbances

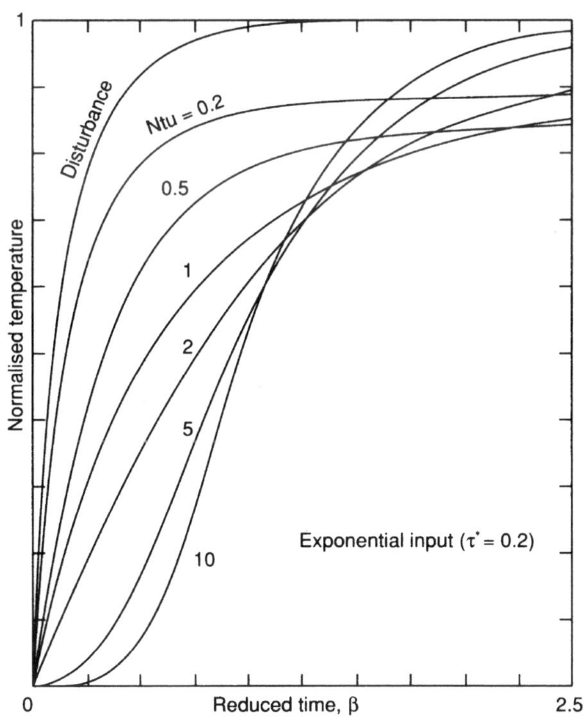

Fig.10.3 Response from exponential input disturbances

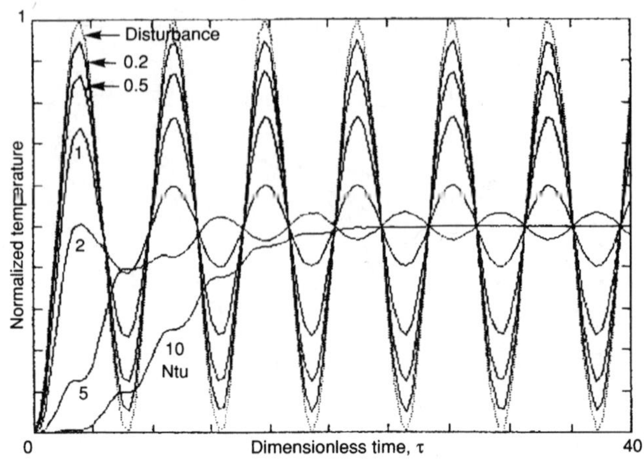

Fig.10.4 Response from harmonic input disturbances

On the precision of the cyclic method, Meek (1962) observed some apparent variation in measured heat-transfer values against frequency, which he attributed to inaccuracies associated with very small downstream temperature amplitudes. Bell & Katz (1949) advised ten heating cycles before measurement of amplitude and phase angle are taken. For a given *Ntu*, Stang & Bush (1972) showed that one frequency exists at which best test results are produced for a given uncertainty in temperature measurement, and they recommend cyclic methods for values of *Ntu* for the difficult range ($0.2 < Ntu < 5.0$). However, from Fig. 10.4 it seems that there would be less difficulty in resolving exponential response curves using complete curve matching in the single-blow technique.

In all these methods involving use of Laplace transforms the experimental input disturbance must follow as near as practicable the form of the mathematical input disturbance. In single-blow testing this can be arranged using thyristor control of the fast-response electrical heater generating the input disturbance (Coombs, 1970). An alternative design of fast-response heater has been developed by Adams *et al.* (2003).

Generating theoretical response curves

In obtaining theoretical response curves, two methods are available to the investigator:

- direct application of numerical procedures to the physical problem
- mathematical derivation of integral expressions which are subsequently evaluated numerically

The direct method starting from the basic differential equations was favoured by Dusinberre (discussion to Coppage & London, 1953). It has also been employed by Johnson (1948), by Howard (1964), and by other workers in an early form. With powerful computers arbitrary shapes of input disturbance can now be handled, and presently this is the preferred approach.

A general solution for the second of the above options has been presented in this chapter in more analytical form to allow comparison of different experimental techniques, without which there would be no benchmark to compare the completely numerical method. Appendix E.2 provides a convenient method for dealing with the integration.

Additional effects which would complicate the canonical solution provided in this paper include:

- longitudinal conduction in the solid (Howard, 1964; Stang & Bush, 1972)
- axial and longitudinal diffusion in the fluid (Amundson, 1956)
- surface losses from the matrix exterior (Dabora, 1957; Dabora *et al.*, 1959)
- internal heat generation (Brinkley, 1947; Clark *et al.*, 1957, 1959, 1961; Chiou & El-Wakil, 1966)
- conduction into the solid interior (Meek, 1961; Smith & Coombs, 1972)

The most important of these is longitudinal conduction in the solid, because high thermal conductivity in the solid matrix is desirable for accurate results. Conduction into the solid interior is less important, as hollow sections can be used. In practice, Smith & Coombs (1972) found that little difference could be detected replacing solid copper rods with hollow copper rods, providing the obvious adjustment is made in calculating mass of the matrix.

10.6 Conclusions on test methods

1. The 'initial rise' method of determining *Ntu* is invalid for any practicable heater.
2. The 'maximum slope' method requires accurate knowledge of heater *exponential* response time constant τ^*, and should be used with caution when determining values of $Ntu < 5.0$.
3. Practical complications exist with the 'periodic method' in that Fourier analysis is required in order to extract the first harmonic from inlet and outlet temperature waves.
4. Single-blow testing with 'complete curve matching' by computer is recommended. Both inlet disturbance and outlet response curves have to be measured accurately whether a fully numerical or an analytical/numerical method is used.
5. A fully numerical approach is desirable to eliminate 'tuning' of the electrical input heater to deliver exponential input disturbances. Both the direct finite-difference method, and the method of characteristics can accommodate arbitrary inlet disturbances close to the exponential form.

10.7 Practical considerations

Earlier sections, directed at selection of the best method of testing, used only the simplified coupled equations (10.1) and (10.2).

Experimental testing requires some additional consideration of physical assumptions made in the analysis. In determining mean heat-transfer coefficients in tube bundles the choice was of materials of construction (copper rods) for the matrix and Tufnol for the test section, to obtain physical properties best matching the mathematical assumptions made in the theory. The test fluid (ambient air) is close to a perfect gas, and the temperature rise introduced into the flow during testing was kept small (e.g. 5–10 K) so as not to change bulk physical properties of the gas. For matrices that are connected in the longitudinal direction, the small temperature rise also helps minimize longitudinal conduction effects.

$$\frac{\partial \theta_g}{\partial t} + u_g \frac{\partial \theta_g}{\partial x} = -\frac{n_g}{\tau_g}(\theta_g - \theta_s) = -\frac{n_g u_g}{L}(\theta_g - \theta_s)$$

$$\frac{\partial \theta_b}{\partial t} = +\frac{n_g}{R_g \tau_g}(\theta_g - \theta_s) = +\frac{n_g}{R_g}\frac{u}{L}(\theta_g - \theta_s)$$

The fuller equations describing the single-blow process for a tube-bundle matrix in crossflow are

$$\frac{\partial \theta_g}{\partial t} + u_g \frac{\partial \theta_g}{\partial x} = -\frac{n_g u_g}{L}(\theta_g - \theta_s) \qquad \text{Fluid}$$

$$\frac{\partial \theta_b}{\partial t} = +\frac{n_g}{R_g}\frac{u_g}{L}(\theta_g - \theta_s) \qquad \text{Solid}$$

$$\frac{\partial \theta_{r,x,t}}{\partial t} = \frac{\lambda_b}{\rho_b C_b}\left[\frac{\partial^2 \theta_{r,x,t}}{\partial r^2} + \frac{1}{r}\frac{\partial \theta_{r,x,t}}{\partial r}\right] \qquad \text{Transient conduction}$$

$$\theta_{s(r,x,t)} = \theta_{a,x,t} \qquad \text{Surface temperature}$$

$$\theta_{b(x,t)} = \frac{1}{\pi a^2}\int_0^a 2\pi r\, \theta_{r,x,t}\, dr \qquad \text{Bulk temperature}$$

The *solid* equation relates the bulk temperature (θ_b) to the surface temperature (θ_s), the *bulk temperature* equation defines (θ_b) itself, while the *surface temperature* equation selects one solid temperature, surface temperature, from all possible solid temperatures ($\theta_{a,x,t}$).

The fuller set of equations is required when poor thermal conductivity in the solid causes bulk temperature to be different from surface temperature (i.e. slow thermal transients in the solid). Don't go down this route if at all possible.

Longitudinal conduction

The governing equations for the case including longitudinal conduction were given earlier as equations (10.1) and (10.2). When the solid material has significant thermal gradients within its mass, then additional equations are required to represent bulk and surface temperatures

$$
\begin{aligned}
\frac{\partial \theta_g}{\partial t} + u_g \frac{\partial \theta_g}{\partial x} &= -n_g\left(\frac{u_g}{L}\right)(\theta_g - \theta_s) \\
\frac{\partial \theta_b}{\partial t} - \kappa_b \frac{\partial^2 \theta_b}{\partial x^2} &= +\frac{n_g}{R_{bg}}\left(\frac{u_g}{L}\right)(\theta_g - \theta_s)
\end{aligned}
$$

Usually wall sections are thin, allowing replacement of (θ_b, θ_s) with the wall temperature (θ_w). The equations may be further simplified by normalizing x and scaling t from where

$$
\left.
\begin{aligned}
\frac{\partial \theta_g}{\partial T} + \frac{\partial \theta_g}{\partial X} &= -n_g(\theta_g - \theta_w) \\
\frac{\partial \theta_w}{\partial T} - \mathrm{Fo}\frac{\partial^2 \theta_w}{\partial X^2} &= +\frac{n_g}{R_{bg}}(\theta_g - \theta_w)
\end{aligned}
\right\} \quad \text{where } \mathrm{Fo} = \kappa\left(\frac{\tau_g}{L^2}\right) \text{ is Fourier number}
$$

Solution by Laplace transforms does not include allowance for temperature-dependent physical properties.

10.8 Solution by finite differences

The more general finite-difference solution approach is capable of handling:

- arbitrary inlet disturbances (in both temperature and mass flowrate)
- temperature-dependent physical properties
- longitudinal conduction terms
- changes in fluid velocity and density (if required)
- direct evaluation of the heat-transfer coefficient (α)

The single-blow equations given below are a subset of the transient contraflow equations described in Appendix A.2. The equations use a slightly different notation to that used in the Laplace transform approach which is looking for *Ntu* values (and includes physical properties) whereas the finite-difference approach is looking for heat-transfer coefficients alone.

Full computation

Subset of the equations for transient response of contraflow heat exchanger

$$
\text{Hot mass flow}
\begin{cases}
\dfrac{\partial \rho_h}{\partial t} + u_h \dfrac{\partial \rho_h}{\partial x} = -\rho_h \dfrac{\partial u_h}{\partial x} & \text{density} \\
\dfrac{\partial (\rho u)_h}{\partial t} + u_h \dfrac{\partial (\rho u)_h}{\partial x} = -(\rho u)_h \dfrac{\partial u_h}{\partial x} - \left[\dfrac{\partial p_h}{\partial x} + \left(\dfrac{f \rho u^2}{2 r_{hyd}}\right)\right] & \\
& \text{density} \times \text{velocity}
\end{cases}
$$

$$\text{Temperatures}\begin{cases} \dfrac{\partial T_h}{\partial t} + u_h\dfrac{\partial T_h}{\partial x} = -\dfrac{u_h}{L}\left(\dfrac{\alpha_h S}{\dot{m}_h C_h}\right)(T_h - T_w) & \text{hot fluid} \\ \dfrac{\partial T_w}{\partial t} - \hat{\kappa}_w\dfrac{\partial^2 T_w}{\partial x^2} = +\left(\dfrac{\alpha_h S}{M_w C_w}\right)(T_h - T_w) & \text{solid wall} \end{cases}$$

Simplification

With constant mass flowrate, if the temperature disturbance is limited to say 6 K compared with ambient at 300 K, this would imply a variation of ± 1 per cent about the mean for change in density ($p = \rho RT$) and hence a variation of ∓ 1 per cent in velocity ($\dot{m} = \rho A u$). When this error level is acceptable, the equations to be solved can be reduced to the set for temperatures alone.

Neglecting longitudinal conduction

When working with tube banks in crossflow, the longitudinal conduction term vanishes, then

$$\text{Temperatures}\begin{cases} \dfrac{\partial T_h}{\partial t} + u_h\dfrac{\partial T_h}{\partial x} = -\dfrac{u_h}{L}\left(\dfrac{\alpha_h S}{\dot{m}_h C_h}\right)(T_h - T_w) & \text{hot fluid} \\ \dfrac{\partial T_w}{\partial t} = +\left(\dfrac{\alpha_h S}{M_w C_w}\right)(T_h - T_w) & \text{solid wall} \end{cases}$$

Solution of any of the above sets of equations is by a subset of the Crank–Nicholson finite-difference equations set out in Appendix A. An alternative route followed by Elliott (1985) in his PhD thesis on single-blow testing was to use forward differences for the fluid flow equations and Crank–Nicholson for the temperature equations. The procedures adopted are further described by Elliott & Rapley (1988). Other solutions may use the MacCormack's finite-difference algorithm, or the method of lines with Runge–Kutta (see Appendix B.8).

10.9 Regenerators

A brief discussion of the regenerator problem is appropriate at this point.

A regenerator is a porous matrix through which hot and cold fluids flow alternately, the objective being to transfer thermal energy from one fluid to the other, the matrix acting as a temporary store for the energy, or rather as a temporary store for exergy, Bejan (1988). The fluids occupy the same porous space in the matrix alternately, and may flow in opposite or in the same directions, or in crossflow.

The matrix prism can be stationary, in which case the hot and cold fluid flows are intermittent, or two matrices can be provided, allowing the hot and cold fluids to flow continuously and be directed by rotary valves. The matrix can be a slowly rotating disc, with hot fluid flowing steadily through one sector, while the cold fluid flows steadily through the other sector. The sectors may not be equal.

Regenerator theory is related to single-blow theory, and to the crossflow recuperator problem, but possesses its own attributes. Hausen's classic (1950) text contains analytical solutions, and the later finite-difference work by Willmott (1964) serves to confirm the accuracy of Hausen's original work.

Excellent reviews of the regenerator problem were prepared by Hausen (1979) and separately by Razelos (1979) in papers which appeared in the same publication. Schmidt & Willmott published a textbook on regenerators in 1981, and more recent papers include those of Evans & Probert (1987), Hill & Willmott (1989), Willmott & Duggan (1980), Romie (1990, 1991), and Foumeny & Pahlevanzadeh (1994). Van den Bulck (1991) and San (1993) have both considered optimal control and performance of crossflow regenerators, while Shen & Worek (1993) have optimized the performance of a rotary regenerator considering both heat transfer and pressure loss. Interest in the Stirling cycle has produced quite a number of papers on regenerators under short-cycle conditions, of which Organ's (1992, 1997) books and other papers are representative.

The fundamental regenerator equations in canonical form are

$$\boxed{\begin{aligned} \frac{\partial \theta_b}{\partial \eta} &= -(\theta_b - \theta_g) \\ \frac{\partial \theta_g}{\partial \xi} &= +(\theta_b - \theta_g) \end{aligned}}$$

where θ_g is dimensionless gas temperature and θ_b is dimensionless solid temperature, and both η and ξ are defined below.

$$\left.\begin{aligned} \theta_g &= \frac{T_g - T_{c1}}{T_{h1} - T_{c1}} \\ \theta_b &= \frac{T_b - T_{c1}}{T_{h1} - T_{c1}} \end{aligned}\right\} \qquad \left.\begin{aligned} \eta &= \frac{\alpha S t}{C_b V} \quad \text{reduced time variable} \\ \xi &= \frac{\alpha S x}{C_g L} \quad \text{reduced length variable} \end{aligned}\right\}$$

The analytical problem may be complicated by a number of factors, viz.:

- allowing for residual fluid in the stationary matrix at the end of a blow, or
- allowing for carry-over in the case of a rotary matrix
- allowing for longitudinal conduction in a matrix
- allowing for the shape of real disturbances
- allowing for two- and three-dimensional effects in the matrix
- allowing for the disturbance not 'breaking-through' the matrix under short cycle times, or partially breaking through (Stirling cycle applications)

The reader is encouraged to set up the fundamental regenerator equations (prior to canonical reduction) to see what terms might have been neglected, and to assess whether a full solution by finite differences may be practicable (see e.g. Organ, 1992, 1993, 1994).

References

Adams, J.C., **Ireland, P.T.**, **Cerza, M.**, and **Oswald, J.** (2003) A detailed investigation of a perforated heat transfer surface applied to gas turbine recuperators. ASME Paper 2003-38740.

Amundson, N.L. (1956) Solid–fluid interactions in fixed and moving beds. I – Fixed beds with small particles. II – Fixed beds with large particles. *Ind. Engng Chem., Ind. Edn*, **48**(1), January, 24–43.

Anzelius, A. (1926) Uber erwärmung vermittels durchströmender Medien. *Zeitschrift angewandte Mathematik und Mechanik*, **6**(4), August, 291–294.

Bejan, A. (1988) *Advanced Engineering Thermodynamics*, John Wiley.

Bell, J.C. and **Katz, E.F.** (1949) A method for measuring surface heat transfer using cyclic temperature variations. In *Proceedings of Heat Transfer and Fluid Mechanics Institute*, 22–24 June 1949, pp. 243–254.

Brinkley, S.R. (1947) Heat transfer between a fluid and a porous solid generating heat. *J. Appl. Physics*, **18**, June, 582–585.

Chiou, J.P. and **El-Wakil, M.M.** (1966) Heat transfer and flow characteristics of porous matrices with radiation as a heat source. *Trans. ASME J. Heat Transfer*, **88**, February, 69–76.

Clark, J.A., **Arpaci, V.S.**, and **Treadwell, K.M.** (1957, 1959, 1961) Dynamic response of heat exchangers having internal heat sources. Part I, *Trans. ASME*, 57-SA-14, p. 612. Part II, *Trans. ASME*, 57-HT-6, p. 625. Part III, *Trans. ASME*, 58-SA-39, p. 253, November 1959. Part IV, *Trans. ASME*, 60-WA-127, p. 321, August 1961.

Close, D.J. (1965) Rock pile thermal storage for comfort air conditioning. *Mech. Chem. Engng Trans., Inst. Engrs, Australia*, **MC1**(1), May 1, 11–22.

Coombs, B.P. (1970) A transient test technique for evaluating the thermal performance of cross-inclined tube bundles. PhD Thesis, University of Newcastle upon Tyne.

Coppage, J.E. and **London, A.L.** (1953) The periodic flow regenerator – a summary of design theory. *Trans. ASME*, **75**, 779–787.

Dabora, E.K. (1957) Regenerative heat exchanger with heat-loss consideration. Air Force Office of Scientific Research, AFOSR Technical Note 57-613. (Also ASME Paper 58-SA-29.)

Dabora, E.K., **Moyle, M.P.**, **Phillips, R.**, **Nicholls, J.A.**, and **Jackson, P.** (1959) Description and experimental results of two regenerative heat exchangers. In *AIChE Chemical Engineering Progress Symposium Series*, vol. 55, No. 29, *Heat Transfer*, pp. 21–28. (Paper from: *2nd National AIChE-ASME Heat Transfer Conference, Chicago 1958.*)

Elliott, M. (1985) The singleblow transient heat transfer technique. PhD Thesis, University of Sunderland, UK.

Elliott, M.P. and **Rapley, C.W.** (1988) The effect of solid conduction on the singleblow experimental method. In *Proceedings of the 2nd UK Heat Transfer Conference,* University of Strathclyde, vol. 2, Paper C155/88, pp. 1623–1634.

Evans, R. and **Probert, S.D.** (1987) Thermal performance of counterflow regenerators: a non-iterative method of prediction. *Appl. Energy*, **26**(1), 9–46.

Foumeny, E.A. and **Pahlevanzadeh, H.** (1994) Performance evaluation of thermal regenerators. *Heat Recovery Systems and CHP*, **14**(1), January, 79–84.

Hart, J.A. and **Szomanski, E.** (1965) Development of the cyclic method of heat transfer measurement at Lucas Heights. *Mech. Chem. Engng Trans., Inst. Engrs, Australia*, **MC1**(1), May, 1–10.

Hausen, H. (1929) Über die Theorie des Wärmeaustauchers in Regeneratoren. *Zeitschrift angewandte Mathematik und Mechanik*, **9**(3), 173–200.

Hausen, H. (1950) *Wärmeübertragung im Gegenstrom, Gleichstrom und Kreuzstrom*, 1st edn, Springer, Berlin (see also under 1976 in Appendix H).

Hausen, H. (1979) Developments of theories on heat transfer in regenerators. *Compact Heat Exchangers – History, Technological and Mechanical Design Problems* (Eds. R.K. Shah, C.F. McDonald, and C.P. Howard), ASME Publication HTD-Vol. 10, pp. 79–89.

Hill, A. and **Willmott, A.J.** (1989) Accurate and rapid thermal regenerator calculations. *Int. J. Heat Mass Transfer*, **32**(3), March, 465–476.

Howard, C.P. (1964) The single-blow problem including the effects of longitudinal conduction. ASME Paper 64-GTP-11.

Johnson, J.E. (1948) Regenerator heat exchangers for gas turbines. Aeronautical Research Council Reports and Memoranda, (UK), R&M No. 2630, May, pp. 1025–1094.

Kohlmayer, G.F. (1968a) Analytical solution of the single-blow problem by a double Laplace transform method. *ASME J. Heat Transfer*, **90**(1), February, 176–178.

Kohlmayer, G.F. (1968b) Properties of the transient heat transfer (single-blow) temperature response function. *AIChE J.*, **14**(3), May, 499–501.

Kohlmayer, G.F. (1971) Implementation of direct curve matching methods for transient matrix heat transfer testing. *Appl. Sci. Res.* **24**, June, 127–148.

Kramers, H. and **Alberda, G.** (1953) Frequency response analysis of continuous flow systems. *Chem. Engng Sci.*, **2**, 173–181.

Liang, C.Y. and **Yang, W.J.** (1975) Modified single-blow technique for performance evaluation on heat transfer surfaces. *ASME J. Heat Transfer, Ser. C*, **97**(1), February, 16–21.

Locke, G.L. (1950) Heat transfer and flow friction characteristics of porous solids. Stanford University Technical Report No. 10, Office of Naval Research NR-035-104.

Meek, R.M.G. (1961) The measurement of heat transfer coefficients in packed beds by the cyclic method. In *Proceedings of the International Heat Transfer Conference*, Boulder, Colorado, 1961–1962, ASME Paper 93, pp. 770–780.

Meek, R.M.G. (1962) Measurement of heat transfer coefficients in randomly packed beds by the cyclic method. NEL Report No. 54, National Engineering Laboratory, East Kilbride, UK.

Mondt, J.R. (1961) Effects of longitudinal thermal conduction in the solid on apparent convective behaviour with data on plate-fin surfaces. In *Proceedings of the International Heat Transfer Conference*, Boulder, Colorado, 1961–1962, ASME Paper 73.

Mondt, J.R. and **Siegla, D.C.** (1972) Performance of perforated heat exchanger surfaces. ASME Paper 72-WA-52.

Nelder, J.A. and **Meade, R.** (1965) A Simplex method for function minimization. *Comp. J.*, **7**, 308–313.

Nusselt, W. (1927) Die Theorie des Winderhitzers. *Zeitschrift Vereines deutscher Ingenieure* **71**(3), January, 85–91.

Organ, A.J. (1992) *Thermodynamics and Gas Dynamics of the Stirling Cycle Machine*, Cambridge University Press, Cambridge.

Organ, A.J. (1993) Flow in the Stirling regenerator characterised in terms of complex conditions. Part 1 – Theoretical developments. *Proc. Inst. Mech Engrs, Part C*, **207**(2), 117–124.

Organ, A.J. (1994) The wire mesh regenerator of the Stirling Cycle machine. *Int. J. Heat Mass Transfer*, **37**(16), November, 2525–2534.

Organ, A.J. (1997) *The Regenerator and the Stirling Engine*, Institution of Mechanical Engineers, London.

Parkingson, J.M. and **Hutchinson, D.** (1972) An investigation into the efficiency of variants on the Simplex method. *Numerical Methods for Non-Linear Optimisation* (Ed. F.A. Lootsma), Academic Press, New York.

Pucci, P.F., **Howard, C.P.**, and **Piersall, C.H.** (1967) The single-blow transient test technique for compact heat exchanger surfaces. *ASME J. Engng Power, Ser. A*, **89**(1), January, 29–40.

Razelos, P. (1979) History and advancement of regenerator thermal design theory. *Compact Heat Exchangers – History, Technological and Mechanical Design Problems.* (Eds. R.K. Shah, C.F. McDonald, and C.P. Howard), ASME Publication HTD-Vol. 10, pp. 91–100.

Romie, F.E. (1990) Response of rotary regenerators to step changes in mass rates. *ASME J. Heat Transfer*, **112**(1), February, 43–48.

Romie, F.E. (1991) Treatment of transverse and longitudinal heat conduction in regenerators. *ASME J. Heat Transfer*, **113**, February, 247–249.

San, J.-Y. (1993) Heat and mass transfer in a two-dimensional crossflow regenerator with a solid conduction effect. *Int. J. Heat Mass Transfer*, **36**(3), 633–643.

Schmidt, F.W. and **Willmott, A.J.** (1981) *Thermal Energy Storage and Regeneration*, Hemisphere, Washington.

Schumann, T.E.W. (1929) Heat transfer. A liquid flowing through a porous prism. *J. Franklin Inst.*, September, 405–416.

Shearer, C.J. (1962) Measurement of heat transfer coefficients in low conductivity packed beds by the cyclic method. NEL Report No. 55, National Engineering Laboratory, East Kilbride, UK.

Shen, C.M. and **Worek, W.M.** (1993) Second-law optimisation of regenerative heat exchangers including the effect of matrix heat conduction. *Energy*, **18**(4), April, 355–363.

Smith, E.M. and **Coombs, B.P.** (1972) Thermal performance of cross-inclined tube bundles measured by a transient method. *J. Mech. Engng Sci.*, **14**(3), 205–220.

Spendley, W., **Hext, G.R.**, and **Hinsworth, F.R.** (1962) Sequential application of Simplex design in optimisation and evolutionary optimisation. *Technometrics*, **4**, 441–461.

Stang, J.H. and **Bush, J.E.** (1972) The periodic method for testing heat exchanger surfaces. ASME Paper 72-WA/HT-57.

Van den Bulck, E. (1991) Optimal thermal control of regenerator heat exchangers. *Design and Operation of Heat Exchangers, Proceedings of the EUROTHERM Seminar No. 18*, Hamburg (Eds. W. Roetzel, P.J. Heggs, and D. Butterworth), 27 February–1 March 1991, pp. 295–304.

Willmott, A.J. (1964) Digital computer simulation of a thermal regenerator. *Int. J. Heat Mass Transfer*, **7**, 1291–1302.

Willmott, A.J. and **Duggan, R.C.** (1980) Refined closed methods for the contraflow thermal regenerator problem. *Int. J. Heat Mass Transfer*, **23**, 655–662.

Bibliography

Baclic, B.S. and **Heggs, P.J.** (1990) Unified regenerator theory and re-examination of the unidirectional regenerator performance. *Adv. Heat Transfer*, **20**, 133–179.

Brown, A. and **Down, W.S.** (1976) Melting and freezing processes as a means of storing heat. In *6th Thermodynamics and Fluid Mechanics Conference*, Durham, Institution of Mechanical Engineers, Thermodynamics and Fluid Mechanics Group, 6–8 April 1976, Paper 57/76, pp. 157–163.

Clenshaw, C.W. (1962) Chebyshev series for mathematical functions. National Physical Laboratory. *Mathematical Tables*, Vol. 5, HMSO.

DeGregoria, A.J. (1991) Modelling the active magnetic regenerator. In *Proceedings of the 1991 Cryogenic Engineering Conference, Advances in Cryogenic Engineering*, vol. 37, Part B, Plenum Press, pp. 867–873.

Furnas, C.C. (1930) Heat transfer from a gas stream to a bed of broken solids – II. *Ind. Engng Chemistry, Ind. Edn*, **22**(7), 721–731.

Hamming, R.W. (1962) *Numerical Methods for Scientists and Engineers*, 2nd edn, Chapter 26, McGraw-Hill, New York, pp. 445–458.

Handley, D. and **Heggs, P.J.** (1969) Effect of thermal conductivity of the material on transient heat transfer in a fixed bed. *Int. J. Heat Mass Transfer*, **12**, 549–570.

Hausen, H. (1937) Feuchitgkeitsablagerung in Regenatoren. *Zeitschrift des Verein deutscher Ingenieures, Beiheft "Verfahrenstechnik"*, **2**, 62–67.

Heggs, P. and **Burns, D.** (1986) Single-blow experimental prediction of heat transfer coefficients: a comparison of four commonly used techniques. *Exp. Thermal Fluid Sci.*, **1**(3), July, 243–252.

Jacquot, R.G., **Steadman, J.W.**, and **Rhodine, C.N.** (1983) The Gaver-Stehfest algorithm for approximate inversion of Laplace transforms. *IEEE Circuits Systems Mag.*, **5**(1), March, 4–8.

Kohlmayer, G.F. (1966) Exact maximum slopes for transient matrix heat transfer testing. *Int. J. Heat Mass Transfer*, **9**, 671–680.

Lowan, A.N., **Davids, N.**, and **Levinson, A.** (1954) Table of the zeros of the Legendre polynomials of order 1-16 and the weight coefficients for Gauss mechanical quadrature formula. *Tables of Functions and Zeros of Functions*, NBS Applied Mathematics Series, No. 37, pp. 185–189.

Organ, A.J. and **Rix, D.H.** (1993) Flow in the Stirling regenerator characterised in terms of complex conditions, Part 2 – Experimental investigation. *Proc. Inst. Mech. Engrs, Part C*, **207**(2), 127–139.

Pfeiffer, S. and **Huebner, H.** (1987) Untersuchung zum Einfrieren von Regeneratur-Wärmeübertragern (Investigation of freezing-up in regenerative heat exchangers), *Ki Klima Kälte Heizung*, **15**(10), October, 449–452.

Rapley, C.W. (1978) Regenerator matrices for automotive gas turbines. In *6th International Heat Transfer Conference*, Toronto, Canada, 7–11 August 1978, vol. 4, Paper HX-3, pp. 201–206.

Rapley, C.W. and **Webb, A.I.C.** (1983) Heat transfer performance of ceramic regenerator matrices with sine-duct shaped passages. *Int. J. Heat Mass Transfer*, **26**(6), 805–814.

Smith, E.M. (1979) General integral solution of the regenerator transient test equations for zero longitudinal conduction. *Int. J. Heat Fluid Flow*, **1**(2), 71–75.

Smith, E.M. and **King, J.L.** (1978) Thermal performance of further cross-inclined in-line and staggered tube banks. In *6th International Heat Transfer Conference*, Toronto, vol. 4, Paper HX-14.

Stehfest, H. (1970) Algorithm 368, Numerical Inversion of Laplace Transforms. *Comm. ACM*, **13**(1), January, 47–49.

Willmott, A.J. and **Hinchcliffe, C.** (1976) The effect of heat storage upon the performance of the thermal regenerator. *Int. J. Heat Mass Transfer*, **19**, 821–826.

Willmott, A.J. and **Scott, D.M.** (1993) Matrix formulation of linear simulation of the operation of thermal regenerators. *Numerical Heat Transfer, Part B – Fundamentals*, **23**(1), January–February, 43–65.

CHAPTER 11
Heat Exchangers in Cryogenic Plant

System development and heat exchanger sizing

11.1 Background

Before discussing step-wise rating of cryogenic heat exchangers it is desirable to understand the procedure employed in arriving at the layout of liquefaction plant. While several textbooks exist on the subject of cryogenics, e.g. Scott (1959), Haselden (1971), Barron (1985), and with many examples of complete plants given in these texts and elsewhere, they lack specific instruction as to how to go about designing liquefaction plant starting from a blank sheet of paper.

The author proceeded to investigate the thermodynamics of the process on his own account, and what follows is a distillation of some of the results of these investigations. Early sections in this chapter will discuss a number of basic considerations as they affect plant design, before examining the design of the heat exchangers themselves. In cryogenic plant emphasis is placed on feasibility, simplicity, and performance, and the difference between desirable and practical approaches will be discussed where appropriate.

Begin by considering Fig. 11.1 which is a representation of Carnot efficiency above and below the 'dead-state' temperature at which all heat may be rejected without any possibility of generating further work. This temperature will vary from place to place on the Earth's surface as it is related to local ambient temperature, but to illustrate the concept we will assume the dead-state temperature to be 300 K.

The 'engine' region exists above 300 K, and in this region thermal energy may be partially converted to work, The Carnot efficiency tends to the asymptote of 1.0 as temperature increases.

Between 300 and 150 K exists the 'heat-pump' region in which it is possible to take energy from one temperature level and reject it at a higher temperature level while doing less work that the energy being shifted. The limit of 150 K is where exactly the same amount of energy is shifted as work is done. If a slightly different 'dead-state' temperature is chosen then the lower temperature limit for the 'heat-pump' region changes accordingly.

Below 150 K we have the true 'cryogenic' region where more work is required to shift energy than the energy itself. At 80 K (just above liquid nitrogen (LN_2) saturation temperature at 1 bar) the Carnot work required is 2.75 times the cooling produced. At 20 K close to liquid hydrogen (LH_2) saturation temperature

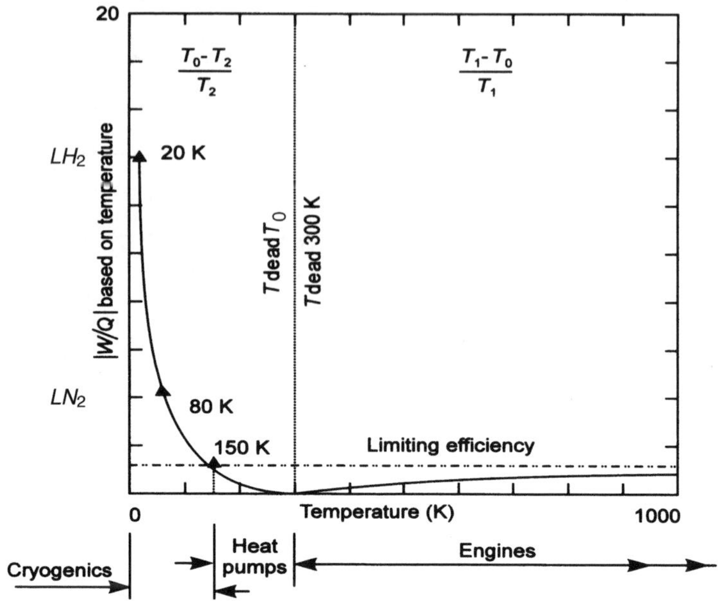

Fig.11.1 Carnot efficiency above and below the dead state

at 1 bar, the Carnot work required is 14 times the cooling produced. There is thus every reason to seek the most efficient thermodynamic means for liquefying gases.

In the region (300 K – 150 K) it is appropriate to consider 'conventional' refrigeration plant using evaporators and condensers as this is the most work-efficient method of cooling available.

Below 150 K two principal means are available for cooling a refrigerating gas. The first involves expansion of high-pressure refrigerating gas in a cryo-turbine with very low frictional losses. By this means, compression work is known to be a principal barrier to improvement in liquefaction performance. The second method involves the use of thermo-magnetic regenerators whose matrix temperature may be changed by application and removal of strong magnetic fields, thus cooling the refrigerating gas at constant pressure. Thermo-magnetic methods can be used with effect at and below liquid helium temperatures, and there have been attempts to extend the method to regions of higher temperature. Peschka (1992) provides a theoretical treatment.

Most commercial plants employ the cryo-turbine method, and this is what will be considered.

11.2 Liquefaction concepts and components

Liquefaction involves cooling a gas below its critical point and in large plant this implies using gas as the refrigerant flowing in contraflow to the product stream. Table 11.1 lists primary cryogens of interest as possible cooling streams.

Table 11.1 Candidate refrigerant fluids

Fluid	*Critical pressure (bar)*	*Critical temp. (K)*	*Saturation temp. @1.0 bar (K)*	*Latent heat (kJ/kg)*	*Gas constant (kJ/kg K)*	*Ratio Cp/Cv (300 K)*
Oxygen	50.9	154.77	90.18	212.3	0.2598	1.396
Argon	50.0	150.86	87.29	159.6	0.2082	1.670
Nitrogen	33.96	126.25	77.35	197.6	0.2968	1.404
Neon	26.54	44.40	27.09	86.1	0.4117	1.640
Hydrogen	12.76	32.98	20.27	434.0	4.157	1.410
Helium	2.3	5.25	4.2	21.0	2.075	1.662

Mixtures of gases with high Joule–Thompson coefficients (e.g. nitrogen–methane–ethane) have produced significant improvements in cooling (Alfeev *et al.*, 1971). In laboratory-scale testing, Little (1984) confirmed Russian claims that cool-down times were reduced from 18 to 2 min and that lower temperatures could be attained with mixtures than by using nitrogen alone. Further work is underway at Stanford University (Paugh, 1990).

Any gas to be liquefied (sometimes a hydrocarbon) will henceforth be referred to as the 'product' stream, and the fluid doing the cooling will be referred to as the 'refrigerating' stream. While establishing the design procedure, we shall restrict ourselves to the gases in Table 11.1.

Forms of hydrogen

Hydrogen has two forms, *ortho*-hydrogen and *para*-hydrogen, which differ in the spins of their protons (Fig. 11.2). These two forms are not isotopes. Above 300 K, the ortho : para concentration ratio remains constant at 75 : 25 and this is known as 'normal' hydrogen. Below 300 K, each temperature level has an equilibrium concentration ratio as shown in Fig. 11.3. As the desired final liquefaction state is 100 per cent para-hydrogen, during cooling of the process stream the objective is to achieve the greatest para concentration at each temperature level. This corresponds to removing the maximum amount of heat at the highest possible temperature levels, i.e. to achieving equilibrium concentration ratio at each temperature level.

From these remarks the reader will appreciate that the two forms of hydrogen have different thermodynamic properties. When consulting data books it might be anticipated that properties would be listed for both ortho- and para-forms. It may come as a mild surprise to find that only normal-hydrogen and para-hydrogen properties are listed. This means that some calculation is required to obtain the properties of equilibrium hydrogen at any temperature level as follows.

The enthalpy of normal hydrogen is given by

$$ {}_{n}h = 0.25\,{}_{p}h + 0.75\,{}_{o}h \tag{11.1}$$

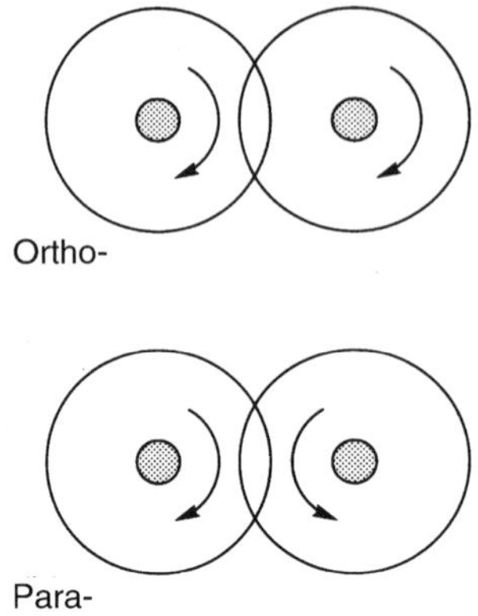

Fig.11.2 Hydrogen molecule configurations

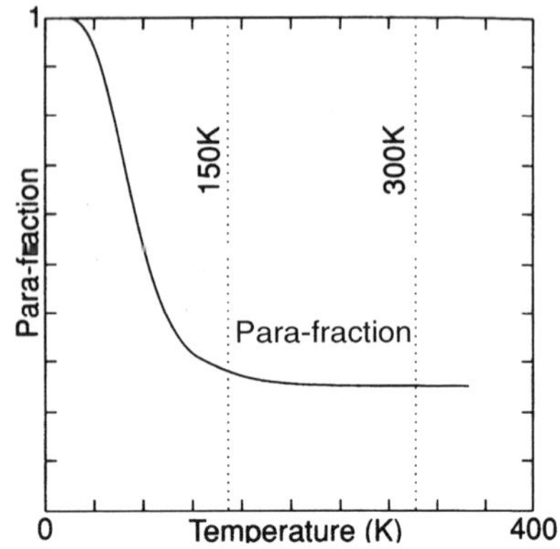

Fig.11.3 Para content versus temperature (K)

Let x be the concentration of para-hydrogen at the desired temperature, the corresponding equilibrium enthalpy is then given by

$$_eh = x_ph + (1 - x)_oh \tag{11.2}$$

Substituting for the enthalpy of ortho-hydrogen from equation (11.1), the enthalpy of equilibrium hydrogen is obtained as

$$\boxed{_eh = y_ph + (1 - y)_nh \quad \text{where } y = (4x - 1)/3} \tag{11.3}$$

Minimum work of liquefaction

This will be illustrated with reference to hydrogen, which is a more complicated case than will be encountered with other gases, but the principles remain the same.

The minimum work of liquefaction of equilibrium hydrogen from 300 K will be compared at different pressure levels of 1, 15, 35, and 50 bar. To make this assessment it is necessary to have values of specific heat at constant pressure. These were obtained by cubic spline-fitting enthalpy data, and then differentiating once to obtain specific heat

$$C = \left(\frac{\partial h}{\partial T}\right)_p$$

The results for four pressure levels at 1, 15, 35, and 50 bar are shown in Fig. 11.4.

The minimum work of liquefaction is evaluated as follows.

In isobaric cooling through δT, the amount of heat removed is $\delta Q = C\delta T$ where C is the specific heat at constant pressure.

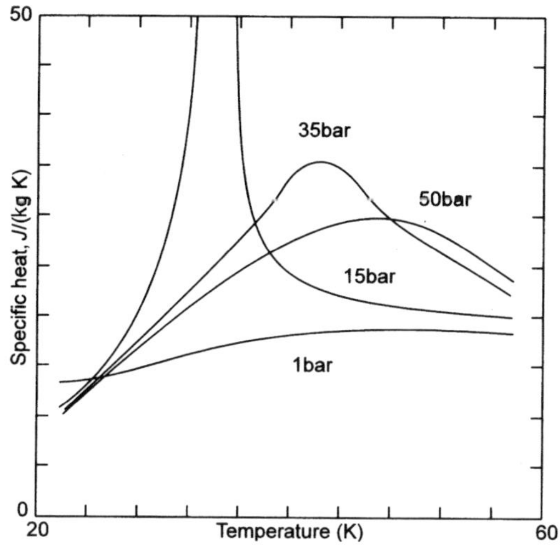

Fig.11.4 Specific heat of equilibrium hydrogen at 1, 15, 35, and 50 bar

The Carnot efficiency is

$$\eta = -\frac{\dot{W}}{\dot{Q}} = -\left(\frac{T_0 - T}{T}\right)$$

Minimum work is then given by

$$\delta W_{\min} = -\int\left(\frac{T_0 - T}{T}\right) C\, dT = -\int\left(\frac{C\,T_0}{T} - C\right) dT \tag{11.4}$$

When specific heat (C) is known as a simple mathematical function of temperature (T), then direct integration of equation (11.4) becomes possible. However, the difficulty of fitting polynomials to the separate curves of Fig. 11.5 is evident, and an alternative method was developed.

If C is constant over a small range $T_a - T_b$, where the subscripts refer to initial and final states, then

$$\Delta W_{\min} = -\bar{C}\left[T_0 \int_a^b \frac{dT}{T} - \int_a^b dT\right] = -\left(\frac{C_b + C_a}{2}\right)[T_0 \ln(T_b/T_a) - (T_b - T_a)]$$

and it is possible to evaluate $W_{\min} = \sum \Delta W_{\min}$ over an extended cooling range (T_1 to T_2) where the mean values of specific heat are taken over

5 K intervals for 300–150 K
2 K intervals for 150–50 K
1 K intervals for 50–20 K

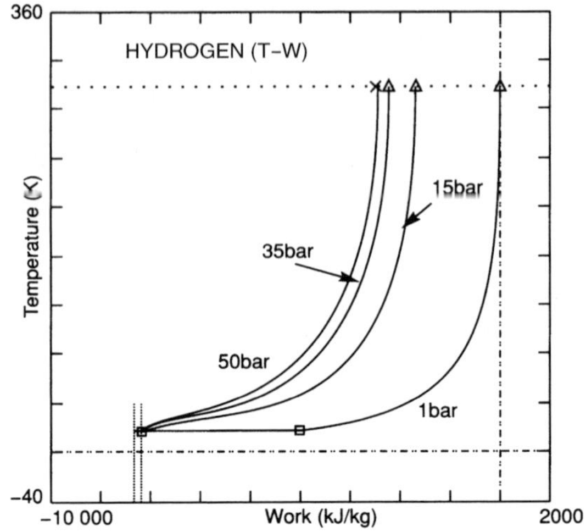

Fig.11.5 Minimum work of liquefaction of hydrogen

For the 1 bar pressure level, a latent heat term

$$-\left(\frac{T_0 - T_2}{T_2}\right)h_{fg}$$

is added to W_{min}, i.e. for the 15, 35, and 50 bar pressure levels, an isothermal work term $-RT_1 \ln(r)$ is added to W_{min}, where r is the compression ratio.

The results are shown in Fig. 11.5 which is a T–W diagram, resembling a T–s diagram, and the two vertical lines shown at the bottom left of the figure represent the maximum and minimum work requirements to liquefy.

While least energy expenditure is achieved by cooling at 1 bar, there is very little difference in work expenditure if the hydrogen is first isothermally compressed to 35 bar and then cooled. A quick look at Fig. 11.1 confirms that any inefficiency in lifting a large amount of latent heat at 1 bar will completely negate the advantage of cooling at 1 bar. Thus in most liquefaction arrangements the product stream is first compressed to supercritical pressure before cooling commences.

Catalysts and continuous conversion

During cooling hydrogen tends to maintain its initial ortho : para ratio, and the conversion ratio can be made rapid enough only by using a catalyst. Ideally the catalyst should be placed inside the heat exchangers used in the cooling, but catalysts can become contaminated. It is presently the practice to provide separate catalyst pots so that the catalyst can be changed if required – (one manufacturer has been brave enough to place catalyst inside the last exchanger, on the basis that any contamination would be caught earlier).

In practice this means that new thermodynamic properties have to be calculated for the constant ortho : para condition between catalyst pots. This is straightforward, once the appropriate scheme for calculating thermodynamic properties has been set up. For other gases, the complication of different molecular forms does not exist.

Catalysts for ortho–para conversion have been described by Newton (1967a,b), Barrick *et al.* (1965), Schmauch & Singleton (1964) and by Keeler & Timmerhaus (1960). Experiments on continuous conversion have been made by Lipman *et al.* (1963), and an arrangement of separate heat exchange and catalytic conversion equipment approximating to this process has been described by Newton (1967a).

The reverse process of para–ortho conversion has also been discussed by Schmauch *et al.* (1963), and the paper provides a list of some 20 candidate catalysts. This is relevant to recovery of maximum cooling effect from the LH_2-vapour return line from the final product storage tank. Substantial amounts of vapour may return via the storage tank while chilling and filling of road-tankers takes place.

Compressors

The product stream must be compressed to supercritical pressure so that cooling may proceed towards the liquid side of the saturation line in the T–s diagram. Refrigerating streams have also to be compressed to suitable pressures.

There is no problem in compressing such gases as oxygen and hydrogen using relatively slow moving reciprocating compressors. Rotary compressors with fast-moving parts may safely be used for inert gases, and may also be used for some hydrocarbons if sufficient care is taken to avoid a high-temperature rub between impellers and casings.

For comparison of prospective compressor arrangements it is practicable to employ an isentropic index of compression to compute the work. When actual machines are constructed then an isentropic efficiency expression $\eta_s = (W_s/W_{real})$ can be used to relate actual performance to the computed value. This avoids having to guess a value for the polytropic index of compression (n), and the isentropic index $\gamma = (C_p/C_v)$ can be used in its place.

Assuming k stages of compression with suction at (p_1, T_1), and final delivery at p_{k+1} with intercooling to T_2, then the expression for minimum work for k stages of compression can be found by standard methods.

$$W_{\min} = -R\left(\frac{\gamma}{\gamma-1}\right)[T_1\{r_1^{(\gamma-1)/\gamma} - 1\} + (k-1)T_2\{r_2^{(\gamma-1)/\gamma} - 1\}]$$

$$\text{where} \qquad r_1 = \left(\frac{p_{k+1}}{p_1}\right)^{\frac{1}{k}} \left(\frac{T_1}{T_2}\right)^{-\left(\frac{\gamma}{\gamma-1}\right)\left(\frac{k-1}{k}\right)}$$

$$\text{and} \qquad r_2 = \left(\frac{p_{k+1}}{p_1}\right)^{\frac{1}{k}} \left(\frac{T_1}{T_2}\right)^{\left(\frac{\gamma}{\gamma-1}\right)\frac{1}{k}} \qquad (11.5)$$

Whenever possible a single-stage compressor is to be preferred (implying restriction of the compression ratio), and plant design may be configured accordingly.

Cryo-expanders

It is not easy to arrange for multi-staging in a single expansion turbine, and the most suitable turbine is the single-stage inward radial flow machine. The limitation on expansion ratio has then to be explored. A relatively crude analysis permits evaluation of comparative pressure expansion ratios for different refrigerant gases, using Fig. 11.6. This suffices for feasibility study of the overall liquefaction system.

For perfect gases:

$$R = (C_p - C_v) = C_p\left(\frac{\gamma - 1}{\gamma}\right)$$

With the following subscript notation:

nozzle inlet, 0
nozzle throat, 1
rotor exhaust, 2

sonic velocity at the throat of the nozzle (c_1) may be expressed as

$$c_1^2 = \gamma R T_1 = C_p(\gamma - 1)T_1 = 2C_p\left(\frac{\gamma - 1}{\gamma + 1}\right)T_0$$

On Fig. 11.6 for an inward radial flow machine having a rotor tip speed U_1, a gas inlet angle α_1, and equal gas velocities before the nozzle and after the diffuser such that entering and leaving losses are the same, then

$$\sin(\alpha_1) = \frac{U_1}{c_1}$$

(see Fig. 11.6). For an isentropic efficiency η_s, the pressure expansion ratio is given by

$$\frac{P_2}{P_0} = \left[\frac{1 - U_1^2}{\eta_s C_p T_0}\right]^{\frac{\gamma}{\gamma - 1}}$$

and the outlet temperature is given by

$$T_2 = T_0 - \frac{U_1^2}{C_p}$$

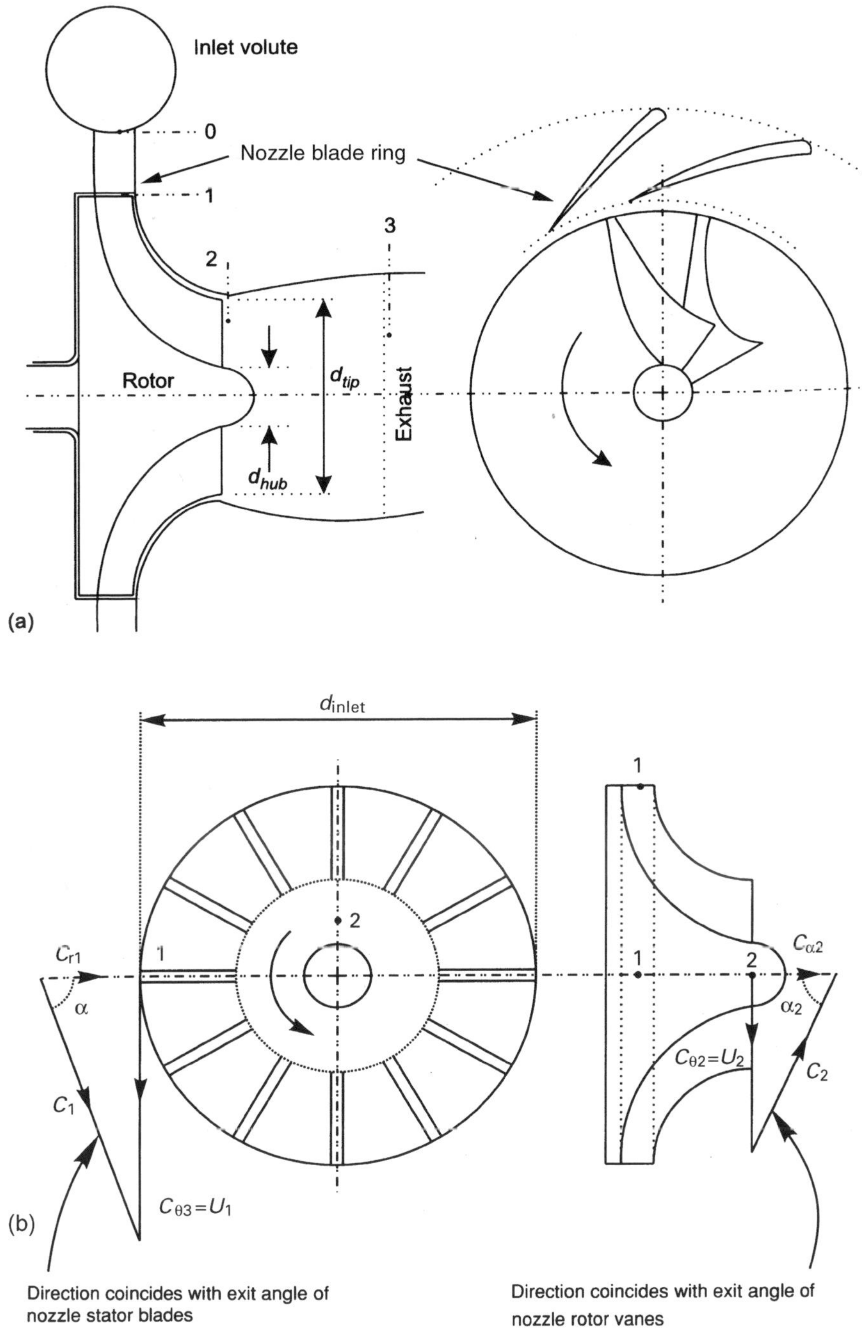

Fig.11.6 Inward radial flow turbine

With substitution it is quickly shown that

$$\frac{P_2}{P_0} = \left[1 - \frac{2}{\eta_s}\left(\frac{\gamma - 1}{\gamma + 1}\right)\sin^2(\alpha_1)\right]^{\frac{\gamma}{\gamma-1}}$$

$$\frac{T_2}{T_0} = \left[1 - 2\left(\frac{\gamma - 1}{\gamma + 1}\right)\sin^2(\alpha_1)\right]$$

and both these relationships depend only on γ and the inlet angle α_1.

For the purpose of comparison, an isentropic efficiency of 0.8 and an inlet angle of $\alpha_1 = 80^\circ$ (see Fig. 11.6) will be assumed. Results for the expansion of five candidate refrigerant gases are presented in Table 11.2 in descending order of the ratio C_p/C_v.

Oxygen is not there because of the very great risk of fire should a high-speed turbine rotor come into contact with its casing, but oxygen is still a possible refrigerant gas as vapour return from the final stages of liquefaction.

The gases clearly fall into two groups, the monatomic group with expansion ratios of about 10/1, and the diatomic group with expansion ratios of about 6/1. The first group achieves the greatest amount of single-stage cooling.

It is desirable to stay away from shock-wave losses whenever possible, and inward radial flow rotor design is eased when incompressible conditions are achieved at below approximately one-third of sonic velocity. Most practical plants try to keep expansion ratios below 3.0, a better choice being 2.5 or less. In maintaining the expansion ratio constant, temperatures will fall in reducing geometric progression. For sequential expansions this gives the optimum expansion ratios for minimization of exergy loss found by Nesselman, which are reported briefly at the end of the paper by Grassmann & Kopp (1957).

Table 11.2 also indicates why there is current interest in mixed refrigerants. Cryo-expansion problems are eased, and some mixtures have been found capable of reaching lower temperatures than those achieved using a single component.

The design of radial inward flow turbines is discussed in the text by Whitfield & Baines (1990), but the later paper by Whitfield (1990) examines cryogenic turbines in more detail.

Table 11.2 Cryo-expansion fluids

Gas	(C_p/C_v)	(P_0/P_2)	(T_2/T_0)
Argon	1.670	10.349	0.5133
Helium	1.662	10.145	0.5176
Neon	1.640	9.688	0.5298
Hydrogen	1.410	6.228	0.6700
Nitrogen	1.404	6.164	0.6740

11.3 Liquefaction of nitrogen

Nitrogen is almost always a first candidate for a refrigerating stream in liquefaction plant because of its abundance, inertness, and low critical pressure. It does not have the properties of a monatomic molecule, but this disadvantage could be mitigated by mixing it with argon.

The present example of a liquefaction plant to produce LN_2 has been chosen so as to illustrate some features of a typical system. In one respect only is the example not typical, for it is possible to mix product and refrigerating streams without affecting the product.

The reader should consider the T–s diagram for nitrogen in her/his mind. We have already decided to compress the product stream to supercritical pressure so that cooling can follow the liquid side of the saturation curve and get close to the final condition before throttling to produce liquid at near ambient pressure. What pressure level should be chosen for the product stream? What governs its selection?

Low-compression work is an important consideration, and thus we seek the lowest practicable pressure level. To do this examine the h–T diagram for nitrogen (Fig. 11.7), and notice that it is possible to construct 'break points' on the h–T curve such that straight lines joining these points provide a near approximation to the curve itself. The existence of these straight segments means that the temperature distributions in the heat exchangers will also be nearly linear.

Only one supercritical curve is shown in Fig. 11.7, but in practice many curves need to be examined, as the choice of pressure level changes the position and spacing of the break points on the product stream curve. Finding break points that produce linear segments is a necessary but not sufficient condition for successful

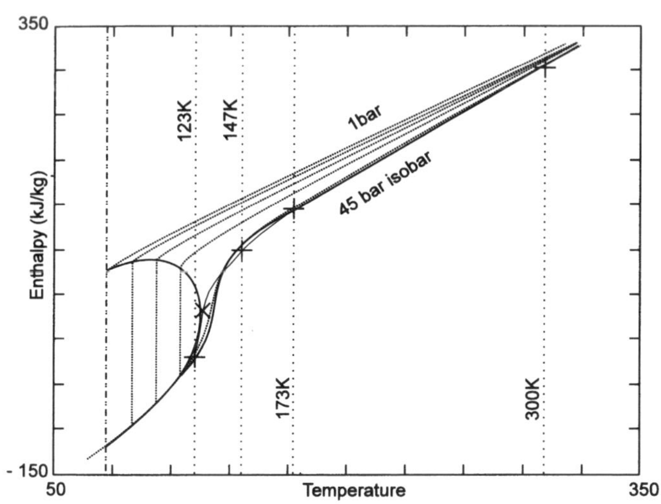

Fig.11.7 Break points on the nitrogen h–T diagram

liquefaction, for consideration has also to be given to expansion ratio and the temperature reduction achievable by cryo-turbines feeding the refrigeration streams.

When the product stream and the refrigeration stream are different gases, there is less incentive to match pressure levels elsewhere in the system. For nitrogen liquefaction using nitrogen as the refrigerant it makes some sense to try to match pressure levels. This matching process is the art of engineering cryogenic plant.

Figure 11.8 shows the T–s diagram for the plant and Fig. 11.9 shows the layout selected. All compressors shown are assumed to include aftercooling to 300 K.

If throttling from station 9 had been directly to 3 bar at station 12, exactly the same fraction of liquid would have been produced. However the much greater gaseous return flow at station 23, would increase the compression work required.

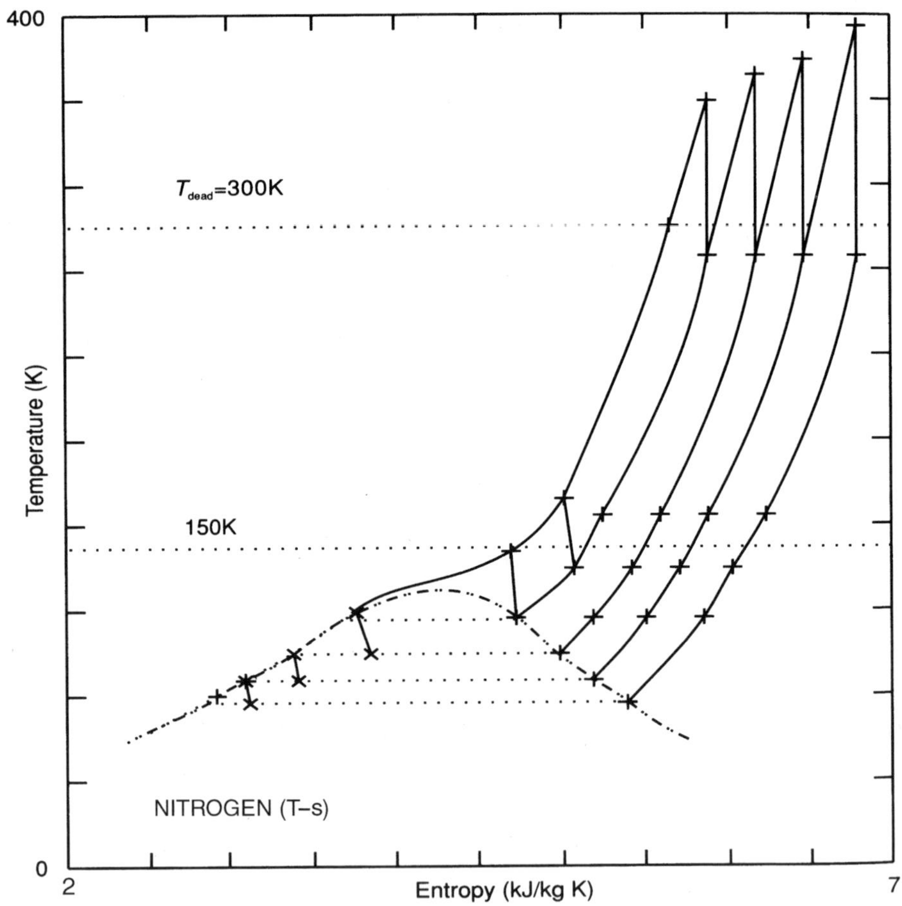

Fig.11.8 ***T–s* diagram for nitrogen liquefaction plant**

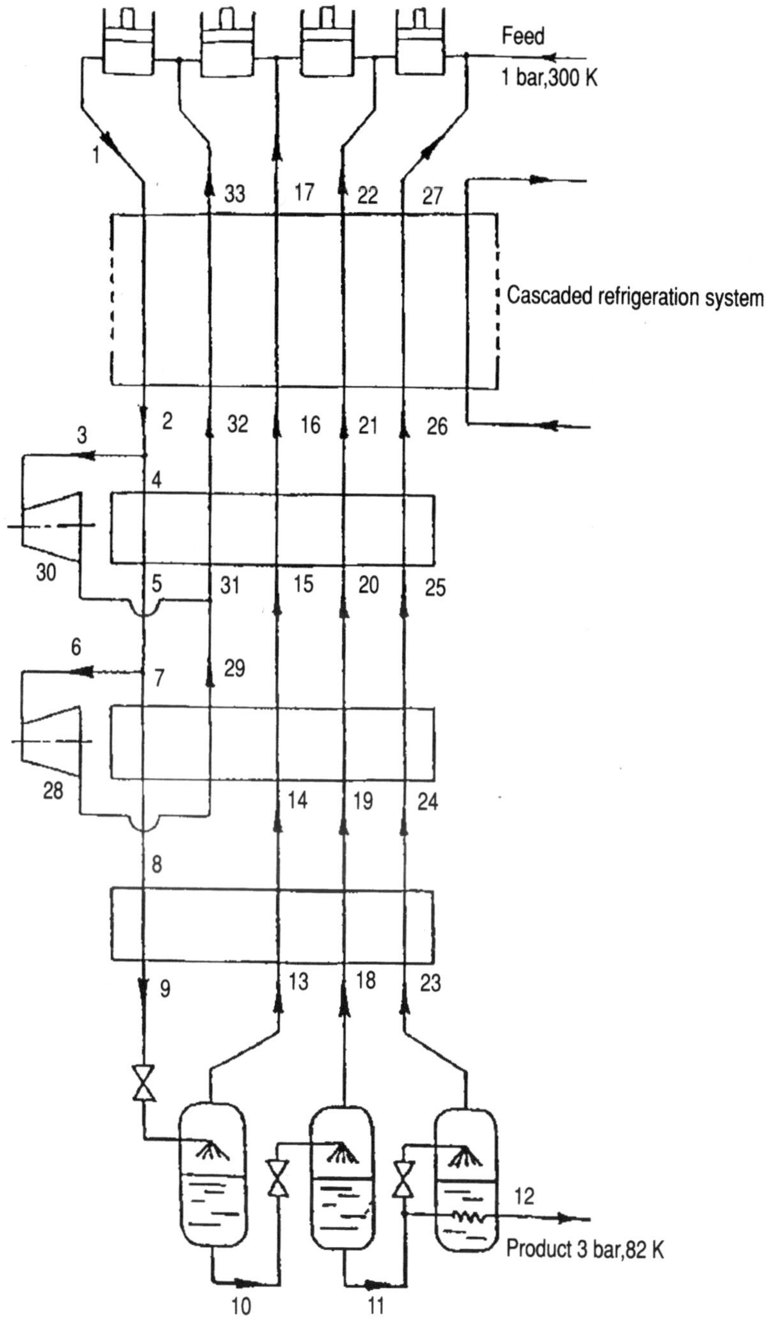

Fig.11.9 Configuration of liquefaction plant to produce LN_2

Essentially the product stream is compressed from 1 bar at around 300 K to 35 bar–300 K, and is then cooled isobarically to (35 bar–123 K). Compression was required to get the product stream to the left-hand side of the saturation line, but the work input has not yet produced any cooling in its own right. To remedy this, a succession of throttling steps are introduced which allow most of the product stream to 'walk-down' the liquid saturation line (Marshall & Oakey, 1985). Any cold vapour produced by throttling the product stream is made to return in parallel with the refrigerating streams. Here careful selection of pressure levels has allowed these return streams to be combined with the refrigerating streams, thus reducing the number of independent streams in the multi-stream heat exchangers – an important simplification.

The product stream is not expanded all the way to 1 bar, as it is better to maintain the product liquid slightly above atmospheric pressure, any leakage then being outwards. However, the liquid product is undercooled as far as is practicable, making use of the last expansion stage for that purpose. This helps counteract 'heat-leak' from the insulated storage tanks. The first refrigeration stage in cooling the product stream is not shown, and this can be a series of cascaded conventional refrigeration plants, using appropriate working fluids, see e.g. Barron (1985). Some refrigerating fluids would be inappropriate for oxygen as a product stream.

For common product and refrigerant streams it makes sense to select compressor pressure levels for the refrigerating system that match those generated by the product system. But which pairs should they be, 1–3 bar, 3–8 bar, 8–20 bar, or 20–45 bar?

Returning to the h–T diagram (Fig. 11.7), the 20-bar isobar has a small curvature which matches the 45-bar product stream slightly better than the isobars at 8, 3, or 1 bar. There is no a priori reason why suction should not be at 20 bar providing the system is pressure-tight, and if necessary canned compressors can be used. The primary refrigeration compressor is smaller as a consequence of higher gas densities. Product return streams are not considered at this stage because these make much smaller lesser contributions to the cooling required.

Figure 11.9 shows the final plant configuration with four compressors, a cascaded refrigeration system, two cryo-turbines, and three throttling stages. The function of the lowest heat exchanger is simply to equalize the temperatures of the returning product vapour streams before serious cooling begins. There have been attempts in the industry to develop liquid expansion machines as a replacement for throttles, but absence of moving parts at cryogenic temperatures leads to plant reliability. It is possible to introduce another throttling stage at 20 bar, but whether this is worthwhile is a matter of economics in plant build. The way the system is shown, there is little opportunity for refrigerating fluid to flow in the wrong direction.

In arriving at *temperatures* for the above configuration, the procedure is from the *top-down*, finding break points that produce linear segments in the h–T curve, matching cryo-turbine expansion ratios with the required break points, matching temperature levels at entry and exit to the heat exchangers, and matching temperatures so that mixing losses do not occur, or are minimized. Grassmann & Kopp temperature profiles are applied in each exchanger to ensure minimum exergy loss – because the compression work penalty is very high in liquefaction systems – consider Fig. 11.1.

Cryo-turbine performance is determined by a simple calculation in which the isothermal efficiency is assumed to be 0.80, and a $T{-}s$ diagram is used to check that the expansion is in the right position (Fig. 11.10). Throttle performance is assessed similarly (Fig. 11.11). In the multi-stream heat exchangers, cooling performance of each return stream is assessed individually, individual component performance allowing *mass flow ratios* to be determined.

Assessment of heat exchanger performance at this stage is restricted to piecewise checking of the enthalpy balance along the exchanger, fixing appropriate low-pressure fluid cold inlet and outlet temperatures, and fixing an appropriate high-pressure fluid warm inlet temperature. Thermodynamic properties of both fluids are obtained from interpolating spline-fits. An appropriate value of pinch point (temperature difference at point of closest approach) is chosen, the mass flow-rate of the cold fluid is set to 1.0 kg/s and the calculation iterated until the pinch point is achieved somewhere in the exchanger. This calculation provides five important items of design information along the exchanger (Figs 11.12 and 11.13), viz.

- shape of ΔT, $T{-}h$ profiles
- shape of the $h{-}T$ profiles
- high-pressure warm fluid outlet temperature

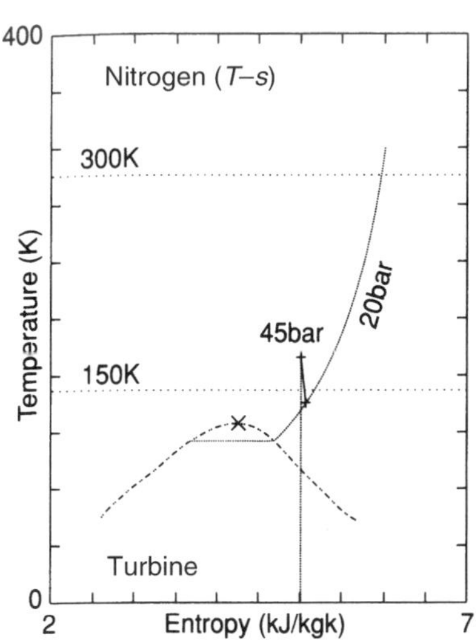

Fig.11.10 Cryo-turbine performance

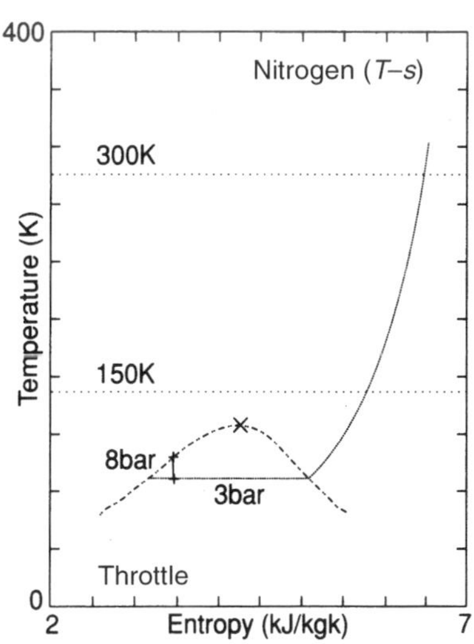

Fig.11.11 Throttle performance

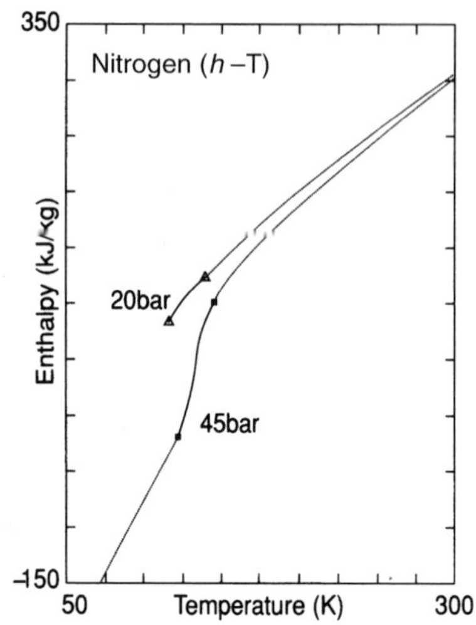

Fig.11.12 Exchanger h–T profiles

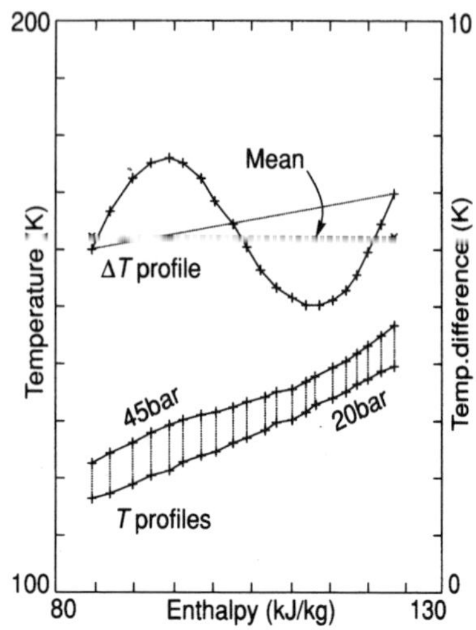

Fig.11.13 Exchanger ΔT, T–h

- mass flowrate of warm fluid for 1.0 kg/s of cold fluid
- position of the temperature pinch point

If the outlet temperature of the high-pressure warm fluid is not the value desired (often a value corresponding to that of another 'mixing' stream, to avoid detrimental mixing losses), then the pinch-point temperature can be adjusted to achieve an outlet temperature match. If no suitable value of outlet temperature becomes available, then it may be necessary to choose new temperature break points (Fig. 11.7), or to reconfigure the plant.

Where more than two fluids are present in an exchanger, then several such calculations have to be made for each possible pair of fluids to determine the best possible combination of energy exchange balances.

Results for one such calculation for the two main fluids in the critical heat exchanger of the nitrogen liquefaction plant are presented in Table 11.3. That this is the critical exchanger can be confirmed by examining corresponding mass flow rates in Table 11.4, but generally it is the exchanger which straddles the critical temperature and most closely approaches the critical point of the fluid being cooled which turns out to be the 'critical' exchanger.

Once correct mass flow ratios for each exchanger have been determined, *true mass flowrates* for the whole plant system can be found starting from the *bottom-up* (Fig. 11.9). This begins with free choice of the desired amount of undercooling of the product stream at 3 bar (noting that it is not possible to cool below the saturation temperature at 1 bar). In working back up a cryogenic system only arithmetic is required, except in a few cases when simple simultaneous algebraic equations may sometimes be needed to determine flowrates. Completion of Table 11.4 is necessary before the design of actual exchangers can proceed.

Table 11.3 Temperature profiles from enthalpy balance

20 bar, T (K)	*20 bar, h (kJ/kg)*	*20 bar, Δh (kJ/kg)*	*45 bar, Δh × R_m (kJ/kg)*	*45 bar, h (kJ/kg)*	*45 bar, T (K)*	*Temp. difference, ΔT (K)*
140.0	123.6	—	—	101.4	147.0	7.00
		3.22	9.78			
137.7	120.4	—	—	92.1	143.7	5.958
		3.30	10.03			
135.4	117.1	—	—	81.7	140.7	5.286
		3.44	10.46			
133.1	113.6	—	—	71.1	138.1	5.030
		3.57	10.86			
130.8	110.1	—	—	60.2	136.0	5.160
		3.68	11.18			
128.5	106.4	—	—	48.7	134.2	5.678
		3.82	11.62			
126.2	102.6	—	—	37.2	132.7	6.459
		4.02	12.21			
123.9	98.6	—	—	25.4	131.1	7.230
		4.28	13.00			
121.6	94.3	—	—	12.2	129.2	7.606
		4.68	14.24			
119.3	89.6	—	—	−2.0	126.5	7.196
		5.12	15.57			
117.0	84.5	—	—	−17.5	123.0	5.996
						Mean 6.236

20–45 bar section of multi-stream heat exchanger with mass flow ratio $R_m = (\dot{m}_{20\,bar}/\dot{m}_{45\,bar}) = 3.0407$.

11.4 Hydrogen liquefaction plant

The same procedures are used in designing other liquefaction plant, except that in the case of hydrogen, care has to be taken to use 'equilibrium' thermodynamic properties where appropriate.

In the very recent industrial-scale hydrogen liquefaction plant described by Bracha *et al.* (1994), liquid nitrogen is used to effect the first ortho : para hydrogen conversion, and the cold gaseous nitrogen is then used to refrigerate the incoming hydrogen streams. The refrigerating nitrogen stream in this plant is not recycled, but is continuously extracted from the air and discharged to atmosphere.

The paper by Bracha *et al.* (1994) provides a good description of a real hydrogen liquefaction system. Syed *et al.* (1998) prepared an economic analysis of three large-scale hydrogen liquefaction systems in which closed-cycle nitrogen precooling is used. They employ the earlier work of Dini & Martorano (1980) for enthalpies at inlet and outlet of heat exchangers.

Table 11.4 Thermodynamic and flow analysis of an LN_2 plant. Data are generated using Vargaftik (1975), except for a value of liquid specific heat which was obtained from Touloukian & Makita (1970)

Station	*Pressure (bar)*	*Temperature (K)*	*Mass flow (kg/s)*	*Enthalpy (kJ/kg)*
1	45.0	300.0	7.2482	302.0
2	45.0	173.0	7.2482	148.5
3	45.0	173.0	2.9726	148.5
4	45.0	173.0	4.2756	148.5
5	45.0	147.0	4.2756	99.51
6	45.0	147.0	2.5357	99.51
7	45.0	147.0	1.7399	99.51
8	45.0	123.0	1.7399	−25.44
9	45.0	120.0	1.7399	−28.00
10	8.0	100.4 sat.	1.2480	−73.60 wet
11	3.0	87.9 sat.	1.0697	−99.87 wet
12	3.0	82.0 u.cool	1.0000*	liq. u.cool
13	8.0	100.4 sat.	0.4919	87.70
14	8.0	117.0	0.4919	110.92
15	8.0	140.0	0.4919	137.3
16	8.0	165.0	0.4919	165.45
17	8.0	285.0	0.4919	293.88
18	3.0	87.9 sat.	0.1783	83.96 dry
19	3.0	117.0	0.1783	118.3
20	3.0	140.0	0.1783	142.3
21	3.0	165.0	0.1783	169.0
22	3.0	285.0	0.1783	295.10
23	1.0	77.4 sat.	0.0697	76.80 dry
24	1.0	117.0	0.0697	120.98
25	1.0	140.0	0.0697	144.2
26	1.0	165.0	0.0697	170.45
27	1.0	285.0	0.0697	295.60
28	20.0	117.0	2.5357	96.94
29	20.0	140.0	2.5357	123.6
30	20.0	140.0	2.9726	123.6
31	20.0	140.0	5.5083	123.6
32	20.0	165.0	5.5083	156.55
33	20.0	285.0	5.5083	291.05

* Indicates the product stream.
sat., saturation; u.cool, undercool.

11.5 Preliminary direct-sizing of multi-stream heat exchangers

Preliminary sizing of exchangers provides a best estimate for the exchanger cross-sections, e.g. edge length in plate–fin designs, and number of tubes and tube spacing in shell-and-tube exchangers. Step-wise rating becomes necessary when the assumption of constant thermophysical properties along the exchanger no longer holds. It is worth summarizing the procedure for a two-stream exchanger, which is in several stages. The most awkward exchanger of the cryogenic plant in Fig. 11.9 is likely to be the multi-stream exchanger associated with the lowest cryo-turbine because conditions for the 20-bar cooling stream and the 45-bar product stream are nearest to the critical point of nitrogen.

Stage one

Table 11.3 is constructed to obtain a first estimate of mean temperature difference. Inlet and outlet temperatures of both fluids are chosen to meet the Grassman & Kopp requirement that $\Delta T = T/20$. A suitable number of intermediate stations is chosen along the temperature span of the cold fluid (usually 20, but 10 is used in Table 11.3 for compactness). Spline-fitted temperature/enthalpy curves for both fluids are then used to calculate and match enthalpy increments on both sides of the exchanger, from which the corresponding temperature increments on the warm side can be found. The calculation requires knowledge of mass flowrates on both sides of the exchanger. It is convenient to set the cold mass flowrate to 1.00 kg/s and the warm fluid mass flowrate is iterated until the desired outlet temperature of the warm fluid is matched.

Usually a match is not obtained at the first trial, and the value of ΔT is then changed until the desired value of the warm fluid outlet temperature is obtained. The mean temperature difference for the exchanger is calculated as the average of the local temperature differences at stations along the exchanger, and this will be different from ΔT.

We now also have the *ratio* of the mass flowrates, and when this procedure is followed for the whole cryogenic plant, then *actual* mass flowrates can be calculated.

Stage two

The mean temperature difference of 6.236 K from Table 11.3 is used in direct-sizing, together with actual mass flowrates, and the assumption of mean thermophysical properties. This procedure is covered in Chapter 4, and includes an adjustment of the mean temperature difference to allow for longitudinal conduction. In this case the adjustment was 0.975, making the mean temperature difference 6.080 K.

Direct-sizing (Fig. 11.14) is carried out for three reasons, first to ensure that the pressure losses are as desired, second to optimize local surface geometries and approach the desired optimum exchanger, and third to obtain the edge length (E) and length (L) of the exchanger.

This is done for each combination-pair of two flow streams in the multi-stream block, and for each combination-pair the design point is chosen at the upper left-hand end of the heat-transfer curve corresponding to the maximum length (L) of that exchanger. If more than one stream is being cooled in the multi-stream block, then more than one selection of sets of combination-pairs will prove possible, and an appropriate selection can be made at this stage.

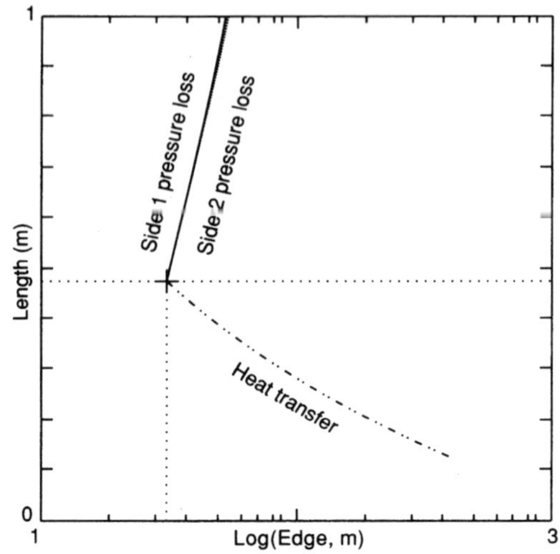

Fig.11.14 Direct-sizing of the two-stream exchanger of Table 11.3 to determine maximum length

Edge lengths (E) have to be whole multiples of block width in the multi-stream exchanger, and a suitable choice is made at this stage. The smallest value of L from the final set of combination-pairs may also be selected as the block length of the multi-stream exchanger. All selected combination-pairs can now be recalculated to determine new pressure losses for the selected E and L values. It is desirable to use the same surface geometry for any stream that is split and serves more than one combination-pair.

Stage three

Stepwise rating of a single combination-pair begins with assembling the necessary thermophysical data against temperature, either as tables or as interpolating spline-fits. When a suitable number of stations are taken along the length of the exchanger in which the enthalpy balances are assured, then each small section of the exchanger can be dealt with as an individual exchanger.

The LMTD of individual sections can be calculated, together with the mean bulk temperatures of both fluids, and corresponding thermophysical properties can be found. Soyars (1991) did not find the ε–Ntu method accurate for this purpose.

Given the edge length E, heat transfer and pressure losses may be determined for each section, and the required length and pressure loss for each section found. The calculation may be checked using the summed values of length and pressure loss for each section; they should be close to those obtained in the previous direct-sizing step.

This produces the actual temperature/length profile for the exchanger of Table 11.3 which is shown in Fig. 11.15.

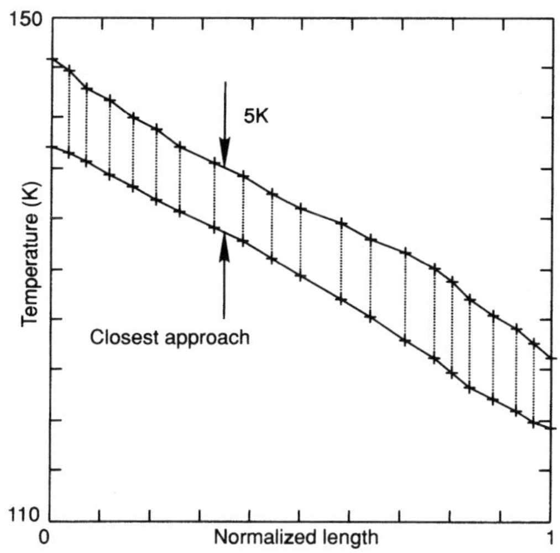

Fig.11.15 Temperature profiles of the two-stream exchanger obtained using step-wise rating

Obviously, each section of the multi-stream exchanger is likely to produce temperature profiles differing slightly from those shown in Fig. 11.15. This introduces further considerations, viz.:

(a) the desirability of using the same surface geometry in each section for the stream that is split
(b) the need to allow for cross-conduction effects
(c) appropriate choice of stacking pattern

11.6 Step-wise rating of multi-stream heat exchangers

For multi-stream exchangers, temperatures of either the hot streams or the cold streams are not usually constant over each cross-section of the exchanger. Then *cross-conduction* effects between adjacent streams may significantly affect the performance. Haseler (1983) analysed this problem for the plate–fin design, using simple fin theory to evaluate cross-conduction effects, and he further showed how to incorporate an allowance for cross-conduction in the design process.

The algebra in Haseler's approach is compact and some assistance in getting quickly into his elegant solution seems appropriate. The differential equation governing heat conduction in a fin is

$$\frac{d^2T_f}{dy^2} - \left(\frac{\alpha P}{\lambda A}\right)\left(T_f - T_c\right) = 0 \tag{11.6}$$

where

T_f = fin temperature
α = heat-transfer coefficient
P = fin perimeter/unit length
λ = fin thermal conductivity
A = fin area for conduction

For a rectangular offset strip fin, equation (11.6) becomes

$$\frac{d^2T_f}{dy^2} - \left(\frac{2\alpha}{\lambda tf}\right)(T_f - T_c) = 0 \qquad \text{where } tf \text{ is fin thickness}$$

Putting $\theta_f = T_f - T_c$

$$\frac{d^2\theta_f}{dy^2} - \left(\frac{2\alpha}{\lambda tf}\right)\theta_f = 0$$

for which the solution is

$$\theta_f = C_1.e^{my} + C_2.e^{-my} \qquad \text{where } m = \sqrt{\left(\frac{2\alpha}{\lambda tf}\right)}$$

Taking the origin at the centre of the fin, mid-way between two plates with spacing $b = 2a$, the boundary conditions become

$$\left.\begin{array}{l} y = -b/2 = -a, \text{ say} \\ \theta_f = \theta_{w1} \end{array}\right\} \qquad \left.\begin{array}{l} y = +b/2 = +a, \text{ say} \\ \theta_f = \theta_{w2} \end{array}\right\}$$

from which

$$C_1 = \left(\frac{e^{ma}\theta_{w2} - e^{-ma}\theta_{w1}}{e^{2ma} - e^{-2ma}}\right), \qquad C_2 = \left(\frac{e^{ma}\theta_{w1} - e^{-ma}\theta_{w2}}{e^{2ma} - e^{-2ma}}\right)$$

and the solution for fin temperature becomes

$$\theta_f = \frac{\theta_{w2}(e^{ma}e^{my} - e^{-ma}e^{-my}) - \theta_{w1}(e^{-ma}e^{my} - e^{ma}e^{-my})}{(e^{2ma} - e^{-2ma})}$$

Digressing at this point, consider the expansion of

$$\left[\frac{A+B}{2}\right]\frac{\cosh(my)}{\cosh(m\,a)} + \left[\frac{A-B}{2}\right]\frac{\sinh(my)}{\sinh(m\,a)}$$
$$= A\left[\frac{e^{ma}e^{my} - e^{-ma}e^{-my}}{e^{2ma} - e^{-2ma}}\right] + B\left[\frac{e^{-ma}e^{my} - e^{ma}e^{-my}}{e^{2ma} - e^{-2ma}}\right]$$

which allows the solution for fin temperature to be written

$$\theta_f = \left[\frac{\theta_{w2} + \theta_{w1}}{2}\right]\frac{\cosh(my)}{\cosh(m\,a)} + \left[\frac{\theta_{w2} - \theta_{w1}}{2}\right]\frac{\sinh(my)}{\sinh(m\,a)} \tag{11.7}$$

Haseler writes heat transfer from the first wall in the y-direction as the sum of direct heat transfer and fin conduction, viz.

$$\dot{Q}_{LT} = \alpha S_1 \theta_{w1} - N\lambda t_f \frac{dT}{dy}\bigg|_{y=-a} \tag{11.8}$$

where

S_1 = primary surface per unit length along the exchanger
N = number of fins across the exchanger
$S_2 = 2aN$, is secondary surface per unit length along the exchanger

Differentiating equation (11.7) and substituting in equation (11.8)

$$\begin{aligned}
\dot{Q}_{LT} &= \alpha S_1 \theta_{w1} + N\lambda tf\, m\left[\left\{\frac{\theta_{w2}+\theta_{w1}}{2}\right\}\tanh(ma) - \left\{\frac{\theta_{w2}-\theta_{w1}}{2}\right\}\coth(ma)\right] \\
&= \alpha S_1 \theta_{w1} + \alpha\left[\frac{\tanh(ma)}{ma}\right]S_2\left[\frac{\theta_{w2}+\theta_{w1}}{2}\right] - \alpha\left[\frac{\coth(ma)}{ma}\right]S_2\left[\frac{\theta_{w2}-\theta_{w1}}{2}\right] \\
&= \alpha S_1 \theta_{w1} + \alpha\left[\frac{\tanh(ma)}{ma}\right]S_2\left[\theta_{w1} + \frac{\theta_{w2}-\theta_{w1}}{2}\right] - \alpha\left[\frac{\coth(ma)}{ma}\right]S_2\left[\frac{\theta_{w2}-\theta_{w1}}{2}\right] \\
&= \alpha S_1 \theta_{w1} + \alpha\left[\frac{\tanh(ma)}{ma}\right]S_2\theta_{w1} + \alpha\left[\frac{\tanh(ma)-\coth(ma)}{2ma}\right]S_2(\theta_{w2}-\theta_{w1})
\end{aligned}$$

The standard expression for fin efficiency is

$$\eta = \frac{\tanh(ma)}{ma}$$

Haseler defines fin 'by-pass' efficiency as

$$\eta' = \frac{\coth(ma) - \tanh(ma)}{2ma} = \frac{\text{cosech}(2ma)}{ma}$$

then Haseler's equations become

$$\boxed{\begin{aligned}
\dot{Q}_{LT} &= \alpha(S_1 + \eta S_2)\theta_{w1} + \alpha\eta' S_2(\theta_{w1} - \theta_{w2}) \\
\dot{Q}_{LT} &= \dot{Q}_L + \dot{Q}_B
\end{aligned}} \tag{11.9}$$

where

$\dot{Q}_{LT}$ = total heat flow from left-hand wall per unit length
$\dot{Q}_L$ = heat flow from left wall to or from fluid stream per unit length
$\dot{Q}_B$ = by-pass heat flow per unit length

A similar set of equations exists for the right-hand wall, viz

$$\dot{Q}_{RT} = \alpha(S_1 + \eta S_2)\theta_{w2} - \alpha\eta' S_2(\theta_{w1} - \theta_{w2})$$
$$\dot{Q}_{RT} = \dot{Q}_R - \dot{Q}_B$$

where

$\dot{Q}_{RT}$ = total heat flow from right-hand wall per unit length
$\dot{Q}_R$ = heat flow from right wall to or from fluid stream per unit length
$\dot{Q}_B$ = by-pass heat flow per unit length

The remainder of the analysis is straightforward, the principal requirement being access to a large computer.

Hasler's analysis highlights the importance of considering *individual channel passages* rather than the number of separate streams in design of multi-stream exchangers, but more importantly it focuses attention on the initial plant configuration stage where there is opportunity to design-out mixing losses and unacceptable temperature profiles, e.g. Fig. 11.9 and Table 11.4.

The full design process involves multi-passage analysis (or perhaps multi-plate analysis), rather than multi-stream analysis, and solution of the simultaneous equations may then require a considerable amount of computational work. In an important paper, Suessman & Mansour (1979) provided a simple method for arriving at a good stacking pattern in the arrangement of individual flow passages. Stacking pattern is often repeated in an exchanger, and this can minimize the amount of computational work required by Haseler's method.

Mollekopf & Ringer (1987) indicate that Linde AG has developed an exact solution of the set of governing differential equations. This scheme assumes constant properties and is valid for incremental steps only, which is sufficient to allow computation of stacking patterns which deviate from the common wall temperature assumption.

Different philosophies are suggested by Haseler (1983) and by Prasad & Gurukul (1992) for carrying out the step-wise rating process for multi-stream exchangers. Prasad & Gurukul prefer to start the computation from the end where the temperature differential between hot and cold fluids is greatest. Haseler prefers to start the computation from the end at which only one stream temperature may be the true unknown, as otherwise he found that instabilities may arise in the calculation.

Papers by Prasad & Gurukul (1987), Paffenbarger (1990) and Prasad (1993, 1996) extend this work, and all papers listed in this Section 11.6 are recommended reading. Designing a multi-stream exchanger is not a fully explicit process.

Feasiblity studies

A reasonable approximation to the final design can be achieved by adopting the stacking pattern of Suessman & Mansour, and then working a step-wise rating design, assuming that hot fluid and cold fluid temperatures in any cross-section are reasonably constant. This should provide a first approximation on which to base cost estimates.

Optimization of multi-stream exchangers

A definitive paper describing optimization of multi-stream exchangers using mathematical techniques of non-linear programming (NLP), especially successive quadratic programming is presented by Reneaume *et al.* (2000).

The design tool is capable of handling the undernoted configurations:

- multi-stream heat exchangers
- multi-phase streams
- plain, perforated, and serrated fins
- counterflow
- re-distribution of streams
- duplex, triplex streams
- design of distributor sections

Either published or proprietary experimental correlations for flow friction and heat transfer of compact surfaces may be used. The classic Kays & London (1984) text is referenced.

11.7 Future commercial applications

The thrust of the present work is towards new engineering developments, particularly in cryogenics. Electricity and hydrogen are the energy vectors of the future, and it is possible that the new energy resource fields will be found in those areas of the world where massive hydropower and geothermal resources exist. It is already projected that hydrogen produced by electrolysis of water will be liquefied, and that bulk liquid hydrogen will then be shipped in sealed tanks mounted on skids to where the energy is required (Petersen *et al.*, 1994).

It is also possible that future transport arrangements will be based on liquid hydrocarbons containing the least amount of carbon, e.g. methanol (CH_3OH). Then conventional tankers can be used for bulk transport, with steam reforming to produce hydrogen and carbon dioxide more locally. Carbon can be regarded simply as a 'carrier' for hydrogen atoms.

Ceramic superconductors embedded in silver have been found which exhibit superconductivity up to 135 K. These have a current-carrying capacity greater than 10^6 amps/cm^2, and are now being spun in lengths of 1000 m (Stansell, 1994).

Liquid nitrogen is a convenient, safe, and inexpensive cryogen at 77 K. It is also the essential refrigerating cryogen in the technology of hydrogen liquefaction (Bracha *et al.*, 1994). Interest in liquid nitrogen is set to increase as its applications grow in importance, including superconducting power generators and electricity storage in superconducting coils.

Pressurized hydrogen gas stored in stainless steel bottles at ambient temperatures is a possible candidate for road vehicles. While liquid hydrogen systems have also been developed these may be technically too complex for general public use.

Fuel cells will ultimately be used for generation of on-board electricity for propulsion. A significant number of alternative propulsion systems are currently being explored by large international companies and the eventual winner may take some time to emerge.

Liquid hydrogen is likely to find its first commercial application as a replacement fuel in aircraft propulsion (Brewer, 1993). The technical advantages of having a fuel with an energy content of 118.6 MJ/kg (which is 2.78 times that of conventional jet fuel) are considerable in the case of aircraft. The wings can be smaller, the landing gear lighter, less-powerful engines are required, resulting in a reduction in gross take-off weigh of 30 per cent. These 'knock-on' advantages do not exist for land-based or sea-based applications.

11.8 Conclusions

1. Considerations underlying the layout and design of a nitrogen liquefaction plant have been set out. This is an essential preliminary stage in obtaining parameters for heat exchanger design.
2. Factors affecting layout of a hydrogen liquefaction plant have been discussed, and an excellent example of such a plant is to be found in the papers by Bracha *et al.* (1994) and Syed *et al.* (1998).
3. In the hydrogen liquefaction plant, coiled-tube heat exchangers have been used in sections in which evaporation of a liquid is employed for ortho : para hydrogen conversion. This allows the evaporating shell-side fluid to equalize across the tube bundle. Multi-stream plate–fin heat exchangers are preferred for heat exchange between single-phase gaseous fluids.
4. A method of arriving at a first estimate of the cross-section of multi-stream exchangers by direct-sizing has been outlined.
5. Papers on step-wise rating of multi-stream exchangers have been indicated, including important aspects of selection of stacking pattern, and of allowance for cross-conduction effects.

References

Alfeev, V.N., **Brodyansky, V.M.**, **Yagodin, V.M.**, **Nikolsky, V.A.**, and **Ivantsov, A.V.** (1971) Refrigerant for a cryogenic throttling unit. British Patent 1336892, Published 14 November 1973.

Barrick, P.L., **Brown, L.F.**, and **Hutchinson, H.L.** (1965) Improved ferric oxide gel catalysts for *ortho-para* hydrogen conversion. *Adv. Cryogenic Engng*, **10**, 181.

Barron, R.F. (1985) *Cryogenic Systems*, 2nd edn, Oxford.

Bracha, M., **Lorenz, G.**, **Patzelt, A.**, and **Wanner, M.** (1994) Large-scale hydrogen liquefaction in Germany. *Int. J. Hydrogen Energy*, **19**(1), 53–59.

Brewer, R.D. (1993) *Hydrogen Aircraft Technology*, CRC Press, Florida.

Dini, D. and **Martorano, L.** (1980) Design of optimised large and small hydrogen liquefaction plants. *Proceedings of the Third World Hydrogen Energy Conference*, Tokyo, *Hydrogen Energy Progress III*, vol. 4, pp. 2393–2421.

Grassmann, P. and **Kopp, J.** (1957) Zur günstigen Wahl der temperaturdifferenz und der Wärmeuberganszahl in Wärmeaustauschern. *Kältetechnik*, **9**(10), 306–308.

Haselden, G.G. (1971) *Cryogenic Fundamentals*, Academic Press.

Haseler, L.E. (1983) Performance calculation methods for multi-stream plate-fin heat exchangers. *Heat Exchangers: Theory and Practice.* (Eds. J. Taborek, G.F. Hewitt, and N. Afgan), Hemisphere/McGraw-Hill, New York, pp. 495–506.

Kays, W.M. and **London, A.L.** (1984) *Compact Heat Exchangers*, 3rd edn, McGraw-Hill, New York.

Keeler, R.N. and **Timmerhaus, K.D.** (1960) Poisioning and reactivation of ortho-para hydrogen converson catalyst. *Advances in Cryogenic Engineering*, vol. 4, Plenum Press, New York, p. 296.

Lipman, M.S., **Cheung, H.**, and **Roberts, O.P.** (1963) Continuous conversion hydrogen liquefaction. *Chem. Engng Prog.*, **59**, August, 49–54.

Little, W.A. (1984) Microminiature refrigeration. *Rev. Scient. Instrum.*, **55**(5), May, 661–680.

Marshall, J. and **Oakey, J.D.** (1985) Gas refrigeration method. European Patent Application no. 85305248.8, Publication no. 0-171-952, International Classification F25 J 1/02.

Mollekopf, N. and **Ringer, D.U.** (1987) Multistream heat exchangers – types, capabilities and limits of design. *Heat and Mass Transfer in Refrigeration and Cryogenics* (Eds. J. Bougard and N. Afgan), Hemisphere/Springer Verlag, New York, pp. 537–546.

Newton, C.L. (1967a) Hydrogen production and liquefaction. *Chem. Process Engng*, December, 51–58.

Newton, C.L. (1967b) Hydrogen production, liquefaction and use. *Cryogenic Engng News*, Part 1, August, 50–60; Part 2, September, 24–30.

Paffenbarger, J. (1990) General computer analysis of multi-stream plate-fin heat exchangers. *Compact Heat Exchangers – a festschrift for A.L. London* (Eds. R.K. Shah, A.D. Kraus, and D. Metzger), Hemisphere, New York, pp. 727–746.

Paugh, R.L. (1990) New class of microminiature Joule–Thompson refrigerator and vacuum package. *Cryogenics*, **30**, December, 1079–1083.

Peschka, W. (1992) *Liquid Hydrogen: Fuel of the Future*, Springer-Verlag.

Petersen, U., **Wursig, G.**, and **Krapp, R.** (1994) Design and safety considrations for large-scale sea borne hydrogen transport. *Int. J. Hydrogen Energy*, **19**(7), 597–604.

Prasad, B.S.V. (1993) The performance prediction of multi-stream plate-fin heat exchangers based on stacking pattern. *Heat Transfer Engng*, **12**(4), 58–70.

Prasad, B.S.V. (1996) The sizing and passage arrangement of multistream plate-fin heat exchangers. *Heat Transfer Engng*, **17**(3), 35–43.

Prasad, B.S.V. and **Gurukul, S.M.K.A.** (1987) Differential methods for sizing multistream plate fin heat exchangers. *Cryogenics*, **27**, 257–262.

Prasad, B.S.V. and **Gurukul, S.M.K.A.** (1992) Differential methods for the performance prediction of multi-stream plate-fin heat exchangers. *Trans. ASME, J. Heat Transfer*, **114**, 41–49.

Reneaume, J.M., **Pingaud, H.**, and **Niclout, N.** (2000) Optimisation of plate fin heat exchangers – a continuous formulation. *Trans. IChemE*, **78**(A), September, 849–859.

Schmauch G.E. and **Singleton, A.H.** (1964) Technical aspects of ortho-para hydrogen conversion. *Ind. Engng Chemistry*, **56**, May, 20.

Schmauch, G.E., **Kucirka, J.F.**, and **Clark, R.G.** (1963) Activity data on improved para-ortho conversion catalysts. *Chem. Engng Prog.*, **59**, August, 55–60.

Scott, R.B. (1959) *Cryogenic Engineering*, Van Nostrand.

Soyars, W.M. (1991) The applicability of constant property analyses in cryogenic helium heat exchangers. In *Proceedings of the 1991 Cryogenic Engineering Conference, Advances in Cryogenic Engineering*, vol. 37, Part A, Plenum Press, New York, pp. 217–223.

Stansell J. (1994) *Sunday Times,* 3rd April 1994, Business Section 3, p. 10.

Suessman, W. and **Mansour, A.** (1979) Passage arrangement in plate-fin exchangers. In *Proceedings of the XV International Congress of Refrigeration*, Venice, vol. 1, pp. 421–429.

Syed, M.T., **Sherif, S.A.**, **Veziroglu, T.N.**, and **Sheffield, J.W.** (1998) An economic analysis of three hydrogen liquefaction systems. *Int. J. Hydogen Energy*, **23**(7), 565–576.

Touloukian, Y.S. and **Makita, T.** (1970) *Specific Heat – Nonmetallic Liquids and Gases, Thermophysical Properties of Matter*, vol. 6, Plenum Press, New York.

Vargaftik, N.B. (1975) *Tables on the Thermophysical Properties of Liquids and Gases*, Hemisphere/Wiley, New York.

Whitfield, A. (1990) The preliminary design of radial inflow turbines. *Trans. ASME, J. Turbomachinery*, **112**, January, 50–57.

Whitfield, A. and **Baines, N.C.** (1990) *Design of Radial Turbomachines*, Longmans.

Bibliography

Bougard, J. and **Afgan, N.** (1987) *Heat and Mass Transfer in Refrigeration and Cryogenics*, Hemisphere/Springer, New York.

Maurice, L.Q., **Leingang, J.L.**, and **Carreiro, L.R.** (1996) Airbreathing space boosters using in-flight oxidiser collection. *J. Propulsion Power*, **12**(2), March–April, 315–321.

Smith, E.M. (1984) A possible method for improving energy efficient *para*-LH_2 production. *Int. J. Hydrogen Energy*, **9**(11), 913–919.

Smith, E.M. (1989a) Slush hydrogen for aerospace applications. *Int. J. Hydrogen Energy*, **14**(3), 201–213.

Smith, E.M. (1989b) Liquid oxygen for aerospace applications. *Int. J. Hydrogen Energy*, **14**(11), 831–837.

Smith, E.M. (1997) Direct-sizing and step-wise rating of compact heat exchangers. *Compact Heat Exchangers for the Process Industries* (Eds. R.K. Shah, K.J. Bell, S. Mochizuki, and V.V. Wadekar), *Proceedings of the International Conference*, Snowbird, Utah, 22–27 June 1997, Begell House Inc., pp. 201–211.

Weimer, R.F. and **Hartzog, D.G.** (1972) Effect of maldistribution on the performance of multistream multipassage heat exchangers. In *Proceedings of the 1972 Cryogenic Engineering Conference, Advances in Cryogenic Engineering*, vol. 18, Plenum Press, New York, Paper B-2, pp. 52–64.

CHAPTER 12
Heat Transfer and Flow Friction in Two-Phase Flow

This chapter provides only an introduction to problems in obtaining and using heat-transfer and flow-friction correlations in two-phase flow

12.1 With and without phase change

Real heat exchangers do not have constant heat-transfer coefficients, even in single-phase designs, because of temperature dependence of thermophysical properties. Some may approximate to the assumption of constant properties, but some may not, e.g. the cryogenic exchanger discussed in Chapter 11.

In single-phase designs the temperature dependence of physical properties is enough to change the values of Reynolds number and Prandtl number, and hence the Nusselt number and ultimately the overall heat-transfer coefficient along the length of the heat exchanger.

Attempts have been made to adjust the expression for overall heat-transfer coefficient (U) allowing for assumed mathematical variation of the overall coefficient along the exchanger (Schack, 1965; Hausen, 1950, 1983), but these analytical methods have less relevance now that computers are generally available. In fact it is necessary to design the exchanger first in order to obtain the variation of U along the length of the exchanger.

For single-phase design it is possible to size an exchanger incrementally, using spline-fits to represent the physical properties involved, and calculate each increment as if it were a small exchanger itself. The approach is no longer that of direct-sizing but direct-sizing can still be used to obtain a good initial feel for the final size of the unit. Most of the earlier material in this text is relevant to designing single-phase heat exchangers by step-wise methods.

Once it has been accepted that step-wise design by computer is the most accurate way to go, it is straightforward to proceed to the more complicated design of heat exchangers involving two-phase flow, but first the method of calculating pressure loss in two-phase flow has to be considered. Second, there is a need to understand the several forms of two-phase flow which will exist in the design so that appropriate correlations may be employed. Third, maldistribution and instability of flow in plate–fin and other exchangers has to be designed out if possible.

The author has been highly selective in the material which follows as the objective is simply to introduce the reader to the computer design approach. The wise reader will read more widely on two-phase flow and consult the several excellent texts now available before proceeding to his/her own design application. Good starting points are the texts by Wallis (1969), Hewitt & Hall-Taylor (1970), Collier (1972), Bergles *et al.* (1981), Chisholm (1983), Smith (1986), Carey (1992), and Hewitt *et al.* (1994). There are also good articles in the *Handbook of Heat Transfer Fundamentals* (Rohsenow *et al.*, 1985), and in the *Handbook of Multiphase Systems* (Hestroni, 1982), and the reader is encouraged to use all journal and database sources to trace other authors and papers. The excellent work of Wadekar (2002) on phase change in compact heat exchangers shows the extent of scatter in predicting heat-transfer coefficients (50 references).

12.2 Two-phase flow regimes

With extreme clarity Rhee (1972) states that . . . 'Knowing the flow patterns of a two-phase flow is as important as knowing whether the flow is laminar or turbulent in single-phase flow.' Flow pattern maps for various tube geometries are to be found, e.g. in Hewitt *et al.* (1994). Here we shall be concerned only with forced-flow evaporation in a horizontal tube, and will use the relatively early work of Rhee (1972) simply to illustrate the computational approach to the problem.

Rhee found for refrigerant 12 that with horizontal tubes the general description of flow pattern in order of increasing vapour quality was nucleate (or bubbly) flow, stratified (sometimes slug or wavy) flow, annular (sometimes with mist, sometimes without mist) flow, and mist flow itself.

Nucleate flow occurs for an extremely short length of tube when vapour bubbles appear as liquid first reaches saturation temperature, quickly changing to stratified flow as more vapour appears with separate streams of liquid and vapour flowing in the tube. Sometimes liquid slugs form during this stage of evaporation, sometimes liquid waves appear.

At a later stage in the evaporation there is a transition from stratified flow to annular flow. This seems to occur when a higher heat-transfer coefficient would result for annular flow than for stratified flow.

After a period of annular flow there is a sudden drop in heat-transfer coefficient with a change to mist flow. This causes a jump in the temperature of the tube wall and is associated with the potentially dangerous condition of 'burnout' which will arise when the heat flux is being produced by external means (e.g. electrical or nuclear heating) and is not reduced immediately. In a normal heat exchanger this is simply a condition to evaluate, and the location of transition seems to be controlled principally by the Weber number. Mist flow then continues until all liquid has evaporated.

For his test fluid refrigerant 12, Rhee found that there was one other condition to be noted which is related to mass velocity.

- Above a critical mass velocity the flow pattern being followed is:
  Nucleate flow ⇒ Stratified flow ⇒ Annular mist flow ⇒ Mist flow
- Below the critical mass velocity the flow pattern is:
  Nucleate flow ⇒ Stratified flow ⇒ Annular (no mist) flow

The critical mass velocity is a parameter which needs to be evaluated before or during the computation so that the correct flow pattern may be computed.

Other flow situations

Obviously there are many other possible two-phase flow design situations, e.g.:

- internal forced-flow condensation in a tube
- external longitudinal forced-flow evaporation on a tube
- external transverse forced-flow evaporation on a tube
- external longitudinal forced-flow condensation on a tube
- annular forced flow between tubes
- flow in plate–fin surfaces
- permanent dropwise condensation on a surface

The reader is encouraged to seek modern methods of design for these other flow situations in the references cited at the end of Section 12.1.

12.3 Two-phase pressure loss

It is necessary to know the saturation temperature at any point along a heat exchanger in order to calculate physical properties. As saturation temperature is dependent on saturation pressure it follows that incremental pressure loss along the exchanger must be evaluated along the exchanger so that correct values of physical properties are obtained.

Several different models have been proposed for calculating pressure loss in two-phase flow, see e.g. Wallis (1969), Collier (1972), Friedel (1979), Bergles *et al.* (1981), and Chisholm (1983). According to Chisholm (1983) the Armand method is the most elegant, and the Lockhart–Martinelli (1949) approach is the most easily applied as it does not explicitly consider flow pattern. Hewitt *et al.* (1994) recommend that the method of Taitel & Dukler (1976) should be used for prediction of the flow pattern on horizontal flow, and they recommend the Friedel (1979) correlation for calculating pressure loss. *These last two approaches are probably now to be preferred and the reader should seek to apply these methods.*

In his 1972 application, Rhee observed that as pressure loss in two-phase flow was small it did not seem to matter much which model was used, and he therefore used a simple *linear* fit of the Lockhart–Martinelli data. We shall stay with the Lockhart–Martinelli correlation so as not to depart from Rhee's calculations. A better *curved* fit for the Lockhart–Martinelli approach was developed by Chisholm & Laird (1958), Chisholm (1967), and Collier (1972).

Using the Lockhart–Martinelli model and defining quality of the vapour as x then

$$x = \frac{\dot{m}_g}{m_g + m_f} \qquad \text{from which} \qquad \frac{\dot{m}_f}{\dot{m}_g} = \frac{1-x}{x}$$

Using frictional pressure loss only (neglecting acceleration loss, and with zero static head loss for a horizontal tube)

$$-\left(\frac{\Delta p}{\Delta \ell}\right) = \frac{4fG^2}{2\rho}\left(\frac{1}{d}\right)$$

and with

$$G = \frac{4\dot{m}}{\pi d^2}, \qquad \mathrm{Re} = \frac{dG}{\eta}, \qquad f = a\mathrm{Re}^{-n}$$

$$-\left(\frac{\Delta p}{\Delta \ell}\right) = \left[\frac{4^{3-n}}{\pi^{2-n}} \cdot \frac{a}{2}\right]\left[\frac{\eta^n}{\rho}\right]\left[\frac{\dot{m}^{2-n}}{d^{5-n}}\right]$$

If the vapour fraction actually flowing, alone occupied the pipe of diameter d, then

$$-\left(\frac{\Delta p}{\Delta \ell}\right)_g = (const.)\left[\frac{\eta_g^n}{\rho_g}\right][\dot{m}_g^{2-n}]$$

Similarly for the liquid fraction,

$$-\left(\frac{\Delta p}{\Delta \ell}\right)_f = (const.)\left[\frac{\eta_f^n}{\rho_f}\right][\dot{m}_f^{2-n}]$$

Defining X where

$$X^2 = \frac{\left(\frac{\Delta p}{\Delta \ell}\right)_f}{\left(\frac{\Delta p}{\Delta \ell}\right)_g} = \left[\frac{\eta_f}{\eta_g}\right]^n \left[\frac{\rho_g}{\rho_f}\right]\left[\frac{\dot{m}_f}{\dot{m}_g}\right]^{2-n}$$

then X^2 provides a measure of the degree to which the two-phase mixture behaves like the liquid rather than like the gas.

Introducing the *two-phase multipliers* relating the pressure loss in each component flow to the same two-phase pressure loss

$$\left(\frac{\Delta p}{\Delta \ell}\right)_{tp} = \left(\frac{\Delta p}{\Delta \ell}\right)_{g} \times (\phi_g)^2$$

$$\left(\frac{\Delta p}{\Delta \ell}\right)_{tp} = \left(\frac{\Delta p}{\Delta \ell}\right)_{f} \times (\phi_f)^2$$

Thus

$$X^2 = \frac{(\phi_g)^2}{(\phi_f)^2} = \frac{\left(\frac{\Delta p}{\Delta \ell}\right)_f}{\left(\frac{\Delta p}{\Delta \ell}\right)_g} = \left[\frac{\eta_g}{\eta_f}\right]^n \left[\frac{\rho_g}{\rho_f}\right] \left[\frac{1-x}{x}\right]^{2-n}$$

Lockhart & Martinelli prepared empirical correlations from experimental data to relate (ϕ_g, ϕ_f, X). Chisholm & Laird (1958), Collier (1972), and Chisholm (1983) report that these curves may be approximated graphically by the following expressions, which are represented in Fig. 12.1,

$$(\phi_g)^2 = 1 + C.X + X^2$$

$$(\phi_f)^2 = \frac{1}{X^2} + \frac{C}{X} + 1$$

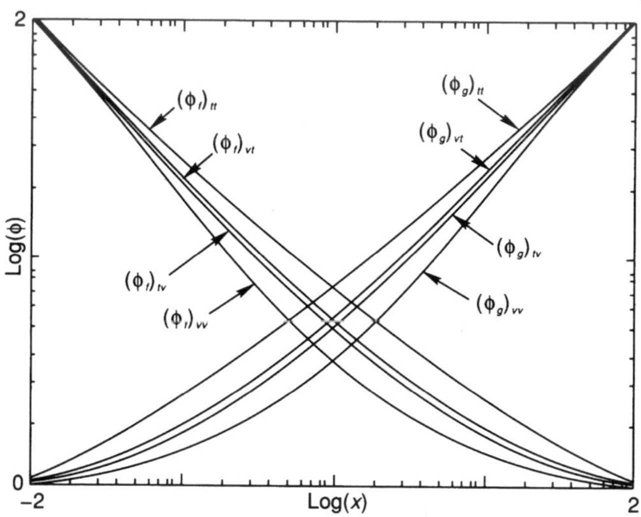

Fig.12.1 Adiabatic friction multipliers for all fluids: ϕ_{tt}, ϕ_{vt}, ϕ_{tv}, and ϕ_{vv} versus x

Table 12.1 Values of constant C (Chisholm, 1983)

Liquid–vapour	C	Re_f	Re_g
Turbulent–turbulent (tt)	21*	>2000	>2000
Viscous–turbulent (vt)	12	>2000	<1000
Turbulent–viscous (tv)	10	<1000	>2000
Viscous–viscous (vv)	5	<1000	<1000

*Some authors may use $C = 20$.

The constant C is to be found from Table 12.1 for the two possible flow conditions in liquid and vapour streams. The turbulent–viscous case is not often encountered.

By the above means, two-phase pressure losses may (approximately) be determined.

For the problem in question

The Lockhart & Martinelli approach provides

$$\left(\frac{\Delta p}{\Delta \ell}\right)_{tp} = (\phi_g)^2 \times \left(\frac{\Delta p}{\Delta \ell}\right)_g$$

$$X = \left[\frac{\eta_f}{\eta_g}\right]^{n/2}\left[\frac{\rho_g}{\rho_f}\right]^{1/2}\left[\frac{1-x}{x}\right]^{1-n/2}$$

For 100 per cent vapour flowing turbulently in a tube

$$\left(\frac{\Delta p}{\Delta \ell}\right)_g = \frac{4fG^2}{2\rho}\left(\frac{1}{d}\right)$$

The Blasius correlation gives

$$f = 0.0791(\mathrm{Re})^{-0.25}$$

from which $n = 0.25$, thus

$$X_{tt} = \left[\frac{\eta_g}{\eta_f}\right]^{0.125}\left[\frac{\rho_g}{\rho_f}\right]\left[\frac{1-x}{x}\right]^{0.875}$$

and the two-phase multiplier for turbulent–turbulent flow is

$$\phi_{tt}^2 = 1 + CX_{tt} + X_{tt}^2$$

where $C = 21$ for $0.01 < X_{tt} < 10.0$. Thus $(\Delta_p/\Delta_\ell)_{tp}$ can be found.

12.4 Two-phase heat-transfer correlations

Correlations presented in this section are those recommended by Rhee for calculating the performance of a double-tube exchanger with refrigerant 12 flowing in the central tube and water in contraflow in the annulus. Hewitt *et al.* (1994) observe that the recommended Chen (1966) correlation was found to be seriously in error for refrigerant 12, and Shah (1976) considered that most correlations were not reliable beyond the range of data for which they were applicable (see Smith, 1986).

Rhee followed a systematic programme of experimentation on a purpose-built test-rig which permitted adjustment of the two-phase inlet condition to any desired vapour quality, and he was thus able to explore each two-phase flow region with some precision and obtain extensive test results on which his correlations are based. The correlations themselves were developed after examining and assessing those of many other workers in the field of two-phase flow.

Rhee & Young (1974) are to be congratulated for seeking to find the solution to a real engineering problem, and for not being satisfied with exploring just one small part of the phenomenon of two-phase flow.

Nucleate boiling

This combines a Dittus–Boelter forced convection correlation with a McNelly pool boiling correlation, following a suggestion by Rohsenow that the two effects could be combined. The nucleate flow heat-transfer coefficient is obtained from Nu.

$$\mathrm{Nu} = 0.023(\mathrm{Re}_f)^{0.8}(\mathrm{Pr}_f)^{0.4} + 0.30(\mathrm{Re}_{th})^{0.69}(\mathrm{Pr}_f)^{0.69}\left(\frac{\rho_f}{\rho_g} - 1\right)^{0.31}\left(\frac{p_i d}{\sigma}\right)^{0.31}$$

where $\mathrm{Re}_{th} = \dfrac{\dot{q}_{flux} d}{\eta_f h_{fg}}$ is the thermal Reynolds number

and σ is the surface tension.

These correlations are valid for very small values of vapour quality, say $x < 0.01$. There is a difficulty in evaluating the correlation as the thermal Reynolds number contains the same heat-transfer coefficient as the Nusselt number on the left-hand side. An iterative approach is required, but this is not a serious impediment to solution as only one value of vapour quality is involved.

Stratified flow

The two-phase heat-transfer coefficient (α_{tp}) is obtained from

$$\frac{\alpha_{tp}}{\alpha_f} = 313.76\left(\frac{1+x}{1-x}\right)^{-0.542}(Fl)^{-0.448}$$

where the heat flux is $Fl = G\alpha_{fg}/\dot{q}_{flux}$ and α_f is obtained from $Nu_f = 0.023(Re)^{0.8} \times (Pr)^{0.4}$.

There will be a numerical problem if attempts are made to evaluate the correlation at very high values of vapour quality (x) However, in design computation it is not necessary to approach the value $x = 1$ too closely, so the difficulty is avoided.

Annular-mist flow

The two-phase heat-transfer coefficient (α_{tp}) is obtained from

$$\frac{\alpha_{tp}}{\alpha_f} = 59.03\left(\frac{1+x}{1-x}\right)^{0.81} (Fl)^{-0.30}$$

where the heat flux is $Fl = G\alpha_{fg}/\dot{q}_{flux}$ and α_f is obtained from $Nu_f = 0.023(Re)^{0.8} \times (Pr)^{0.4}$.

Annular (no mist) flow

The two-phase heat-transfer coefficient (α_{tp}) is obtained from

$$\frac{\alpha_{tp}}{\alpha_f} = 140.8\left(\frac{1+x}{1-x}\right)^{0.565} (Fl)^{-0.41}$$

where the heat flux is $Fl = G\alpha_{fg}/\dot{q}_{flux}$ and α_f is obtained from $Nu_f = 0.023(Re)^{0.8} \times (Pr)^{0.4}$.

Mist flow

$$Nu_f = 0.023(Re_g)^{0.8}(Pr_g)^{0.4}(x)^{-1.23}$$

where x is the vapour quality.

Transition from annular-mist flow to mist flow

This depends on the Weber number obtained from the undernoted correlation

$$We = 1.89 \times 10^{-7} Re_g \left(\frac{\rho_f}{\rho_g}\right)^{0.6} \left(\frac{d}{18.5318}\right)^{-0.48}$$

The form of the Weber number below was derived by Groothuis & Hendal (1959) for the case of heat addition to fluid flowing in a tube, when annular flow breaks down into mist flow. The Weber number is obtained using a velocity profile near the wall defined by the Prandtl universal profile, with the Blasius equation for wall friction,

$$\text{We} = \frac{x^2 G \eta_f}{\sigma \rho_g}\left[\frac{dG}{\eta_f}\right]^{0.125}$$

from which x can be determined.

Demarcation mass velocity

Evaporation of refrigerant 12 in a horizontal tube proceeds in different ways depending on a critical mass velocity (G_{crit}):

- above G_{crit} the flow regimes being followed to 100 per cent dry vapour are:
  Nucleate ⇒ Stratified ⇒ Annular (Mist) ⇒ Mist
- below G_{crit} the flow regimes being followed to 100 per cent dry vapour are:
  Nucleate ⇒ Stratified ⇒ Annular (no mist)

Rhee found that the log-linear correlation $G_{crit} = B\exp(mT_{tp} + c)$ based on boiling temperature T_{tp} provided a good fit for refrigerant 12. For T_{tp} in K and G_{crit} in kg/(m^{2}s) the constants take the following values

$$B = 1.3563 \times 10^{-3}, \qquad m = 0.037\,71, \qquad c = 1.215\,69$$

All the above correlations are for forced-flow evaporation of refrigerant 12 in a horizontal tube, and should not be used in any other circumstances without first checking their validity.

12.5 Two-phase design of a double-tube exchanger

The design exercise tackled was that of a double-tube heat exchanger with refrigerant 12 evaporating in the central tube, being heated in contraflow by water flowing in the annulus. The undernoted flow parameters are for mass velocity $G = 221.40$ kg/(m^2 s).

Tube parameters

Inner tube bore, m	$d_i = 0.011\,887$
Inner tube o.d., m	$d = 0.012\,700$
Inner tube wall thermal conductivity, J/(m s K)	$\lambda_t = 386.0$
Outer tube bore, m	$D = 0.019\,050$

Refrigerant 12

Mass rate of flow of refrigerant, kg/s	$\dot{m}_r = 0.024\,570$
Inlet pressure of refrigerant, bar	$p_1 = 4.102\,21$
Outlet pressure of refrigerant, bar	$p_2 = 3.943\,625$

Water

Mass flow rate of water, kg/s	$\dot{m}_w = 0.653\,94$
Inlet temperature of water, K	$T_1 = 287.37$
Outlet temperature of water, K	$T_2 = 288.67$

All physical properties were obtained from polynomial fits of data in a region close to the design conditions. In repeating the exercise the author converted all data to SI units before proceeding. In this exercise it was found that some of the data-fits used by Rhee were not adequate and new data-fits were produced, thus the results presented here may differ somewhat from those of the original work by Rhee.

The first task is to determine the evaporative duty of the exchanger. A good approximation to this is obtained using the latent heat of refrigerant 12 at the inlet condition. This, however, is not the correct duty because pressure loss due to friction and acceleration produces a different saturation condition at exit.

A good approximation to the other end conditions of both fluids is now available and the design can proceed. After a first design pass the mass flowrate of water can be adjusted proportionately until the thermal duty on both sides becomes the same.

First design pass

The numerical procedure is by increments of vapour dryness fraction (x), and it is recommended that not less than 100 increments be used so that dryness increments in steps of 0.01.

Design proceeds by first evaluating G_{crit} to determine which correlations are to be used after stratified flow. It may be more accurate to evaluate G_{crit} at the end of stratified flow but this is more easily done in a second design pass.

The correlation for nucleate flow is evaluated for only one very small increment of dryness, say 0.0001, as the end of this two-phase flow region is extremely short.

All other correlations are to be evaluated separately for dryness increments of 0.01 over as much of the range as seems necessary, remembering that there is a numerical restriction in evaluating the term $(1+x)/(1-x)$ which appears in stratified flow, annular-mist flow and annular (no mist) flow. It is convenient to stop short of reaching 100 per cent dryness as this does not affect the computation.

Each two-phase flow correlation and its associated Lockhart–Martinelli pressure loss correlation is placed inside a separate 'procedure body' together with the heat-transfer and pressure loss correlations for flow of water in the annulus. In each procedure body the dryness increment is used to calculate the following parameters starting from the inlet end for refrigerant 12:

heat transferred in length increment $d\ell$
heat flux in length increment $d\ell$
overall heat-transfer coefficient in length increment $d\ell$
pressure loss in length increment $d\ell$
pressure in refrigerant 12 at exit from length increment $d\ell$
water inlet temperature to length increment $d\ell$
cumulative length of tube $\Sigma(d\ell)$

Thus different curves can be produced over almost the whole length of the exchanger showing how the two-phase heat-transfer coefficient changes for each flow regime during evaporation (Figs 12.2 and 12.3).

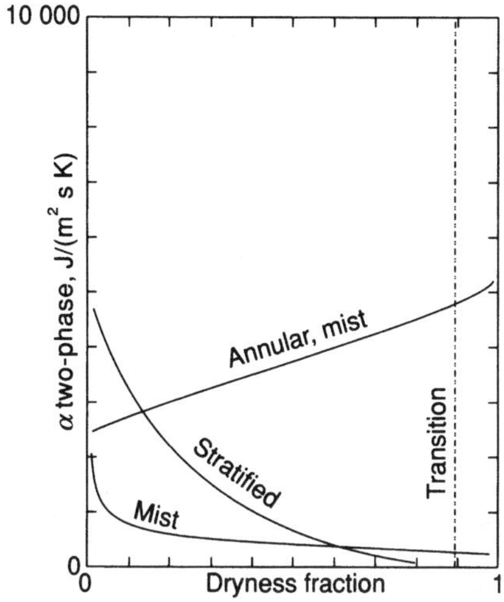

Fig.12.2 Individual curves for two-phase flow above G_{crit} to base of dryness x

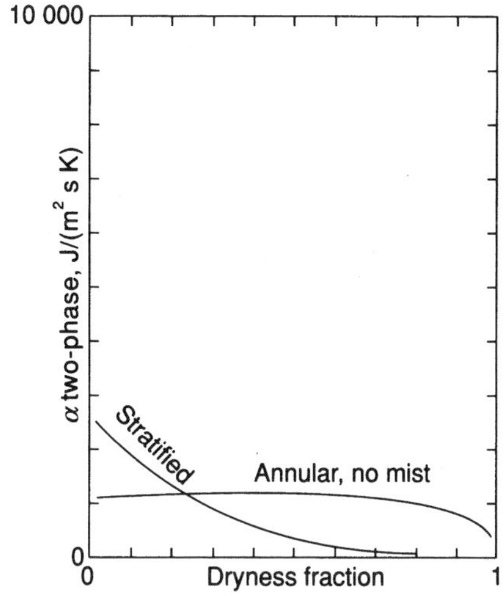

Fig.12.3 Individual curves for two-phase flow below G_{crit} to base of dryness x

This information can be used to construct actual behaviour of the evaporating fluid. Nucleate flow is the first point on the curve at $x = 0$. Stratified flow proceeds until its heat-transfer coefficient is exceeded by either annular-mist flow or annular (no mist) flow. If refrigerant 12 mass velocity is below G_{crit} then annular (no mist) flow continues

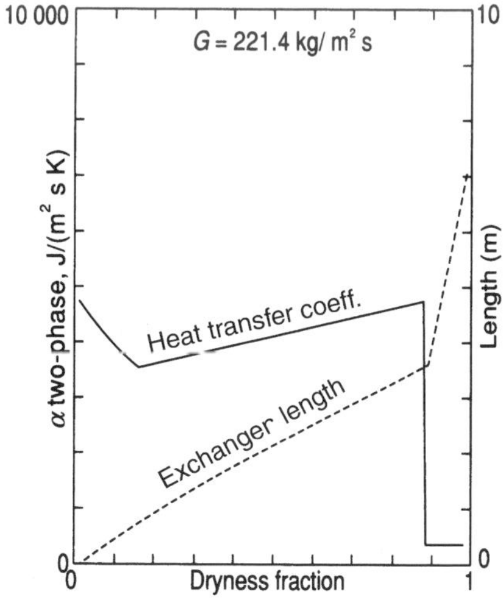

Fig.12.4 Composite curve for two-phase flow above G_{crit} to base of dryness x

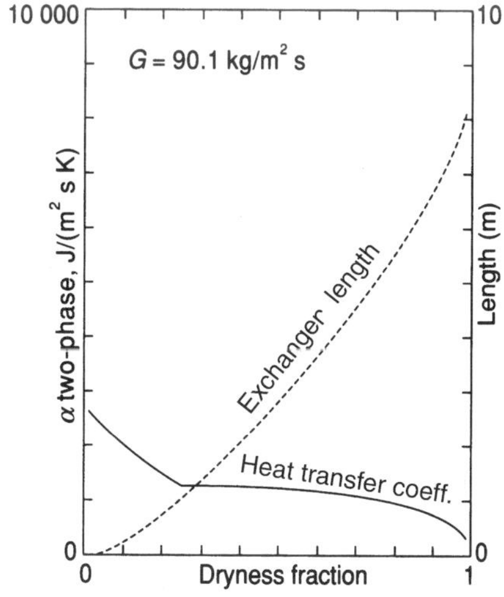

Fig.12.5 Composite curve for two-phase flow below G_{crit} to base of dryness x

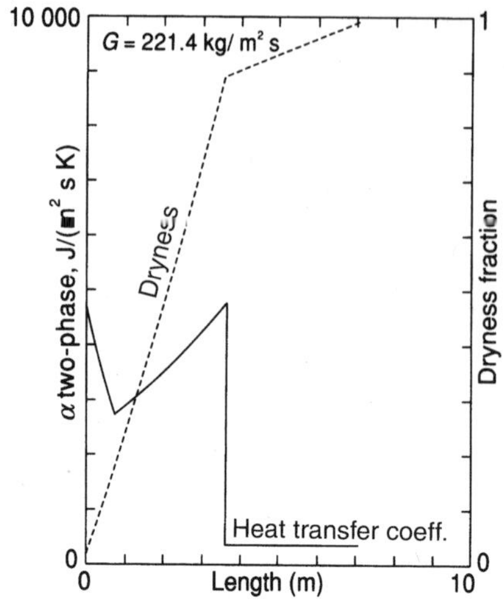

Fig.12.6 Composite curve for two-phase flow above G_{crit} to base of length ℓ

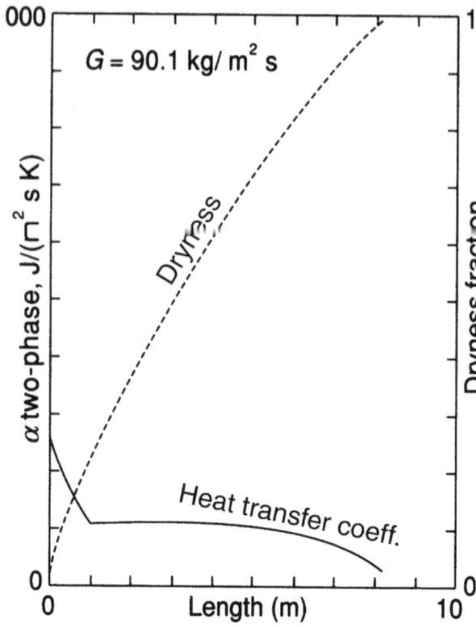

Fig.12.7 Composite curve for two-phase flow below G_{crit} to base of length ℓ

to 100 per cent dryness. If refrigerant mass velocity is above G_{crit} then annular-mist flow continues to the dryness value determined by the Weber number, and after that point, mist flow continues to 100 per cent dryness (Figs 12.4–12.7).

12.6 Discussion

The software was written from scratch in SI units by the author, following guidelines provided by Rhee. The physical properties were not spline-fitted which is the recommended procedure but were included as polynomial fits of data so as to follow as closely as possible the method used by Rhee. Rhee's data-fits were not used, instead the best available data were refitted by polynomials, and in the process some serious discrepancies were found in the two representations of refrigerant 12.

Rhee used Du Pont data for Freon 12, while the author used ICI data for Arcton 12. Rhee admitted that there were disturbing inconsistencies with the Freon 12 data. In this light it cannot be certain that the computational predictions of Rhee are absolutely correct, and in consequence the author's computations cannot be compared exactly with those of Rhee.

This is really not a serious problem as world-wide production of refrigerant 12 has now ceased because of damage to the ozone layer, and the above results are not likely to be used in anger.

However, the curves in Figs 12.2–12.7 correspond very well in form to the test results obtained by Rhee (1972) and also to the independent experimental data of Chawla (1967) on refrigerant 11 boiling.

It will be noticed that the two-phase heat-transfer coefficient in stratified flow always decreases as vapour dryness increases. Rhee reports that this effect was also experimentally noticed by other investigators in low mass flowrate studies (Chawla, 1967; Zahn, 1964), but that its explanation is straightforward, viz.

> '... As the stratified flow develops, volume of the vapour on the top of the tube increases, lowering the value of the heat-transfer coefficient in the upper part of the tube to that close to the heat-transfer coefficient of the pure vapour. The more vapour generation, the larger the area covered by the vapour until it reaches the point where annular flow develops and the tube wall is again wetted with liquid.'

It might be further remarked that as annular-mist flow develops, liquid adheres to the tube wall because its higher viscosity allows a better match of slow-moving fluid to the stationary tube wall. In the core, the still higher speed vapour is happier to match speed with the faster-moving liquid interface on its perimeter.

The 'dryout' transition in two-phase heat-transfer coefficient going from annular-mist flow to mist flow is not so sharp in the experimental data of Chawla (1967), but this could be the result of other effects such as longitudinal thermal conduction in the tube wall affecting experimental results.

In particular it is worth noting in Figs 12.4–12.7 how the last small increment in dryness fraction requires a disproportionate length of the exchanger. It is clear that the much lower heat-transfer coefficient on the water-side is controlling this design during existence of the very high two-phase heat-transfer coefficients, but when mist flow occurs there is closer correspondence with the heat-transfer coefficients for water and refrigerant 12.

For those who may be despairing that no correlations yet exist for the two-phase fluid and horizontal surface geometry of their interest, it may be worth first trying to establish the Weber number that provides the transition between high and low overall heat-transfer coefficients. This will ease design as an inaccurate value for two-phase flow heat-transfer coefficient before transition will not much affect design of the exchanger.

Where the problem may become more difficult is when both fluids in the exchanger change phase together. The computational problem becomes more complex, and the final heat exchanger will undoubtedly be short. This makes it easier for a slight change in operating conditions to perhaps move one fluid partially out of a short exchanger as regards two-phase flow conditions. Caution is necessary as this is a situation to be avoided.

In general, the work of Rhee and Young is a valuable contribution to design for two-phase flow, for it established a methodology of experimentation and also of heat exchanger design procedures on which future work may be based.

However, there is scope for reworking Rhee's data using the later paper of Friedel (1979) which provides the two-phase pressure loss correlations. There is little point in applying the Taitel & Dukler (1976) two-phase flow pattern map because Rhee's experimental technique has already identified each flow regime.

Friedel two-phase pressure-loss correlation

This is based on evaluating friction factors for the pipe either totally filled with liquid (quality $x = 0$) or totally filled with dry vapour (quality $x = 1$).

Evaluate Reynolds numbers for flow with full mass velocity $G = (\dot{m}/A)$

$$\text{Reynolds number for liquid only, } \mathrm{Re}_f = \frac{dG}{\eta_f}$$

$$\text{Reynolds number for gas only, } \mathrm{Re}_g = \frac{dG}{\eta_g}$$

and determine the friction factors (f_f, f_g). Either the Fanning ('16/Re'), or the Moody ('64/Re') definitions for friction factor will do, as the ratio will subsequently be taken of the two values.

The Friedel two-phase friction correlation is

$$\boxed{(\phi_f)^2 = E + \frac{3.24FH}{(\mathrm{Fr}_{tp})^{0.045}(\mathrm{We}_{tp})^{0.035}}}$$

where

$$E = (1-x)^2 + x^2\left(\frac{\rho_f}{\rho_g}\right)\left(\frac{f_g}{f_f}\right)$$

$$F = x^{0.78}(1-x)^{0.24}$$

$$H = \left(\frac{\rho_f}{\rho_g}\right)^{0.91}\left(\frac{\eta_g}{\eta_f}\right)^{0.19}\left[1 - \frac{\eta_g}{\eta_f}\right]^{0.7}$$

two-phase density,

$$\rho_{tp} = \left[\frac{x}{\rho_g} + \frac{(1-x)}{\rho_f}\right]^{-1}$$

Froude number, where g is acceleration due to gravity ($\mathrm{m/s^2}$)

$$\boxed{\mathrm{Fr} = \frac{G^2}{gd(\rho_{tp})^2}}$$

Weber number, where σ is surface tension (N/m)

$$\boxed{\mathrm{We} = \frac{G^2 d}{\sigma \rho_{tp}}}$$

The two-phase pressure gradient is determined with the same expression as used in the Martinelli treatment, viz.

$$\boxed{\left(\frac{\Delta p}{\Delta \ell}\right)_{tp} = \left(\frac{\Delta p}{\Delta \ell}\right)_f \times (\phi_f)^2}$$

where the liquid-only pressure gradient (Fanning definition) is

$$-\left(\frac{\Delta p}{\Delta \ell}\right)_f - \frac{4 f_f G^2 (1-x)^2}{2\rho_f}\left(\frac{1}{d}\right)$$

A full numerical example is to be found in Hewitt *et al.* (1994).

Later supporting work

New experimental results for two-phase boiling of n-pentane published by Kandlbinder *et al.* (1997) exhibit very similar heat-transfer coefficient trends to those predicted by Rhee. The data of Kandlbinder *et al.* extends well into nucleate boiling, a region which was not covered by Rhee's experimental work. New analytical correlations are thus awaited with some interest.

Judge & Radermacher (1997) examined ten different heat-transfer correlations for condensation and evaporation and compared their predictions with experimental data. Of the five flow evaporation and five flow condensation correlations tested, the best evaporation correlation was due to Jung & Radermacher (1989), and the best boiling correlation was due to Dobson *et al.* (1994).

For refrigerant 22, the Jung & Radermacher correlation produced a smooth curve without discontinuities resembling the general form shown in Fig. 12.4 for refrigerant 12. This is encouraging, and further examination of these correlations would be appropriate, except that chlorofluorocarbon and hydrochlorofluorocarbon refrigerants are being phased out in favour of hydrofluorocarbons and natural CO_2.

Readers with an interest in condensation should consult Chu & McNaught (1992), McNaught (1982, 1985), Bergles *et al.* (1981), and Collier (1972).

Plate–fin surfaces

Recent two-phase work with plate–fin surfaces is to be found in the paper by Wadekar (1991) who considers vertical flow boiling of heptane, with earlier work on cyclohexane. Chen *et al.* (1981) have also studied boiling in plate–fin exchangers. Clarke & Robertson (1984) investigated convective boiling of liquid nitrogen in plate–fin heat exchanger passages and found that there were regions of superheated liquid in the exchanger where boiling would have been expected but the onset of boiling was delayed. This produced a considerable length of exchanger in which very low heat flux conditions existed and little heat transfer took place.

It is further remarked by Clarke & Robertson that the point of onset of evaporation appears to be affected by the method by which the desired operating conditions were achieved, and that both stable and meta-stable onset conditions are possible.

Two-phase flow in compact heat exchangers is becoming better understood through work by Kew & Cornwell (1997), by Thonon *et al.* (1997), and by a good number of other workers referenced in these two papers. It may not yet be well understood in multi-stream plate–fin exchangers, and presently multi-start coil helical-tube heat exchangers might still be preferred for commercial evaporating service, because the shell-side is fully interconnected. It seems desirable that compact plate–fin exchangers should also be configured to interconnect evaporating or condensing passages. This may involve the use of surface geometries like the rectangular offset strip-fin configuration which is everywhere connected, plus transverse interconnection between all identical channels in the exchanger to equalize pressures in the evaporating or condensing stream. This last concept will require reworking of the manufacturing process. Further experimental work is necessary on compact plate–fin exchangers to resolve the situation and demonstrate stability in two-phase operation.

This effect may have similarities to that of 'roll-over' in cryogenic tanks where the temperature of liquid at the bottom of the tank may be higher than saturation at the evaporating surface. In vertical boiling in a channel the column of liquid may exert sufficient pressure to suppress evaporation until explosive evaporation takes place. With vertical boiling it seems that the presence of gravitational forces in theoretical correlations may be anticipated.

12.7 Aspects of air conditioning

Air conditioning

Air dehumidification using plate–fin and tube heat exchangers is discussed by Seshimo *et al.* (1989), with extension to frosting conditions by Ogawa *et al.* (1993). Related work is reported by Kondepudi & O'Neal (1989, 1993), and by Machielson & Kershbaumer (1989). McQuiston & Parker (1994), Jones (1985), and Threlkeld (1970) are good textual references on heating, ventilating, and air conditioning.

Vardhan & Dhar (1998) describe an approach to design of air conditioning tube-and-fin coiled heat exchangers in which the finned coil is split into equal geometric blocks each of which 'contains' a single section of coolant tube. Each block is then analysed as a separate heat exchanger.

Condensation

Readers with an interest in condensation should consult Chu & McNaught (1992), McNaught (1982, 1985), Bergles *et al.* (1981), and Collier (1972). The international Keynote lectures of Rose (1997 to date) may be traced on the Internet, and some of his other papers are listed in the Bibliography.

Contact resistance

Critoph *et al.* (1996) report important work on the attachment of plain aluminium fins to tube coils. The traditional pressed fit was found to be less satisfactory than aluminium brazed fins using a commercial process. Heat transfer with pressed fins was found to be around 12 per cent of the air-side resistance, and this was almost completely eliminated by use of brazing (see also Sheffield *et al.*, 1989). When ice formation is likely, brazed fins would seem to offer less likelihood of water freezing and ice expanding between tubes and fins.

Fin-and-tube heat exchangers

Such crossflow exchangers are frequently used as condensers and evaporators in refrigeration or air conditioning plant, and they require their own design procedures. An exchanger with some flow depth in the tube bank may have three or more hairpin tubes to be traversed by the air flow. Full thermal design of tube-and-fin heat exchangers may require the approach developed by Vardhan & Dhar (1998).

The definitive paper by Kim *et al.* (1999) provides universal heat-transfer and pressure loss correlations for the fin-side of staggered tube arrangements (in-line configurations are not recommended).

Plain fin-and-tube surfaces

Wang *et al.* (1996) tested 15 plate fin-and-tube surfaces, and found good agreement with the heat-transfer correlations of Gray & Webb (1986). The recommended correlations for flow friction and heat transfer in the range $800 < \mathrm{Re}_{D_c} < 7500$ are

$$f = 1.039(\mathrm{Re}_{D_c})^{-0.418}\left(\frac{t}{D_c}\right)^{-0.104} N^{-0.0935}\left(\frac{F_p}{D_c}\right)^{-0.197}$$

$$j = 0.394(\mathrm{Re}_{D_c})^{-0.392}\left(\frac{t}{D_c}\right)^{-0.0449} N^{-0.0897}\left(\frac{F_p}{D_c}\right)^{-0.212}$$

where

D_c = fin collar outside diameter (m)
F_p = fin pitch (m)
N = number of tube row
t = fin thickness (m)

The j-correlation could describe 97 per cent of the experimental data within 10 per cent

The f-correlation could describe 88 per cent of the experimental data within 10 per cent

The tube size was given as 9.52 mm o.d

Slit fin-and-tube surfaces

Wang *et al.* (1999) further investigated the performance of slit-fin surfaces, which regenerate the boundary layer at each new leading edge. The correlations are more complicated than those for plain fins and readers are referred to the paper.

The j-correlation could describe 83 per cent of the experimental data within 10 per cent

The f-correlation could describe 93 per cent of the experimental data within 10 per cent

The tube size was given as 9.52 mm i.d. before expansion

An inconsistency appears to exist about whether tube diameter is for inside or outside surfaces between plain- and slip-fin surfaces, but this should hardly affect the correlations.

A considerable 'reference' collection of other sources is to be found in the two papers published by Wang *et al.*, including correlations for wavy-fin and convex louvre-fin surfaces.

Frosting and defrosting

Heat transfer involving the growth of ice on refrigerated surfaces is a moving boundary or Stefan problem discussed in the text edited by Ockendon & Hodgkins (1975). More thorough analysis of heat transfer during frost formation on circular cylinders is contained in the papers by Raju & Sherif (1993) and by Ismail *et al.* (1997). Modelling of frost formation on a flat plate has been presented by Sherif *et al.* (1993). A paper on defrosting of evaporators has been presented by Radenco *et al.* (1995).

When icing may be encountered, it may also be advantageous to omit every second fin in the bank for the first tube hairpin, so that a new fin leading edge becomes available for ice formation deeper into the exchanger (Ogawa *et al.*, 1993). With icing, the attachment of fins to tubes may also require brazing instead of press-fitting to ensure maintenance of good thermal contact (Critoph *et al.*, 1996). The formation of ice under a bad press-fit simply makes a bad fit more loose.

Heat transfer during icing becomes complex for air flow over iced surfaces when simple geometries can become modified by selective ablation, making the numerical problem more difficult (Date, 1991). With air conditioning equipment the best philosophy is probably to 'burn off' the frost before it gets too thick. One of the best ways of defrosting an evaporator is to use hot gas directly from the compressor discharge. It may be necessary to control the short time duration of defrost, and to fit a throttle to the hot-gas by-pass line to prevent loss of compressor head pressure.

Ice harvesting

Off-peak operation of gas turbine plant has been used to make ice used later during peak power operation to cool compressor inlet air. This is one way of increasing system capacity without increasing system operating costs. The same principle can be used for air conditioning of buildings.

Harvesting of the ice sheet is done by switching refrigerant fluid from liquid to gas inside the bayonet tube or vertical hollow plate. Once the ice has fallen into a storage tank a new cycle of ice production and harvesting can begin. Transient modelling of an ice-bank system has been presented by Finer *et al.* (1993).

Mixtures of ice crystals and glycol are now of interest as secondary circuit fluids in refrigeration systems.

12.8 Rate processes

It would not be proper to close without brief mention of the many correlations used in heat-transfer design. It was never the intention to cover this topic fully, yet it would be remiss not to mention the contribution of Churchill (1988, 1992) in explaining rate processes, and the assessment of his work by Kabel (1992).

Condensation

Readers with an interest in condensation should consult Collier (1972), Bergles *et al.* (1981), McNaught (1982, 1985), and Chu & McNaught (1992).

References

Bergles, A.E., **Collier, J.G.**, **Delhaye, J.M.**, **Hewitt, G.F.**, and **Mayinger, F.** (1981) *Two-Phase Flow and Heat Transfer in the Power and Process Industries*, Hemisphere, New York.

Carey, V.P. (1992) *Liquid-Vapor Phase-Change Phenomena*, Hemisphere, New York.

Chawla, J.M. (1967) Wärmeübergang and Druckabfall in waagerechten Rohren bei der Strömung von Verdampfenden Kältemitteln. (Local heat transfer and pressure drop for refrigerants evaporating in horizontal tubes.) *Kaltetechnik*, **19**(8), August, 246–252.

Chen, J.C. (1966) Correlation for boiling heat transfer to saturated fluids in convective flow. *Ind. Engng Chemistry, Process Des. Dev.*, **5**(3), 322–333.

Chen, C.C., **Loh, J.U.**, and **Westwater, J.W.** (1981) Prediction of boiling heat transfer duty in a compact plate-fin heat exchanger using the improved local assumption. *Int. J. Heat Mass Transfer*, **24**(12), December, 1907–1912.

Chisholm, D. (1967) A theoretical basis for the Lockhart–Martinelli correlation for two-phase flow. *Int. J. Heat Mass Transfer*, **10**(12), 1767–1778.

Chisholm, D. (1983) *Two-Phase Flow in Pipelines and Heat Exchangers*, Longman.

Chisholm, D. and **Laird, A.D.K.** (1958) Two-phase flow in rough tubes. *Trans. ASME*, **80**(2), February, 276–286.

Chu, C.M. and **McNaught, J.M.** (1992) Condensation on bundles of plain and low-finned tubes – effects of vapour shear and condensate inundation. *IChemE Symposium Series*, No. 129, vol. 1, pp. 225–232.

Churchill, S.W. (1988) The role of mathematics in heat transfer. *AIChE Symposium Series*, No. 263 – *Heat Transfer*, Houston, 1988, vol. 84, pp. 1–13.

Churchill, S.W. (1992) The role of analysis in the rate processess. *Ind. Engng Chemistry Res.*, **31**, 643–658.

Clarke, R.H. and **Robertson, J.M.** (1984) Investigations into the onset of convective boiling with liquid nitrogen in plate-fin heat exchanger passages under constant wall temperature boundary conditions. *AIChE Symposium Series – Heat Transfer*, Niagara Falls, 1984, pp. 98–103.

Collier, J.G. (1972) *Convective Boiling and Condensation*, McGraw-Hill, New York.

Collier, J.G. and **Thome, J.R.** (1994) *Convective Boiling and Condensation*, 3rd edn, McGraw-Hill, New York.

Critoph, R.E., **Holland, M.K.**, and **Turner, L.** (1996) Contact resistance in air-cooled plate fin-tube air-conditioning condensers. *Int. J. Refrigeration*, **19**(6), 400–406.

Date, A.W. (1991) A strong enthalpy formulation for the Stefan problem. *Int. J. Heat Mass Transfer*, **34**(9), 2231–2235.

Dobson, M.K., **Chato, J.C.**, **Wattelet, J.P.**, **Gaibel, J.A.**, **Poncher, M.**, **Kenney, P.J.**, **Shimon, R.L.**, **Villaneuva, T.C.**, **Rhines, N.L.**, **Sweeney, K.A.**, **Allen, D.G.**, and **Hershberger, T.T.** (1994) Heat transfer and flow regimes during condensation in horizontal tubes, ACRC Project 37.

Finer, S.I., **Cleland, A.C.**, and **Lovatt, S.J.** (1993) Simple mathematical model for prediciting the transient behaviour of an ice-bank system. *Int. J. Refrigeration*, **16**(5), 312–320.

Friedel, L. (1979) New friction pressure drop correlations for upward, horizontal and downward two-phase pipe flow. Presented at the *HTFS Symposium*, Oxford, 1979 (Hoechst AG Reference 372217/24 698).

Gray, D.L. and **Webb, R.L.** (1986) Heat transfer and friction correlations for plant fin-and-tube exchangers having plain fins. In *Proceedings of the 9th International Heat Transfer Conference*, San Francisco.

Groothuis, H. and **Hendal, W.P.** (1959) Heat transfer in two-phase flow. *Chem. Engng Sci.*, **11**(3), November, 212–220.

Hausen, H. (1950) *Wärmeübertragung im Gegenstrom, Gleichstrom und Kreuzstrom*, 1st edn, Springer-Verlag.

Hausen, H. (1983) *Heat Transfer in Counterflow, Parallel flow and Cross flow*, 2nd edn, McGraw-Hill, New York.

Hestroni, G. (Ed.) (1982) *Handbook of Multiphase Systems*, Hemisphere, New York (see Hewitt, G.F., Chapter 2, Gas-liquid flow).

Hewitt, G.F. and **Hall-Taylor, M.S.** (1970) *Annular Two-Phase Flow*, Pergamon.

Hewitt, G.F., **Shires, G.L.**, and **Bott, T.R.** (1994) *Process Heat Transfer*, Chapter 10, CRC Press, Florida, pp. 391–421.

Ismail, K.A.R., **Salinas, C.**, and **Gonçlaves, M.M.** (1997) Frost growth around a cylinder in a wet air stream. *Int. J. Refrigeration*, **20**(2), 106–119.

Jones, W.P. (1985) *Air Conditioning Engineering*, 3rd edn, Edward Arnold.

Judge, J. and **Radermacher, R.** (1997) A heat exchanger model for mixtures and pure refrigerant cycle simulations. *Int. J. Refrigeration*, **20**(4), 244–255.

Jung, D.S. and **Radermacher, R.** (1989) Prediction of pressure drop during horizontal annular flow boiling of pure and mixed refrigerants. *Int. J. Heat Mass Transfer*, **32**, 2435–2446.

Kabel, R.L. (1992) Reflections on rates. *Ind. Engng Chemistry*, **31**, 641–643.

Kandlbinder, T., **Wadekar, V.V.**, and **Hewitt, G.F.** (1997) Flow boiling of a single component and binary mixture hydrocarbon. In *5th UK National Conference on Heat Transfer*, Imperial College of Science and Technology, 17–18 September 1997, Session D – Boiling and Gas Liquid Flow.

Kew, P.A. and **Cornwell, K.** (1997) Correlations for the prediction of boiling heat transfer in small-diameter channels. *Appl. Thermal Engng*, **17**(8–10), 705–715.

Kim, N.H., **Youn, B.**, and **Webb, R.L.** (1999) Air-side heat transfer and friction correlations for plain fin-and-tube heat exchangers with staggered tube arrangements. *Trans. ASME, J. Heat Transfer*, **121**, August, 662–667.

Kondepudi, S.N. and **O'Neal, D.L.** (1989) Effect of frost formation on the performance of louvred finned tube heat exchangers. *Int. J. Refrigeration*, **12**, May, 151–158.

Kondepudi, S.N. and **O'Neal, D.L.** (1993) Performance of finned-tube heat exchangers under frosting conditions. Part 1 – simulation model. Part 2 – Comparison of experimental data with model. *Int. J. Refrigeration*, **16**(3), 175–180 and 181–184.

Lockhart, R.W. and **Martinelli, R.C.** (1949) Proposed correlation of data for isothermal two-phase, two component flow in pipes. *Chem. Engng Prog.*, **45**(1), January, 39–48.

Machielsen, C.H.M. and **Kerschbaumer, H.G.** (1989) Influence of frost formation and defrosting on the performance of air coolers: standards and dimensionless coefficients for the system designer. *Int. J. Refrigeration*, **12**(5), September, 283–290.

McNaught, J.M. (1982) Two-phase forced convection heat transfer during condensation on a horizontal tube bundle. In *Proceedings of the 7th International Heat Transfer Conference*, Munich, 1982 (Ed. U. Grigull), vol. 5, Paper CS21, Hemisphere, New York, pp. 125–131.

McNaught, J.M. (1985) Condenser design. *Process Engng*, **66**(1), January, 38–39, and 42–43.

McQuiston, F.C. and **Parker, J.D.** (1994) *Heating, Ventilating and Air Conditioning – Analysis and Design*, 4th edn, John Wiley.

Ockendon, J.R. and **Hodgkins, W.R.** (Eds.) (1975) *Moving Boundary Problems in Heat Flow and Diffusion*, Clarendon Press, Oxford.

Ogawa, K., **Tanaka, N.**, and **Takeshita, M.** (1993) Performance improvement of plate fin-and-tube heat exchangers under frosting conditions. *ASHRAE Trans.: Symposia*, **99**(1), 762–771.

Radenco, V., **Vargas, J.V.C.**, **Bejan, A.**, and **Lim, J.S.** (1995) Two design aspects of defrosting refrigerators. *Int. J. Refrigeration*, **18**(2), 76–86.

Raju, S.P. and **Sherif, S.A.** (1993) Frost formation and heat transfer on circular cylinders in crossflow. *Int. J. Refrigeration*, **16**(6), 390–402.

Rhee, B.-W. and **Young, E.H.** (1974) Heat transfer to boiling refrigerants flowing inside a plain copper tube. *Heat Transfer-Research and Design, AIChE Symposium Series* No. 138, vol. 70, pp. 64–70.

Rhee, B.-W. (1972) Heat transfer to boiling refrigerants R-12 and R-22 flowing inside a plain copper tube, PhD thesis, Department of Chemical Engineering, University of Michigan.

Rohsenow, W.M., **Hartnett, J.P.**, and **Ganic, E.N.** (1985) Two-phase flow. *Handbook of Heat Transfer Fundamentals*, 2nd revised edn, Chapter 13, McGraw-Hill, New York.

Schack, A. (1965) *Industrial Heat Transfer*, Chapman & Hall, London.

Seshimo, Y., **Ogawa, K.**, **Marumoto, K.**, and **Fujii, M.** (1989) Heat and mass transfer performances of plate-fin and tube heat exchangers with dehumidification. *Heat Transfer, Japan. Res.*, **18**(5), September–October, 79–94.

Shah, M.M. (1976) A new correlation for heat transfer during boiling flow through pipes. *ASHRAE Trans.*, **82**(2), 66–86.

Sheffield, J.W., **Wood, R.A.**, and **Sauer, H.J.** (1989) Experimental investigation of thermal conductance of finned tube contacts. *Exp. Thermal Fluid Sci.*, 23–31.

Sherif, S.A., **Raju, S.P.**, **Padki, M.M.**, and **Chan, A.B.** (1993) A semi-empirical transient method for modelling frost formation on a flat plate. *Int. J. Refrigeration*, **16**(5), 321–329.

Smith, R.A. (1986) *Vaporisers, Selection Design and Operation*, Longman, pp. 190–192.

Taitel, Y. and **Dukler, A.E.** (1976) A model for predicting flow regime transitions in horizontal and near horizontal gas-liquid flow. *AIChE J.*, **22**, 47–55.

Thonon, B., **Feldman, A.**, **Margat, L.**, and **Marvillet, C.** (1997) Transition from nucleate boiling to convective boiling in compact heat exchangers. *Int. J. Refrigeration*, **20**(8), 592–597.

Threlkeld, J.L. (1970) *Thermal Environmental Engineering*, 2nd edn, Prentice-Hall.

Vardhan, A. and **Dhar, P.L.** (1998) A new procedure for performance prediction of air conditioning coils. *Int. J. Refrigeration*, **21**(1), 77–83.

Wadekar, V.V. (1991) Vertical slug flow heat transfer with nucleate boiling. *27th National Heat Transfer Conference*, Minneapolis, July 1991, *Phase Change Heat Transfer, ASME HTD-159*, pp. 157–161.

Wadekar, V.V. (2002) Compact exchangers for phase change. *Int. J. Heat Exchangers*, **III**(2), December.

Wallis, G.B. (1969) *One-Dimensional Two-Phase Flow*, McGraw-Hill, New York.

Wang, C.-C., **Chang, Y.-J.**, **Hsieh, Y.-C.**, and **Lin, Y.-T.** (1996) Sensible heat and friction characteristics of plate fin-and-tube heat exchangers having plane fins. *Int. J. Refrigeration*, **19**(4), 223–230.

Wang, C.-C., **Tao, W.-H.**, and **Chang, C.-J.** (1999) An investigation of the airside performance of the slip fin-and-tube heat exchangers. *Int. J. Refrigeration*, **22**, 595–603.

Zahn, W.R. (1964) A visual study of two-phase flow while evaporating in a horizontal tube. *ASME J. Heat Transfer, Ser. C*, **86**, 417–429.

Bibliography

Bell, G.E. (1981) Solidification of a liquid about a cylindrical pipe. *Int. J. Heat Mass Transfer*, **22**, 1981–1986.

Bredesen, A.M., **Pettersen, J.**, **Hafner, A.**, **Neskå, P.**, and **Skaugen, G.** (1997) Studies on CO_2 heat exchangers and heat transfer. In *CO_2 Technology on Refrigeration, Heat Pump and Air Conditioning Systems*, Trondheim, Norway, 13–14 May 1997, pp. 329–358. IEA Heat Pump Centre, Sittard, The Netherlands, Report HPC-WR-19.

Churchill, S.W. and **Gupta, J.P.** (1977) Approximation for conduction with freezing or melting. *Int. J. Heat Mass Transfer*, **20**, 1251–1253.

Clarke, R.H. (1992) Condensation heat transfer characteristics of liquid nitrogen in serrated plate-fin passages. In *3rd UK National Conference incorporating 1st European Conference on Thermal Sciences, Institution of Chemical Engineers, Symposium Series*, No. 129, vol. 2, pp. 1301–1309.

Foumeny, E.A. and **Heggs, P.J.** (1991) *Heat Exchange Engineering*, vols. 1 and 2. Vol. 2, *Compact Heat Exchangers: Techniques of Size Reduction*, Ellis Horwood, New York & London.

Frivik, P.E. and **Pettersen, J.** (1997) CO_2 as a working fluid in perspective. In *CO_2 Technology on Refrigeration, Heat Pump and Air Conditioning Systems*, Trondheim, Norway, 13–14 May 1997, pp. 3–31. IEA Heat Pump Centre, Sittard, The Netherlands, Report HPC-WR-19.

Fuchs, P.H. (1975) Heat transfer and pressure drop during flow of evaporating liquid in horizontal tubes and bends. PhD thesis, Norwegian Institute of Technology (in Norwegian).

Gungar, K.E. and **Winterton, R.H.S.** (1986) General correlation for flow boiling in tubes and annuli. *Int. J. Heat Mass Transfer*, **29**(3), March, 351–358.

Hahne, E., **Spindler, K.**, and **Skok, N.** (1993) A new pressure drop correlation for subcooled flow boiling of refrigerants (R12, R134a). *Int. J. Heat Mass Transfer*, **36**(17), November, 4267–4274.

Haseler, L.E. (1980) Condensation of nitrogen in brazed aluminium plate-fin heat exchangers. In *ASME/AIChE National Heat Transfer Conference*, Orlando 1980, Paper 80-HT-57.

Hwang, Y. and **Radermacher, R.** (1997) Carbon dioxide refrigeration system. In *CO_2 Technology in Refrigeration, Heat Pump and Air Conditioning Systems*, Workshop Proceedings, Trondheim, Norway, 13–14 May 1997, The IEA Heat Pump Centre, Sittard, The Netherlands, p. 71–78.

Jung, D.S. (1988) Mixture effects on horizontal convective boiling heat transfer. PhD thesis, Department of Mechanial Engineering, University of Maryland, College Park, Maryland.

Kandlikar, S.G., **Bijlani, C.A.**, and **Sukhatme, S.P.** (1975) Predicting properties of mixtures of R22 and R12 - Part II. Transport Properties. *ASHRAE Trans.* 2343.

Kayansayan, N. (1994) Heat transfer characterisation of plate fin-tube heat exchangers. *Int. J. Refrigeration*, **17**(1), 49–57.

Kreissig, G. and **Muller-Steinhagen, H.M.** (1992) Frictional pressure drop for gas/liquid two-phase flow in plate heat exchangers. *Heat Transfer Engng*, **13**(4), 42–52.

Martinelli, R.C. and **Nelson, D.B.** (1948) Prediction of pressure drop during forced-circulation boiling of water. *Trans. ASME*, **70**, 695–702.

McLinden, M.O., **Lemmon, E.W.**, and **Jacobsen, R.T.** (1998) Thermodynamic properties of alternative refrigerants (includes mixtures). *Int. J. Refrigeration*, **21**(4), 322–338.

Morrison, G. and **McLinden, M.O.** (1986) Application of hard sphere equation of state to refrigerants and refrigerant mixtures. NBS Technical Note 1226, National Bureau of Standards, Gaithersburg, Maryland.

Neskå, P., **Rekstad, H.**, **Reza Zakeri, G.**, and **Schiefloe, P.A.** (1998) CO_2-heat pump water heater: characteristics, system design and experimental results. *Int. J. Refrigeration*, **21**(3), 172–179.

Paliwoda, A. (1989) Generalised method of pressure drop and tube length calculation with boiling and condensing refrigerants within the entire zone of saturation. *Int. J. Refrigeration*, **12**, November, 314–322.

Paliwoda, A. (1992) Generalised method of pressure drop calculation across components containing two-phase flow of refrigerants. *Int. J. Refrigeration*, **15**(2), 119–125.

Pettersen, J., **Hafner, A.**, and **Skaugen, G.** (1998) Development of compact heat exchangers for CO_2 air-conditioning systems. *Int. J. Refrigeration*, **21**(3), 180–193.

Polyakov, A.F. (1991) Heat transfer under supercritical pressures. *Adv. Heat Transfer*, **21**, 1–53. See also curves on pp. 251–252 in *CO_2 technology in refrigeration, heat pump and air conditioning systems*, Workshop proceedings Trondheim, Norway 13–14 May 1997, The IEA Heat Pump Centre, Sittard, The Netherlands.

Ramos, M., **Cerrato, Y.**, and **Gutierrez, J.** (1994) An exact solution of the finite Stefan problem with temperature dependent thermal conductivity and specific heat. *Int. J. Refrigeration*, **17**(2), 130–134.

Reid, R.C., **Prausnitz, J.M.**, and **Poling, B.E.** (1987) *The Properties of Gases and Liquids*, 4th edn, McGraw-Hill, New York.

Robertson, J.M. (1979) Boiling heat transfer with liquid nitrogen in brazed aluminium plate-fin heat exchangers. *AIChE Symposium Series*, No. 189, vol. 75, pp. 151–164.

Rohsenow, W.M. and **Choi, H.Y.** (1961) *Heat, Mass and Momentum Transfer*, Prentice-Hall, Englewood Cliffs, New Jersey.

Rose, J.W. (1998a) Condensation heat transfer fundamentals. *Trans. Inst. Chem. Engrs*, **76**, Part A, 143–152.

Rose, J.W. (1998b) Interphase matter transfer, the condensation coefficient and dropwise condensation, Keynote lecture. *Proc 11th International Heat Transfer Conference*, Korea, Vol. 1, pp. 89–104.

Rose, J.W. (1999) Condensation heat transfer. *Heat and Mass Transfer Journal*, Vol. 35, Springer-Verlag, pp. 479–485.

Rose, J.W. (2002) Dropwise condensation theory and experiment: a review. *Proc. Instn Mech. Engrs, Part A, J. of Power and Energy*, **216**, 115–128.

Savkin, N., **Mesarkishvili, Z.**, and **Bartsch, G.** (1993) Experimental study of void fraction and heat transfer under upward and downward flow boiling conditions (water in an annulus). *Int. Comm. Heat Mass Transfer*, **20**(6), November–December, 783–792.

Schlager, L.M., **Pate, M.B.**, and **Bergles, A.E.** (1989) Heat transfer and pressure drop during evaporation and condensation of R22 in horizontal micro-fin tubes. *Int. J. Refrigeration*, **12**, January, 6–14.

Schlunder, E.U. (Ed. in Chief) (1990) *Heat Exchanger Design Handbook*, Hemisphere, Washington.

Shah, R.K. (1988) Plate-fin and tube-fin heat exchanger design procedures. *Heat Transfer Equipment Design*, (Eds, R.K. Shah, E.C. Subbarao, and R.A. Mashelkar) Hemisphere, New York, pp. 256–266.

Span, R. and **Wagner, W.** (1996) A new equation of state for carbon dioxide covering the fluid region from the triple point to 1100 at pressures up to 800 MPa. *J. Phys. Chem. Reference Data*, **26**, 1509–1596.

Wadekar, V.V. (1992) Flow boiling of heptane in a plate-fin heat exchanger passage. *Compact Heat Exchangers for Power and Process Industries, 28th National Heat Transfer Conference*, San Diego, California, 1992, ASME HTD-201, pp. 1–6.

Walisch, T. and **Trepp, Ch.** (1997) Heat transfer to supercritical carbon dioxide in tubes with mixed convection. Presented at the *IEA Annex 22, Workshop on Compression Systems with Natural Working Fluids*, Gatlinsburg, Tennessee, 2–3 October, 1997.

Wang, C.-C., **Hsieh, Y.-C.**, and **Lin, Y.-T.** (1997) Performance of plate-finned tube heat exchangers under dehumidifying conditions. *ASME J. Heat Transfer*, **119**, February, 109–117.

Wang, L.K., **Sundén, B.**, and **Yang, Q.S.** (1999) Pressure drop analysis of steam condensation in a plate heat exchanger. *Heat Transfer Engng*, **20**(1), 71–77.

Weimer, R.F. and **Hartzog, D.G.** (1972) Effects of maldistribution on the performance of multistream multipassage heat exchangers. In *Proceedings of the 1972 Cryogenic Engineering Conference, Advances in Cryogenic Engineering*, vol. 18, Plenum Press, Paper B-2, pp. 52–64.

Willatzen, M., **Pettit, N.B.O.L.**, and **Ploug-Sørensen, L.** A general dynamic simulation model for evaporators and condensers in refrigeration. Part 1 – Moving boundary formulation of two-phase flows with heat exchange. Part 2 – Simulation and control of an evaporator. *Int. J. Refrigeration*, **21**(5), 398–403 and 404–414.

APPENDIX A
Transient Equations with Longitudinal Conduction and Wall Thermal Storage

Temperature and velocity fields

A.1 Mass flow and temperature transients in contraflow

The complete set of equations to be solved are presented as equations (A.1), they were obtained using a continuum approach. Thermal diffusivity κ_w is defined in the notation. Their validity may be checked using Schlichting (1960), and further development is summarized in Appendix A.2.

$$\text{Hot fluid}\begin{cases} \dfrac{\partial\rho}{\partial t}+\dfrac{\partial(\rho u)}{\partial x}=0 & \text{mass} \\ \dfrac{\partial u}{\partial t}+u\dfrac{\partial u}{\partial x}=-\dfrac{1}{\rho}\dfrac{\partial p}{\partial x}-\dfrac{fu^2}{2r_{hyd}} & \text{momentum} \\ \rho\dfrac{Dh}{Dt}-\dfrac{Dp}{Dt}=+\lambda_h\dfrac{\partial^2 T_h}{\partial x^2}+\left(-\dfrac{\partial\dot{q}_i}{\partial x_i}\right)+\eta\Phi & \text{energy} \end{cases}$$

$$\text{Solid wall}\left\{\dfrac{\partial T_w}{\partial t}-\hat{\kappa}_w\dfrac{\partial^2 T_w}{\partial x^2}=+\left(\dfrac{\alpha_h S}{M_w C_w}\right)(T_h-T_w)-\left(\dfrac{\alpha_c S}{M_w C_w}\right)(T_w-T_c)\right. \quad \text{energy}$$

$$\text{Cold fluid}\begin{cases} \rho\dfrac{Dh}{Dt}-\dfrac{Dp}{Dt}=+\lambda_c\dfrac{\partial^2 T_c}{\partial x^2}+\left(+\dfrac{\partial\dot{q}_i}{\partial x_i}\right)+\eta\Phi & \text{energy} \\ \dfrac{\partial u}{\partial t}-u\dfrac{\partial u}{\partial x}=+\dfrac{1}{\rho}\dfrac{\partial p}{\partial x}+\dfrac{fu^2}{2r_{hyd}} & \text{momentum} \\ \dfrac{\partial\rho}{\partial t}-\dfrac{\partial(\rho u)}{\partial x}=0 & \text{mass} \end{cases} \tag{A.1}$$

In the fluid energy equations, Φ is the Rayleigh dissipation function. The energy terms $\partial\dot{q}_i/\partial x_i$ are volumetric heat-transfer rates, excluding longitudinal conduction which is accounted for separately. For the hot fluid the term can be expressed as

$$\left(\frac{\partial \dot{q}_i}{\partial x_i}\right) = \frac{\alpha_h \delta S(T_h - T_w)}{\delta V_h} = \frac{\alpha_h}{(A_h \delta x)}\left(\frac{S\delta x}{L}\right)(T_h - T_w) = \frac{u_h}{L}\left(\frac{\alpha_h S}{A_h u_h}\right)(T_h - T_w)$$

where the flow area is constant. When divided by $(\rho_h C_h)$, the last expression provides the familiar convective heat-transfer terms

$$\frac{u_h}{L}\left(\frac{\alpha_h S}{\dot{m}_h C_h}\right)(T_h - T_w)$$

In the above equations the constitutive equation for an isotropic viscous (Stokes) fluid is used

$$\sigma_{ij} = -pI + 2\eta \dot{e}_{ij} - \frac{2}{3}\eta \delta_{ij}\dot{e}_{kk} \tag{A.2}$$

which in expanded form becomes

$$\sigma_{ij} = \frac{1}{3}\delta_{ij}\sigma_{kk} + 2\eta\left\{\frac{1}{2}\left(\frac{\partial u_i}{\partial x_j} + \frac{\partial u_j}{\partial x_i}\right)\right\} - \frac{2}{3}\eta\left(\frac{\partial u_1}{\partial x_1} + \frac{\partial u_2}{\partial x_2} + \frac{\partial u_3}{\partial x_3}\right)$$

This form is already incorporated in the fluid energy equations (A.1).

The state equation for a perfect gas may be interpreted locally.

$$p = \rho RT \tag{A.3}$$

The balance of linear momentum equation can be recast to bring it into a more convenient form.

Pressure gradient due to friction

Fanning friction factor

$$f = \frac{\tau_w}{\rho u^2/2}$$

Frictional resistance to flow

$$\tau_w \delta S = \left(\frac{f\rho u^2}{2}\right)\frac{A\delta x}{r_{hyd}} \qquad \text{where } r_{hyd} = \frac{AL}{S}$$

Force balance

$$[p - (p - \delta p)]A = \left(\frac{f\rho u^2}{2}\right)\frac{A\delta x}{r_{hyd}}$$

Pressure gradient due to friction

$$\frac{\partial p}{\partial x} = \frac{f\rho u^2}{2r_{hyd}}$$

Alternative form of balance of linear momentum

The balance of linear momentum equation for the hot fluid is

$$\frac{\partial u}{\partial t} + u\frac{\partial u}{\partial x} = -\frac{1}{\rho}\frac{\partial p}{\partial x} - \frac{fu^2}{2r_{hyd}}$$

Adding $u(\partial\rho/\partial t)$ to both sides

$$\rho\frac{\partial u}{\partial t} + u\frac{\partial \rho}{\partial t} = -\frac{\partial p}{\partial x} - \rho u\frac{\partial u}{\partial x} + u\frac{\partial \rho}{\partial t} - \frac{f\rho u^2}{2r_{hyd}}$$

$$\frac{\partial(\rho u)}{\partial t} = -\frac{\partial p}{\partial x} - \rho u\frac{\partial u}{\partial x} + u\left[\frac{\partial \rho}{\partial t}\right] - \frac{f\rho u^2}{2r_{hyd}}$$

Substituting for $\partial\rho/\partial t$ from balance of mass equation

$$\frac{\partial(\rho u)}{\partial t} = -\frac{\partial p}{\partial x} - \rho u\frac{\partial u}{\partial x} - u\frac{\partial(\rho u)}{\partial x} - \frac{f\rho u^2}{2r_{hyd}}$$

and the hot fluid equation becomes

$$\frac{\partial(\rho u)_h}{\partial t} = -\frac{\partial(\rho u^2)_h}{\partial x} - \left[\frac{\partial p_h}{\partial x} + \left(\frac{f\rho u^2}{2r_{hyd}}\right)_h\right] \tag{A.4}$$

with an identical equation for the cold fluid, but in the other direction.

Balance of energy

For the balance of energy equations, we neglect fluid longitudinal conduction terms. These are extremely small for gases and very small for many liquids, but can be significant for liquid metals. Expand the remaining terms and collect fluid work contributions together

$$\left\{\begin{aligned}
\frac{\partial T_h}{\partial t} + u_h\frac{\partial T_h}{\partial x} &= -\frac{u_h}{L}\left(\frac{\alpha_h S}{\dot{m}_h C_h}\right)(T_h - T_w) + \frac{1}{\rho_h C_h}\left(\frac{\partial p_h}{\partial t} + u_h\frac{\partial p_h}{\partial x} + \eta_h\Phi_h\right)\\
\frac{\partial T_w}{\partial t} - \hat{\kappa}_w\frac{\partial^2 T_w}{\partial x^2} &= +\left(\frac{\alpha_h S}{M_w C_w}\right)(T_h - T_w) - \left(\frac{\alpha_c S}{M_w C_w}\right)(T_w - T_c)\\
\frac{\partial T_c}{\partial t} - u_c\frac{\partial T_c}{\partial x} &= +\frac{u_c}{L}\left(\frac{\alpha_c S}{\dot{m}_c C_c}\right)(T_w - T_c) + \frac{1}{\rho_c C_c}\left(\frac{\partial p_c}{\partial t} - u_c\frac{\partial p_c}{\partial x} + \eta_c\Phi_c\right)
\end{aligned}\right.$$

Remnants of the Rayleigh dissipation terms (Φ) may not be significant, viz.

$$\Phi = 2\left(\frac{\partial u}{\partial x}\right)^2 - \frac{2}{3}\left(\frac{\partial u}{\partial x}\right)^2 = \frac{4}{3}\left(\frac{\partial u}{\partial x}\right)^2$$

as order of magnitude arguments show that it is the transverse velocity terms which contribute most to dissipation.

Lumping together pressure and dissipation terms as (W_h, W_c), and rewriting

$$\begin{cases} \dfrac{\partial T_h}{\partial t} + u_h \dfrac{\partial T_h}{\partial x} = -\dfrac{u_h}{L}\left(\dfrac{\alpha_h S}{\dot{m}_h C_h}\right)(T_h - T_w) + W_h \\ \dfrac{\partial T_w}{\partial t} - \hat{\kappa}_w \dfrac{\partial^2 T_w}{\partial x^2} = +\left(\dfrac{\alpha_h S}{M_w C_w}\right)(T_h - T_w) - \left(\dfrac{\alpha_c S}{M_w C_w}\right)(T_w - T_c) \\ \dfrac{\partial T_c}{\partial t} - u_c \dfrac{\partial T_c}{\partial x} = +\dfrac{u_c}{L}\left(\dfrac{\alpha_c S}{\dot{m}_c C_c}\right)(T_w - T_c) + W_c \end{cases} \tag{A.5}$$

Response of the temperature field for both hot and cold fluids and the wall is coupled, and simultaneous solution of the finite-difference energy balance equations is straightforward by matrix inversion, there being no missing values.

The balance of energy equations (A.5) can be written

$$\begin{cases} \dfrac{\partial T_h}{\partial t} + u_h \dfrac{\partial T_h}{\partial x} = -E(T_h - T_w) + W_h \\ \dfrac{\partial T_w}{\partial t} - \hat{\kappa}_w \dfrac{\partial^2 T_w}{\partial x^2} = +F(T_h - T_w) - G(T_w - T_c) \\ \dfrac{\partial T_c}{\partial t} - u_c \dfrac{\partial T_c}{\partial x} = +H(T_w - T_c) + W_c \end{cases} \tag{A.6}$$

where (E, F, G, H) may be regarded as constants which vary with space and time, and are to be evaluated numerically at each grid station. Initially the wall longitudinal conduction term involving the second derivative of temperature will be omitted, but it can be allowed for in the numerical computation. The pressure and dissipation contributions (W_h, W_c) are available as numerical contributions from solution of the balance of mass and balance of linear momentum equations if desired. In the present case these corrections were omitted.

Alternative numerical solution routes for these equations are discussed in Chapter 9 and Appendix B.8.

A.2 Summarized development of transient equations for contraflow

Fundamental

One-dimensional form of continuum equations in which no rotational velocities exist

$$\text{Hot fluid}\begin{cases} \dfrac{\partial \rho}{\partial t} + \dfrac{\partial(\rho u)}{\partial x} = 0 & \text{mass} \\ \dfrac{\partial u}{\partial t} + u\dfrac{\partial u}{\partial x} = -\dfrac{1}{\rho}\dfrac{\partial p}{\partial x} - \dfrac{fu^2}{2r_{hyd}} & \text{momentum} \\ \rho\dfrac{Dh}{Dt} - \dfrac{Dp}{Dt} = +\lambda_h \dfrac{\partial^2 T_h}{\partial x^2} + \left(-\dfrac{\partial q_i}{\partial x_i}\right) + \eta\Phi & \text{energy} \end{cases}$$

$$\text{Solid wall}\begin{cases} \dfrac{\partial T_w}{\partial t} - \hat{\kappa}_w \dfrac{\partial^2 T_w}{\partial x^2} = +\left(\dfrac{\alpha_h S}{M_w C_w}\right)(T_h - T_w) \\ \qquad\qquad -\left(\dfrac{\alpha_c S}{M_w C_w}\right)(T_w - T_c) \qquad \text{energy} \end{cases}$$

$$\text{Cold fluid}\begin{cases} \rho \dfrac{Dh}{Dt} - \dfrac{Dp}{Dt} = +\lambda_c \dfrac{\partial^2 T_c}{\partial x^2} + \left(+\dfrac{\partial q_i}{\partial x_i}\right) + \eta\Phi & \text{energy} \\ \dfrac{\partial u}{\partial t} - u\dfrac{\partial u}{\partial x} = +\dfrac{1}{\rho}\dfrac{\partial p}{\partial x} + \dfrac{fu^2}{2r_{hyd}} & \text{momentum} \\ \dfrac{\partial \rho}{\partial t} - \dfrac{\partial(\rho u)}{\partial x} = 0 & \text{mass} \end{cases}$$

Expanded and rearranged

Use same algorithms for cold mass flow as those for hot mass flow.

$$\text{Hot fluid}\begin{cases} \dfrac{\partial \rho_h}{\partial t} = -\dfrac{\partial(\rho u)_h}{\partial x} & \text{mass} \\ \dfrac{\partial(\rho u)_h}{\partial t} = -\dfrac{\partial(\rho u^2)_h}{\partial x} - \left[\dfrac{\partial p_h}{\partial x} + \left(\dfrac{f\rho u^2}{2r_{hyd}}\right)_h\right] & \text{momentum} \\ \dfrac{\partial T_h}{\partial t} + u_h \dfrac{\partial T_h}{\partial x} - \hat{\kappa}_h \dfrac{\partial^2 T_h}{\partial x^2} = -\dfrac{u_h}{L}\left(\dfrac{\alpha_h S}{\dot{m}_h C_h}\right)(T_h - T_w) & \\ \qquad + \dfrac{1}{\rho_h C_h}\left(\dfrac{\partial p_h}{\partial t} + u_h \dfrac{\partial p_h}{\partial x} + \eta_h \Phi_h\right) & \text{energy} \end{cases}$$

$$\text{Solid wall}\begin{cases} \dfrac{\partial T_w}{\partial t} - \hat{\kappa}_w \dfrac{\partial^2 T_w}{\partial x^2} = +\left(\dfrac{\alpha_h S}{M_w C_w}\right)(T_h - T_w) - \left(\dfrac{\alpha_c S}{M_w C_w}\right)(T_w - T_c) \\ \qquad\qquad\qquad\qquad\qquad \text{energy} \end{cases}$$

$$\text{Cold fluid}\begin{cases} \dfrac{\partial T_c}{\partial t} - u_c \dfrac{\partial T_c}{\partial x} - -\hat{\kappa}_c \dfrac{\partial^2 T_c}{\partial x^2} = +\dfrac{u_c}{L}\left(\dfrac{\alpha_c S}{\dot{m}_c C_c}\right)(T_w - T_c) & \\ \qquad + \dfrac{1}{\rho_c C_c}\left(\dfrac{\partial p_c}{\partial t} - u_c \dfrac{\partial p_c}{\partial x} + \eta_c \Phi_c\right) & \text{energy} \\ \text{(same as hot fluid, but with reversed stations)} & \text{momentum} \\ \text{(same as hot fluid, but with reversed stations)} & \text{mass} \end{cases}$$

Cleaned up

Rearrangement of terms to permit solution, with neglect of some minor contributions.

$$\text{Hot fluid}\begin{cases}\dfrac{\partial \rho_h}{\partial t} = -\rho_h \dfrac{\partial u_h}{\partial x} - u_h \dfrac{\partial \rho_h}{\partial x} & \text{mass} \\ \dfrac{\partial (\rho u)_h}{\partial t} = -u_h \dfrac{\partial (\rho u)_h}{\partial x} - (\rho u)_h \dfrac{\partial u_h}{\partial x} - \left[\dfrac{\partial p_h}{\partial x} + \left(\dfrac{f \rho u^2}{2 r_{hyd}}\right)\right] & \text{momentum} \\ \dfrac{\partial T_h}{\partial t} + u_h \dfrac{\partial T_h}{\partial x} = -\dfrac{u_h}{L}\left(\dfrac{\alpha_h S}{\dot{m}_h C_h}\right)(T_h - T_w) & \text{energy}\end{cases}$$

$$\text{Solid wall}\begin{cases}\dfrac{\partial T_w}{\partial t} - \hat{\kappa}_w \dfrac{\partial^2 T_w}{\partial x^2} = +\left(\dfrac{\alpha_h S}{M_w C_w}\right)(T_h - T_w) & \\ \qquad -\left(\dfrac{\alpha_c S}{M_w C_w}\right)(T_w - T_c) & \text{energy}\end{cases}$$

$$\text{Cold fluid}\begin{cases}\dfrac{\partial T_c}{\partial t} - u_c \dfrac{\partial T_c}{\partial x} = +\dfrac{u_c}{L}\left(\dfrac{\alpha_c S}{\dot{m}_c C_c}\right)(T_w - T_c) & \text{energy} \\ \text{(same as hot fluid, but with reversed stations)} & \text{momentum} \\ \text{(same as hot fluid, but with reversed stations)} & \text{mass}\end{cases}$$

Simplified for computation

Pressure terms omitted from this set, with subsequent adjustment for frictional loss only. Note: pressure gradient terms may be important in adjusting flow velocities.

$$\text{Hot mass flow}\begin{cases}\dfrac{\partial \rho_h}{\partial t} = -\rho_h \dfrac{\partial u_h}{\partial x} - u_h \dfrac{\partial \rho_h}{\partial x} & \text{density} \\ \dfrac{\partial (\rho u)_h}{\partial t} = -(\rho u)_h \dfrac{\partial u_h}{\partial x} - u_h \dfrac{\partial (\rho u)_h}{\partial x} & \\ \qquad -[\text{pressure field}] & \text{density} \times \text{velocity}\end{cases}$$

$$\text{Temperatures}\begin{cases}\dfrac{\partial T_h}{\partial t} + u_h \dfrac{\partial T_h}{\partial x} = -\dfrac{u_h}{L}\left(\dfrac{\alpha_h S}{\dot{m}_h C_h}\right)(T_h - T_w) & \text{hot fluid} \\ \dfrac{\partial T_w}{\partial t} - \hat{\kappa}_w \dfrac{\partial^2 T_w}{\partial x^2} = +\left(\dfrac{\alpha_h S}{M_w C_w}\right)(T_h - T_w) & \\ \qquad -\left(\dfrac{\alpha_c S}{M_w C_w}\right)(T_w - T_c) & \text{solid wall} \\ \dfrac{\partial T_c}{\partial t} - u_c \dfrac{\partial T_c}{\partial x} = +\dfrac{u_c}{L}\left(\dfrac{\alpha_c S}{\dot{m}_c C_c}\right)(T_w - T_c) & \text{cold fluid}\end{cases}$$

$$\text{Cold mass flow}\begin{cases}\text{(same as hot fluid, but with reversed stations)} & \text{density}\\ \text{(same as hot fluid, but with reversed stations)} & \text{density} \times \text{velocity}\end{cases}$$

Steady-state solution

The set of steady-state equations is obtained by omitting time-dependent terms.

A.3 Computational approach

The approach will be considered in three stages.

- First, numerical considerations – the Crank–Nicholson finite-difference approach will be used on the simplified set of equations with no pressure-field terms, to assess stability of the solution, and highlight any computational difficulties.
- Second, the introduction of pressure-field terms involves both use of a state equation plus a friction factor correlation for each fluid, making the full set of transient equations unique for the Stokesian fluids chosen.
- Third, consideration of contraflow and crossflow design under transients.

Numerical considerations

Splitting the problem

With inlet disturbances $(\dot{m}, T)$ instantaneous mass flow rates can be expressed either as $\dot{m}$ or as ρuA. The unknowns ρu and $\dot{m}$ thus correspond for prismatic flow. However, the fluid momentum equations contain unknowns ρu, while the fluid energy equations contain unknowns $\dot{m}^{-1}$, which would create difficulty when attempting to solve for all seven unknowns simultaneously.

Because of this difficulty, the approach adopted is sequential solution of the mass flow and linear momentum fields for each fluid stream independently, followed by separate and simultaneous solution of the set of the energy-field equations for wall plus hot and cold fluid temperatures. This de-coupling is permissible because Mach numbers are normally less than 0.1, well below compressible heating levels (see Appendix L).

Selection of time intervals

Time intervals for high-speed flow are constrained by the modified Courant–Friedrichs–Lewy (CFL) stability condition, see e.g. Fletcher (1991), which simply requires that mechanical disturbances travelling at the speed of sound in the fluid shall not leave the space interval under consideration.[1]

$$\Delta t \leq \frac{\Delta x}{(|u| + c)} \tag{A.7}$$

where c is the speed of sound.

---

[1]Thermal disturbances travel at much slower speeds than mechanical disturbances.

At each time interval velocity values are found for both hot and cold fluids along the length of the exchanger. The largest absolute value is inserted in equation (A.7) to find the maximum permissible time increment for the next time step. Events during the next time step are not yet known, but the definition (A.7) should ensure that a sufficient gap exists between x stations.

Algorithms that allow for convective transport are the MacCormack predictor–corrector scheme (Appendix B.8), and the method of lines with Runge–Kutta. An alternative not yet explored is to first transform the independent variables (x, t) as was done for Laplace transformation in Section 10.3, then solve by Crank–Nicholson and transform back to the original (x, t) coordinates.

Fluid flow equations

The fluid flow equations are first recast in a form suitable for numerical work, and then arranged so that for each fluid the balance of mass and balance of linear momentum equations may be solved in turn. Only the numerical solution approach for the hot fluid is required, as the cold fluid can make use of the same solution method providing the renumbering of grid points is carried out before inversion. The cold fluid solution so found is then reverse renumbered to match the physical problem. This provides new velocity fields, and the information thus obtained is used to update mass flowrates in the energy equations.

Solution of the balance of mass equation is for unknown values of ρ. Solution of the balance of linear momentum equation is for unknown values of ρu. In both cases the velocity coefficients (u) are of first order only and approximate numerical values can be taken from the last time interval to permit solutions [very small time intervals (Δt) are desirable]. The new velocity fields are obtained from $u = (\rho u)/\rho$.

Initial conditions along the full length of the exchanger are required for ρ and ρu to start the computation – this usually requires an initial steady state. Thereafter only inlet boundary conditions are known at each subsequent time interval. Using the Crank–Nicholson finite-difference formulation, outlet boundary conditions might be handled either by assuming that gradients of ρ and ρu are zero at outlet [equation (A.8)], and extrapolating for an external fictitious value one step beyond the real outlet condition using data from the previous time interval [equation (A.9)], or more simply by using equation (A.10).

Either

$$\left(\frac{\partial y}{\partial x}\right)_n^t = \frac{y_{n+1}^t - y_{n-1}^t}{2\Delta x} = 0 \qquad \text{from where } y_{n+1}^t = y_{n-1}^t \tag{A.8}$$

and

$$y_{n+1}^t = 4y_n^{t-1} - 6y_{n-1}^{t-1} + 4y_{n-2}^{t-1} - y_{n-3}^{t-1} \tag{A.9}$$

or, better

$$y_{n+1}^{t} = y_{n-1}^{t-1} \tag{A.10}$$

Energy equations

The energy equations are solved simultaneously for the transient temperature field using the same finite-difference scheme as was found effective in determining the effects of longitudinal conduction in steady-state flow, see Appendix B.5. Using this approach there are no problems with boundary conditions.

Updated temperatures (T) are used first to calculate new physical properties from interpolating cubic spline fits, and updated values incorporated into the complete set of equations ready for the next time interval, and so on.

A feature of Crank–Nicholson solution of the energy-field equations is that it handles the interaction of all unknowns simultaneously, and can present snapshots of the temperature fields.

Completing the solution

Pressure-field terms

In the pressure-field group both pressure gradient and flow-friction terms were omitted from the balance of linear momentum equations during a first solution. This was done because symmetry of the mass and momentum equations simplified development of algorithms. Pressure-field terms introduce additional effects.

For an ideal gas the equation of state is equation (A.3), and for this special case we reset the pressure-field expression, and observe the additional effect due to temperature transients

$$\frac{\partial p}{\partial x} + \left(\frac{f\rho u^2}{2r_{hyd}}\right) \Rightarrow (RT)\frac{\partial \rho}{\partial x} + \left(\frac{uf}{2r_{hyd}}\right)(\rho u) \qquad \text{for perfect gas}$$

Distributed values of (ρ, T) are available prior to solution of the balance of linear momentum equation allowing the pressure gradient term to be evaluated numerically for placement on the right-hand side of the solution matrix. The importance of the pressure gradient term is because large changes in absolute temperature (T) can affect flowrates significantly. Extreme values of the pressure gradient term could lead to flow 'choking'.

In the friction loss term the velocity squared term is split to improve accuracy. Distributed values of coefficient u from the last time interval are used to evaluate Reynolds numbers and obtain frictional coefficients (f) from an interpolating cubic spline fit. The distributed coefficients ($uf/2r_{hyd}$) are now known and terms involving the unknown ρu can be worked into the left-hand side of the solution matrix.

Reflection of transients in contraflow

A feature of contraflow exchangers subject to transients is that disturbances carried by the first fluid will be felt by the second fluid and transported backwards, causing knock-on effects in the first fluid. Reflections will continue to pass backwards and forwards until the system settles down. Fortunately, in most cases thermal capacity of the intervening wall is high and damping should be sufficient to permit control.

Transients travelling against the flow in contraflow

A transient generated by one fluid flowing in one direction may be experienced by the second fluid as a transient travelling in a direction opposite to its flow. This was the reason for suggesting a modification to the CFL rule discussed in Chapter 9, Section 9.2, viz.

$$\Delta t = \frac{\Delta x/2}{(|u| + c)} \tag{A.11}$$

However, this cautionary adjustment need only be tried if computational difficulties should arise.

Change in sign of velocity

When predictor–corrector algorithms of the MacCormack type are in use, problems may arise if the disturbance is travelling in the opposite direction to the flow velocity. The sign of the correction term may then depend on whether the flow velocity is less than the disturbance velocity. Inspection of velocities pertaining in the simple illustrative examples of crossflow and contraflow given in Chapter 4 do indicate the possibility of transients running in a direction opposing the flow.

The computational problem should not exist with the implicit Crank–Nicholson type of finite-difference solution (although slow in execution), particularly as the disturbances would be slowed by the thermal inertia of the wall and longitudinal conduction effects.

Potential problems with crossflow

For crossflow under steady-state conditions, the initial assumption of equal mass flowrate in each channel is incorrect. Equal pressure loss in each channel is the proper constraint, particularly when pressure loss in the headers can be made zero by following Dow's method outlined in the second half of Chapter 8. Pressure loss is controlled by density and viscosity and both of these values are temperature dependent. Inspection of the temperature sheets in Section 3.5 shows that the additional effect of cross-conduction may also have to be taken into account. The work of Haseler outlined in Section 11.6 seems a good starting point.

Further departure from the initial assumption of equal mass flowrate in each channel is likely to occur under transient conditions, but there may be compensation in the absence of temperature end-reflection effects. Prediction of crossflow transients thus requires channel-by-channel analysis. Experimentation with an exchanger core deliberately configured to emphasize parasitic losses might be used to check theoretical predictions.

Development of algorithms

Software for development of algorithms for solution of the transient equations in contraflow will be found in the supplement to Appendix B.

References

Mostly contraflow – see Chapter 9 for more extensive collection.

Fletcher, C.A.J. (1991) *Computational Techniques for Fluid Mechanics*, 2nd edn, Springer, Berlin.

Schlichting, H. (1960) *Boundary Layer Theory*, 4th edn, McGraw-Hill, New York, p. 291.

Bibliography

Acklin, L. and **Laubli, F.** (1960) Calculation of the dynamic behaviour of heat exchangers with the aid of analogue computers. *Sulzer Tech. Rev.*, **4**, 13–20.

Acrivos, A. (1956) Method of characteristics technique – application to heat and mass transfer problems. *Ind. Engng Chemistry*, **48**(4), April, 703–710.

Ames, W.F. (1965) *Nonlinear Partial Differential Equations in Engineering*, Academic Press, New York, pp. 453–454.

Anderson, D.A., **Tannehill, J.C.**, and **Pletcher, R.H.** (1984) *Computational Fluid Mechanics and Heat Transfer*, Chapter 9, Hemisphere, New York.

Cima, R.M. and **London, A.L.** (1958) The transient response of a two-fluid counterflow heat exchanger. *Trans. ASME*, **80**, July, 1169–1179.

Clark, J.A. and **Arpaci, V.S.** (1958) Dynamic response of heat exchangers having internal heat sources – Part II. *Trans. ASME*, **80**, April, 623–634.

Clark, J.A. and **Arpaci, V.S.** (1959) Dynamic response of heat exchangers having internal heat sources – Part III. *ASME J. Heat Transfer*, **81**, November, 233–266.

Clark, J.A., **Arpaci, V.S.**, and **Treadwell, K.M.** (1958) Dynamic response of heat exchangers having internal heat sources – Part I. *Trans. ASME, Ser. C.*, **80**, April, 612–624.

Cohen, W.C. and **Johnson, E.F.** (1956) Dynamic characteristics of double-pipe heat exchangers. *Ind. Engng Chemistry*, **48**, June, 1031–1034.

Dzubenko, B.V., **Dreitser, G.A.**, and **Ashmantas, L.V.-A.** (1990) *Unsteady Heat and Mass Transfer*.

Gartner, J.R. and **Harrison, H.L.** (1965) Dynamic characteristics of water-to-air crossflow heat exchanger. *Trans. ASHRAE*, **71**, Part 1, 212–224.

Gvozdenac, D.D. (1990) Transient response of the parallel flow heat exchanger with finite wall capacitance. *Ingenieure Arch.*, **60**, 481–490.

Harmon, R.W. (1966) Forced-flow heat exchanger dynamics. *Ind. Engng Chemistry, Fundamentals*, **5**(1), February, 138–139.

Jaswon, M.A. and **Smith, W.** (1954) Countercurrent transfer processes in the non-steady state. *Proc. R. Soc., Ser. A*, **225**, 226–244.

Koppel, L.B. (1962) Dynamics of a forced-flow heat exchanger. *Ind. Engng Chemistry, Fundamentals*, **1**(2), May, 131–134.

London, A.L., **Biancardi, F.R.**, and **Mitchell, J.W.** (1959) The transient response of gas-turbine plant heat exchangers – recuperators, intercoolers, pre-coolers, and ducting. *ASME J. Engng Power*, **81**, October, 433–448.

Masubuchi, M. (1960) Dynamic response and control of multipass heat exchangers. *ASME J. Basic Engng, Ser. D*, **82**, March, 51–55.

Mitchell, A.R. and **Griffiths, D.F.** *The Finite Difference Method in Partial Differential Equations*, John Wiley, pp. 184–185 and 250.

Mozley, J.M. (1956) Predicting dynamics of concentric pipe heat exchangers. *Ind. Engng Chemistry*, **48**, June, 1035–1041.

Ontko, J.S. (1989) A parametric study of counterflow heat exchanger transients. Report IAR 89-10, Institute for Aviation Research The Wichita State University, Wichita, Kansas.

Paynter, H.M. and **Takahashi, Y.** (1956) A new method of evaluating dynamic response of counterflow and parallel flow heat exchangers. *Trans. ASME*, **78**, 749–758.

Peyret, R. and **Taylor, T.D.** (1982) *Computational Methods for Fluid Flow*, Springer, Berlin.

Press, W.H., **Flannery, B.P.**, **Teukolsky, S.A.**, and **Vettering, W.T.** (1989) *Numerical Recipies in Pascal: The Art of Scientific Computing*, Chapter 17, Cambridge University Press, Cambridge, pp. 691–693

Stermole, F.J. and **Larson, M.A.** (1963) Dynamic response of heat exchangers to flow rate changes. *Ind. Engng Chemistry, Fundamentals*, **2**(1), February, 62–67.

Tan, K.S. and **Spinner, I.H.** (1978) Dynamics of a shell and tube heat exchanger with finite tube-wall heat capacity and finite shell-side resistance. *Ind. Engng Chemistry, Fundamentals*, **17**(4), 353–358.

Tan, K.S. and **Spinner, I.H.** (1984) Numerical methods of solution for continuous countercurrent processes in the non-steady state. Part I – Model equations and development of numerical methods and algorithms. Part II – Application of numerical methods. *AIChE J.*, **30**(5), September, 770–779 and 780–786.

Tan, K.S. and **Spinner, I.H.** (1991) Approximate solution for transient response of a shell and tube heat exchanger. *Ind. Engng Chemistry Res.*, **30**, 1639–1646.

Todo, I. (1976) Dynamic response of bayonet type heat exchangers. Part I – Response to inlet temperature changes. *Bull. Japan. Soc. Mech. Engrs*, **19**(136), October, 1135–1140.

Todo, I. (1978) Dynamic response of bayonet-tube heat exchangers. Part II – Response to flow-rate changes. *Bull. Japan. Soc. Mech. Engrs*, **21**(154), April, 644–651.

Yamashita, H., **Izumi, R.**, and **Yamaguchi, S.** (1978) Analysis of dynamic characteristics of crossflow heat exchangers with both fluids unmixed. (On the transient response to a step change in inlet temperature.) *Bull. Japan. Soc. Mech. Engrs*, **21**(153), March, paper 153-17, 479–480.

Yang, W.J. (1964) Transient heat transfer in a vapour-heated heat exchanger with time-wise varient flow disturbance. *ASME J. Heat Transfer, Ser. C*, **86**, May, 133–142.

Yang, W.J., **Clark, J.A.**, and **Arpaci, V.S.** (1958) Dynamic response of heat exchangers having internal heat sources – Part IV. *Trans. ASME, Ser. C*, **83**, April, 321–338.

Appendix B
Algorithms And Schematic Source Listings

Development of Pascal listings

B.1 Algorithms for mean temperature distribution in one-pass unmixed crossflow

Alternative algorithms for generating temperature sheets and the temperature-difference sheet for crossflow are presented below. These produced mean temperature difference for parameters of local *Ntu* numbers [equation (3.27)].

The approach was explicit finite difference, accuracy being affected by error propagation, particularly from the steepest parts of the temperature fields. For equally spaced intervals, at least a 50×50 mesh should be used. Improvements can be expected using a smaller mesh near the starting corner, but it is simpler to increase the size of the full mesh.

The algorithm employs the modified Euler–Cauchy method which obtains an estimate of the slope at the new point, and takes the mean of this estimated slope and the known slope at the old point to obtain a better estimate for the slope in the interval.

More recent developments in symbolic mathematical solution of equations may present other approaches to solution.

```
{first origin square}
  k:=0;
  islTg:=-Tg[k,k+1]+Tf[k,k+1];                   {initial slope}
  islTf:=+Tg[k+1,k]-Tf[k+1,k];                   {initial slope}
  eTg:=Tg[k,k+1]+islTg*Ng/50;                    {estimated Tg}
  eTf:=Tf[k+1,k]+islTf*Nf/50;                    {estimated Tf}
  fslTg:=-eTg+eTf;                               {final slope}
  fslTf:=+eTg-eTf;                               {final slope}
  mslTg:=(islTg+fslTg)/2;                        {mean slope}
  mslTf:=(islTf+fslTf)/2;                        {mean slope}
  Tg[k+1,k+1]:=Tg[k,k+1]+mslTg*Ng/50;            {new Tg}
  Tf[k+1,k+1]:=Tf[k+1,k]+mslTf*Nf/50;            {new Tf}
  FOR k:=1 TO 49 DO
BEGIN
  n:=k;
  FOR m:=n TO 49 DO                              {in X-direction}
```

```
BEGIN
  islTg:=-Tg[m,n]+Tf[m,n];                      {initial slope}
  islTf:=+Tf[m+1,n-1]-Tf[m+1,n-1];              {initial slope}
  eTg:=Tg[m,n]+islTg*Ng/50;                     {estimated Tg}
  eTf:=Tf[m+1,n-1]+islTf*Nf/50;                 {estimated Tf}
  fslTg:=-eTg+eTf;                              {final slope}
  fslTf:=+eTg-eTf;                              {final slope}
  mslTg:=(islTg+fslTg)/2;                       {final slope}
  mslTf:=(islTf+fslTf)/2;                       {final slope}
  Tg[m+1,n]:=Tg[m,n]+mslTg*Ng/50;               {new Tg}
  Tf[m+1,n]:=Tf[m+1,n-1]+mslTg*Nf/50;           {new Tf}
END;

  m:=k;
  FOR n:=m TO 49 DO                             {in Y-direction}
BEGIN
  islTg:=-Tg[m-1,n+1]+Tf[m-1,n+1];              {initial slope}
  islTf:=+Tg[m,n]-Tf[m,n];                      {initial slope}
  eTg:=Tg[m-1,n+1]+islTg*Ng/50;                 {estimated Tg}
  eTf:=Tf[m,n]+islTf*Nf/50;                     {estimated Tf}
  fslTg:=-eTg+eTf;                              {final slope}
  fslTf:=+eTg-eTf;                              {final slope}
  mslTg:=(islTg+fslTg)/2;                       {final slope}
  mslTf:=(islTf+fslTf)/2;                       {final slope}
  Tg[m,n+1]:=Tg[m-1,n+1]+mslTg*Ng/50;           {new Tg}
  Tf[m,n+1]:=Tf[m,n]+mslTg*Nf/50;               {new Tf}
END;

{next origin square}
  islTg:=-Tg[k,k+1]+Tf[k,k+1];                  {initial slope}
  islTf:=+Tg[k+1,k]-Tf[k+1,k];                  {initial slope}
  eTg:=Tg[k,k+1]+islTg*Ng/50;                   {estimated Tg}
  eTf:=Tf[k+1,k]+islTf*Nf/50;                   {estimated Tf}
  fslTg:=-eTg+eTf;                              {final slope}
  fslTf:=+eTg-eTf;                              {final slope}
  mslTg:=(islTg+fslTg)/2;                       {mean slope}
  mslTf:=(islTf+fslTf)/2;                       {mean slope}
  Tg[k+1,k+1]:=Tg[k,k+1]+mslTg*Ng/50;           {new Tg}
  Tf[k+1,k+1]:=Tf[k+1,k]+mslTf*Nf/50;           {new Tf}
END;                                            {for k-loop}

{other thermal parameters}
  {sum over inside temperatures}
  sum:=0.0;
  FOR m:=1 TO 49 DO
BEGIN
  FOR n:=1 TO 49 DO
```

```
  sum:=sum+(Tg[m,n]-Tf[m,n]);
END;
  {sum over edge temps}
  FOR n:=0 TO 50 DO
  sum:=sum+((Tg[0,n]-Tf[0,n])+(Tg[50,n]-Tf[50,n]))/2;
  FOR m:=0 TO 50 DO
  sum:=sum+((Tg[m,0]-Tf[m,0])+(Tg[m,50]-Tf[m,50]))/2;

{mean temperature difference}
  meanTDiff:=sum/(50*50);

  Tginn:=1.0;                          {mixed warm inlet temp}
  Tfinn:=0.0;                          {mixed cold inlet temp}
  Tspan:=Tginn-Tfinn;                  {temp.span in core}

{mixed warm outlet temperature}
  sum:=0.0;
  m:=50;                               {at warm fluid outlet}
  FOR n:=1 TO 49 DO
  sum:=sum+Tg[m,n];                    {sum over inner temps}
  sum:=sum+(Tg[m,0]+Tg[m,50])/2;       {add mean outer temps}
  Tgout:=sum/50;                       {mixed warm outlet}

{mixed cold outlet temperature}
  sum:=0.0;
  n:=50;                               {cold fluid outlet}
  FOR m:=1 TO 49 DO
  sum:=sum+Tf[m,n];                    {over inner temps}
  sum:=sum+(Tf[0,n]+Tf[50,n])/2; {add mean outer temps}
  Tfout:=sum/50;                       {mixed cold outlet}

{effectiveness}
  TdiffG:=Tginn-Tgout;
  TdiffF:=Tfout-Tfinn;
  IF (TdiffG>TdiffF)
    THEN Tdiff:=TdiffG
    ELSE Tdiff:=TdiffF;
  Eff:=Tdiff/Tspan;                    {effectiveness}
```

In using this algorithm it is to be recognized that the explicit type of solution always produces some error propagation, which affects both temperature sheets more or less equally. Consequently the value of ‘meanTdiff’ is more reliable than the computed outlet temperatures ‘Tgout’ and ‘Tfout’.

The value of ‘meanTdiff’ is used in design, and it is better to calculate the mean outlet temperature for each side using the energy balance equation, viz.

```
Q:=mg*Cg*(Tginn-Tgout);
Q:=mf*Cf*(Tfout-Tfinn);
```

B.2 Schematic source listing for direct-sizing of compact one-pass crossflow exchanger

This schematic algorithm is given below.

```
{one-pass unmixed-unmixed crossflow}
refMR=m1/m2                          {desired mass flow ratio}
iterate Re1                          {until Q matches Qduty}
| f1=correlation                     {interpolating splinefit}
| G1=Re*mu1/D1                       {mass velocity side-1}
| Lp1=dp1*2*rho1*D1/(4*f1*G1^2)      {length of channel}
| E2=Lp1                             {edge length, side-2}
| iterate 'aspect'                   {until newMR matches refMR}
| | Lp2=aspect*Lp1                   {plate aspect=E1/E2}
| | E1=Lp2                           {edge length side-1}
| | Splate=E1*E2                     {area of single plate}
| | given dp2                        {pressure loss on side-2}
| | iterate Re2                      {until dp matches dp2}
| | | f2=correlation                 {interpolating splinefit}
| | | G2=Re2*mu2/D2                  {mass velocity side-2}
| | | dp=4*f2*G2^2*L2/(2*rho2*D2)     {estimate for dp2}
| | until dp=dp2                     {Re known on both sides}
| | Afront1=E1*(b1/2)                {Afront, 1/2 plate spacing}
| | Aflow1=sigma1*Afront1            {Aflow, 1/2 plate spacing}
| | mP1=G1*Aflow1                    {Mflow, 1/2 plate spacing}
| | repeat last 3 lines for side-2
| | newMr=mP1/mP2                    {estimate of refMR}
| until newMR=refMR                  {'aspect' for Re1}
| Nw=TRUNC(m1/MP1)+1                 {number of plates}
| wide=Nw*(b1/2+tp+b2/2)             {exchanger width}
| vol=E1*E2*wide                     {exchanger core volume}
| Sexchr=Nw*Splate                   {total plate surface}
| Stotal1=Sexchr*kappa1              {total surface side-1}
| Stotal2=Sexchr*kappa2              {total surface side-2}
| Pr1=Cp1*mu1/k1                     {at mean bulk temperatures}
| St1=j-correlation                  {interpolating splinefit}
| h1=St1*Cp1*G1                      {heat trans.coeff, side-1}
| Y1=b1/2                            {approx.fin height}
| mY1=Y1*SQRT(2*h1/(kf1*tf1))        {fin parameter}
| phi1=TANH(mY1)/mY1                 {fin performance ratio}
| eta1=1-gamma1*(1-phi1)             {correct to total surface}
| u1=h1*eta1*kappa1                  {heat trans.coeff @ plate}
| u2=similarly for side-2
| u3=kp/tp                           {plate coefficient}
```

```
| U=1/(1/u1+1/u2+1/u3)          {overall coeff.at plate}
| Ntu1=U*Sexchr/(m1*Cp1)        {Ntu, side-1}
| Ntu2=U*Sexchr/(m2*Cp2)        {Ntu, side-2}
| find meanTD                   {using T1,t1,Ntu1,Ntu2}
| Q=U*Sexchr*meanTD             {exchanger duty at Re1}
until Q=Qduty                   {Q is desired peformance}
```

B.3 Schematic source listing for direct-sizing of compact contraflow exchanger

This schematic algorithm includes separate procedure bodies.

```
{main program}
Re1=2500                                  {mid-range value}
fric(Re1,loRe1F,hiRe1F,f1)                {f1, corr.limits}
heat(Re1,loRe1H,hiRe1H,StPr^2/3)          {StPr^2/3, corr.limits}
test (max-loRe1 < Re1 < min-hiRe1)        {inside valid range}
surf1(fins,b,c,x,D,tf,ts,beta,gamma)      {surface params}
surf2(fins,b,c,x,D,tf,ts,beta,gamma)      {surface params}
scan over valid range of Re1              {loRe1->hiRe1 ...
|                                           over 100 steps}
| heatrans(Re1,forcedRe2,Lh,Edge)         {PROC.find Edge}
| pdrop1(Re1,dp1,Lp1)                     {PROC.find Lp1}
| pdrop2(Re2,dp2,Lp2)                     {PROC.find Lp2}
until scan complete                       {full validity range}
plot curves (Lh,Lp1,Lp2) vs Edge          {visual check}
iterate for rh-intersection & L           {design point, Re1,Re2}
if Lp1 rh-curve, calc.NEWdp2              {forced pressure loss}
if Lp2 rh-curve, calc.NEWdp1              {forced pressure loss}
design                                    {PROC.exchanger block}

PROCEDURE heatrans(Re1,forced-Re2,Lh,Edge)
G1=Re1*mu1/D1                             {mass velocity]
Aflow1=m1/G1                              {total flow area}
Afront1=Aflow1/sigma1                     {total frontal area}
E=Afront1/(b1/2)                          {edge length}
z1=E/c1                                   {no.cells on side-1}
z2=E/c2                                   {no.cells on side-2}
Afront2=E*(b2/2)                          {total frontal area}
Re2=D2*m2/(eta2*Aflow2)                   {forced Re2}
G2=Re2*mu2/D2                             {forced mass velocity}
PrX=EXP(2/3*LN*(Pr1))                     {side-1, PrX=Pr1^2/3}
heat(Re1,loRe1H,hiRe1H,StPr^2/3)          {splinefitted corr.}
St1=(StPr^2/3)/PrX                        {side-1, Stanton no.}
h1=St1*Cp1*G1                             (cell h.t.coeff}
Y1=b1/2                                   {approx. fin height}
mY1=Y1*SQRT(2*h1/(kf1*tf1))               {fin parameter}
```

```
phi1=TANH(mY1)/mY1                      {fin performance ratio}
eta1=1-gamma1*(1-phi1)                  {correct to Stotal}
u1=h1*eta1*kappa1                       {h.trans.coeff.@ plate}
u2 similarly for side-2                 {h.trans.coeff.@ plate}
u3=kp/tp                                {plate coeffcient}
U=1/(1/u1+1/u2+1/u3)                    {overall coeff.@ plate}
Splate=Q/(U*LMTD)                       {design surface plate}
Lh=Splate/E                             {length for Q}

PROCEDURE pdrop(Re1,dp1,Lp1)
fric(Re1,loRe1F,hiRe1F,f1)              {splinefitted corr.}
G1=Re1*mu1/D1                           {mass velocity}
Lp1=dp1*2*rho1*D1/(4*f1*G1^2)           {length for dp1}

PROCEDURE pdrop(Re2,dp2,Lp2)            {same as for side-1}

PROCEDURE design
Splate=E*L                              {total plate surface}
Ntu1=U*S/(m1*Cp1)                       {whole exchanger}
Ntu2=U*S/(m2*Cp2)                       {whole exchanger}
Stotal1=Splate*kappa1                   {total surface, side-1}
Stotal2=Splate*kappa2                   {total surface, side-2}
V=L*E*(b1/2+tp+b2/2)                    {volume exchanger}
zR1=TRUNC(z1)+1                         {no.cells, side-1}
zR2=TRUNC(z2)+1                         {no.cells, side-2}
Ac=X-sect.for long.condn.               {depends on surface}
Am=X-sect.for mass evaluation           {depends on surface}
Mblock=rhoM*Am*L                        {mass exchr.core}
Py=1/Mblock/(rhoM*V)                    {porosity exchr.core}
```

B.4 Parameters for rectangular offset strip fins

In running software for both 'rating' and 'direct-sizing' careful attention must be paid to accurate definition of the surface geometry. Quite small deviations from correct values may cause significant change in exchanger performance or in final dimensions.

It was found that significant errors existed in some published data. For the rectangular offset strip fin it is practicable to proceed from basic dimensions and compute consistent values. This procedure is recommended as the best way of avoiding data entry problems. Definitions of parameters are provided in Table 4.11 of Chapter 4, but the reader may find the following explanations helpful in understanding the generation of values.

Single-cell geometries

(Terminators 1 and 2 to be added to identifiers to designate side-1 and side-2, except where already indicated for parameter 'alpha'.)

Flow-cell characteristic dimension, flow area, and effective perimeter

Parameters under this section are required for one complete flow cell, so that Reynolds numbers can be evaluated.

```
Cell:=(b-tf)*(c-tf);                {cell Aflow}
Per:=2*(b-tf) + 2*(c-tf);           {cell perimeter}
```

We need to take cell ends into account, as the extra surface area will contribute. Fin ends are considered to be half thickness on each side of a single cell. Half of each base end is attached to the next cell, thus only the other half contributes surface area.

```
Perx:=Per*x +  4*(b-tf)*(tf/2) + 2*(c/2)*tf;
              {4 half-fin ends} {2 half-base ends}
```

Recover effective perimeter

```
Per:=Perx/x;                          {effective perimeter}
```

before evaluating cell hydraulic radius and hydraulic diameter.

```
rh:=Cell/Per;                         {hydraulic radius}
D:=4*rh;                              {hydraulic diameter}
```

Values of single-cell parameters per unit length

The following parameters are evaluated for the cell spaces between two separating plates. In design of the heat exchanger only half-cells on either side of one plate are used, and adjustment for this effect is made later. Half of each base end is attached to the next cell, thus only the other half contributes surface area.

Total surface area (heat transfer/strip length) – i.e. total surface per unit length

```
{1 cell}   {2 sides}     {2 bases}     {4 half-fin ends}
Stotal:=   2*(b-tf) +    2*(c-tf)  +    4*(b-tf)*(tf/2)/x
                                   +    2*(c/2)*tf/x;
                                       {2 half-base ends}
```

For fin surface area the difficulty lies in deciding what to do with the fin ends, as these are attached to the plate, and for heat transfer might well be lumped with the separating plate. However, the fin ends act as steps in the flow direction, and thus contribute to enhanced heat transfer.

```
{1-cell}   {2 sides}                  {4 half-fin ends}
Sfins:=     2*(b-tf)              +    4*(b-tf)*(tf/2)/x
                                  +    2*(c/2)*tf/x;
                                      {2 half-base ends}
```

The undernoted parameters are for a complete cell space between two plates.

```
Vtotal:=b*c;                          {single cell-level}
Splate:=2*c;                          {single cell-level}
Sbase:=2*(c-tf);                      {single cell-level}
```

Values of ratios valid for half single-cell heights

Here ratios are taken that apply to both full and half-height surfaces.

```
beta:=Stotal/Vtotal;                  {Stotal/Vtotal}
alpha1:=b1*beta1/(b1+2*tp+b2);        {Stotal/Vexchr, side-1}
alpha2:=b2*beta2/(b1+2*tp+b2);        {Stotal/Vexchr, side-2}
gamma:=Sfins/Stotal;                  {Sfins/Stotal}
kappa:=Stotal/Splate;                 {Stotal/Splate}
lambda:=Sfins/Splate;                 {Sfins/Splate}
sigma:=Cell/(b*c+c*tp);               {Aflow/Afront}
tau:=Sbase/Splate;                    {Sbase/Splate}
omega:=alpha/kappa;                   {Splate/Vexchr}
```

Partial CHECK – omega should be the same for both sides.

Double-cell geometries

(Terminators 1 and 2 to be added to identifiers to designate side-1 and side-2, except where already indicated for parameter 'alpha'.)

Flow-cell characteristic dimension, flow area and effective perimeter

Parameters under this section are required for one complete flow cell, so that Reynolds numbers can be evaluated.

```
Cell:=((b-ts)/2-tf)*(c-tf);          {cell Aflow}
Per:=2*((b-ts)/2-tf) + 2*(c-tf);     {cell perimeter}
     {2 sides -----}    {2 bases}
```

Cell perimeter needs to take cell ends into account, as the extra surface area will contribute. Fin ends are taken as half thickness on each side of a single cell. Half of each base end is attached to the next cell, thus only the other half contributes surface area.

```
             {4 half-fin ends ------}
Perx:=Per*x + 4*((b-ts)/2-tf))*(tf/2)
            + 2*(c/2)*tf;
             {2 half-base ends}
```

We recover effective perimeter

```
Per:=Perx/x;                          {effective perimeter}
```

before evaluating cell hydraulic radius and hydraulic diameter.

```
rh:=Cellx/Perx;                     {hydraulic radius}
D:=4*rh;                            {hydraulic diameter}
```

Values of double-cell parameters per unit length

The following parameters are evaluated for the cell spaces between two separating plates. In design of the heat exchanger only half-cells on either side of one plate are used and adjustment for this effect is made later. Half of each base end is attached to the next cell, thus only the other half contributes surface area.

Total surface area (heat transfer/strip length) – i.e. total surface per unit length

```
{2-cells}  {4 sides -----}  {splitter}  {2 plates}
Stotal:=    4*((b-ts)/2-tf)   +2*(c-tf)  + 2*(c-tf)
          + 8*((b-ts)/2-tf)*(tf/2)/x + 4*(c/2)*tf/x;
           {8 half-fin ends ------}  {4 half-base ends}

{2-cells}  {4 sides -----}  {splitter}
Sfins:=     4*((b-ts)/2-tf)   +2*(c-tf)
            8*((b-ts)/2-tf)*(tf/2)/x +   4*(c/2)*tf/x
           {8 half-fin ends ------}   {4 half-base ends}
```

The undernoted parameters are for a complete cell space between two plates.

```
Vtotal:=b*c;                        {double cell-level}
Splate:=2*c;                        {double cell-level}
Sbase:=2*(c-tf);                    {double cell-level}
```

Values of ratios valid for half double-cell heights

Here ratios are taken that apply to both full- and half-height surfaces.

```
beta:=Stotal/Vtotal;                {Stotal/Vtotal}
alpha1:=b1*beta1/(b1+2*tp+b2);      {Stotal/Vexchr, side-1}
alpha2:=b2*beta2/(b1+2*tp+b2);      {Stotal/Vexchr, side-2}
gamma:=Sfins/Stotal;                {Sfins/Stotal}
kappa:=Stotal/Splate;               {Stotal/Splate}
lambda:=Sfins/Splate;               {Sfins/Splate}
sigma:=2*Cell/(b*c+c*tp);           {Aflow/Afront}
tau:=Sbase/Splate;                  {Sbase/Splate}
omega:=alpha/kappa;                 {Splate/Vexchr}
```

Partial CHECK – omega should be the same for both sides

A contribution from fin base thickness on both sides of the plate should be added to plate thickness for both single- and double-cell fins. This is not done, simply because fluid heat-transfer coefficients are very much smaller than plate heat-transfer coefficients.

The plate coefficient is, however, evaluated in computation simply to provide an immediate indication that fin base thicknesses, and indeed the plate itself, may be ignored in evaluating overall heat-transfer coefficients.

B.5 Longitudinal conduction in contraflow

Finite-difference layout

Wall temperatures are evaluated at stations intermediate to the fluid stations, and the assumption is made that zero wall temperature gradient exists at both ends.

Schematic 5-cell layout

	known											
	H_0		H_1		H_2		H_3		H_4		H_5	
...	+	−	+	−	+	−	+	−	+	−	+	...
W_0	$\vert$	W_0	$\vert$	W_1	$\vert$	W_2	$\vert$	W_3	$\vert$	W_4	$\vert$	W_4
...	+	−	+	−	+	−	+	−	+	−	+	...
	C_0		C_1		C_2		C_3		C_4		C_5	
											known	

Equations (3.25)

$$+\frac{\partial H}{\partial x} = -P(H - W)$$

$$-\frac{\partial^2 W}{\partial x^2} = +Q(H - W) - R(W - C)$$

$$-\frac{\partial C}{\partial x} = \qquad\qquad +S(W - C)$$

where

$$P = \left(\frac{\alpha_h S}{\dot{m}_h C_h}\right)\frac{1}{L}, \qquad Q = \left(\frac{\alpha_h S}{M_w C_w}\right)\frac{1}{\hat{\kappa}}, \qquad R = \left(\frac{\alpha_c S}{M_w C_w}\right)\frac{1}{\hat{\kappa}},$$

$$S = \left(\frac{\alpha_c S}{\dot{m}_c C_c}\right)\frac{1}{L}$$

Hot fluid equation

The finite-difference form becomes

$$\frac{H_{j+1} - H_j}{\Delta x} = -P_j\left[\frac{H_{j+1} + H_j}{2} - W_j\right] \qquad \text{where } P_j \text{ refers to } W_j \text{ position}$$

Simplifying

$$H_{j+1}\left[1+\frac{\Delta x}{2}P_j\right]+H_j\left[-1+\frac{\Delta x}{2}P_j\right]+W_j[-\Delta xP_j]=0$$

At hot fluid inlet

$$H_1\left[1+\frac{\Delta x}{2}P_0\right]+W_0[-\Delta xP_0]=H_0\left[1-\frac{\Delta x}{2}P_0\right]$$

For the preliminary computer solution assume $P_j = P = constant$.

Solid wall equation

The finite-difference form becomes

$$-\frac{(W_{j+1}-2W_j+W_{j-1})}{(\Delta x)^2}=+Q_j\left[\frac{(H_{j+1}+H_j)}{2}-W_j\right]+R_j\left[W_j-\frac{(C_{j+1}+C_j)}{2}\right]$$

where Q_j and R_j are evaluated at W_j stations.

Simplifying and dividing though by $(\Delta x)^2$

$$W_{j+1}\left[\frac{-1}{(\Delta x)^2}\right]+W_j\left[\frac{2}{(\Delta x)^2}+Q+R\right]+W_{j-1}\left[\frac{-1}{(\Delta x)^2}\right]$$
$$+\left\{\begin{array}{l}+H_{j+1}\left[\dfrac{-Q_j}{2}\right]+H_j\left[\dfrac{-Q_j}{2}\right]\\ +C_{j+1}\left[\dfrac{-R_j}{2}\right]+C_j\left[\dfrac{-R_j}{2}\right]\end{array}\right\}=0$$

At hot fluid inlet, $j = 0$ and $W_{-1} = W_0$

$$W_1\left[\frac{-1}{(\Delta x)^2}\right]+W_0\left[\frac{1}{(\Delta x)^2}+Q_0+R_0\right]$$
$$+\left\{\begin{array}{l}+H_1\left[\dfrac{-Q_0}{2}\right]\\ +C_1\left[\dfrac{-R_0}{2}\right]+C_0\left[\dfrac{-R_0}{2}\right]\end{array}\right\}=H_0\left[\frac{Q_0}{2}\right]$$

Table B.1 Matrix for longitudinal conduction in contraflow (position of terms)

	Unknown Equation	1 H_1	2 H_2	3 H_3	4 H_4	5 H_5	6 W_0	7 W_1	8 W_2	9 W_3	10 W_4	11 C_0	12 C_1	13 C_2	14 C_3	15 C_4	16 RHS
1	**1**	#					#										#
2	**2**	#	#					#									
	3		#	#					#								
n − 1	**4**			#	#					#							
n	**5**				#	#					#						
n + 1	**6**	#					#	#				#	#				#
n + 2	**7**						#	#	#								
	8							#	#	#							
2n − 1	**9**								#	#	#						
2n	**10**				#	#				#	#					#	#
2n + 1	**11**						#					#	#				
2n + 2	**12**							#					#	#			
	13								#					#	#		
3n − 1	**14**									#					#	#	
3n	**15**										#					#	#
		1	2		n − 1	n	n + 1	n + 2		2n − 1	2n	2n + 1	2n + 2		3n − 1	3n	
3n + 1																	

RHS, right-hand side.

At cold fluid inlet, $j = n$ and $W_{n+1} = W_n$

$$W_n\left[\frac{1}{(\Delta x)^2} + Q_n + R_n\right] + W_{n-1}\left[\frac{-1}{(\Delta x)^2}\right] + \left\{\begin{array}{r} +H_{n+1}\left[\dfrac{-Q_n}{2}\right] + H_n\left[\dfrac{-Q_n}{2}\right] \\ +C_n\left[\dfrac{-R_n}{2}\right] \end{array}\right\} = C_{n+1}\left[\frac{R_n}{2}\right]$$

For preliminary computer calculations assume $Q_j = Q = const.$ and $R_j = R = const.$

Cold fluid equation

The finite-difference form becomes

$$-\frac{(C_{j+1} - C_j)}{\Delta x} = S_j\left[W_j - \frac{(C_{j+1} + C_j)}{2}\right] \qquad \text{where } S_j \text{ is evaluated at } W_j$$

Simplifying

$$C_{j+1}\left[-1 + \frac{\Delta x}{2}S_j\right] + C_j\left[+1 + \frac{\Delta x}{2}S_j\right] + W_j[-\Delta x S_j] = 0$$

At cold fluid inlet, $j = n$ and C_{n+1} is inlet temperature

$$C_j\left[+1 + \frac{\Delta x}{2}S_j\right] + W_j[-\Delta x S_j] = C_{j+1}\left[+1 - \frac{\Delta x}{2}S_j\right]$$

For the preliminary computer solution assume $S_j = S = const.$

Positioning of terms in matrix

Before writing the general algorithm it is helpful to get some idea of the shape of the matrix to be solved, as symmetries can be identified, compactness in writing the algorithm achieved, and useful checks can be carried out The matrix layout for n=5 is shown in Table B.1.

Solution algorithm

```
{clear matrix}
       FOR j:=1 TO 3*n DO              {j rows, k cols}
BEGIN  FOR k:=1 TO 3*n+1 DO
       coeff[j,k]:=0.0;      {left & right hand sides}
END;

{known inlet temperatures}
       H[0]:=Th1;
       C[n]:=Tc2;
```

```
{load matrix}

{hot fluid equation}
FOR j:=1 TO n-1 DO coeff[j     +1,j       ]:=-1+dX*P/2;
FOR j:=1 TO n   DO coeff[j       ,j       ]:=+1+dX*P/2;
FOR j:=1 TO n   DO coeff[j       ,j   +n  ]:=-dX*P;

coeff[1+0*n,3*n+1]:=(1-dX*P/2)*H[0];

{wall equation}
FOR j:=1 TO n   DO coeff[j  +n   ,j       ]:=-Q/2;
FOR j:=2 TO n   DO coeff[j  +n   ,j     -1]:=-Q/2;

FOR j:=1 TO n-1 DO coeff[j  +n+1,j   +n  ]:=-1/(dX*dX);
FOR j:=1 TO n   DO coeff[j  +n  ,j   +n  ]:=+2/(dX*dX)+Q+R;
FOR j:=1 TO n-1 DO coeff[j  +n  ,j   +n+1]:=-1/(dX*dX);

coeff[n+1,n+1]:=coeff[n+1,n+1]-1/(dX*dX);
coeff[2*n,2*n]:=coeff[2*n,2*n]-1/(dX*dX);

FOR j:=1 TO n   DO coeff[j  +n   ,j+2*n  ]:=-R/2;
FOR j:=2 TO n   DO coeff[j  +n-1,j+2*n  ]:=-R/2;

coeff[1+1*n,3*n+1]:=+(Q/2)*H[0];
coeff[n+1*n,3*n+1]:=+(R/2)*C[n];

{cold fluid equation}
FOR j:=1 TO n   DO coeff[j+2*n  ,j   +n  ]:=-dX*S;

FOR j:=1 TO n   DO coeff[j+2*n  ,j+2*n  ]:=+1+dX*S/2;
FOR j:=2 TO n   DO coeff[j+2*n-1,j+2*n  ]:=-1+dX*S/2;

coeff[n+2*n,3*n=1]:=(+1-dX*S/2)*C[n];

{invert matrix}
llgauss(3*n,delta,coeff,soln); {delta is pivot error}
{temperature field solution}
FOR j:=1     TO n   DO H[j      ]:=soln[j]; {H[1] to H[n]}
FOR j:=n+1   TO 2*n DO W[j  -n-1]:=soln[j]; {W[0] To W[n-1]}
FOR j:=2*n+1 TO 3*n DO C[j-2*n-1]:=soln[j]; {C[0] to C[n-1]}
```

This provides the three temperature profiles in an exchanger with longitudinal conduction, and allows temperature differences at each station along the exchanger to be evaluated, except at each end where there is no longitudinal conduction. The treatment for obtaining the missing end-wall temperatures is outlined in Section 3.2, equation (3.11).

It remains to calculate temperature profiles without longitudinal conduction using equations derived in Section 3.2, and again to evaluate temperature differences at each station along the exchanger.

The difference between values of temperature difference with and without longitudinal conduction is then summed and a mean taken to obtain the reduction in LMTD due to longitudinal conduction to be applied in design.

B.6 Spline-fitting of data

Cubic spline-fitting is the preferred method for representing temperature-dependent physical properties plus both flow-friction and heat-transfer data. When using interpolating spline-fits there is no need to worry whether data are being used outside their range of validity in design, as extrapolation is not built in.

Original data are required in the form of tables of values, which may contain experimental errors. The spline-fitting algorithm of Woodford (1970) allows for experimental errors by including an estimated standard deviation of each ordinate. Woodford's method also allows the smoothing spline to be an arbitrary polynomial, but the author found the cubic polynomial to be adequate for most applications. The exception is when the curve being fitted goes through a point of infinite gradient, but this can be fitted by two spline-fits with points adjacent to the infinite gradient being fitted by more elementary means.

de Boor (1978) examined a number of spline-fitting procedures in his book, and in assessing cubic splines, exponential splines, and taut splines, he observed that performance of the simple cubic spline is often difficult to improve upon. When oscillations are found, the trick is simply to include additional knot points.

When data are very sparse and considerable changes in ordinates are involved with sharp changes in direction, then the variable power spline of Soanes (1976) is capable of providing a smooth fit. With variable power splines a possible technique is to calculate sufficient intermediate points from the variable power spline and then use this new data to fit the standard interpolating cubic spline.

The author tested both taut splines and variable power splines as alternatives to the cubic spline-fit for the representation of data. With sufficient knots, no significant advantage over cubic splines was found in comparison with taut splines. There are, however, other engineering applications in which taut splines are preferred.

When difficulty is experienced with the straightforward cubic spline-fit, it is very possible that using logarithmic data will produce a good fit. When interpolating the logarithmic spline-fit care is then necessary to recover the original linear form. With cubic spline-fits, and with variable power splines as back-up, most datasets can be fitted.

One of the thrusts in looking at different methods of fitting data was to find a twice-differentiable representation that may be useful in certain other applications, see e.g. Young (1988) who calculated the thermodynamic properties of steam from a few fundamental properties.

B.7 Extrapolation of data

This section is concerned with simple extrapolation over one space step only. The basis for extrapolation is comparison of an extrapolated cubic fit of data with the finite-difference expression for a second derivative. A requirement is equally spaced abscissae.

Cubic fit

A cubic fit to four equally spaced points can be represented by the polynomial

$$y = Ax^3 + Bx^2 + Cx + D \tag{B.1}$$

With the coordinates of four points (x_0,y_0), (x_1,y_1), (x_2,y_2), and (x_3,y_3), the coefficients of the cubic curve through these points may be obtained by solution of a set of four simultaneous equations obtained by substituting algebraic pairs of values in equation (B.1).

Retaining strict symmetry in the solution, and whenever possible putting

$$(x_3 - x_2) = (x_2 - x_1) = (x_1 - x_0) = h$$

it is possible to solve for constants (A,B,C) in sequence, leading to

$$A = \frac{y_3 - 3y_2 + 3y_1 - y_0}{6h^3} \tag{B.2}$$

$$B = \frac{(y_3 - y_2 - y_1 + y_0)}{4h^2} - \frac{3A}{2}(x_2 + x_1) \tag{B.3}$$

$$C = \left[\frac{y_3 - y_0}{3h}\right] - 3A[x_1^2 + hx_1 + h^2] - B[2x_1 + h] \tag{B.4}$$

The value for D can be obtained directly by substituting back into equation (B.1).

Extrapolation

Assume that the extrapolation is in the direction

$$(x_0,y_0),\ (x_1,y_1),\ (x_2,y_2),\ (x_3,y_3) \Longrightarrow (x_4,y_4)$$

Equation (B.1) can be differentiated twice to give

$$\frac{d^2y}{dx^2} = 6Ax + 2B = 6Ax + 2\left[\frac{(y_3 - y_2 - y_1 + y_0)}{4h^2} - \frac{3A}{2}(x_2 + x_1)\right]$$

If the abscissae are taken as $x_0 = 0$, $x_1 = h$, $x_2 = 2h$, and $x_3 = 3h$, then the second derivative can be evaluated at the known end point x_3, thus

$$\begin{aligned}\left(\frac{d^2y}{dx^2}\right)_{x_3} &= 6A_3 + 2\left[\frac{(y_3 - y_2 - y_1 + y_0)}{4h^2} - \frac{3A}{2}(x_2 + x_1)\right] \\ &= 9Ah + 2\left[\frac{(y_3 - y_2 - y_1 + y_0)}{6h^3}\right] \\ &= 9\left[\frac{y_3 - 3y_2 + 3y_1 - y_0}{6h^3}\right]h + 2\left[\frac{(y_3 - y_2 - y_1 + y_0)}{4h^2}\right] \\ &= \frac{(2y_3 - 5y_2 + 4y_1 - y_0)}{h^2}\end{aligned} \tag{B.5}$$

By finite differences, the second derivative at (x_3,y_3) is

$$\left(\frac{d^2y}{dx^2}\right)_{x_3} = \frac{y_4 - 2y_3 + y_2}{h^2} \tag{B.6}$$

Equating equations (B.5) and (B.6)

$$\left(\frac{d^2y}{dx^2}\right)_{x_3} = \frac{y_4 - 2y_3 + y_2}{h^2} = \frac{2y_3 - 5y_2 + 4y_1 - y_0}{h^2}$$

from which the extrapolated value of y_4 can be obtained as

$$\boxed{y_4 = 4y_3 - 6y_2 + 4y_1 - y_0} \tag{B.7}$$

B.8 Finite-difference solution schemes for transients

In a simple contraflow heat exchanger two fluids flow in opposite directions. The direction of algorithmic propagation for the disturbance in the second fluid is differerent from that in the first fluid. This is not a problem for the balance of mass and linear momentum equations which can be separated from the balance of energy equations for the low Mach numbers involved.

The energy equations provide the coupling between the two fluids. If the approach is other than by simultaneous solution by inversion of a Crank–Nicholson matrix, then the direction of solution plus the direction of propagating disturbances makes the situation very much more complicated.

Crank–Nicholson approach

This is a first option for solution of partial differential equations as the method is unconditionally stable, but time steps are restricted by the Courant–Friedrichs–Lewy (CFL) condition which ensures that disturbances stay within each space step.

At each time-step, physical properties are adjusted using interpolating cubic spline-fits for each temperature-dependent parameter. Interpolating cubic spline-fits are also used to prepare heat-transfer and flow-friction correlations from raw data–the data–fits often being better than those published with the data.

Balance of mass

Solution for density (ρ). Densities are replaced by their subscripts below.

Schematic 5-cell layout

known												*fictitious*
H_0		H_1		H_2		H_3		H_4		H_5		H_6
+	–	+	–	+	–	+	–	+	–	+	···	+
|		|		|		|		|		|		⋮
+	–	+	–	+	–	+	–	+	–	+	···	+

The scheme shown is that for the hot fluid. Only the algorithm for the hot fluid is required, as the identical algorithm can be used for the cold fluid providing the input data is reverse numbered before solution and the output results reverse re-numbered after solution.

The balance of mass equation to be solved is

$$\frac{\partial \rho}{\partial t} = -\rho \frac{\partial u}{\partial x} - u \frac{\partial \rho}{\partial x}$$

We use values of velocity (u) from the previous time interval. Unknowns in the finite-difference form are H_1 to H_5, but Crank–Nicholson formulation requires a value at H_6. This is not a problem at start-up from an isothermal condition. Beyond the first time-step, zero density gradient at H_5 is assumed, and since

$$\left(\frac{\partial \rho}{\partial x}\right)_n^t = \frac{(\rho_{n+1}^t - \rho_{n-1}^t)}{2\Delta x} = 0$$

there results $\rho_{n+1}^t = \rho_{n-1}^t$.

Thus at time $t+1$ we may put $\rho_{n+1}^{t+1} = \rho_{n-1}^t$ without serious error followed by matrix inversion using Gaussian elimination. If necessary, ρ_{n+1}^{t+1} can be iterated until $\rho_{n+1}^{t+1} = \rho_{n-1}^{t+1}$.

Balance of linear momentum

Solution for the product of density × velocity ρu.

The balance of linear momentum equation to be solved is

$$\frac{\partial(\rho u)}{\partial t} = (\rho u)\frac{\partial u}{\partial x} - u\frac{\partial(\rho u)}{\partial x} \quad \text{in conservative form, and}$$

$$\frac{\partial(\rho u)}{\partial t} = (\rho u)\frac{\partial u}{\partial x} - u\frac{\partial(\rho u)}{\partial x} - \left[\frac{\partial p}{\partial x} + \left(\frac{f\rho u^2}{2r_{hyd}}\right)\right] \quad \text{in dissipative form, and}$$

$$\frac{\partial(\rho u)}{\partial t} = (\rho u)\frac{\partial u}{\partial x} - u\frac{\partial(\rho u)}{\partial x} - \left[(RT)\frac{\partial \rho}{\partial x} + \left(\frac{uf}{2r_{hyd}}\right)(\rho u)\right] \quad \text{for a perfect gas.}$$

Values of velocity (u) and temperature (T) are from the previous time interval are used. New values for density (ρ) are obtained directly from solution of the balance of mass equation.

The approach to solution is the same as for the balance of mass equation, now providing distributed values of ρu, from which new values of velocity can be obtained.

As before, the identical algorithm can be used for the cold fluid providing the input data are reverse numbered before solution, and reverse re-numbered after solution.

Balance of energy

Solution for the absolute temperatures T. Temperatures are replaced by their subscripts below.

Schematic 5-cell layout

	known											
	H_0		H_1		H_2		H_3		H_4		H_5	
···	+	−	+	−	+	−	+	−	+	−	+	···
W_0	|	W_0	|	W_1	|	W_2	|	W_3	|	W_4	|	W_4
···	+	−	+	−	+	−	+	−	+	−	+	···
	C_0		C_1		C_2		C_3		C_4		C_5	
											known	

$$
\left.\begin{aligned}
&\frac{\partial H}{\partial t} + u_h \frac{\partial H}{\partial x} = -P(H - W) \\
&\frac{\partial W}{\partial t} - \kappa \frac{\partial^2 W}{\partial x^2} = +Q(H - W) - R(W - C) \\
&\frac{\partial C}{\partial t} - u_c \frac{\partial C}{\partial x} = \qquad\qquad\qquad + S(W - C)
\end{aligned}\right\}
$$

where

$$
P = \left(\frac{\alpha_h S}{\dot{m}_h C_h}\right)\frac{u_h}{L}, \qquad Q = \left(\frac{\alpha_h S}{M_w C_w}\right), \qquad R = \left(\frac{\alpha_c S}{M_w C_w}\right), \qquad S = \left(\frac{\alpha_c S}{\dot{m}_c C_c}\right)\frac{u_c}{L}
$$

We set up the solution along the lines adopted for longitudinal conduction in Appendix B.5 except that time-dependent terms are now involved. This requires the assumption of zero wall temperature gradients at each end of the exchanger, but wall temperatures at each end of the exchanger remain unknown.

The Crank–Nicholson transient solution is set up at the mid-point of cells. This requires the average of *fluid* temperatures at each end of a cell in developing the algorithms.

For the hot fluid equation, the time-wise temperature gradient at mid-points for the hot fluid equation may be written

$$\frac{1}{2}\left[\left(\frac{\partial H}{\partial t}\right)_{j-mid}^{t+1}+\left(\frac{\partial H}{\partial t}\right)_{j-mid}^{t}\right]=\frac{(H_{j+1}^{i+1}+H_{j}^{i+1})-(H_{j+1}^{t}+H_{j}^{t})}{2\Delta t}$$

$$=\frac{1}{2}\left\{(RHS)_{j}^{t+1}+(RHS)_{j}^{t}\right\}$$

where j refers to the wall stations. The (*RHS*) terms are forward and backward differences for the right-hand side of the hot fluid equation given in the set above.

The process is repeated for wall and cold fluid equations. By this means we get exactly the same number of unknowns as there are equations, and solution of the temperature-field matrix can be by Gaussian inversion. Note that the solution provides fluid temperatures at the cell boundaries, but wall temperatures at the mid-point of cells.

Once temperatures are known, updated physical properties for the next time interval can be obtained from interpolating cubic spline-fits. The maximum speed of sound found in the exchanger allows the next time interval to be estimated from the modified CFL condition. New values of Reynolds numbers allow heat-transfer coefficients and flow-friction factors to be obtained for each fluid station.

Solution of the balance of mass, balance of linear momentum, and balance of energy equations can now be repeated for the next time interval. The essential requirement is a very fast computer.

Alternative approaches

Alternatives include the two-step Lax–Wendroff scheme, for which an easily understood graphical representation is given by Press *et al.* (1989). Various extensions of the Lax–Wendroff method exist (see Mitchell & Griffiths, 1980), including several different versions of the two-step MacCormack algorithm. Accuracy of numerical prediction of transient response of heat exchangers needs to be demonstrated against experimental results.

Solution of coupled systems of conservation equations by the method due to MacCormack is described by Mitchell & Griffiths (1980), by Peyret & Taylor (1982), by Anderson *et al.* (1984), and by Fletcher (1991). Explicit solution of individual equations involves solution by predictor–corrector algorithms. Anderson *et al.* (1984) observed that best results are obtained when differences in the predictor are taken in the direction of the flow disturbance, and differences in the corrector are taken in the opposite direction.

Ontko (1989) adopted a modified MacCormack approach for solution of transients in a contraflow heat exchanger. He solved for a single-step inlet disturbance with constant physical properties, however, it is perfectly possible to allow for temperature-dependent physical properties.

It may be optimal to solve the balance of mass and linear momentum equations by MacCormack, or by the method of lines plus Runge–Kutta, while solving the balance of energy equations simultaneously using Crank–Nicholson and matrix inversion. Stability of the explicit and implicit MacCormack schemes is discussed by Fletcher (1991, vol. II, Chapter 18, Sections 18.2 and 18.3). The implicit scheme is an extension to the explicit scheme and as an example it is applied to a particular one-dimensional transport equation.

The method of lines involves reducing the partial differential equation to a system of ordinary differential equations for the nodal values (Fletcher, 1991, vol. I, Chapter 7, Section 7.4). One of the best ways of solving this sequence of discrete problems is by using the Runge–Kutta method. See also Fletcher (1991, vol. II, Chapter 18, Section 18.2.2).

The best confirmation of accuracy of these alternative solution methods would be experimental, involving construction of a small laboratory test-rig to produce measured transients in model contraflow exchangers (see Chapter 4, Fig. 8.4).

References

Anderson, D.A., Tannehill, J.C., and **Fletcher, R.H.** (1984) *Computational Techniques for Fluid Mechanics*, Chapter 9, Hemisphere, New York.

de Boor, C. (1978) *A Practical Guide to Splines, Applied Mathematical Sciences*, 27, Springer, p. 303.

Fletcher, C.A.J. (1991) *Computational Techniques for Fluid Dynamics*, vols. I and II, 2nd edn, Springer, Berlin.

Mitchell, A.R. and **Griffiths, D.F.** (1980) *The Finite Difference Method in Partial Differential Equations*, John Wiley, Chichester.

Ontko, J.S. (1989) A parametric study of counterflow heat exchanger transients. Report IAR 89-10, Institute for Aviation Research, Wichita State University, Witchita, Kansas.

Peyret, R. and **Taylor, T.D.** (1982) *Computational Methods for Fluid Flow*, Springer, Berlin.

Press, W.H., Flannery, B.P., Teukolsky, S.A., and **Vetterling, W.T.** (1989) *Numerical Recipies in Pascal*, Cambridge University Press, Cambridge.

Soanes, R.V. (1976) VP-splines, an extension of twice differentiable interpolation. In *Proceedings of the 1976 Army Numerical Analysis and Computer Conference*, ARD Report 76-3, US Army Research Office, Research Triangle Park, North Carolina, pp. 141–152.

Woodford, C.H. (1970) An algorithm for data smoothing using spline functions. *B.I.T.*, **10**, June, 501–510.

Young, J.B. (1988) An equation of state for steam for turbomachinery and other flow calculations. *ASME J. Engng Gas Turbines Power*, **110**, January, 1–7.

Bibliography

Anon. (1960–1978) Index by Subject to Algorithms. *Comm. Assoc. Comp. Mach. (ACM)*, 1976, December 1977, 1978 (subsequently as loose-leaf *Collected Algorithms from ACM*).

Cheney, W. and **Kincaid, D.** (1985) *Numerical Mathematics and Computing*, 2nd edn, Brooks/Cole Publishing Co.

Jeffrey, A. (1989) *Mathematics for Engineers and Scientists*, 4th edn, Van Nostrand Reinhold (International).

Noble, B. (1964) *Numerical Methods: 2 – Differences Integration and Differential Equations*, Oliver and Boyd, Edinburgh and London.

Ontko, J.S. and **Harris, J.A.** (1990) Transients in counterflow heat exchanger. *Compact Heat Exchangers – a Festschrift for A.L. London* (Eds., R.K. Shah, A.D. Kraus, and D. Metzger), Hemisphere, New York, pp. 531–548.

Spencer, A.J.M., Parker, D.F., Berry, D.S., England, A.H., Faulkner, T.R., Green, W.A., Holden, J.T., Middleton, D., and **Rogers, T.G.** (1981) *Engineering Mathematics*, vols. I and II, Van Nostrand Reinhold Co. Ltd.

Supplement to Appendix B
Transient Algorithms

Crank–Nicholson finite-difference formulation

Mass flow and temperature transients in contraflow

A finite-difference solution involves a set of seven simultaneous partial differential equations. As the Mach number for flow does not exceed 0.1 the solution process may be arranged as separate and sequential solution of mass flow and temperature fields. The approach to solving the transient problem can be found in Chapter 9 and Appendices A and B. Automatic selection of time intervals is essential and is controlled by the software.

Time interval solution has to accommodate hot and cold transients travelling in opposite directions, and it may be appropriate to use a modified version of the Courant–Friedrichs–Lewy condition, viz $\Delta t \leq (\Delta x/2)/(|u| + c)$, where the speed of sound for a perfect gas is $c = \sqrt{\gamma RT}$ and the largest value of $(|u| + c)$ is taken.

Because physical properties are temperature dependent it is recommended that the computation is started from initially isothermal conditions.

S.1 Preparation of algorithms

The finite-difference approach used is the stable Crank–Nicholson solution. Disturbances are prevented from passing into the next space interval by invoking the CFL constraint (Chapter 9, Section 9.2). At the beginning of each time interval it is required that the temperature profiles be known throughout the exchanger for the hot fluid, the separating wall and the cold fluid. Thus transients must initiate from a known steady-state. Evaluation of physical properties is by use of interpolating cubic spline-fits. In a single time interval the first stage is to solve the balancc of mass and balancc of lincar momentum equations for density (ρ) and velocity (u).

These equations were solved first to understand any computational problems before particularizing the solution.

$$\left.\begin{aligned}\frac{\partial\rho}{\partial t} &= -\frac{\partial(\rho u)}{\partial x}\\ \frac{\partial(\rho u)}{\partial t} &= -\frac{\partial(\rho u^2)}{\partial x}\end{aligned}\right\} \tag{S.1}$$

Neither pressure (p) nor friction loss enter into the conservative continuum equation for linear momentum, but pressure can be evaluated from density (ρ) using the state equation for the fluid, and the pressure loss due to friction can be evaluated using both density (ρ) and velocity (u) from the last time step. Pressure group terms added to the balance of linear momentum equation will particularize the solution, and in the process will then introduce further transient effects.

$$\left.\begin{aligned} \frac{\partial\rho}{\partial t} &= -\frac{\partial(\rho u)}{\partial x} \\ \frac{\partial(\rho u)}{\partial t} &= -\frac{\partial(\rho u^2)}{\partial x} - \left[\frac{\partial p}{\partial x} + \left(\frac{f\rho u^2}{2r_{hyd}}\right)\right] \end{aligned}\right\} \tag{S.2}$$

Once the balance of mass and balance of linear momentum equations have been solved then, for the same time interval, the three simultaneous equations for the temperature fields (hot fluid, solid wall, cold fluid) can be solved. The temperature-field solutions then provides new temperature profiles from which new physical properties for the next time interval are obtained from interpolating spline-fits. The above procedures are repeated until the transient is complete.

S.2 Balance of mass

The governing equation is first expanded

$$\frac{\partial\rho}{\partial t} = -\rho\frac{\partial u}{\partial x} - u\frac{\partial\rho}{\partial x}$$

Stations along the exchanger are numbered ($0 \cdots n+1$) where (0) is the inlet condition, (n) is the outlet condition and ($n+1$) is a fictitious condition. Forward and backward finite differences are taken as

$$\left[\frac{\partial\rho}{\partial t}\right]_j^t = -\rho_j^t\left[\frac{u_{j+1}^t - u_{j-1}^t}{2\Delta x}\right] - u_j^t\left[\frac{\rho_{j+1}^t - \rho_{j-1}^t}{2\Delta x}\right]$$

$$\left[\frac{\partial\rho}{\partial t}\right]_j^{t+1} = -\rho_j^{t+1}\left[\frac{u_{j+1}^{t+1} - u_{j-1}^{t+1}}{2\Delta x}\right] - u_j^{t+1}\left[\frac{\rho_{j+1}^{t+1} - \rho_{j-1}^{t+1}}{2\Delta x}\right]$$

from which the standard Crank–Nicholson finite-difference form is produced.

$$\frac{\rho_j^{t+1} - \rho_j^t}{\Delta t} = \frac{1}{2}\left\{\begin{aligned} &\rho_{j-1}^t\left(\frac{+u_j^t}{2\Delta x}\right) + \rho_j^t\left(\frac{-u_{j+1}^t + u_{j-1}^t}{2\Delta x}\right) + \rho_{j+1}^t\left(\frac{-u_j^t}{2\Delta x}\right) \\ &\rho_{j-1}^{t+1}\left(\frac{+u_j^t}{2\Delta x}\right) + \rho_j^{t+1}\left(\frac{-u_{j+1}^{t+1} + u_{j-1}^{t+1}}{2\Delta x}\right) + \rho_{j+1}^{t+1}\left(\frac{-u_j^t}{2\Delta x}\right) \end{aligned}\right\}$$

Rearranging unknowns on the left-hand side (LHS)

$$\rho_{j-1}^{t+1}\left\{\frac{-u_j^{t+1}}{4\Delta x}\right\} + \rho_j^{t+1}\left\{\frac{1}{\Delta t} + \frac{(u_{j+1}^{t+1} - u_{j-1}^{t+1})}{4\Delta x}\right\} + \rho_{j+1}^{t+1}\left\{\frac{+u_j^{t+1}}{4\Delta x}\right\}$$

$$= \rho_{j-1}^{t}\left\{\frac{+u_j^{t}}{4\Delta x}\right\} + \rho_j^{t}\left\{\frac{1}{\Delta t} - \frac{(u_{j+1}^{t} - u_{j-1}^{t})}{4\Delta x}\right\} + \rho_{j+1}^{t}\left\{\frac{-u_j^{t}}{4\Delta x}\right\} = [RHS_j]$$

allows a compact notation to be adopted

$$[A_j]\rho_{j-1}^{t+1} + [B_j]\rho_j^{t+1} + [C_j]\rho_{j+1}^{t+1} = [RHS_j]$$

where

$$A_j = \left\{\frac{-u_j^{t+1}}{4\Delta x}\right\}, \qquad B_j = \left\{\frac{1}{\Delta t} + \frac{(u_{j+1}^{t+1} - u_{j-1}^{t+1})}{4\Delta x}\right\}, \qquad C_j = \left\{\frac{+u_j^{t+1}}{4\Delta x}\right\}$$

and

$$RHS_j = \rho_{j-1}^{t}\left\{\frac{+u_j^{t}}{4\Delta x}\right\} + \rho_j^{t}\left\{\frac{1}{\Delta t} - \frac{(u_{j+1}^{t} - u_{j-1}^{t})}{4\Delta x}\right\} + \rho_{j+1}^{t}\left\{\frac{-u_j^{t}}{4\Delta x}\right\}$$

The solution matrix may now be loaded as follows:

- first equation in matrix

$$[B_1]\rho_1^{t+1} + [C_1]\rho_2^{t+1} = [RHS_1] - [A_1]\rho_0^{t+1}$$

- intermediate equations in matrix

$$[A_j]\rho_{j-1}^{t+1} + [B_j]\rho_j^{t+1} + [C_j]\rho_{j+1}^{t+1} = [RHS_j]$$

- last equation in matrix

$$[A_n]\rho_{n-1}^{t+1} + [B_n]\rho_n^{t+1} = [RHS_n] - [C_n]\rho_{n+1}^{t+1}$$

The compact notation (A_j, B_j, C_j) requires knowledge of velocities at time $t+1$, which are unknown. However, we can use known velocities at time t to load and solve the matrix, which will not introduce too much error if time intervals are small. The fictitious value $\rho_{n+1}^{t+1} \equiv (\rho)_6$ is not known for the last equation, and two approaches to approximating an acceptable value are by extrapolation or by zero gradient.

Extrapolation

One approach is to extrapolate for ρ_{n+1}^{t} from the solution of the matrix for the previous time interval, and use this as an approximate value for ρ_{n+1}^{t+1}. The fictitious value ρ_{n+1}^{t} may be found using the extrapolation rule explained in Appendix A,

Section A.3 viz.

$$(\rho)_{n+1} = 4(\rho)_n - 6(\rho)_{n-1} + 4(\rho)_{n-2} - (\rho)_{n-3}$$

Zero gradient

A simpler and better approach is to assume zero gradient at outlet, allowing the fictitious value $\rho_{n+1}^{t+1} = \rho_{n-1}^{t}$.

Schematic 5-cell layout

known												*fictitious*
H_0		H_1		H_2		H_3		H_4		H_5		H_6
+	–	+	–	+	–	+	–	+	–	+	···	+
|		|		|		|		|		|		⋮
+	–	+	–	+	–	+	–	+	–	+	···	+

Cold fluid equation

The matrix Table S.1 has been set up for the hot fluid, but the solution method serves equally well for the cold fluid providing input data stations are renumbered appropriately. It is only necessary to reverse-renumber the cold fluid solution when it emerges.

Table S.1 Matrix for density

	Unknown	**1**	**2**	**3**	**4**	**5**	
	Equation	ρ_1	ρ_2	ρ_3	ρ_4	ρ_5	*RHS*
1	**1**	B_1	C_1				$RHS_1 - A_1(\rho)_0$
2	**2**	A_2	B_2	C_2			RHS_2
	3		A_3	B_3	C_3		RHS_3
n − 1	**4**			A_4	B_4	C_4	RHS_4
n	**5**				A_5	B_5	$RHS_5 - C_5(\rho)_6$
		1	2		n − 1	n	n + 1

Known inlet condition $(\rho)_0$, estimated fictitious condition $(\rho)_6$, $(\rho)_6$ adjusted and matrix solution iterated until $(\rho)_4 = (\rho)_6$.

S.3 Balance of linear momentum

The conservative form of the governing equation is

$$\frac{\partial(\rho u)}{\partial t} = (\rho u)\frac{\partial u}{\partial x} - u\frac{\partial(\rho u)}{\partial x}$$

This has the same form as the balance of mass equation, and thus provides the same compact notation

$$[A_j](\rho u)_{j-1}^{t+1} + [B_j](\rho u)_j^{t+1} + [C_j](\rho u)_{j+1}^{t+1} = [RHS_j]$$

where

$$A_j = \left\{ \frac{-u_j^{t+1}}{4\Delta x} \right\}, \qquad B_j = \left\{ \frac{1}{\Delta t} + \frac{(u_{j+1}^{t+1} - u_{j-1}^{t+1})}{4\Delta x} \right\}, \qquad C_j = \left\{ \frac{+u_j^{t+1}}{4\Delta x} \right\}$$

and

$$RHS_j = +(\rho u)_{j-1}^t \left[\frac{+u_j^t}{4\Delta x} \right] + (\rho u)_j^t \left[\frac{1}{\Delta t} - \frac{(u_{j+1}^t - u_{j-1}^t)}{4\Delta x} \right] + (\rho u)_{j+1}^t \left[\frac{-u_j^t}{4\Delta x} \right]$$

Inclusion of pressure terms (see Appendix A.3)

Following evaluation of Reynolds numbers $\mathrm{Re}_j^{t-1} = (\rho u d_{hyd}/\eta)_j^{t-1}$ to obtain friction factors $(f)_j^{t-1}$ from an interpolating cubic spline-fit:

1. ... add $\left(\dfrac{uf}{2r_{hyd}} \right)_j^{t-1}$ to expression for B_j
2. ... add $(RT)_j^t \left(\dfrac{p_{j+1}^t - p_{j-1}^t}{2\Delta x} \right)$ to expression for RHS_j

Pressure gradient terms at entry and exit are zero, and should be replaced by numerical expressions for losses due to entry and exit effects.

The solution matrix may now be loaded as follows:

- first equation in matrix

$$[B_1](\rho u)_1^{t+1} + [C_1](\rho u)_2^{t+1} = [RHS_1] - [A_1](\rho u)_0^{t+1}$$

- intermediate equations in matrix

$$[A_j](\rho u)_{j-1}^{t+1} + [B_j](\rho u)_j^{t+1} + [C_j](\rho u)_{j+1}^{t+1} = [RHS_j]$$

- last equation in matrix

$$[A_n](\rho u)_{n-1}^{t+1} + [B_n](\rho u)_n^{t+1} = [RHS_n] - [C_n](\rho u)_{n+1}^{t+1}$$

For the density $\times$ velocity matrix, finding values of (A_j, B_j, C_j) and $(\rho u)_{n+1}^{t+1}$ follows the same route as employed for the (density) matrix.

Table S.2 Matrix for density × velocity

	Unknown Equation	1 $(\rho u)_1$	2 $(\rho u)_2$	3 $(\rho u)_3$	4 $(\rho u)_4$	5 $(\rho u)_5$	*RHS*
1	**1**	B_1	C_1				$RHS_1 - A_1(\rho u)_0$
2	**2**	A_2	B_2	C_2			RHS_2
	3		A_3	B_3	C_3		RHS_3
n − 1	**4**			A_4	B_4	C_4	RHS_4
n	**5**				A_5	B_5	$RHS_5 - C_5(\rho u)_5$
		1	2		n − 1	n	n + 1

Known inlet condition $(\rho u)_0$, estimated fictitious condition $(\rho u)_6$, $(\rho u)_6$ adjusted and matrix solution iterated until $(\rho u)_4 = (\rho u)_6$.

Schematic 5-cell layout

known												*fictitious*
H_0		H_1		H_2		H_3		H_4		H_5		H_6
+	−	+	−	+	−	+	−	+	−	+	⋯	+
I		I		I		I		I		I		⋮
+	−	+	−	+	−	+	−	+	−	+	⋯	+

Cold fluid equations

The matrix Table S.2 has been set up for the hot fluid, but this serves equally well for the cold fluid providing input data stations are renumbered appropriately. It is only necessary to reverse-renumber the cold fluid solution when it emerges.

S.4 Balance of energy

The use of Crank–Nicholson method for the three coupled temperature equations does not involve any extrapolation in the solution. The first step is to settle the finite-difference layout for solution. End temperatures for the solid wall remain unknown, but we might assume that there may be zero temperature gradient at the ends.

This requires setting up hot and cold fluid temperatures in the range $(0 \cdots n)$, and the solid wall temperatures in the range $(0 \cdots n - 1)$. The mid-point of cells is used as the basis for the algorithm.

Schematic 5-cell layout

	known											
	H_0		H_1		H_2		H_3		H_4		H_5	
⋯	+	−	+	−	+	−	+	−	+	−	+	⋯
W_0	I	W_0	I	W_1	I	W_2	I	W_3	I	W_4	I	W_4
⋯	+	−	+	−	+	−	+	−	+	−	+	⋯
	C_0		C_1		C_2		C_3		C_4		C_5	
											known	

Simplification of the governing equations

$$\left.\begin{aligned}
&\frac{\partial T_h}{\partial t} + u_h \frac{\partial T_h}{\partial x} = -\frac{u_h}{L}\left(\frac{\alpha_h S}{\dot{m}_h C_h}\right)(T_h - T_w) \\
&\frac{\partial T_w}{\partial t} - \kappa \frac{\partial^2 T_w}{\partial x^2} = +\left(\frac{\alpha_h S}{M_w C_w}\right)(T_h - T_w) - \left(\frac{\alpha_c S}{M_w C_w}\right)(T_w - T_c) \\
&\frac{\partial T_c}{\partial t} - u_c \frac{\partial T_c}{\partial x} = +\frac{u_c}{L}\left(\frac{\alpha_c S}{\dot{m}_c C_c}\right)(T_w - T_c)
\end{aligned}\right\} \tag{S.3}$$

Replace temperatures with their subscripts, and replace coefficients with (P, Q, R, S), noticing that these values are different from those defined for the steady-state.

$$\left.\begin{aligned}
&\frac{\partial H}{\partial t} + u_h \frac{\partial H}{\partial x} = -P(H - W) \\
&\frac{\partial W}{\partial t} - \kappa \frac{\partial^2 W}{\partial x^2} = +Q(H - W) - R(W - C) \\
&\frac{\partial C}{\partial t} - u_c \frac{\partial C}{\partial x} = +S(W - C)
\end{aligned}\right\}$$

where

$$P = \left(\frac{\alpha_h S}{\dot{m}_h C_h}\right)\frac{u_h}{L}, \qquad Q = \left(\frac{\alpha_h S}{M_w C_w}\right), \qquad R = \left(\frac{\alpha_c S}{M_w C_w}\right), \qquad S = \left(\frac{\alpha_c S}{\dot{m}_c C_c}\right)\frac{u_c}{L}$$

Each equation is now converted to Crank-Nicholson finite-difference form separately.

Hot fluid equation

$$\frac{\partial H}{\partial t} + u_h \frac{\partial H}{\partial x} = -P(H - W)$$

Forward differences evaluated at W_j

$$\begin{aligned}
\left(\frac{\partial H}{\partial t}\right)_j^t &= -(u_h)_j^t\left[\frac{H_{j+1}^t - H_j^t}{\Delta x}\right] - P_j^t\left[\frac{(H_{j+1}^t + H_j^t)}{2} - W_j^t\right] \\
&= H_{j+1}^t\left[\frac{-(u_h)_j^t}{\Delta x}\right] + H_j^t\left[\frac{+(u_h)_j^t}{\Delta x}\right] + H_{j+1}^t\left[\frac{-P_j^t}{2}\right] + H_j^t\left[\frac{-P_j^t}{2}\right] + W_j^t[P_j^t] \\
&= H_{j+1}^t\left[-\frac{(u_h)_j^t}{\Delta x} - \frac{P_j^t}{2}\right] + H_j^t\left[+\frac{(u_h)_j^t}{\Delta x} - \frac{P_j^t}{2}\right] + W_j^t[P_j^t]
\end{aligned}$$

and backward differences at W_j

$$\left(\frac{\partial H}{\partial t}\right)_j^{t+1} = -(u_h)_j^{t+1}\left[\frac{H_{j+1}^{t+1} - H_j^{t+1}}{\Delta x}\right] - P_j^t\left[\frac{(H_{j+1}^{t+1} + H_j^{t+1})}{2} - W_j^{t+1}\right]$$

$$= H_{j+1}^{t+1}\left[\frac{-(u_h)_j^{t+1}}{\Delta x}\right] + H_j^{t+1}\left[\frac{+(u_h)_j^{t+1}}{\Delta x}\right] + H_{j+1}^{t+1}\left[\frac{-P_j^{t+1}}{2}\right]$$

$$+ H_j^{t+1}\left[\frac{-P_j^{t+1}}{2}\right] + W_j^{t+1}[P_j^{t+1}]$$

$$= H_{j+1}^{t+1}\left[-\frac{(u_h)_j^{t+1}}{\Delta x} - \frac{P_j^{t+1}}{2}\right] + H_j^{t+1}\left[+\frac{(u_h)_j^{t+1}}{\Delta x} - \frac{P_j^{t+1}}{2}\right] + W_j^{t+1}[P_j^{t+1}]$$

Crank–Nicholson is mean of forward and backward differences.

The time-wise temperature gradient between mid-points (both hot and cold fluids) is given by

$$\frac{\begin{pmatrix}\text{mid temp}\\ @\ t+1\end{pmatrix} - \begin{pmatrix}\text{mid temp}\\ @\ t\end{pmatrix}}{\Delta t} = \frac{\left(\dfrac{H_{j+1}^{t+1} + H_j^{t+1}}{2}\right) - \left(\dfrac{H_{j+1}^{t} + H_j^{t}}{2}\right)}{\Delta t}$$

$$= \frac{\left(H_{j+1}^{t+1} + H_j^{t+1}\right) - \left(H_{j+1}^{t} + H_j^{t}\right)}{2\Delta t} = \begin{pmatrix}\text{mean of time wise}\\ \text{temperature gradients}\\ \text{at each end}\end{pmatrix}$$

$$= \frac{1}{2}\left[\left(\frac{\partial H}{\partial t}\right)_{j-mid}^{t+1} + \left(\frac{\partial H}{\partial t}\right)_{j-mid}^{t}\right] = \frac{1}{2}\left\{(RHS)_j^{t+1} + (RHS)_j^t\right\}$$

where j refers to wall stations

from which

$$\frac{(H_{j+1}^{t+1} + H_j^{t+1}) - (H_{j+1}^{t} + H_j^{t})}{2\Delta t}$$

$$= \frac{1}{2}\left\{\begin{aligned} & H_{j+1}^{t}\left[-\frac{(u_h)_j^t}{\Delta x} - \frac{P_j^t}{2}\right] + H_j^t\left[+\frac{(u_h)_j^t}{\Delta x} - \frac{P_j^t}{2}\right] + W_j^t[P_j^t] \\ & + H_{j+1}^{t+1}\left[-\frac{(u_h)_j^{t+1}}{\Delta x} - \frac{P_j^{t+1}}{2}\right] + H_j^{t+1}\left[+\frac{(u_h)_j^{t+1}}{\Delta x} - \frac{P_j^{t+1}}{2}\right] + W_j^{t+1}[P_j^{t+1}]\end{aligned}\right\}$$

then

$$
(H_{j+1}^{t+1} + H_j^{t+1}) - (H_{j+1}^{t} + H_j^{t})
$$

$$
= \left\{
\begin{array}{l}
H_{j+1}^{t}\left[-\dfrac{\Delta t}{\Delta x}(u_h)_j^t - \dfrac{\Delta t\ P_j^t}{2}\right] + H_j^t\left[\dfrac{\Delta t}{\Delta x}(u_h)_j^t - \dfrac{\Delta t\ P_j^t}{2}\right] \\
+ W_j^t[\Delta t\ P_j^t] + H_{j+1}^{t+1}\left[-\dfrac{\Delta t}{\Delta x}(u_h)_j^{t+1} - \dfrac{\Delta t\ P_j^{t+1}}{2}\right] \\
+ H_j^{t+1}\left[\dfrac{\Delta t}{\Delta x}(u_h)_j^{t+1} - \dfrac{\Delta t\ P_j^{t+1}}{2}\right] + W_j^{t+1}[\Delta t\ P_j^{t+1}]
\end{array}
\right\}
$$

Collect unknown $t+1$ terms on LHS, and known t terms on the RHS

$$
\begin{aligned}
&H_{j+1}^{t+1}\left[1 + \frac{\Delta t}{\Delta x}(u_h)_j^{t+1} + \frac{\Delta t\ P_j^{t+1}}{2}\right] + H_j^{t+1}\left[1 - \frac{\Delta t}{\Delta x}(u_h)_j^{t+1} + \frac{\Delta t\ P_j^{t+1}}{2}\right] \\
&\quad + W_j^{t+1}[-\Delta t\ P_j^{t+1}] \\
&\quad = \left\{H_{j+1}^{t}\left[1 - \frac{\Delta t}{\Delta x}(u_h)_j^t - \frac{\Delta t\ P_j^t}{2}\right] + H_j^t\left[1 + \frac{\Delta t}{\Delta x}(u_h)_j^t - \frac{\Delta t\ P_j^t}{2}\right] + W_j^t[+\Delta t\ P_j^t]\right\}
\end{aligned}
$$

If all coefficients are evaluated at time t, and mass flow rates remain unchanged so that $\left\{(u_h)_j^{t+1} = (u_h)_j^t\right\}$ then putting

$$
\left[A = \frac{\Delta t}{\Delta x}(u_h)_j^t, \quad B = \frac{\Delta t\ P_j^t}{2}\right],
$$

$$
\begin{aligned}
&H_{j+1}^{t+1}[1 + A + B] + H_j^{t+1}[1 - A + B] + W_j^{t+1}[2B] \\
&\quad = \{H_{j+1}^{t}[1 - A - B] + H_j^t[1 + A - B] + W_j^t[2B]\} = [RHS_j]
\end{aligned}
$$

At $j = 0$ the hot inlet temperature H_0^{t+1} is known

$$
\begin{aligned}
&H_1^{t+1}[1 + A + B] + H_0^{t+1}[1 - A + B] + W_0^{t+1}[2B] \\
&\quad = \{H_1^t[1 - A - B] + H_0^t[1 + A - B] + W_0^t[2B]\} = [RHS_0]
\end{aligned}
$$

thus

$$
H_1^{t+1}[1 + A + B] + W_0^{t+1}[-2B] = [RHS_0] - H_0^{t+1}[1 - A + B]
$$

Solid wall equation

$$\frac{\partial W}{\partial t} - \kappa \frac{\partial^2 W}{\partial x^2} = +Q(H - W) - R(W - C)$$

Forward differences give

$$\left(\frac{\partial W}{\partial t}\right)_j^t = +\kappa_j^t \left[\frac{W_{j+1}^t - 2W_j^t + W_{j-1}^t}{(\Delta x)^2}\right] + Q_j^t \left[\frac{(H_{j+1}^t + H_j^t)}{2} - W_j^t\right] - R_j^t \left[W_j^t - \frac{(C_{j+1}^t + C_j^t)}{2}\right]$$

$$\left(\frac{\partial W}{\partial t}\right)_j^t = W_{j+1}^t \left[\frac{\kappa_j^t}{(\Delta x)^2}\right] + W_j^t \left[\frac{-2\kappa_j^t}{(\Delta x)^2}\right] + W_{j-1}^t \left[\frac{\kappa_j^t}{(\Delta x)^2}\right] + \left\{\begin{matrix} H_{j+1}^t \left[\frac{Q_j^t}{2}\right] + H_j^t \left[\frac{Q_j^t}{2}\right] + W_j^t[-Q_j^t] \\ C_{j+1}^t \left[\frac{R_j^t}{2}\right] + C_j^t \left[\frac{R_j^t}{2}\right] + W_j^t[-R_j^t] \end{matrix}\right\}$$

$$\left(\frac{\partial W}{\partial t}\right)_j^t = W_{j+1}^t \left[\frac{\kappa_j^t}{(\Delta x)^2}\right] + W_j^t \left[-\frac{2\kappa_j^t}{(\Delta x)^2} - Q_j^t - R_j^t\right] + W_{j-1}^t \left[\frac{\kappa_j^t}{(\Delta x)^2}\right] + \left\{\begin{matrix} H_{j+1}^t \left[\frac{Q_j^t}{2}\right] + H_j^t \left[\frac{Q_j^t}{2}\right] \\ C_{j+1}^t \left[\frac{R_j^t}{2}\right] + C_j^t \left[\frac{R_j^t}{2}\right] \end{matrix}\right\}$$

Backward differences give

$$\left(\frac{\partial W}{\partial t}\right)_j^{t+1} = W_{j+1}^{t+1} \left[\frac{\kappa_j^{t+1}}{(\Delta x)^2}\right] + W_j^{t+1} \left[-\frac{2\kappa_j^{t+1}}{(\Delta x)^2} - Q_j^{t+1} - R_j^{t+1}\right] + W_{j-1}^{t+1} \left[\frac{\kappa_j^{t+1}}{(\Delta x)^2}\right] + \left\{\begin{matrix} H_{j+1}^{t+1} \left[\frac{Q_j^{t+1}}{2}\right] + H_j^{t+1} \left[\frac{Q_j^{t+1}}{2}\right] \\ C_{j+1}^{t+1} \left[\frac{R_j^{t+1}}{2}\right] + C_j^{t+1} \left[\frac{R_j^{t+1}}{2}\right] \end{matrix}\right\}$$

Crank–Nicholson is average of forward and backward differences, thus time-wise temperature gradient between mid-points (for the solid wall) is given by

$$\frac{W_j^{t+1} - W_j^t}{\Delta t} = \frac{1}{2}\left\{\left(\frac{\partial W}{\partial t}\right)_j^{t+1} + \left(\frac{\partial W}{\partial t}\right)_j^t\right\} = \frac{1}{2}\{(RHS)_j^{t+1} + (RHS)_j^t\}$$

where j refers to wall stations.

$$\frac{W_j^{t+1} - W_j^t}{\Delta t}$$

$$= \frac{1}{2}\left\{\begin{array}{l} W_{j+1}^t\left[\dfrac{\kappa_j^t}{(\Delta x)^2}\right] + W_j^t\left[\dfrac{-2\kappa_j^t}{(\Delta x)^2} - Q_j^t - R_j^t\right] + W_{j-1}^t\left[\dfrac{\kappa_j^t}{(\Delta x)^2}\right] \\ +W_{j+1}^{t+1}\left[\dfrac{\kappa_j^{t+1}}{(\Delta x)^2}\right] + W_j^{t+1}\left[\dfrac{-2\kappa_j^{t+1}}{(\Delta x)^2} - Q_j^{t+1} - R_j^{t+1}\right] + W_{j-1}^{t+1}\left[\dfrac{\kappa_j^{t+1}}{(\Delta x)^2}\right] \\ +H_{j+1}^t\left[\dfrac{Q_j^t}{2}\right] + H_j^t\left[\dfrac{Q_j^t}{2}\right] + C_{j+1}^t\left[\dfrac{R_j^t}{2}\right] + C_j^t\left[\dfrac{R_j^t}{2}\right] \\ +H_{j+1}^{t+1}\left[\dfrac{Q_j^{t+1}}{2}\right] + H_j^{t+1}\left[\dfrac{Q_j^{t+1}}{2}\right] + C_{j+1}^{t+1}\left[\dfrac{R_j^{t+1}}{2}\right] + C_j^{t+1}\left[\dfrac{R_j^{t+1}}{2}\right] \end{array}\right\}$$

Then

$$W_j^{t+1}\left[\frac{2}{\Delta t}\right] - W_j^t\left[\frac{2}{\Delta t}\right] = \{\quad\}$$

Unknown $t+1$ terms are now on LHS and known t terms are on RHS

$$\left\{\begin{array}{l} W_{j+1}^{t+1}\left[-\dfrac{\kappa_j^{t+1}}{(\Delta x)^2}\right] + W_j^{t+1}\left[\dfrac{2}{\Delta t} + \dfrac{2\kappa_j^{t+1}}{(\Delta x)^2} + (Q_j^{t+1} + R_j^{t+1})\right] + W_{j-1}^{t+1}\left[-\dfrac{\kappa_j^{t+1}}{(\Delta x)^2}\right] \\ +H_{j+1}^{t+1}\left[-\dfrac{Q_j^{t+1}}{2}\right] + H_j^{t+1}\left[-\dfrac{Q_j^{t+1}}{2}\right] + C_{j+1}^{t+1}\left[-\dfrac{R_j^{t+1}}{2}\right] + C_j^{t+1}\left[-\dfrac{R_j^{t+1}}{2}\right] \end{array}\right\}$$

$$= \left\{\begin{array}{l} W_{j+1}^t\left[+\dfrac{\kappa_j^{t+1}}{(\Delta x)^2}\right] + W_j^t\left[\dfrac{2}{\Delta t} - \dfrac{2\kappa_j^t}{(\Delta x)^2} - (Q_j^t + R_j^t)\right] + W_{j-1}^t\left[+\dfrac{\kappa_j^t}{(\Delta x)^2}\right] \\ +H_{j+1}^t\left[+\dfrac{Q_j^t}{2}\right] + H_j^t\left[+\dfrac{Q_j^t}{2}\right] + C_{j+1}^t\left[+\dfrac{R_j^t}{2}\right] + C_j^t\left[+\dfrac{R_j^t}{2}\right] \end{array}\right\}$$

If all coefficients for $t+1$ and t are evaluated at time interval t, then putting

$$D = \frac{\kappa_j^t}{(\Delta x)^2}, \qquad E = \frac{2}{\Delta t}, \qquad F = \frac{Q_j^t}{2}, \qquad G = \frac{R_j^t}{2} \quad \text{it follows that}$$

$$\left\{ \begin{array}{l} W_{j+1}^{t+1}[-D] + W_j^{t+1}[E + 2(D + F + G)] + W_{j-1}^{t+1}[-D] \\ +H_{j+1}^{t+1}[-F] + H_j^{t+1}[-F] + C_{j+1}^{t+1}[-G] + C_j^{t+1}[-G] \end{array} \right\}$$

$$= \left\{ \begin{array}{c} W_{j+1}^t[D] + W_j^t[E - 2(D + F + G)] + W_{j-1}^t[D] \\ +H_{j+1}^t[F] + H_j^t[F] + C_{j+1}^t[G] + C_j^t[G] \end{array} \right\} = [RHS_j]$$

where j is in the range 1 to $(n - 1)$

- *Hot end equation.* At $j = 0$ inlet temperature H_0^{t+1} is known, and $W_{-1}^{t+1} = W_0^{t+1}$ because hot-end wall temperature gradient is zero

$$\left\{ \begin{array}{l} W_1^{t+1}[-D] + W_0^{t+1}[E + 2(D + F + G)] + W_0^{t+1}[-D] \\ +H_1^{t+1}[-F] + H_0^{t+1}[-F] + C_1^{t+1}[-G] + C_0^{t+1}[-G] \end{array} \right\}$$

$$= \left\{ \begin{array}{c} W_1^t[D] + W_0^t[E - 2(D + F + G)] + W_0^t[D] \\ +H_1^t[F] + H_0^t[F] + C_1^t[G] + C_0^t[G] \end{array} \right\}$$

collecting W_0^{t+1} terms and moving inlet temperature H_0^{t+1} to RHS

$$\left\{ \begin{array}{l} W_1^{t+1}[-D] + W_0^{t+1}[E + D + 2(F + G)] \\ +H_1^{t+1}[-F] + C_1^{t+1}[-G] + C_0^{t+1}[-G] \end{array} \right\}$$

$$= \left\{ \begin{array}{c} W_1^t[D] + W_0^t[E - D - 2(F + G)] \\ +H_1^t[F] + H_0^t[F] + C_1^t[G] + C_0^t[G] \end{array} \right\} - H_0^{t+1}[-F] = [RHS_0] - H_0^{t+1}[-F]$$

- *Cold-end equation.* At $(j = n)$ inlet temperature C_{n+1}^{t+1} is known, and $(W_{n+1}^{t+1} = W_n^{t+1})$ because end-wall temperature gradient is zero

$$\left\{ \begin{array}{l} W_n^{t+1}[-D] + W_n^{t+1}[E + 2(D + F + G) + W_{n-1}^{t+1}[-D] \\ +H_{n+1}^{t+1}[-F] + H_n^{t+1}[-F] + C_{n+1}^{t+1}[-G] + C_n^{t+1}[-G] \end{array} \right\}$$

$$= \left\{ \begin{array}{c} W_n^t[D] + W_n^t[E - 2(D + F + G)] + W_{n-1}^t[D] \\ +H_{n+1}^t[F] + H_n^t[F] + C_{n+1}^t[G] + C_n^t[G] \end{array} \right\}$$

collecting W_n^{t+1} terms and moving inlet temperature C_{n+1}^{t+1} to RHS

$$\left\{\begin{array}{c} W_n^{t+1}[E+D+2(F+G)]+W_{n-1}^{t+1}[-D] \\ +H_{n+1}^{t+1}[-F]+H_n^{t+1}[-F]+C_n^{t+1}[-G] \end{array}\right\}$$

$$=\left\{\begin{array}{c} W_n^t[E-D-2(F+G)]+W_{n-1}^t[D] \\ +H_{n+1}^t[F]+H_n^t[F]+C_{n+1}^t[G]+C_n^t[G] \end{array}\right\}-C_{n+1}^{t+1}[-G]$$

$$=[RHS_n]-C_{n+1}^{t+1}[-G]$$

Cold fluid equation

$$\frac{\partial C}{\partial t}-u_c\frac{\partial C}{\partial x}=+S(W-C)$$

Forward differences evaluated at W_j

$$\left(\frac{\partial C}{\partial t}\right)_j^t=+(u_c)_j^t\left[\frac{C_{j+1}^t-C_j^t}{\Delta x}\right]+S_j^t\left[W_j^t-\frac{(C_{j+1}^t+C_j^t)}{2}\right]$$

$$=C_{j+1}^t\left[\frac{+(u_c)_j^t}{\Delta x}\right]+C_j^t\left[\frac{-(u_c)_j^t}{\Delta x}\right]+C_{j+1}^t\left[\frac{-S_j^t}{2}\right]+C_j^t\left[\frac{-S_j^t}{2}\right]+W_j^t[S_j^t]$$

$$=C_{j+1}^t\left[+\frac{(u_c)_j^t}{\Delta x}-\frac{S_j^t}{2}\right]+C_j^t\left[-\frac{(u_c)_j^t}{\Delta x}-\frac{S_j^t}{2}\right]+W_j^t[S_j^t]$$

and backward differences at W_j

$$\left(\frac{\partial C}{\partial t}\right)_j^{t+1}=+(u_c)_j^{t+1}\left[\frac{C_{j+1}^{t+1}-C_j^{t+1}}{\Delta x}\right]+S_j^{t+1}\left[W_j^{t+1}-\frac{(C_{j+1}^{t+1}+C_j^{t+1})}{2}\right]$$

$$=C_{j+1}^{t+1}\left[\frac{+(u_c)_j^{t+1}}{\Delta x}\right]+C_j^{t+1}\left[\frac{-(u_c)_j^{t+1}}{\Delta x}\right]$$

$$+C_{j+1}^{t+1}\left[\frac{-S_j^{t+1}}{2}\right]+C_j^{t+1}\left[\frac{-S_j^{t+1}}{2}\right]+W_j^{t+1}[S_j^{t+1}]$$

$$=C_{j+1}^{t+1}\left[+\frac{(u_c)_j^{t+1}}{\Delta x}-\frac{S_j^{t+1}}{2}\right]+C_j^{t+1}\left[-\frac{(u_c)_j^{t+1}}{\Delta x}-\frac{S_j^{t+1}}{2}\right]+W_j^{t+1}[S_j^{t+1}]$$

Crank–Nicholson is mean of forward and backward differences

$$\frac{(C_{j+1}^{t+1}+C_j^{t+1})-(C_{j+1}^{t}+C_j^{t})}{2\Delta t}$$

$$=\frac{1}{2}\left\{\begin{aligned}&C_{j+1}^{t}\left[+\frac{(u_c)_j^t}{\Delta x}-\frac{S_j^t}{2}\right]+C_j^t\left[-\frac{(u_c)_j^t}{\Delta x}-\frac{S_j^t}{2}\right]+W_j^t[S_j^t]\\&+C_{j+1}^{t+1}\left[+\frac{(u_c)_j^{t+1}}{\Delta x}-\frac{S_j^{t+1}}{2}\right]+C_j^{t+1}\left[-\frac{(u_c)_j^{t+1}}{\Delta x}-\frac{S_j^{t+1}}{2}\right]+W_j^{t+1}[S_j^{t+1}]\end{aligned}\right\}$$

$$(C_{j+1}^{t+1}+C_j^{t+1})-(C_{j+1}^{t}+C_j^{t})$$

$$=\left\{\begin{aligned}&C_{j+1}^{t}\left[+\frac{\Delta t}{\Delta x}(u_c)_j^t\quad-\frac{\Delta tS_j^t}{2}\right]+C_j^t\left[-\frac{\Delta t}{\Delta x}(u_c)_j^t\quad-\frac{\Delta tS_j^t}{2}\right]+W_j^t[\Delta tS_j^t]\\&+C_{j+1}^{t+1}\left[+\frac{\Delta t}{\Delta x}(u_c)_j^{t+1}-\frac{\Delta tS_j^{t+1}}{2}\right]+C_j^{t+1}\left[-\frac{\Delta t}{\Delta x}(u_c)_j^{t+1}-\frac{\Delta tS_j^{t+1}}{2}\right]\\&+W_j^{t+1}[\Delta tS_j^{t+1}]\end{aligned}\right\}$$

Collect unknown $t+1$ terms on LHS and known t terms on RHS

$$C_{j+1}^{t+1}\left[1-\frac{\Delta t}{\Delta x}(u_c)_j^{t+1}+\frac{\Delta tS_j^{t+1}}{2}\right]+C_j^{t+1}\left[1+\frac{\Delta t}{\Delta x}(u_c)_j^{t+1}+\frac{\Delta tS_j^{t+1}}{2}\right]$$

$$+W_j^{t+1}[-\Delta tS_j^{t+1}]$$

$$=\left\{C_{j+1}^{t}\left[1+\frac{\Delta t}{\Delta x}(u_c)_j^t-\frac{\Delta tS_j^t}{2}\right]+C_j^t\left[1-\frac{\Delta t}{\Delta x}(u_c)_j^t-\frac{\Delta tS_j^t}{2}\right]+W_j^t[+\Delta tS_j^t]\right\}$$

$$=\{RHS\}$$

If all coefficients are evaluated at time t, and mass flow rates remain unchanged so that $\{(u_h)_j^{t+1}=(u_h)_j^t\}$, then putting

$$\left[Y=\frac{\Delta t}{\Delta x}(u_c)_j^t,\quad Z=\frac{\Delta tS_j^t}{2}\right]$$

$$C_{j+1}^{t+1}[1-Y+Z]+C_j^{t+1}[1+Y+Z]+W_j^{t+1}[-2Z]$$

$$=\{C_{j+1}^{t}[1+Y-Z]+C_j^t[1-Y-Z]+W_j^t[2Z]\}=[RHS_j]$$

At $j = n - 1$ the cold inlet temperature C_n^{t+1} is known

$$C_n^{t+1}[1 - Y + Z] + C_{n-1}^{t+1}[1 + Y + Z] + W_{n-1}^{t+1}[-2Z]$$
$$= \{C_n^t[1 + Y - Z] + C_{n-1}^t[1 - Y - Z] + W_{n-1}^t[2Z]\} = [RHS_{n-1}]$$

thus

$$C_{n-1}^{t+1}[1 + Y + Z] + W_{n-1}^{t+1}[-2Z] = [RHS_{n-1}] - C_n^{t+1}[1 - Y + Z]$$

The temperature-field matrix is large, even in compact form. To simplify the notation still further the following symbol key table is to be used together with the main matrix.

Table S.3 Symbol key for temperature matrix

a = 1 + A + B	c = −2B	
b = 1 − A + B		
	e = E + D + 2(F + G)	
d = −F	f = E + 2(D + F + G)	h = −G
	g = −D	
		k = 1 + Y + Z
	j = −2Z	m = 1 − Y + Z

RHS entries vary depending on location. Hend and Cend entries each include multiplying coefficents.

Preparation of Table S.4 which follows is the necessary prior step to writing a finite-difference algorithm for a $(1 \cdots 3n,\ 1 \cdots 3n + 1)$ matrix. The matrix may be solved by Gaussian elimination.

S.5 Coding of temperature matrix

See pages 400–403.

Table S.4 Transient temperatures in contraflow with longitudinal conduction

Unknown Equation		1 H_1	2 H_2	3 H_3	4 H_4	5 H_5	6 W_0	7 W_1	8 W_2	9 W_3	10 W_4	11 C_0	12 C_1	13 $C_2{}^{\cdot}$	14 C_3	15 C_4	16 *RHS*
1	**1**	a					c										RHS-Hend
2	**2**	b	a					c									RHS
	3		b	a					c								RHS
n − 1	**4**			b	a					c							RHS
n	**5**				b	a					c						RHS
n + 1	**6**	d					e	g				h	h				RHS-Hend
n + 2	**7**	d	d				g	f	g				h	h			RHS
	8		d	d				g	f	g				h	h		RHS
2n − 1	**9**			d	d				g	f	g				h	h	RHS
2n	**10**				d	d				g	e					h	RHS-Cend
2n + 1	**11**						j					k	m				RHS
2n + 2	**12**							j					k	m			RHS
	13								j					k	m		RHS
3n − 1	**14**									j					k	m	RHS
3n	**15**										j					k	RHS-Cend
		1	2		n − 1	n	n + 1	n + 2		2n − 1	2n	2n + 1	2n + 2		3n − 1	3n	3n + 1

```
{------|*******^*******^*******^*******^*******|---------------------------------}
{Filename TMATRIX.TEXT - loads coefficients for temperature matrix in TRANS}
{------|*******^*******^*******^*******^*******|---------------------------------}
       PROCEDURE tmatrix(Hend,Cend:real);
       TYPE      vectorL=ARRAY[0..n] OF real;
       VAR       j,k:integer;
                 RHS:real;
                 A,B,D,E,F,G,Y,Z:vectorL;
BEGIN
{matrix coordinates .....
       (n) is the number of space cells
       (p=n+1) is the number of stations between cells
       By solving the temperature field matrix at the mid point of cells we w have
       (n) unknowns for each of the hot fluid (H)
                                      the wall (W)
                             and the cold fluid (C)

             known
   H =>        0       1       2       3       4       5
          + ----- + ===== + ===== + ===== + ===== + ===== + ----- +
          :       |       |       |       |       |       |       :
   W =>   :   0   |   0   |   1   |   2   |   3   |   4   |   4   :
          :       |       |       |       |       |       |       :
          + ----- + ===== + ===== + ===== + ===== + ===== + ----- +
   C =>        0       1       2       3       4       5
                                                       known

The resulting matrix is [3*n,3*n+1] ..... }

       {thermal diffusivity of wall at cell boundaries}
       FOR j:=0 TO n DO
       kappaW[j]:=lamW[j]/(rhoW*CpW[j]);

{------|*******^*******^*******^*******^*******|---------------------------------}
```

```
{writeln('begin Tmatrix components A,B,D,E,F,G,Y,Z');}
        {coeffs for solution of hot fluid temp.field at wall stations}
        FOR j:=0 TO n DO
BEGIN   av_Vh:=(velH[j+1]+velH[j])/2;
        WmH[j]:=(mH[j+1]+mH[j])/2;
        WCpH[j]:=(CpH[j+1]+CpH[j])/2;
        A[j]:=(dT/dX)*av_Vh;
        B[j]:=(dT*Surf/(2*L)) * (alphaH[j]*av_Vh)/(WmH[j]*WCpH[j]);
END;    {writeln('A,B complete');}
        {-------------------------------------------}
        {coeffs for solution of solid wall temp.field at wall stations}
        FOR j:=0 TO n DO
BEGIN   D[j]:=kappaW[j]/(dX*dX);
        E[j]:=2/dT;
        F[j]:=(Surf/(2*Mw))*(alphaH[j]/CpW[j]);
        G[j]:=(Surf/(2*Mw))*(alphaC[j]/CpW[j]);
END;    {writeln('D,E,F,G complete');}
        {-------------------------------------------}
        {coeffs for solution of cold fluid temp.field at wall stations}
        FOR j:=0 TO n DO
BEGIN   av_Vc:=(velC[j+1]+velC[j])/2;
        WmC[j]:=(mC[j+1]+mC[j])/2;
        WCpC[j]:=(CpC[j+1]+CpC[j])/2;
        Y[j]:=(dT/dX)*av_Vc;
        Z[j]:=(dT*Surf/(2*L)) * (alphaC[j]*av_Vc)/(WmC[j]*WCpC[j]);
END;    {writeln('Y,Z complete');}

{------|*******^*******^*******^*******^*******|----------------------------------}

{clear matrix}
        writeln('clear Tmatrix')
        FOR j:=1 TO 3*n DO                       {j rows, k cols}
BEGIN   FOR k:=1 TO 3*n+1 DO
        coeff^[j,k]:=0.0;                        {left and right hand sides}
END;

{------|*******^*******^*******^*******^*******|----------------------------------}
```

```
{load matrix}
{################################################################################}

{top section}
        {writeln('top section of Tmatrix');}

                                       coeff^[     +1,     +1]:=1+A[0]+B[0];           {H[1] - a}
        FOR j:=2 TO n-1 DO coeff^[    +j   ,    +j  ]:=1+A[j-1]+B[j-1];         {H[1] to H[n-1]}
                                       coeff^[ +n     , +n    ]:=1+A[n-1]+B[n-1];         {H[n] - a}

                                   {... coeff^[     +1,     +0]:=1-A[0]+B[0]; ........... H[0] on RHS}
        FOR j:=2 TO n-1 DO coeff^[    +j   ,    +j-1]:=1-A[j-1]+B[j-1];         {H[1] to H[n-1]}
                                       coeff^[ +n     , +n  -1]:=1-A[n-1]+B[n-1];         {H[n] - b}

        {--------------------------------------------}

                                       coeff^[     +1, +n  +1]:=-2*B[0];               {W[0] - c}
        FOR j:=2 TO n-1 DO coeff^[    +j   , +n+j  ]:=-2*B[j-1];             {W[1] to W[n-1]}
                                       coeff^[ +n     ,2*n    ]:=-2*B[n-1];             {W[n] - c}

        {--------------------------------------------}

        FOR j:=0 TO n-1 DO
BEGIN   RHS:=  Th[j+1]*(1-A[j]-B[j])
             + Th[j  ]*(1+A[j]-B[j]) + (Tw[j]*(+2*B[j]));                            {known}
        coeff^[  +j+1,3*n+1]:=RHS;                                                   {rhs}
END;    {adjust first value, j=0}
        coeff^[1,3*n+1]:=coeff^[1,3*n+1]-Hend*(1-A[0]+B[0]);                          {rhs-H*(1-A+B)}

{################################################################################}

{mid-section}
        {writeln('mid-section of Tmatrix');}

                                       coeff^[ +n  +1,     +1]:=-F[0];                 {H[1] - d}
        FOR j:=2 TO n-1 DO coeff^[ +n+j   ,    +j  ]:=-F[j-1];               {H[2] to H[n-1]}
                                       coeff^[2*n     , +n    ]:=-F[n-1];               {H[n] - d}

                                   {... coeff^[ +n  +1,     +0]:=-F[0]; ................. H[0] on RHS}
        FOR j:=2 TO n-1 DO coeff^[ +n+j   ,    +j-1]:=-F[j-1];               {H[1] to H[n-2]}
                                       coeff^[2*n     , +n  -1]:=-F[n-1];               {H[n-1] - d}

        {------------------------------------}
```

```
                   {... coeff^[ +n  +1, +n    ]:=-D[0]; ................ not valid}
      FOR j:=2 TO n-1 DO coeff^[ +n+j   , +n+j-1]:=-D[j-1];                {W[0] to W[n-2]}
                         coeff^[2*n    ,2*n  -1]:=-D[n-1];                 {W[n-1] - g}

                         coeff^[ +n  +1, +n  +1]:=E[0]+D[0]+2*(F[0]+G[0]);          {W[0] - e}
      FOR j:=2 TO n-1 DO coeff^[ +n+j   , +n+j  ]:=E[j-1]+2*(D[j-1]+F[j-1]+G[j-1]); {W[1] to W[n-2]}
                         coeff^[2*n    ,2*n    ]:=E[n-1]+D[n-1]+2*(F[n-1]+G[n-1]); {W[n-1] - e}

                         coeff^[ +n  +1, +n  +2]:=-D[0];                   {W[1] - f}
      FOR j:=2 TO n-1 DO coeff^[ +n+j   , +n+j+1]:=-D[j-1];                {W[2] to W[n-2]}
                   {... coeff^[2*n    ,2*n  +1]:=-D[n-1]; .............. not valid}

      {-----------------------------------}

                         coeff^[ +n  +1,2*n  +1]:=-G[0];                   {C[0] - h}
      FOR j:=2 TO n-1 DO coeff^[ +n+j   ,2*n+j  ]:=-G[j-1];                {C[0] to C[n-2]}
                         coeff^[2*n    ,3*n    ]:=-G[n-1];                 {C[n-1] - h}

                         coeff^[ +n  +1,2*n  +2]:=-G[0];                   {C[1] - h}
      FOR j:=2 TO n-1 DO coeff^[ +n+j   ,2*n+j+1]:=-G[j-1];                {C[2] to C[n-2]}
                   {... coeff^[2*n    ,3*n  +1]:=-G[n-1]; .............. C[n] on RHS}

      {zero end gradient}
      Tw[-1]:=Tw[0];
      Tw[n]:=Tw[n-1];

      FOR j:=0 TO n-1 DO
BEGIN RHS:=  Th[j+1]*(F[j]  + Th[j]*(F[j])
           + Tw[j+1]*(D[j]  + Tw[j]*(E[j]-2*(D[j]+F[j]+G[j]+G[j])) + Tw[j-1]*(D[j])
           + Tc[j+1]*(G[j]  + Tc[j]*(G[j]);                               {known}
      coeff^[ +n+j+1,3*n+1]:=RHS;                                         {mid RHS}
END;

      {adjust  hot end, j=0}
      coeff^[n+1,3*n+1]:=coeff^[n+1,3*n+1]-Hend*(-F[0]);                  {top RHS}
      {adjust cold end, j=n-1}
      coeff^[2*n,3*n+1]:=coeff^[2*n,3*n+1]-Cend*(-G[n-1]);                {bot RHS}

{##########################################################################}
```

```
{bottom section}
         {writeln('bot section of Tmatrix');}

                             coeff^[2*n  +1, +n  +1]:=-2*Z[0];            {W[0] - j}
         FOR j:=2 TO n-1 DO coeff^[2*n+j  , +n+j  ]:=-2*Z[j-1];           {W[1] to W[n-2]}
                             coeff^[3*n    ,2*n    ]:=-2*Z[n-1];           {W[n-1] - j}

         {--------------------------------------}

                             coeff^[2*n  +1,2*n  +1]:=1+Y[0]+Z[0];        {C[0] - k}
         FOR j:=2 TO n-1 DO coeff^[2*n+j  ,2*n+j  ]:=1+Y[j-1]+Z[j-1];     {C[1] to C[n-2]}
                             coeff^[3*n    ,3*n    ]:=1+Y[n-1]+Z[n-1];     {C[n-1] - k}

                             coeff^[2*n  +1,2*n  +2]:=1-Y[0]+Z[0];        {C[1] - n}
         FOR j:=2 TO n-1 DO coeff^[2*n+j  ,2*n+j+1]:=1-Y[j-1]+Z[j-1];     {C[2] to C[n-2]}
                        {... coeff^[3*n    ,3*n  +1]:=1-Y[n-1]+Z[n-1]: ....... C[n-1] on RHS}

         FOR j:=0 TO n-1 DO
BEGIN    RHS:=  Tc[j+1]*(1+Y[j]-Z[j])
              + Tc[j  ]*(1-Y[j]-Z[j]) + Tw[j]*(+2*Z[j]);          {known}
         coeff^[2*n+j+1,3*n+1]:=RHS;
END;     {adjust last value j=n-1}
         coeff^[3*n,3*n+1]:=coeff^[3*n,3*n+1]-Cend*(1-Y[j]+Z[j]);  {cold end}

{###############################################################################}

END;     {PROCEDURE tmatrix}
{------|*******^*******^*******^*******^*******|-----------------------------------}
```

WARNING. The author give no assurance that this algorithm is correct. Potential users should check every line of the analysis before committing themselves to computational predictions.

S.6 Conclusions

In the solution approach presented, two assumptions were proposed neither of which may correctly match the actual situation of zero temperature gradient at flow exit in the balance of mass and balance of linear momentum equations. Rapidly rising or falling temperatures to the exchanger side of the flow exit must match with constant temperature levels outside the exchanger after the flow exit. The two numerical solutions proposed both assumed fictitious temperature gradients external to the exchanger.

The problem is to ensure that

- for $(x < L)$ temperatures and flow rates are functions of time and position
- for $(x \geq L)$ the temperature and flow rate gradients are zero

and this mathematical end condition is not easily modelled. The computation may thus become unstable once a flow transient reaches the end of the exchanger, or until heat transfer from the other fluid penetrates the solid wall.

The most reliable route to investigating the problem is then to study transient behaviour on a test rig such as Fig. 8.5, using a representative section of the actual exchanger.

Bibliography

Cheney, W. and **Kincaid, D.** (1985) *Numerical Mathematics and Computing*, 2nd edn, Brooks/Cole Publishing Co.

Fletcher, C.A.J. (1997) *Computational Techniques for Fluid Dynamics, Vols I and II*, 2nd edn, Springer, Berlin.

Jeffrey, A. (1989) *Mathematics for Engineers and Scientists*, 4th edn, Van Nostrand Reinhold (International).

Noble, B. (1964) *Numerical Methods: 2-Differences Integration and Differential Equations*, Oliver and Boyd, Edinburgh and London.

Press, W.H., **Flannery, B.P.**, **Teukolsky, S.A.**, and **Vetterling, W.T.** (1989) *Numerical Recipes in Pascal*, Cambridge University Press, Cambridge.

Spencer, A.J.M., **Parker, D.F.**, **Berry, D.S.**, **England, A.H.**, **Faulkner, T.R.**, **Green, W.A.**, **Holden, J.T.**, **Middleton, D.**, and **Rogers, T.G.** (1981) Engineering Mathematics, Vols. I & II, Van Nostrand Reinhold Co. Ltd.

APPENDIX C
Optimization of Rectangular Offset-Strip, Plate–Fin Surfaces

Directions in which to move

C.1 Fine-tuning of rectangular offset-strip fins

The generalized Manglik & Bergles (1990) correlations for heat transfer and flow friction allow exploration of the effect of varying surface geometries on final core size. For the same thermodynamic performance, the optimum surface geometry is sought for the following geometric parameters:

- minimum block volume (overall dimensions)
- minimum block length
- minimum frontal area
- minimum plate surface

Choice of exchanger

A two-stream compact plate–fin heat exchanger with single-cell rectangular offset-strip fin surfaces on each sides was chosen as the model. In operational mode 1, the hot fluid was made the high-pressure fluid (corresponding to a cryogenic exchanger). In operational mode 2, the hot fluid was made the low-pressure fluid (corresponding to a gas turbine recuperator). The only change in operational parameters between the two modes will be to swap the inlet pressure levels of the fluids. Pressure loss on one side of the exchanger was kept constant while the pressure loss on the other side was allowed to float and find its correct level at the design point. This search arrangement was applied to both sides of the exchanger.

In design it is best to seek coincidence of pressure loss curves on the direct-sizing plot, which makes both pressure losses controlling. Hence only performance characteristics for controlling sides are given in the figures which follow.

LMTD reduction for longitudinal conduction was not applied as interest is for trends at this time.

Exchanger specifications

Thermal parameters

A 200 kW contraflow exchanger with an effectiveness of 0.86 was chosen with hot inlet temperature $T_{h1} = 410.0\,\text{K}$ and cold inlet temperature $T_{c2} = 340.0\,\text{K}$.

Pressure levels (which were swapped to complete the investigation) were 1.1 and 6.0 bar.

One outlet temperature was forced using the effectiveness of 0.86, and a forced mean specific heat was obtained from spline-fits of physical property data. One forced mass flowrate was then found using the thermal duty $Q = 200\,\text{kW}$. Then an arbitrary mass flowrate ratio of 1.15 was selected, to produce the missing mass flowrate.

Parameters still required were an outlet temperature and mean specific heat of one fluid. The outlet temperature was iterated and an estimated mean specific heat obtained from a spline-fit until the required thermal duty of 200 kW was matched.

Surface geometries

The effects of changing fin thickness might be explored, but it was thought that the credibility of the Manglik & Bergles correlations might be pushed too far. Keeping cell width flow area constant it was found that varying high-pressure fin thickness had virtually no effect on surface area. Small low-pressure fin thickness helped minimize surface area. The result is inconclusive because the work of Kelkar & Patankar (1989) and Hesselgreaves (1993) needs further study, however, it is to be noted that thin fins also cause less longitudinal conduction.

Surface geometry was varied according to the following scheme (Table c.1).

Nominal sizes for both sides: $b = 5.00$ (mm), $c = 2.0$ (mm), $x = 6.00$ (mm)

Plate material

Plate thickness, mm	$tp = 2.00$
Fin thickness, mm	$tf = 0.15$
Density Al alloy, kg/m^3	$\rho = 2770.0$

Observations concerning all dimensional parameters stem from validity of the Manglik & Bergles correlations, and the scatter of data should be noted in Figs 4.5 and 4.6. Also, the approach of varying one parameter at a time and then selecting a combination of these to optimize against a particular requirement may find the

Table C.1 Range of geometrical parameters, variation about nominal – (one dimension at a time)

Plate spacing b (mm)	*Cell pitch c (mm)*	*Strip length x (mm)*
2	1	2
3	1.5	3
4	2	4
5	2.5	5
6	3	6
7	3.5	7
8	4	8

general area of best performance, but may miss a true optimum configuration. Automatic optimization techniques can encounter the problem of changing limits on Reynolds number validity during iteration, which may cause problems. Here a manual search was used.

C.2 Trend curves

Primary design parameters of interest are block volume, block length and block frontal area. Secondary parameters include block mass, block porosity, plate surface area and total surface area. The objective is to indicate the most profitable direction in which changes in the local geometry of rectangular offset-strip fins may be made when optimizing thermal performance of an exchanger.

The computational scheme employed covered both single- and double-cell ROSF geometries. Graphs were generated by changing the dimensions of plate spacing 'b', cell width 'c' and strip length 'x', one at a time while the other values remained at a median position. Thus the reader should not expect to find that selection of three individually-optimized parameters will lead to a fully-optimized design. To obtain a complete picture of the situation, readers should refer to the set of 20 figures presented in Smith (1997, 99). Cool *et al.* (1999) provide a complete set of search parameters using generic algorithms, but their results are presented in scatter plots.

There has been no attempt to explore the effect of varying fin thickness on exchanger performance. Although this could have been done, it might have pushed the Manglik & Bergles correlations just a little too far. However, there is no reason why such work should not be done so that results obtained can be compared with other papers directly concerned with the effect of fin thickness on exchanger performance (Xi *et al.*, 1989).

Minimization of block volume

In the study of block volume there emerged from the complete set of four figures (Smith, 1997) clear-cut evidence that cell width, c, could be minimized on both sides without affecting other parameters. This implied a minor pressure loss penalty, which could easily be accommodated through larger values in cell-heights b and strip-lengths x.

Minimization of block length

The same situation applies to block length as with block volume, if we ignore the behaviour of strip-length x.

Minimization of block mass

Here the situation is less clear, for a reduction in cell width c with a modest increase in cell-height b on one side, is coupled with an increase in cell-width c and a decrease of cell-height b on the other side. There is also an indication that a marginally higher value is required for cell-widths ($c = 1.5$ mm instead of 1.0 mm).

Minimization of frontal area

Increasing the value of length L would reduce block frontal area. There is evidence that cell-width c can be reduced on both sides, with the option of decreasing cell-height b on one side while maintaining a more or less constant b on the other side.

For all of the above options, constraints in the selection of plain rectangular surfaces may be seen in Fig. 4.11.

How to use the graphs

Select the rectangular offset-strip, plate–fin geometry that you think may be suitable, and plot the values of (b, c, x) on the graphs. Now examine slopes of the graphs and from the ordinate and abscissae scales determine the direction in which it would be beneficial to alter the original surface specification.

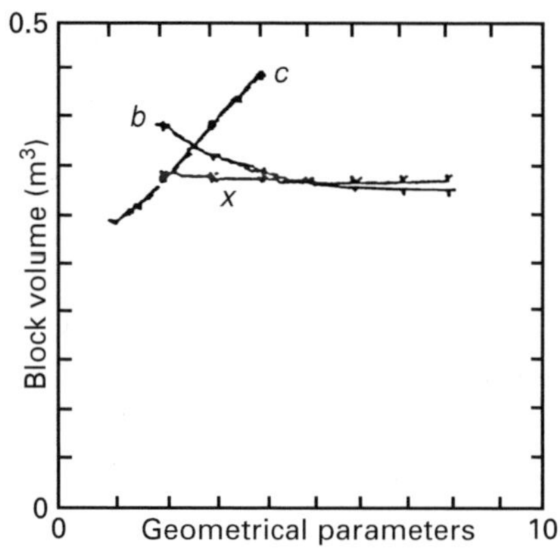

Fig.C.1 Block volume versus (b, c, x)

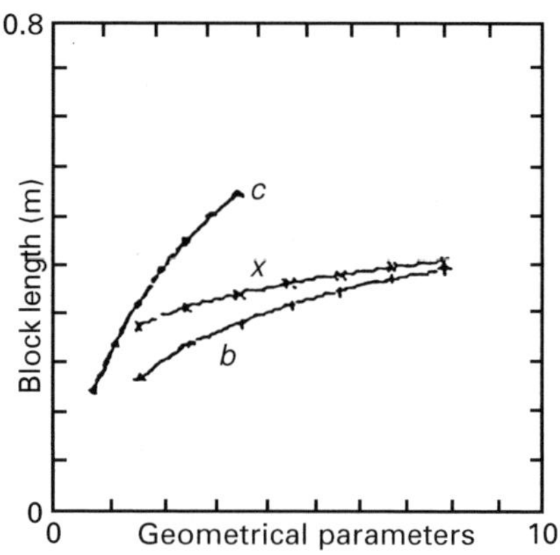

Fig.C.2 Block length versus (b, c, x)

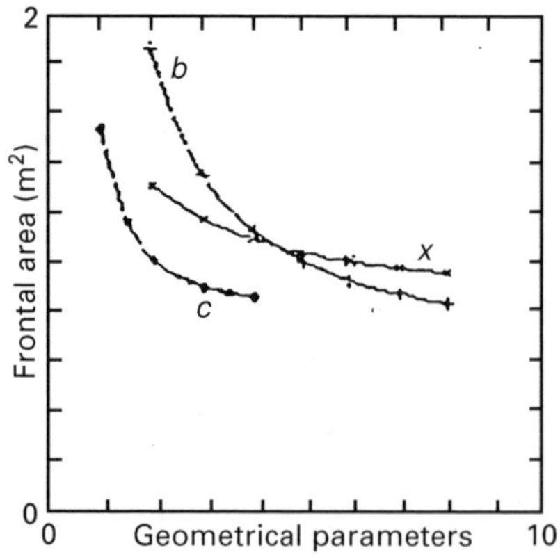

Fig.C.3 Frontal area versus (b, c, x)

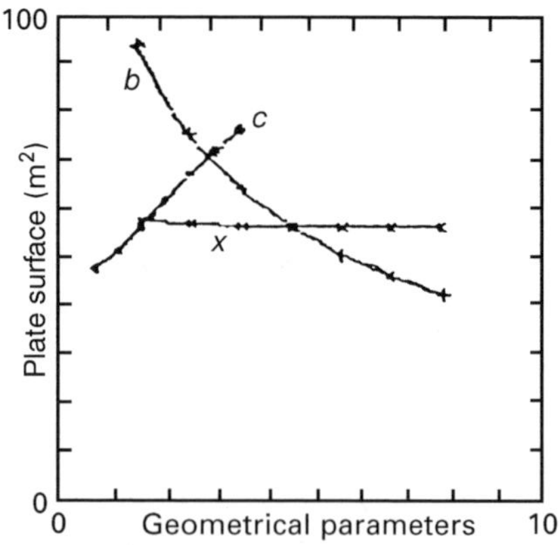

Fig.C.4 Plate surface area versus (b, c, x)

C.3 Optimization graphs

Sample trend curves (without pressure loss levels) are presented showing how block volume, block length, frontal area, and plate surface change as rectangular offset-strip-fin parameters (b, c, x) are varied.

It is somewhat unexpected that changing strip length (x) hardly affects block volume or plate surface area, although it does affect block length and frontal area. For minimum block volume large values of plate spacing (b) and small values of cell pitch (c) are appropriate. More detailed discussion of optimization of plate spacing and cell pitch is to be found in Chapter 4 and Appendix J.

Analysis of laminar flow heat transfer along a flat plate predicts infinite heat-transfer coefficients at the leading edge, and a mean value of heat transfer over the plate to be twice that calculated at the trailing edge. Further investigation of ROSF geometries might be worthwhile.

C.4 Manglik & Bergles correlations

In the notation of this text:

$$\text{Manglik \& Bergles } \alpha = \frac{\text{cell pitch}}{\text{plates pacing}} = \left(\frac{c}{b}\right)$$

$$\text{Manglik \& Bergles } \delta = \frac{\text{fin thickness}}{\text{strip length}} = \left(\frac{tf}{x}\right)$$

$$\text{Manglik \& Bergles } \gamma = \frac{\text{fin thickness}}{\text{cell pitch}} = \left(\frac{tf}{c}\right)$$

Flow friction:

$$f = 9.6243(\mathrm{Re})^{-0.7422}(\alpha)^{-0.1856}(\delta)^{0.3053}(\gamma)^{-0.2659} \times \left[1 + 7.669 \times 10^{-8}(\mathrm{Re})^{4.429}(\alpha)^{0.920}(\delta)^{3.767}(\gamma)^{0.236}\right]^{0.1}$$

Heat transfer

$$j = 0.6522(\mathrm{Re})^{-0.5403}(\alpha)^{-0.1541}(\delta)^{0.1499}(\gamma)^{-0.0678} \times \left[1 + 5.269 \times 10^{-5}(\mathrm{Re})^{1.340}(\alpha)^{0.504}(\delta)^{0.456}(\gamma)^{-1.055}\right]^{0.1}$$

where the Colburn j-factor is $j = \mathrm{St}\,\mathrm{Pr}^{2/3}$.

References

Cool, T., Stevens, A., and **Adderley, C.I.** (1999) Heat exchanger optimisation using genetic algorithms. In *6th UK National Heat Transfer Conference*, Institution of Mechanical Engineers, London.

Hesselgreaves, J. (1993) Optimising size and weight of plate-fin heat exchangers. In *Proceedings of the 1st International Conference on Heat Exchanger Technology*, Palo Alta, California, 15–17 February 1993. (Eds R.K. Shah, and A. Hashemi), Elsevier, Oxford pp. 391–399.

Kelkar, K.M. and **Patankar, S.V.** (1989) Numerical prediction of heat transfer and fluid flow in rectangular offset-fin arrays. *Int. J. Comp. Method., Part A, Applied; Numerical Heat Transfer*, **15**(2), March, 149–164. (Also, ASME Publication HTD-Vol.52, pp. 21–28.)

Manglik, R.M. and **Bergles, A.E.** (1990) The thermal hydraulic design of the rectangular offset-strip fin compact heat exchanger. *Compact Heat Exchangers – a Festschrift for A.L. London* (Eds R.K. Shah, A.D. Kraus, and D. Metzger), Hemisphere, New York, pp. 123–149.

Smith, E.M. (1997, 99) *Thermal Design of Heat Exchangers*, John Wiley & Sons, Ltd, Chichester. Reprinted with corrections 1999.

Xi, G., Suzuki, K., Hagiwara, Y., and **Murata, T.** (1989) Basic study on the heat transfer characteristics of offset fin arrays. (Effect of fin thickness on the middle range of Reynolds number.) *Trans. Japan Soc. Mech. Engrs, Part B*, **55**(519), November, 3507–3513.

Bibliography

Manglik, R.M. and **Bergles, A.E.** (1995) Heat transfer and pressure drop correlations for the rectanglar offset strip-fin compact heat exchanger. *Exp. Thermal Fluid Sci.* **10**, 171–180.

APPENDIX D
Performance Data for RODbaffle Exchangers

Extra correlations

D.1 Further heat-transfer and flow-friction data

Towards completion of this text, the writer received a number of experimental data-sets for the RODbaffle geometries from Dr C.C. Gentry of the Phillips Petroleum Company, Oklahoma. It seemed useful to plot these for comparison, and to generate a set of smoothed data for the geometry 02WARA, the geometry of which is not identical to that used in Chapter 7.

The RODbaffle codes (e.g. 02WARA) do not refer to dimensions of the RODbaffle geometry. The first two digits (e.g. 02, 03, 04) denote the specific test sequence. The letter symbols (O, W) denote the test fluid, O for oil and W for water. The final three letters identify baffle ring geometry.

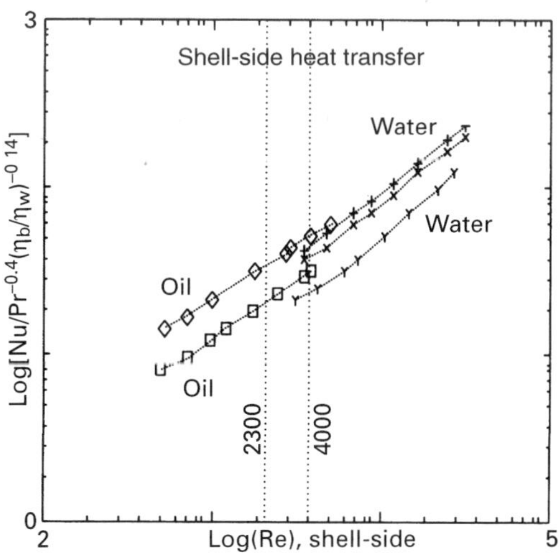

Fig.D.1 Heat-transfer correlations for RODbaffle geometries (experimental data, courtesy of C.C. Gentry, Phillips Petroleum Company). Light oil: 02OARA (◇), 04OARE (□). Water: 02WARA (+), 03WARB (×), 02WARE (Y)

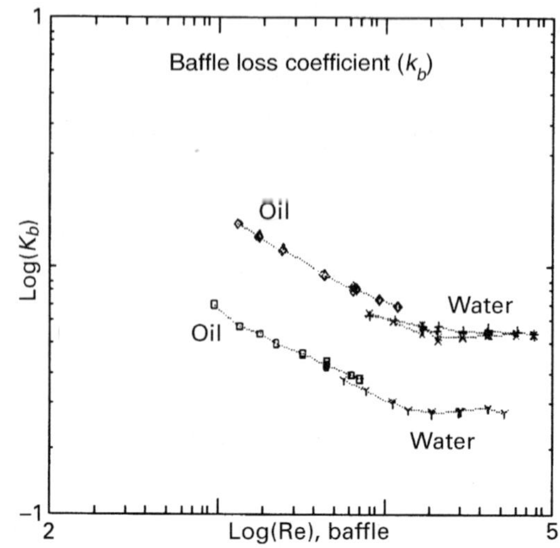

Fig.D.2 Baffle loss coefficient for RODbaffle geometries (experimental data, courtesy of C.C. Gentry, Phillips Petroleum Company). Light oil: 02OARA (◇), 04OARE (□). Water: 02WARA (+), 03WARB (×), 02WARE (Y)

Figures D.1 and D.2 correspond to Figs 7.2 and 7.4 of Chapter 7. Curves at lower Reynolds numbers with open symbols are for oil, while curves at higher Reynolds numbers are for water. While the curves suggest the possibility of a unified correlation for shell-side heat-transfer and baffle loss coefficients which might be useful in optimization (c.f. Manglik & Bergles, 1990, Chapter 4), it is evident from the consistency of individual datasets that better designs would always result when individual correlations are used, as recommended for plate–fin designs by Kays & London (see Chapter 4).

It is usually known in advance as to whether the shell-side fluid is to be water or oil, and universal correlations may perhaps be more easily sought for correlations generated using the same fluid.

Table D.1 Geometries for RODbaffle exchangers (courtesy of C.C. Gentry)

Bundle geometry	*Tube o.d., d (mm)*	*Tube pitch, p (mm)*	*Baffle spacing, L_b (mm)*	*p/d*	*L_b/d*
Chapter 7	38.10	44.45	150	1.1666	3.937
02WARA	12.70	17.4625	124.46	1.375	9.80
02OARA	12.70	17.4625	124.46	1.375	9.80
03WARB	12.70	17.4625	248.92	1.375	19.60
02WARE	15.875	19.050	76.2	1.200	4.80
04OARE	15.875	19.050	76.2	1.200	4.80

Table D.2 Shell-side heat transfer for 02WARA (cubic spline-fit smoothed data)

Reynolds no., shell-side	$\dfrac{Nu}{Pr^{0.4}(\eta_b/\eta_w)^{0.14}}$
30 580	232.207
30 000	228.041
25 000	193.808
20 000	161.215
15 000	128.626
12 000	107.946
10 000	93.109
8000	78.180
6000	63.275
5000	55.426
4000	47.309
3500	43.153
3292	41.408

Table D.1 provides a comparison of the ARA geometry used in Chapter 7 with the additional five sets of data provided separately by Gentry. The geometries are quite different.

Tables D.2 and D.3 are smoothed datasets for configuration 02WARA. Two tables with differing Reynolds numbers are provided because:

1. Regular values of shell-side Reynolds number are useful in setting up an interpolation scheme for the group containing Nu, Pr, and η_b/η_w.

Table D.3 Baffle loss coefficient for 02WARA (cubic spline-fit smoothed data)

Reynolds no., baffle flow	*Baffle loss coeff. (k_b)*
77 936	0.54795
60 000	0.55479
50 000	0.55697
40 000	0.55554
30 000	0.55904
25 000	0.55904
20 000	0.57113
15 000	0.59493
12 000	0.61511
10 000	0.62995
8391	0.64224

2. Regular values of baffle flow Reynolds number are useful in setting up an interpolation scheme for baffle loss coefficient (k_b).

A relationship between these two Reynolds numbers exists for the test data, but as this depends on geometry, mass flowrate, and thermodynamic conditions it was not set up in Tables D.2 and D.3.

D.2 Baffle-ring by-pass

Shell-side by-pass flow degrades exchanger performance. In the RODbaffle exchanger it should be possible to make a reasonable estimate of the by-pass mass flowrate, and thus improve the calculation of exchanger performance. The pressure losses are specified for the RODbaffle bundle, and the shell-side loss must be the same for by-pass flow. Knowing the number of baffles, the approximate pressure loss across a single baffle may be calculated.

Bell & Berglin (1957) researched a method for calculating by-pass mass flowrate for both 'concentric' and 'tangential' baffles.

When a baffle is concentric with the shell of the exchanger, the by-pass flow is named 'concentric'. When a baffle touches the shell at one point, the by-pass flow is named 'tangential'. In practice many by-pass flow situations should lie between these two limiting cases. The tangential case produces the greatest by-pass flow.

In its simplest form, three equations would be used to calculate by-pass mass flowrate, viz.

By-pass flow area shell i.d. $= D$, baffle-ring o.d. $= d$

$$A = \frac{\pi}{4}(D^2 - d^2) \tag{D.1}$$

By-pass Reynolds number ($G = \dot{m}/A$)

$$\mathrm{Re} = \frac{(D - d)G}{\eta} \tag{D.2}$$

By-pass pressure loss

$$\Delta p = \frac{1}{2\rho}\left(\frac{\dot{m}}{CA}\right)^2 \tag{D.3}$$

The actual by-pass pressure loss is found by dividing the shell-side tube-bundle pressure loss by the number of baffles (n), viz.

$$\Delta p = \frac{\Delta p_s}{n} = \frac{11305.46}{76} = 148.76\,\mathrm{N/m^2}$$

This is the value to be matched.

For a 'concentric' baffle, the mass flowrate is first guessed to obtain a Reynolds number. The value of the by-pass coefficient (C) is obtained from an experimentally determined plot of $C = f(\mathrm{Re})$, and the corresponding pressure loss is evaluated from equation (D.3). Iteration can be used until the actual and calculated values of by-pass pressure loss are the same, although solution by plotting a curve of guessed $\dot{m}$ versus calculated Δp is safer.

For a 'tangential' baffle, the process is a little more complicated. The exchanger shell is divided into suitable small segments, such that each segment may be considered as part of a 'concentric' baffle arrangement with by-pass coefficient C' and flow area ΔA. The 'tangential' coefficient (C) is obtained from the relationship

$$C = \frac{\sum (C'\Delta A)}{A} \tag{D.4}$$

Recognizing the possible existence of laminar, transitional, and turbulent flow in the by-pass, Bell & Bergelin realized that more detailed allowances may have to be made to cover such items as

- prior existence of a developed boundary layer on the exchanger shell wall between baffles
- re-creation of boundary layer on baffle ring
- flow acceleration nearing entry
- flow contraction (and possible existence of a vena contracta, or of a flow recirculation cell)
- flow friction in a short duct
- dissipation of kinetic energy loss on expansion from the duct
- partial mixing of leakage flow with main shell-side flow between baffles

For thin sharp-edged baffles, a plot of kinetic energy loss parameter $K = f(\mathrm{Z}/\mathrm{Re})$ was used to estimate kinetic energy losses where,

By-pass length-to-width ratio

$$Z = \frac{\text{baffle thickness}}{\text{mean radial gap}} = \frac{2L}{(D-d)} \tag{D.5}$$

For thick square-edged baffles, a friction allowance was introduced. For thick round-edged entry baffles, both friction and kinetic energy allowances were made. Gentry (1990) provides dimensions for RODbaffle baffle rings and for longitudinal slide bars.

Experimental and practical geometries

The internal diameter of the test shell was 133.45 mm, with mean baffle clearance gaps in the range 0.6900–2.915 mm. Baffle thickness (L) lay in the range 1.460–6.4500 mm with one exceptional value at 22.60 mm. Values for the dimensionless geometrical parameter Z lie in the range 0.1179–9.6209 with one exceptional value at 33.272. Only a single baffle was used in testing, which may not be fully representative of actual conditions.

The industrial exchanger of Chapter 7 has an internal shell diameter of 1217.0 mm, and a mean baffle-ring clearance gap of 3.0 mm. Baffle-ring thickness

(L) may lie in the range 10–50 mm, and would be around $2.5 \times d_r = 15$ mm in this case. The corresponding value of the dimensionless geometrical parameter Z would be 0.6. Some 76 baffles with a spacing of 150 mm are used.

Calculation of by-pass flow

Examination of TEMA (1988) recommendations for clearance between shell inside diameter (D) and baffle-ring outside diameter (d) showed that the expression

$$1 - \left(\frac{d}{D}\right)^2 = 0.01 \tag{D.6}$$

held for shell inside diameters (D) greater than 1000 mm.

Below $D = 1000$ mm the constant 0.01 increased progressively to about 0.04, as radial gaps reduced progressively from 2.5 mm to a lower limit of 1.5 mm.

Assuming a thick square-edged concentric baffle-ring for the exchanger of Chapter 7,

shell inside diameter, m	$D = 1.217$
baffle ring outside diameter, m	$d = 1.210$
by-pass flow area, m^2	$A = \pi/4(D^2 - d^2) = 0.007\,845$
pressure loss per baffle, N/m^2	$\Delta_p = 148.76$
absolute viscosity, J/(m s K)	$\eta = 0.000\,0245$
fluid density, kg/m^3	$\rho = 4.1026$

Apply equations (D.2) and (D.3), and guess a by-pass mass flowrate of 0.15 kg/s. Using equation (D.2)

$$\text{Re} = \frac{(D-d)}{\eta}\left(\frac{\dot{m}}{A}\right) = \frac{0.007}{0.000\,0245}\left(\frac{0.15}{0.007\,845}\right) = 5463.0$$

Using the graph published by Bell & Berglin (1957), $C = f(\text{Re}) = 0.56$. For computer calculation an interpolating spline-fit of this relationship would be preferable.

Using equation (D.3)

$$\Delta p = \frac{1}{2\rho}\left(\frac{\dot{m}}{CA}\right)^2 = \frac{1}{2 \times 4.1026}\left(\frac{0.15}{0.56 \times 0.007\,845}\right)^2 = 142.08\ \text{N/m}^2$$

This is close enough to the required value of 148.76 N/m^2. A more accurate result can be obtained by programming the calculation.

Bell and Berglin provide further corrections to be made when calculating the loss coefficient (C), and study of the published papers listed in Chapter 7 is recommended. In the light of improved experimental and computational methods there might be a case for re-examining the problem to model exactly what is happening in by-pass flow. With the presently available results an immediate advance can be achieved by using interpolative spline-fitting on empirical relationships.

Once the by-pass flowrate is found, the whole exchanger has to be sized again, because the shell-side mass flow was initially assumed to be the total mass flowrate. The new shell-side flowrate will be the total shell-side flowrate minus the by-pass flowrate.

The final shell-side outlet temperature is a result of mixing by-pass flow and shell-side core flow over the tube bundle. A simple enthalpy balance is made at shell core outlet after sizing calculations and by-pass flowrate calculations are complete, viz

$$(\dot{m}_{shellcore} + \dot{m}_{by\text{-}pass})h_{mix} = \dot{m}_{shellcore}h_{shellcore} + \dot{m}_{by\text{-}pass}h_{by\text{-}pass} \tag{D.7}$$

References

Bell, K.J. and **Bergelin, O.P.** (1957) Flow through annular orifices. *Trans. ASME*, **79**, 593–601.

Gentry, C.C. (1990) RODbaffle heat exchanger technology. *Chem. Engng Prog.*, July, 48–57.

Manglik, R.M. and **Bergles, A.E.** (1990) The thermal hydraulic design of the rectangular offset strip-fin compact heat exchanger. *Compact Heat Exchangers – A festschrift for A.L. London*. (Eds R.K. Shah, A.D. Kraus, and D. Metzger), Hemisphere, New York.

TEMA (1988) Standard of Tubular Exchanger Manufacturers' Association, 7th edn, TEMA, Tarrytown, New York.

Bibliography

Bell, K.J. (1955) Annular orifice coefficients with application to heat exchanger design. PhD thesis, Department of Chemical Engineering, University of Delaware, Newark, Delaware.

Appendix E
Proving the Single-Blow Test Method – Theory and Experimentation

The analytical approach

E.1 Analytical approach using Laplace transforms

The required inverse Laplace transforms

$$L^{-1}\left\{\exp\left(\frac{n}{s-a}\right)\right\} = \delta(t) + e^{at}\frac{nI_1(2\sqrt{nt})}{\sqrt{nt}}$$

$$L^{-1}\left\{\frac{1}{s-a}\exp\left(\frac{n}{s-a}\right)\right\} = e^{at}I_0(2\sqrt{nt})$$

may be obtained by series expansion and term-by-term inversion. While deriving these inversions it was considered that a gap existed in published tables of inverse transforms.[1] Tables E.1, E.2, and E.3 provide a sequence of inversions in which those of interest above are to be found. I_0 and I_1 are modified Bessel functions.

Table E.1 Laplace transforms – elementary

Transform $\hat{f}(s)$	*Inversion* $f(t)$
1	$\delta(t)$
$1/s$	1
$1/(s-a)$	$\exp(at)$
$1/s^2$	t
$1/[s(s-a)]$	$(1/a)[\exp(at)-1]$

[1] Dr Jeffrey Lewins, in later private correspondence, referred the author to some inversions in Carlslaw & Jaeger (1948, 2nd edn) which the author had not seen.

Table E.2 Laplace transforms involving exp(n/s)

Transform $\hat{f}(s)$	*Inversion* $f(t)$
$\exp\left(\frac{n}{s}\right)$	$\delta(t)+\frac{nI_1(2\sqrt{nt})}{\sqrt{nt}}$
$\frac{1}{s}\exp\left(\frac{n}{s}\right)$	$I_0(2\sqrt{nt})$
$\frac{1}{s-a}\exp\left(\frac{n}{s}\right)$	$I_0(2\sqrt{nt})+a\int_0^t e^{-a\sigma}I_0(2\sqrt{n\sigma})d\sigma$
$\frac{1}{s^2}\exp\left(\frac{n}{s}\right)$	$\int_0^t I_0(2\sqrt{n\sigma})d\sigma$
$\frac{1}{s(s-a)}\exp\left(\frac{n}{s}\right)$	$e^{at}\int_0^t e^{-at}I_0(2\sqrt{n\sigma})d\sigma$

Table E.3 Laplace transforms involving exp[$n/(s-a)$]

Transform $\hat{f}(s)$	*Inversion* $f(t)$
$\exp\left(\frac{n}{s-a}\right)$	$\delta(t)+e^{at}\frac{nI_1(2\sqrt{nt})}{\sqrt{nt}}$
$\frac{1}{s}\exp\left(\frac{n}{s-a}\right)$	$1+\int_0^t e^{a\sigma}\frac{nI_1(2\sqrt{n\sigma})}{\sqrt{n\sigma}}d\sigma$
$\frac{1}{s-a}\exp\left(\frac{n}{s-a}\right)$	$e^{at}I_0(2\sqrt{nt})$
$\frac{1}{s^2}\exp\left(\frac{n}{s-a}\right)$	$t-te^{at}+e^{at}\int_0^t I_0(2\sqrt{n\sigma})d\sigma$
$\frac{1}{s(s-a)}\exp\left(\frac{n}{s-a}\right)$	$\int_0^t e^{a\sigma}I_0(2\sqrt{n\sigma})d\sigma$

E.2 Numerical evaluation of Laplace outlet response

The following procedure minimizes the computational requirement. Assume the inlet disturbance D to be exponential in form (see Fig. E.1), corresponding closely in shape to that obtained from a fast-response electrical heater. Then

$$D(0,t)=1-\exp\left(\frac{-\tau}{\tau^*}\right)$$

with non-dimensional time

$$\tau=\frac{Ntu}{W}\frac{(ut-x)}{L}$$

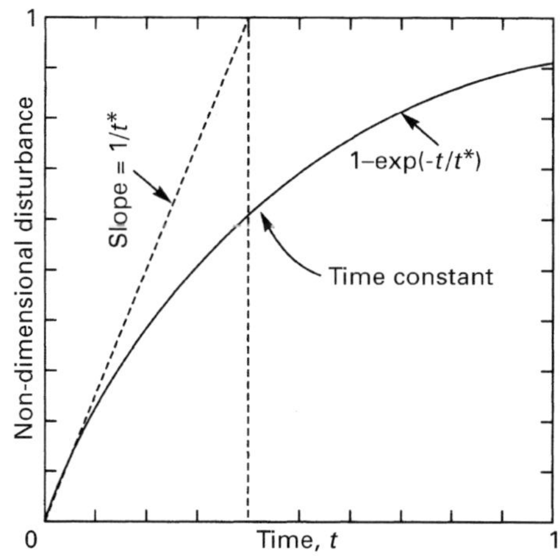

Fig.E.1 Non-dimensional disturbance and time constant

non-dimensional time constant

$$\tau^* = \frac{Ntu}{W}\frac{ut^*}{L} \qquad \text{at } x = 0$$

The outlet fluid temperature response then becomes

$$G^{\#} = \exp(-n)\left[1 - \exp\left(-\frac{\tau}{\tau^*}\right) + \int_0^{\tau}\left\{1 - \exp\left(-\frac{\tau-\sigma}{\tau^*}\right)\right\}R(\sigma)d\sigma\right]$$

$$= \exp(-n)\left[1 - \exp\left(-\frac{\tau}{\tau^*}\right) + \int_0^{\tau} R(\sigma)d\sigma - \exp\left(-\frac{\tau}{\tau^*}\right)\int_0^{\tau}\exp\left(\frac{\sigma}{\tau^*}\right)R(\sigma)d\sigma\right]$$

The expected response is of the form shown in Fig. E.2.

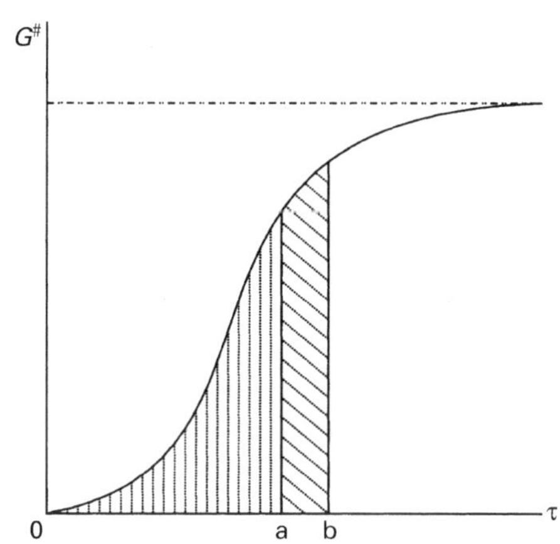

Fig.E.2 Outlet temperature response

Suppose the value of each integral

$$\int_0^{\tau} (\ )d\sigma$$

is known up to $\tau - a$, then to continue evaluation of the $G^{\#}$ τ curve the increment (*cross-hatched area*) is required to continue the summation.

The two integrals to be evaluated are

$$\int_0^{\tau} R(\sigma)d\sigma \qquad \text{and} \qquad \int_0^{\tau} \exp\left(\frac{\sigma}{\tau^*}\right)R(\sigma)d\sigma$$

Let us consider evaluation of the first of these between limits $\tau = a$ and $\tau = b$

$$\int_a^b R(\sigma)d\sigma = \int_a^b e^{-\sigma}\frac{nI_1(2\sqrt{n\sigma})}{(2\sqrt{n\sigma})}d\sigma$$

To avoid difficulties in the denominator when $a = 0$, we change the variable. Putting $(\sigma = n\alpha^2, d\sigma = 2n\alpha \cdot d\alpha)$ the integral becomes

$$\int_{\alpha=\sqrt{a/n}}^{\alpha=\sqrt{b/n}} e^{-n\alpha^2}\frac{nI_1(2n\alpha)}{n\alpha}2n\alpha d\alpha = 2n\int_{\sqrt{a/n}}^{\sqrt{b/n}} e^{-n\alpha^2}I_1(2n\alpha)d\alpha$$

At a new value of $\tau = b$, i.e. $\alpha = \sqrt{b/n}$, the new value of the integral is given by

New value = Old value + Increment

Each *increment* of integral may be evaluated using Legendre polynomials in four-point Gaussian quadrature

$$\int_a^b y d\alpha = \frac{1}{2}(b-a)\int_{-1}^{+1} y dA$$

$$= \frac{1}{2}(b-a)[w_1 y(A_1) + w_2 y(A_2) + w_3 y(A_3) + w_4 y(A_4)]$$

where A is abscissae value, and w is weighting value, given in Table E.4 for four-point Gaussian quadrature described in the paper by Lowan *et al.* (1954). Values of the modified Bessel function, $I_1(2n\alpha) \equiv y(A)$ are computed using an algorithm given by Clenshaw (1962).

In present computations a top limit of *Ntu* around 75.0 was obtained before machine overflow occurred within the program. Curves for values of *Ntu* up to 500.0 have been obtained by Furnas (1930) using graphical methods. In testing it is seldom that values exceeding 20.0 will be encountered, while in real cryogenic practice values of *Ntu* over 40.0 may be encountered.

Table E.4 Gaussian four-point quadrature

Position	*Abscissae*	*Weighting*
1	−0.861 136 311 594 053	0.347 854 845 137 454
2	−0.339 981 043 584 856	0.652 145 154 862 546
3	+0.339 981 043 584 856	0.652 145 154 862 546
4	+0.861 136 311 594 053	0.347 854 845 137 454

E.3 Experimental test equipment

Detailed descriptions of a precision single-blow test-rig are to be found in the theses of Coombs (1970) and of King (1976). A shorter description can also be found in the paper by Smith & Coombs (1972).

Although this test-rig was used for evaluation of the thermal performance of tube bundles only, its design and construction and its instrumentation were state of the art at that time. The identical hardware could be used today, but with improved data logging and computational equipment.

The once-through open tunnel had a flared inlet and contraction with honeycomb flow straightener leading to a 150 mm × 150 mm square duct, based on a UK National Physics Laboratory design by Cheers (1945). After velocity profile flattening by wire mesh, the air passed over two electrical heaters – the first was used to adjust for variation in ambient temperatures during the extensive test programmes, and the second was used to generate a rapid exponential increase in air temperature for testing. The rise was restricted to about 6 K, which with an ambient absolute temperature of around 300 K meant that flow velocities and densities would remain within ± 1 per cent of mean temperature.

The fast-response in-plane heaters were constructed of 0.1 mm nichrome wire coils, supported on hollow elliptical alumina insulators (1 mm × 2.5 mm), each insulator being arranged so that its major axis was parallel to the flow stream. The coils were thus virtually free in the air stream, having point contact with the ellipse only at leading and trailing edges.

The inlet temperature disturbance could be tuned. The fast-response heater was controlled by thyristor, so that higher input power could be adjusted over the first 10 cycles of 50 Hz supply to allow for thermal storage requirements of the heater wire and the supporting ceramic insulators, to the point where close approximations to exponential inputs were produced. This is preferable to assuming step change disturbances that are physically impracticable.

Following the fast-response heater, a square inlet section, square test section and square outlet pressure recovery section were constructed from smooth tufnol sheet to minimize thermal storage effects. The pressure recovery section had a number of longitudinal tapping points so that the point of maximum pressure recovery from the test exchanger core could be determined. Each tapping point has to have a small enough diameter so as not to disturb the flow pattern, but the flexible tubing connecting tapping points to the manometers needs to be large enough so as not to dampen response.

Beyond the tufnol sections there was a sheet steel transition section from square to circular section leading to an orifice plate for flow measurement to British Standard 1042 : 1943. This also incorporated thermocouples for temperature measurement. The suction compressor was placed at exit from the orifice plate pipework.

Velocity and temperature profiles were taken in front of the test section at right angles to prove flatness. These probes were removed before thermal testing commenced.

Inlet temperature disturbance and outlet temperature response measurements were made by in-plane platinum resistance thermometers consisting of 0.025 mm bare wire strung in zig-zag arrangement across the duct. Each response was measured by Kelvin double-resistance bridge units, designed to compensate for lead resistances, and to balance automatically.

Test results showed a variation under ± 10 per cent over the complete laminar and turbulent test regions explored. One particular geometry tested produced an unusual result for pressure drop only, in that on slowly increasing the mass flowrate the transition to turbulence was at the upper end of transition, and on slowly decreasing the flowrate the transition to laminar flow was at the lower end of transition. This 'hysteresis loop' was considered to show the quality of flow stability achieved within the test-rig. However, heat-transfer testing wiped out the hysteresis loop completely.

Before constructing any single-blow testing facility, it is strongly recommended that the reader consult as many sources as possible before deciding on the features of his/her test-rig. More references are to be found at the end of Chapter 10.

References

Carlslaw, H.S. and **Jaeger, J.C.** (1948) *Operational Methods in Applied Mathematics*, 2nd edn, Oxford University Press, Oxford.

Cheers, F. (1945) Note on wind tunnel contractions. Aeronautical Research Council, Report & Memorandum, No. 2137.

Clenshaw, C.W. (1962) Chebyshev Series for Mathematical Functions. *National Physical Laboratory, Mathematical Tables*, vol. 5, HMSO.

Coombs, B.P. (1970) A transient technique for evaluating the thermal performance of cross-inclined tube bundles. PhD thesis, University of Newcastle upon Tyne.

Furnas, C.C. (1930) Heat transfer from a gas stream to a bed of broken solids – II. *Ind. Engng Chemistry, Industrial edn.*, **22**(7), 721–731.

King, J.L. (1976) Local and overall thermal characterisitics of tube banks in cross flow. PhD thesis, University of Newcastle upon Tyne.

Lowan, A.N., Davids, N., and **Levinson, A.** (1954) Table of the zeros of the Legendre polynomials of order 1-16 and the weight coefficients for Gauss mechanical quadrature formula. *Tables of Functions and Zeros of Functions, NBS Applied Mathematical Series*, No. 37, pp. 185–189.

Smith, E.M. and **Coombs, B.P.** (1972) Thermal performance of cross-inclinded tube bundles measured by a transient method. *J. Mech. Engng Sci.*, **14**(3), 205–220.

Appendix F
Most Efficient Temperature Difference in Contraflow

Formal mathematics

F.1 Calculus of variations

A clear exposition of the theory for the *calculus of variations* is given in Hildebrand (1976). Other texts are those by Courant & Hilbert (1989), Mathews & Walker (1970), and Rektorys (1969).

The basic problem concerns a function

$$F\left(y, \frac{dy}{dx}, x\right)$$

and the finding of a maximum or minimum of the integral of this function

$$J[y] = \int_{x_0}^{x_1} F\left(y, \frac{dy}{dx}, x\right) dx$$

where end values $x_0, x_1, y(x_0), y(x_1)$ are known. Conditions concerning continuity of functions and of their derivatives are covered in the reference texts, and the required solution reduces to solving the Euler equation

$$\frac{\partial F}{\partial y} - \frac{d}{dx}\left(\frac{\partial F}{\partial y'}\right) = 0$$

Euler equation

Generalization

The problem can be extended to include a constraint in minimization or maximization of the integral

$$J[y] = \int_{x_0}^{x_1} F(y, y', x) dx \tag{F.1}$$

where $y(x)$ is to satisfy the prescribed end conditions

$$\left.\begin{aligned} y(x_0) &= C \\ y(x_1) &= D \end{aligned}\right\} \tag{F.2}$$

as before, but a constraint condition is also imposed in the form

$$\int_{x_0}^{x_1} G(y, y', x)dx = K \tag{F.3}$$

where K is a prescribed constant, then the appropriate Euler equation is found to be the result of replacing F in equation (F.1) by the auxiliary function

$$H = F + \lambda \cdot G \tag{F.4}$$

where λ is an unknown constant. This constant, which is in the nature of a Lagrange multiplier, will generally appear in the Euler equation and in its solution, and is to be determined together with the two constants of integration in such a way that all three conditions are satisfied.

F.2 Optimum temperature profiles

From definition of LMTD (Chapter 2, Section 2.4)

$$N_h - N_c = \ln\left(\frac{\Delta\theta_1}{\Delta\theta_2}\right)$$

$$\exp(N_h - N_c) = \left(\frac{\Delta\theta_1}{\Delta\theta_2}\right)$$

From optimum contraflow exchanger (Chapter 2, Section 2.12)

$$\frac{\Delta\theta}{T_h} = \frac{a}{1+a} \qquad \frac{\Delta\theta}{T_c} = \frac{a}{1}$$

$$\frac{\Delta\theta_1}{\Delta\theta_2} = \frac{aT_{c1}}{aT_{c2}} = \frac{T_{c1}}{T_{c2}} = \frac{T_{h1}}{T_{h2}}$$

From general contraflow temperature profiles [Chapter 3, Section 3.2, equation (3.8)]

$$\frac{T_{h1} - T_h}{T_{h1} - T_{h2}} = \frac{1 - \exp\left[-(N_h - N_c)\dfrac{x}{L}\right]}{1 - \exp[-(N_h - N_c)]} = \frac{T_{c1} - T_c}{T_{c1} - T_{c2}}$$

$$= \frac{1 - \dfrac{1}{\exp\left[\dfrac{x}{L}\{N_h - N_c\}\right]}}{1 - \dfrac{1}{\exp(N_h - N_c)}} = \frac{1 - \dfrac{1}{\exp\left[\dfrac{x}{L}\{\ln(\Delta\theta_1/\Delta\theta_2)\}\right]}}{1 - \dfrac{1}{(\Delta\theta_1/\Delta\theta_2)}}$$

$$= \frac{1 - \dfrac{1}{\exp\left[\ln(T_{h1}/T_{h2})^{\frac{x}{L}}\right]}}{1 - \dfrac{1}{(T_{h1}/T_{h2})}} = \frac{1 - \dfrac{1}{(T_{h1}/T_{h2})^{\frac{x}{L}}}}{1 - \dfrac{1}{(T_{h1}/T_{h2})}}$$

$$= \frac{1 - (T_{h2}/T_{h1})^{\frac{x}{L}}}{1 - (T_{h2}/T_{h1})} = \frac{1 - (T_{c2}/T_{c1})^{\frac{x}{L}}}{1 - (T_{c2}/T_{c1})}$$

$$= \frac{T_{h1} - T_h}{T_{h1} - T_{h2}} = \frac{T_{h1} - T_{h1}\left(\frac{T_{h2}}{T_{h1}}\right)^{\frac{x}{L}}}{T_{h1} - T_{h2}} = \frac{T_{c1} - T_{c1}\left(\frac{T_{c2}}{T_{c1}}\right)^{\frac{x}{L}}}{T_{c1} - T_{c2}} = \frac{T_{c1} - T_c}{T_{c1} - T_{c2}}$$

Hot fluid profile	Cold fluid profile
$\frac{T_h}{T_{h1}} = \left(\frac{T_{h2}}{T_{h1}}\right)^{\frac{x}{L}}$	$\frac{T_c}{T_{c1}} = \left(\frac{T_{c2}}{T_{c1}}\right)^{\frac{x}{L}}$

The log mean temperature difference for these profiles depends on choice of the value for constant a.

References

Courant, R. and **Hilbert, D.** (1989) *Mathematical Methods of Physics*, vol. I, John Wiley, p. 184.

Hildebrand, F.B. (1976) *Advanced Calculus for Applications*, 2nd edn, Prentice Hall, New Jersey, p. 360.

Mathews, J. and **Walker, R.L.** (1970) *Mathematical Methods of Physics*, 2nd edn, Addison-Wesley, p. 322 (based on course given by R.P. Feynman at Cornell).

Rektorys, K. (Ed.) (1969) *Survey of Applicable Mathematics*, MIT Press, Cambridge, Massachusetts, p. 1020.

APPENDIX G
Physical Properties of Materials and Fluids

Where to find and how to fit data

G.1 Sources of data

Over the years the author encountered many delays in attempting to source information on the physical properties of materials of construction. The data are scattered, and are often presented in units not generally used by engineers. Some data need conversion to appropriate engineering SI units, viz. J/(kg K) for specific heat, J/(m s K) for thermal conductivity and m^2/s for thermal diffusivity. Useful conversion factors are listed in Appendix M. Density in kg/m^3 can be obtained from thermal diffusivity.

G.2 Fluids

Particularly near the critical points of fluids, property values tend to change significantly with both temperature and pressure – this behaviour being instanced in later

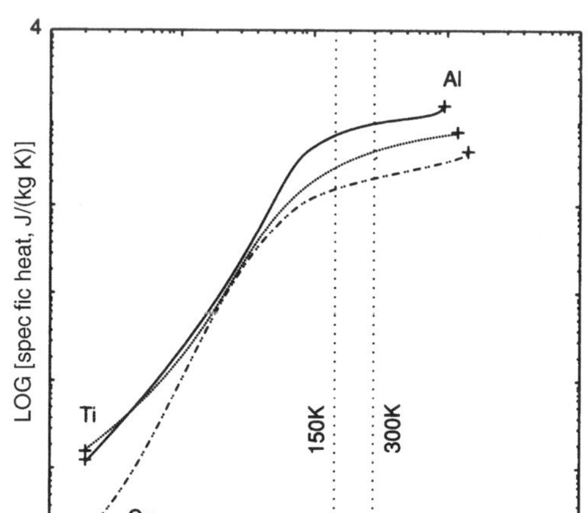

Fig.G.1 Specific heat of aluminium, copper, and titanium, J/(kg K)

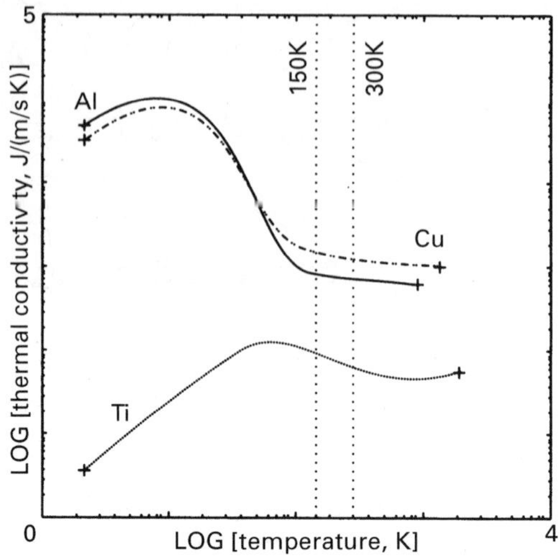

Fig.G.2 Thermal conductivity of aluminium, copper, and titanium, J/(m s K)

examples of steam tables, e.g. the UK Steam Tables in SI Units (1975). For other fluids the reader may wish to consult Vargaftik (1983), Touloukian *et al.* (1970), and the IUPAC Series of which the representative volume on oxygen (Wagner & de Reuck, 1987) is listed below. Other references can be obtained by consulting the *Journal of Physical and Chemical Reference Data* (ACS), a recent issue of the *Chemical Engineers Handbook*, or by seeking information from the

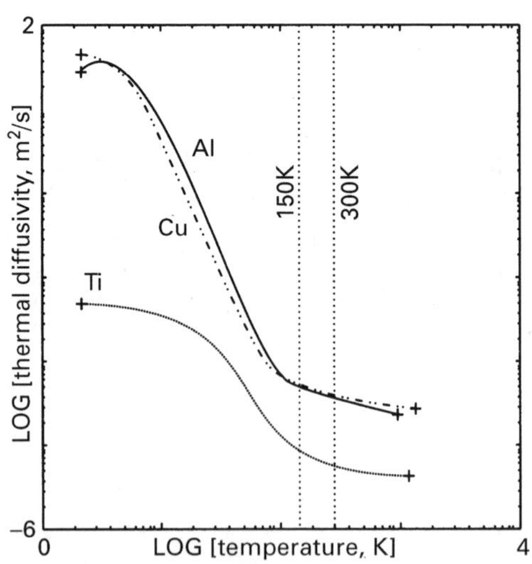

Fig.G.3 Thermal diffusivity of aluminium, copper, and titanium, m^2/s

manufacturers of working fluids, e.g. the *KLEA Refrigerants* from ICI Chemicals & Polymers Division.

G.3 Solids

For aluminium, copper, and titanium the properties of specific heat, thermal conductivity, and thermal diffusivity are presented so that the engineer may see what kind of behaviour exists. These curves are not necessarily typical for other solids and the series of volumes on *Thermophysical Properties of Matter* by Touloukian and others (1970) should be consulted.

One point of including these three graphs is to encourage the use of interpolating cubic spline-fits to fit data. In particular it can be time-saving to fit the complete set of data available, even though the current design requirement needs data only over a limited range. This avoids extra work involved in re-fitting data for another range.

References

American Chemical Society (1971 to date) *J. Phys. Chem. Reference Data.*

IUPAC Thermodynamic Tables Properties Centre, Physical Properties Data Service, Department of Chemical Engineering and Chemical Technology, Imperial College of Science, Technology and Medicine, Prince Consort Road, London SW7 2BY.

Touloukian, Y.S. *et al.* (1970 onwards) *Thermophysical Properties of Matter*, vols. 1–11, IFI/Plenum Press, Washington.

UK Committee on Properties of Steam (1975) *UK Steam Tables in SI Units 1970*, Arnold.

Vargaftik, N.B. (1983) *Handbook of Physical Properties of Liquids and Gases*, Hemisphere/Springer.

Wagner, W. and **de Reuck, K.M.** (1987) *Oxygen, International Thermodynamic Tables of the Fluid State – 9*, IUPAC Series, Blackwell. (See also other volumes in the Series published by Blackwell, by Oxford, and by Pergamon.)

Bibliography

McCarty, R.D. (1977) *Hydrogen Properties.* In *Hydrogen its Technology and Implications*, vol. 3. (Eds, Cox, K.E. and Williamson, K.D.) CRC Press, Florida.

APPENDIX H
Source Books on Heat Exchangers

Read more than the present text

H.1 Texts in chronological order

The undernoted texts should provide excellent sources for tracing other published work on heat exchangers. The landmark texts have added commentary to indicate their importance to this author's work. All books included in the following list are here on merit.

1950

Hausen, H. (1950) *Wärmeübertragung im Gegenstrom, Gleichstrom und Kreuzstrom*, 1st edn, Springer, Berlin (see also 1976). (This is the first definitive text which treats heat exchanger design with imagination and thoroughness. The work is largely analytical, and its relevance and permanence is emphasized by the appearance of an English edition 26 years later.)

Kern, D.Q. (1950) *Process Heat Transfer*, McGraw-Hill. (Engineers involved in chemical plant design will welcome this text as a source of essential information on the configuration and sizing of heat exchangers for different industrial applications.)

1957

Jakob, M. (1949) and (1957) *Heat Transfer*, vol. I (1949) and especially vol. II (1957), John Wiley. (The first volume appeared in 1949, and should by rights be listed before Hausen. The wide ranging thoroughness of the treatment of topics in heat transfer does not detract from the chapters on heat exchangers in volume II.)

1964

Kays, W.M. and **London, A.L.** (1964) and (1984) *Compact Heat Exchangers*, 2nd edn (1964) and 3rd edn (1984), McGraw-Hill, New York. (The first edition was published in 1955. The second and third editions are recommended for their thoroughness in the treatment of plate–fin exchangers, and especially for the heat-transfer and flow-friction correlations used in design today.)

1969

Wallis, G.B. (1969) *One-Dimensional Two-Phase Flow*, McGraw-Hill, New York.

1970

Hewitt, G.F. and **Hall-Taylor, M.S.** (1970) *Annular Two-Phase Flow*, Pergamon.

1972

Collier, J.G. (1972) *Convective Boiling and Condensation* (see 3rd edn 1994), McGraw-Hill, New York.

Kern, D.Q. and **Kraus, A.D.** (1972) *Extended Surface Heat Transfer*, McGraw-Hill, New York.

1973

Gregorig, R. (1973) *Wärmeaustausch und Wärmeaustaucher, Grundlagen der chemishen Technik*, Sauerländer, Aarau & Frankfurt um Main.

1974

Afgan, N.H. and **Schlünder, E.U.** (1974) *Heat Exchangers – Design and Theory Sourcebook*, McGraw-Hill, New York.

1976

Hausen, H. (1976) *Heat Transfer in Counterflow, Parallel Flow and Crossflow*, (English edition of 1950 text) McGraw-Hill, New York.

1978

Shah, R.K. and **London, A.L.** (1978) *Laminar Forced Flow Convection in Ducts*, Supplement 1 to *Advances in Heat Transfer*, Academic Press, New York. (The analytical data provided in this volume extend the experimental data of Kays & London (1964), and have been found valuable in the optimization of plate–fin exchangers.)

1980

Shah, R.K., **McDonald, C.F.**, and **Howard, C.P.** (1980) *Compact Heat Exchangers – History, Technological Advances and Mechanical Design Problems*, ASME Heat Transfer Division, HTD vol. 10, ASME, New York.

Walker, G. (1980) *Stirling Engines* (see bibliography therein), Oxford University Press, Oxford.

1981

Kakaç, S., **Bergles, A.E.**, and **Mayinger, F.** (1981) *Heat Exchangers – Thermal Hydraulic Fundamentals and Design*, Hemisphere, Washington.

Palen, J. (Ed.) (1981) *Heat Exchanger Sourcebook*, Hemisphere, Washington.

Schmidt, F.W. and **Willmott, A.T.** (1981) *Thermal Energy Storage and Regeneration*, Hemisphere, Washington.

1982

Hestroni, G. (Ed.) (1982) *Handbook of Multiphase Systems*, Hemisphere, Washington.

1983

Chisholm, D. (1983) *Two-Phase Flow in Pipelines and Heat Exchangers*, Longmans.

Hausen, H. (1983) *Heat Transfer in Counterflow, Parallel Flow and Cross Flow*, 2nd edn, McGraw-Hill, New York.

Kakaç, S., **Shah, R.K.**, and **Bergles, A.E.** (Eds) (1983) *Low Reynolds Number Heat Exchangers*, Hemisphere, Washington.

Schlünder, E.U. (Ed.) (1983) *Heat Exchanger Design Handbook*, Hemisphere, New York.

Taborek, J., **Hewitt, G.F.**, and **Afgan, N.H.** (Eds) (1983) *Heat Exchangers – Theory and Practice*, Hemisphere, Washington.

1984

Kays, W.M. and **London, A.L.** (1984) *Compact Heat Exchangers*, 3rd edn, McGraw-Hill, New York. (Refer to 2nd edn, 1964.)

1985

Kotas, T.J. (1985) *The Exergy Method of Thermal Plant Analysis*, Butterworths.

Rohsenow, W.M. and **Hartnett, J.P.** (1985) *Handbook of Heat Exchanger Applications*, McGraw-Hill, New York.

Rohsenow, W.M., **Hartnett, J.P.**, and **Ganic, E.N.** (1985) *Handbook of Heat Exchanger Fundamentals*, McGraw-Hill, New York.

1986

Smith, R.A. (1986) *Vaporisers – Selection, Design and Operation*, Longmans, UK.

1987

Kakaç *et al.* (Eds) (1987) *Evaporators – Thermal Hydraulic Fundamentals and Design of Two-phase Flow Heat Exchangers*, NATO Advanced Study Institute, Porto, Portugal.

Kakaç, S., **Shah, R.K.**, and **Aung, W.** (Eds) (1987) *Handbook of Single-phase Convective Heat Transfer*, John Wiley, New York.

Vilemas, J., **Cesna, B.**, and **Survila, V.** (1987) *Heat Transfer in Gas-cooled Annular Channels*, Hemisphere/Springer Verlag.

Wang, B.-X. (Ed.) (1987) *Heat Transfer Science and Technology*, Hemisphere, Washington.

1988

Bejan, A. (1988) *Advanced Engineering Thermodynamics*, John Wiley. [An essential text on recent thermodynamics – to be read for insight, together with Kestin's two volumes entitled *A Course in Thermodynamics*, McGraw-Hill (1978).]

Chisholm, D. (1988) *Heat Exchanger Technology*, Elsevier, Oxford.

Kakaç, S., **Bergles, A.E.**, and **Fernandes, E.O.** (Eds) (1988) *Two-Phase Flow Heat Exchangers – Thermal Hydraulic Fundamentals and Design*, NATO ASI Series E, vol. 143, Kluwer Academic Publishers, Dordrecht.

Minkowycz, W.J., **Sparrow, E.M.**, **Schneider, G.E.**, and **Pletcher, R.H.** (1988) *Handbook of Numerical Heat Transfer*, John Wiley.

Saunders, E.A.D. (1988) *Heat Exchangers – Selection, Design and Construction*, Longmans, UK.

Shah, R.K., **Subbarao, E.C.**, and **Mashelkar, R.A.** (Eds) (1988) *Heat Transfer Equipment Design*, Hemisphere, Washington.

Stasiulevicius, J. and **Skrinska, A.** (1988) (English edition, G.F. Hewitt) *Heat Transfer of Finned Tubes in Crossflow*, Hemisphere, Washington.

TEMA, Tubular Exchanger Manufacturers' Association (1988) *Standard of Tubular Exchanger Manufacturers' Association*, 7th edn TEMA, Tarrytown, New York.

Zukauskas, A. and **Ulinskas, R.** (1988) *Heat Transfer in Tube Banks in Crossflow*, Hemisphere/Springer Verlag.

1989

Zukauskas, A. (1989) *High-Performance Single-Phase Heat Exchangers*, Hemisphere, Washington.

1990

Dzyubenko, B.V., **Dreitser, G.A.**, and **Ashmantas, L.-V.A.** (1990) *Unsteady Heat and Mass Transfer in Helical Tube Bundles*, Hemisphere, New York. (Note: the 'helical tubes' are actually 'twisted flattened tubes'.)

Hewitt, G. (Coordinating Ed.) (1990) *Hemisphere Handbook of Heat Exchanger Design*, Hemisphere, Washington.

Ievlev, V.M., **Danilov, Yu.N.**, **Dzyubenko, B.V.**, **Dreitser, G.A.**, **Ashmantas, L.A.** (T.F. Irvine, editor of English edition) (1990) *Analysis and Design of Swirl-augmented Heat Exchangers*, Hemisphere, Washington.

Shah, R.K., **Kraus, A.D.**, and **Metzger, D.** (1990) *Compact Heat Exchangers – a Festshrift for A.L. London*, Hemisphere, Washington. [Contains the paper by Manglik & Bergles which provides the universal heat-transfer and flow-friction correlations for rectangular offset-strip fins (ROSF) surfaces.]

Thome, J.R. (1990) *Enhanced Boiling Heat Transfer*, Hemisphere, Washington.

1991

Foumeny, E.A. and **Heggs, P.J.** (1991) *Heat Exchange Engineering*. Vol. 1 – Design of heat exchangers. Vol. 2 – Compact heat exchangers: techniques of size reduction. Ellis Horwood, New York and London.

Roetzel, W., **Heggs, P.J.**, and **Butterworth, D.** (Eds) *Design and Operation of Heat Exchangers*, Springer, Berlin.

1992

Carey, V.P. (1992) *Liquid-Vapor Phase-change Phenomena*, Hemisphere, Washington.

McKetta, J.J. (1992) *Heat Transfer Design Methods*, Marcell Dekker, Inc.

Martin, H. (1992) *Heat Exchangers*, Hemisphere, Washington.

Organ, A.J. (1992) *Thermodynamics and Gas Dynamics of the Stirling Cycle Machine*, Cambridge University Press, Cambridge.

Stephan, K. (1992) *Heat Transfer in Condensation and Boiling*, Springer Verlag, Berlin.

1993

Shah, R.K. and **Hashem, A.** (Eds) (1993) *Aerospace Heat Exchanger Technology 1993, Proceedings of 1st International Conference on Aerospace Heat Exchanger Technology*, Palo Alto, California, 15–17 February 1993, Elsevier, Oxford.

1994

Collier, J.G. and **Thome, J.R.** (1994) *Convective Boiling and Condensation*, Oxford University Press, Oxford.

Hewitt, G.F., **Shires, G.L.**, and **Bott, T.R.** (1994) *Process Heat Transfer*, CRC Press, Florida. [An excellent modern treatment and successor to Kern (1950).]

Lock, G.S.H. (1994) *Latent Heat Transfer*, Oxford University Press, Oxford.

Webb, R.L. (1994) *Principles of Enhanced Heat Transfer*, John Wiley, New York. (Where finning is involved this is an excellent modern treatment.)

1995

Sekulic, D.P. and **Shah, R.K.** (1995) Thermal Design of Three Fluid Heat Exchangers. *Adv. Heat Transfer*, **26**, 219–324. (Substantial article, should be a monograph.)

1996

Afgan, N., **Carvalho, M.**, **Bar-Cohen, A.**, **Butterworth, D.**, and **Roetzel, W.** (Eds) (1996) *New Developments in Heat Exchangers*, Gordon & Breach.

1997

Shah, R.K., **Bell, K.J.**, **Mochizuki, S.**, and **Wadekar, V.V.** (Eds) (1997) *Compact Heat Exchangers for the Process Industries, Proceedings of an International Conference*, Snowbird, Utah, 22–27 June 1997, Begell House, New York.

Smith, E.M. (1997) *Thermal Design of Heat Exchangers – A Numerical Approach: Direct Sizing and Stepwise Rating*, 1st edn, John Wiley, Chichester.

1998

Hewitt, G.F. (1998) *Heat Exchanger Design Handbook*, 3 vols., Begell House, New York.

Hewitt, G.F., **Shires, G.L.**, and **Polezhaev, Y.V.** (Eds) (1998) *International Encyclopaedia of Heat and Mass Transfer*, CRC Press, Florida, p. 1344.

Kakaç, S. and **Liu, H.** (1998) *Heat Exchangers: Selection, Rating and Thermal Design*, CRC Press, Florida, p. 448.

Sundén, B. and **Faghri, M.** (Eds) (1998) *Computer Simulations in Compact Heat Exchangers, Developments in Heat Transfer*, Vol. 1, WIT Press, Southampton & Boston.

Sundén, B. and **Heggs, P.J.** (Eds) (1998) *Recent Advances in Analysis of Heat Transfer for Fin Type Surfaces, Developments in Heat Transfer*, Vol. 2 WIT Press, Southampton & Boston.

1999

Bejan, A. and **Mamut, E.** (1999) *Thermodynamic Optimisation of Complex Energy Systems, Proceedings of the NATO Advanced Study Institute*, Neptun, Romania, 13–14 July 1998, Kluwer Academic Publishers.

Kakaç, S., **Bergles, A.E.**, **Mayinger, F.**, and **Yüncü, H.** (Eds) (1999) *Heat Transfer Enhancement of Heat Exchangers, Proceedings of the NATO Advanced Study Institute*, Çesme-Izmir, Turkey, 25 May–5 June 1998, NATO Science Series E, Applied Sciences 355.

Reay, D. (1999) *Learning from Experiences with Compact Heat Exchangers*, Centre for the Analysis and Dissemination of Demonstration Energy Technologies CADDET Analysis Support Unit, Series No. 25, CADDET, Sittard, The Netherlands.

Roetzel, W. and **Xuan, Y.** (1999) *Dynamic Behaviour of Heat Exchangers, Developments in Heat Transfer*, vol. 3, WIT Press, Southampton.

Shah, R.K. with others (Eds) (1999) *Compact Heat Exchangers and Enhancement Technology for the Process Industries, Proceedings of an International Conference*, Banff, Canada, July 1999, Begell House, New York.

2000

Dzyubenco, B.-V., Ashmantas, L.-V., and **Segal, M.D.** (2000) *Modelling and Design of Twisted Tube Heat Exchangers*, Begell House, New York.

Kuppan, T. (2000) *Heat Exchanger Design Handbook*, Marcell Dekker, Inc.

Sundén, B. and **Manglik, R.M.** (2000) *Thermal-Hydraulic Analysis of Plate-and-Frame Heat Exchangers, Developments in Heat Transfer*, WIT Press, Southampton.

Hesselgreaves, J. (2000) *Compact Heat Exchangers, Selection, Design, Operation*, Elsevier, Oxford.

2001

Kraus, A.D., Aziz, A., and **Welty, J.** (2001) *Extended Surface Heat Transfer*, John Wiley.

2003

Shah, R.K. and **Sekulic, D.P.** (2003) *Fundamentals of Heat Exchanger Design*, John Wiley.

Nee, M.J. (2003) *Heat Exchanger Engineering Techniques*, ASME Technical Publishing.

H.2 Exchanger types not already covered

Plate-and-frame exchangers

Design of plate-and-frame heat exchangers is related to direct-sizing of plate–fin heat exchangers, but specific papers and articles provide a better introduction. A few references are provided below, and the reader is encouraged to widen the search, not omitting the texts listed in Section H.1.

Bassiouny, M.F. and **Martin, H.** (1984) Flow distribution and pressure drop in plate heat exchangers. Part 1 – U-type arrangement. Part 2 – Z-type arrangement. *Chem. Engng Sci.*, **39**(4), 693–700 and 701–704.

Hewitt, G.F., Shires, G.L., and **Bott, T.R.** (1994) *Process Heat Transfer*, 2nd edition, CRC Press, Florida.

Kakaç, S. and **Liu, H.** (2002) *Heat Exchangers: Selection, Rating and Thermal Design*, CRC Press, Florida.

Martin, H. (1992) *Heat Exchangers*, Hemisphere, New York.

Muley, A. and **Manglik, R.M.** (1999) Experimental study of turbulent flow heat transfer and pressure drop in a plate heat exchanger with chevron plates. *Trans. ASME, J. Heat Transfer*, **121**, February, 110–117.

Sundén, B. and **Manglik, R.M.** (Eds) (2000) *Thermal–Hydraulic Analysis of Plate-and-Frame Heat Exchangers, Developments in Heat Transfer*, WIT Press, Southampton.

Fin-and-tube heat exchangers

Such crossflow exchangers are frequently used as condensers and evaporators in refrigeration or air-conditioning plant, and they require their own design procedures. An exchanger with some flow depth in the tube bank may have three or more hairpin tube returns to be traversed by the air flow. Full thermal design of tube-and-fin heat exchangers may require the approach developed by Vardhan & Dhar (1998).

There is also the definitive paper by Kim *et al.* (1999) which provides universal heat-transfer and pressure loss correlations for the fin-side of staggered tube arrangements. In-line configurations are not recommended.

The spacing of fins may be twice the developed boundary layer thickness, plus some allowance for core flow. The theory is likely to be more complicated than that for laminar flow between flat plates, and development of design procedures would need to be supported by experimental results.

When icing may be encountered, it may also be advantageous to omit every second fin in the bank for the depth of the first tube hairpin, so that a new leading edge becomes available for ice formation deeper into the exchanger (Ogawa *et al.*, 1993). With icing the attachment of fins to tubes may also require brazing instead of press-fitting to ensure maintenance of good thermal contact (Critoph *et al.*, 1996). The formation of ice under a bad press-fit simply makes a bad fit more loose.

An effective solution for icing is to take hot gas from the compressor discharge and throttle it directly to the evaporator intake. A short timed blast of no more than one or two minutes is sufficient to burn the ice off.

The reader is encouraged to widen the search for papers.

Ataer, Ö.E., **Heri, A.**, and **Gögüs, Y.** (1995) Transient behaviour of finned-tube cross-flow heat exchangers. *Int. J. Refrigeration*, **18**(3), 153–160.

Critoph, R.E., **Holland, M.K.**, and **Turner, L.** (1996) Contact resistance in air-cooled plate fin-and-tube air conditioning condensers. *Int. J. Refrigeration*, **19**(6), 400–406.

Glöckner, G., **Haussmann, B.**, **Heinritz, S.**, **Nowotny, S.**, and **Thiele, K.** (1993) Computer-assisted design of plate-fin heat exchangers – example of an evaporator (in French). *Int. J. Refrigeration*, **16**(1), 40–44.

Kayansayan, N. (1994) Heat transfer characteristics of plate fin-tube heat exchangers. *Int. J. Refrigeration*, **17**(1), 49–57.

Kim, N.H., **Youn, B.**, and **Webb, R.L.** (1999) Air-side heat transfer and friction correlations for plain fin-and-tube heat exchangers with staggered tube arrangements. *Trans. ASME, J. Heat Transfer*, **121**, August, 662–667.

Kondepundi, S.N. and **O'Neal, D.L.** (1989) Effect of frost growth on the performance of louvered finned tube heat exchangers. *Int. J. Refrigeration*, **12**, May, 151–158.

Kondepundi, S.N. and **O'Neal, D.L.** (1993) Performance of finned-tube heat exchangers under frosting conditions. Part 1 – Simulation model. Part 2 – Comparison of experimental data with model. *Int. J. Refrigeration*, **16**(3), 175–180 and 181–184.

Machielson, C.H.M. and **Kershbaumer, H.G.** (1989) Influence of frost formation and defrosting on the performance of air coolers: standards and dimensionless coefficients for the system designer. *Int. J. Refrigeration*, **12**, September, 283–290.

McQuiston, F.C. and **Parker, J.D.** (1994) *Heating, Ventilation and Air Conditioning – Analysis and Design*, 4th edn, John Wiley, New York.

Ogawa, K., **Tanaka, N.**, and **Takeshita, M.** (1993) Performance improvement of plate fin-and-tube heat exchangers under frosting conditions. *ASHRAE Trans.: Symposia*, pp. 762–771.

Paliwoda, A. (1992) Generalised method of pressure drop calculation across components containing two-phase flow of refrigerants. *Int. J. Refrigeration*, **15**(2), 119–125.

Shah, R.K. (1988) Plate-fin and tube-fin heat exchanger design procedures. *Heat Transfer Equipment Design* (Eds R.K. Shah, E.C. Subbarao, and R.A. Mashelkar), Hemisphere, New York, pp. 256–266.

Vardhan, A. and **Dhar, P.L.** (1998) A new procedure for performance prediction of air conditioning coils. *Int. J. Refrigeration*, **21**(1), 77–83.

Willatzen, M., **Pettit, N.B.O.L.**, and **Ploug-Sørensen, L.** (1998) A general dynamic simulation model for evaporators and condensers in refrigeration. Part 1 – Moving boundary formulation for two-phase flows with heat exchange. Part 2 – Simulation and control of an evaporator. *Int. J. Refrigeration*, **21**(5), 398–403 and 404–414.

Micro-channel heat transfer and flow friction, fuel cells

Miniaturization of process plant equipment is the driving force, but other applications exist.

Webb, R.L. and **Zhang, M.** (1988) Heat transfer and friction in small diameter channels. *J. Micro-scale Engng*, **2**(3), 189–202.

Welty, J.R. (1998) Experimental study of flow and heat transfer behaviour of single-phase flow of fluids in rectangular micro-channels. Work in progress, *Oregon State University, Corvalis.*

Recent conferences

Kandlikar, S.G., **Stephan, P.**, **Celata, G.P.**, **Nishio, S.**, and **Thonon, B.** (2003) In *First International Conference on Microchannels and Minichannels*, 24–26 April 2003, Rochester, New York, sponsored by ASME and Rochester Institute of Technology.

Shah, R.K., **Kandlikar, S.G.**, **Beale, S.B.**, **Cheng, P.**, **Djilali, N.**, **Giorgi, L.**, **Hernandez-Guerrero, A.**, **Lee, A.**, **Leo, A.**, **Ma, C.-F.**, **Mukerjee, S.**, **Müller-Steinhagen, H.**, **Onda, K.**, **Ota, K.-I.**, **Penny, T.**, **Shyu, R.-J.**, **Sing, P.**, **Sundén, B.**, **Thonon, B.**, **Toghiani, H.**, **Virkar, A.V.**, and **Voecks, G.** (2003) In *First International Conference on Fuel Cell Science, Engineering and Technology* (Sessions: Proton exchange membrane fuel cell – technology advances and opportunities; General topics related to fuel cells; Micro fuel cells – science and applications; Fuels and fuel reforming technology; Solid oxide fuel cells – prospects in auto and stationary applications; Heat/mass transfer/flow phenomena in fuel cells; Proton exchange membrane fuel cell advanced studies; Novel fuel cells, Molten carbonate fuel cells – a promising stationary power generation technology; Panel on codes and standards for fuel cell systems; Fuels and fuel processing – a success for fuel cell technology; Thermodynamic analysis, modelling and simulation in fuel cells; Molten carbon fuel cells; Balance of power plant of fuel cell systems; Basic research needs in fuel cell technology – challenges and opportunities; Automotive fuel cell applications; Heat/water/temperature balance in PEM fuel cells; Advances in solid oxide fuel cell technology; Industry, government, and academia partnership and funding opportunities.) 21–23 April 2003, Rochester, New York, sponsored by ASME and Rochester Institute of Technology.

Shah, R.K., **Deakin, A.W.**, **Honda, H.**, and **Rudy, T.M.** (2003) In *Fourth International Conference on Compact Heat Exchangers and Enhancement Technology for the Process Industries*, 28 September–3 October 2003, Crete, Greece, Engineering Conferences International, Brooklyn, New York.

H.3 Fouling – some recent literature

This field does not form part of the main theme of the present text, but it is an important subject, particularly in industrial processing. The literature is considerable, and sampling of a few recent international conferences is undertaken below. The author list is substantial in every case.

The reader may locate some textbooks on the subject (e.g. author Bott, T.R. and author Walker, J.) but more often the subject of fouling is kept to one chapter in a more general text on heat exchangers/process heat transfer, or reduced to one session in conference proceedings.

Recent conferences

Panchal, C.B., **Bott, T.R.**, **Somerscales, E.F.C.**, and **Toyama, S.** (1997) *Fouling Mitigation of Industrial Heat Exchange Equipment, Proceedings of an International Conference*, June 1995, San Luis Obispo, California, Begell House, p. 612.

Panchal, C.B., **Bott, T.R.**, **Melo, L.F.**, and **Somerscales, E.F.C.** (1999) *Understanding Heat Exchanger Fouling and its Mitigation* (Sessions: Fundamentals of fouling mechanisms and design; Aqueous systems – cooling water; Fouling in the food industry; Aqueous systems – scaling; Gas systems – combustion; Chemical reaction fouling – refineries; Monitoring; Data evaluation and applications.) *Proceedings of an International Conference*, 11–16 May 1997, Castelvecchio Pascoli (near Barga), Italy, Begell House, p. 418.

Müller-Steinhagen, H., **Watkinson, P.**, and **Malayeri, M.R.** (2001) *Heat Exchanger Fouling, Fundamental Approaches and Technical Solutions* (Sessions: Introduction; Surface treatment; Crystallisation and scaling; Modelling; Fouling in the food industry; Industrial fouling problems and solutions; Fouling in the oil industry; Fouling in power plants; Fouling mitigation.) 8–13 July 2001, Davos, Switzerland, United Engineering Foundation, New York.

Watkinson, P., **Muller-Steinhagen. H.**, and **Malayeri, M.R.** (2003) *Conference Heat Exchanger Fouling and Cleaning Fundamentals and Applications* (Sessions: Water and aqueous systems fouling; Surface modification and modelling of fouling processes; Fouling and cleaning in food and related industries; Petroleum and organic fluid fouling; Fouling in the power industries and in boiling systems; Fouling mitigation and cleaning.) 18–22 May 2003, Santa Fe, New Mexico, Engineering Conferences International, Brooklyn, New York.

Appendix I
Creep Life of Thick Tubes

Operation in the creep/fatigue region. Isotropic creep produces anisotropic damage

I.1 Applications

Conditions being considered for the helium-cooled very-high temperature reactor (VHTR) nuclear reactor are maximum gas temperatures of 1000 °C and pressures in the range 7–15 MPa. Solid oxide fuel-cell systems may operate with temperatures up to 850 °C at 5 bar. Supercritical water-cooled nuclear reactors are proposed for conditions of 375 °C at 25 MPa. Each of these applications may involve heat exchangers operating in the creep/fatigue field.

The most appropriate form of containment is then a tube which may be described as 'thick' or 'thin' in engineering terms, but the distinction is whether the tube may be thin enough to make approximations in the theory without significant error. Under purely elastic conditions tubes with a radial aspect ratio of less than 1.10 might be regarded as thin. Under creep conditions deformations occur which progressively change the stress distributions in the component, and thick tube theory will be outlined to ensure that both thick and thin cases are properly covered.

I.2 Fundamental equations

The nine basic equations for stress readjustment in the wall of a thick tube under internal pressure with closed ends were given by the author (Smith, 1964a), viz.

Radial equilibrium of force

$$\frac{d(r\sigma_r)}{dr} - \sigma_t = 0 \tag{I.1}$$

Radial compatibility of *total* strain

$$\frac{d(re_t)}{dr} - e_r = 0 \tag{I.2}$$

Axial equilibrium of force

$$\int_{r_1}^{r_2} 2\pi r\sigma_a dr = -\pi(r_1^2\sigma_{r1} - r_2^2\sigma_{r2}) \tag{I.3}$$

Axial compatibility of *total* strain

$$e_a = const., \text{ independent of } r \tag{I.4}$$

Total strain

$$\left.\begin{aligned} e_t &= e_{et} + e_{\theta t} + e_{ct} \\ e_r &= e_{er} + e_{\theta r} + e_{cr} \\ e_a &= e_{ea} + e_{\theta a} + e_{ca} \end{aligned}\right\} \tag{I.5}$$

(the author's 1964a paper allowed for plastic strains, but see below).
Constitutive elastic strains due to stresses

$$\left.\begin{aligned} e_{et} &= \frac{1}{E}[\sigma_t - \nu(\sigma_a + \sigma_r)] \\ e_{er} &= \frac{1}{E}[\sigma_r - \nu(\sigma_t + \sigma_a)] \\ e_{ea} &= \frac{1}{E}[\sigma_a - \nu(\sigma_r + \sigma_t)] \end{aligned}\right\} \tag{I.6}$$

Constitutive thermal strains due to temperature

$$\left.\begin{aligned} e_{\theta t} &= \alpha\theta \\ e_{\theta r} &= \alpha\theta \\ e_{\theta a} &= \alpha\theta \end{aligned}\right\} \tag{I.7}$$

Constitutive creep strains (temperature and stress dependent)

$$\left.\begin{aligned} &e_{ct} \\ &e_{cr} \\ &e_{ca} \end{aligned}\right\} \text{requiring constitutive equations} \tag{I.8}$$

Constitutive temperature distribution

$$\theta = f(t, r) \qquad \text{dependent on heat flow} \tag{I.9}$$

These equations have to be solved numerically. By substitution, the first eight equations are reduced to the modified equation set (I.1–I.4), giving two ordinary differential equations, one integral equation, and one algebraic equation to be solved simultaneously for the stress field by matrix inversion. Solution of the temperature field, equation (I.9) is handled separately, which is permitted when energy and linear momentum equations do not involve speeds approaching ballistic impact.

The reader may wonder why plastic strains are not included. The answer is that any form of creep (and indeed plasticity) involves irreversibility which by definition is time dependent, and time-independent creep or plasticity is a thermodynamic impossibility.

To illustrate this point, the author predicted tensile ramp loading behaviour for Nimonic 90 at both ambient and high temperatures using only steady load creep data. The results were compared with commercially quoted data for 0.1

and 0.2 per cent tensile proof strain (Anon. 1961, 1966) and the results were quite close (Ellison & Smith, 1973). Straining from ambient temperatures involves different metallurgical damage from that encountered under creep conditions, so the findings were encouraging, but not definitive. They were also relevant for one material only.

The massive contributions of workers in low-temperature plasticity are not to be ignored, as many valuable predictions have been made assuming time independence. This is however an approximation, as it is well understood that different rates of straining produce different tensile stress–strain curves.

I.3 Early work on thick tubes

In an outstandingly comprehensive treatise on several aspects of creep design, Bailey's (1935) treatment of the thick cylinder problem made the simplifying assumption of zero axial creep, which permitted an explicit solution of the problem. However, it involved a flawed assumption, which was to be repeated time and again by many other workers.

Bailey showed that the assumption of zero axial creep was consistent with the requirements of axial equilibrium. This carries the condition that the axial stress is always the mean of the radial and tangential stresses, and follows from the general expression for creep rate given by

$$\dot{e}_a = f(J_2')S_a\phi(t) = 0$$

then

$$S_a = \frac{2\sigma_a - \sigma_t - \sigma_r}{3}$$

must be zero, whence

$$\sigma_a = \frac{\sigma_t + \sigma_r}{2} \tag{I.10}$$

Equation (I.10) also holds for purely elastic stresses in thick tubes, the axial elastic deformation being not zero, but constant over the cross-section.

Because axial deformation was assumed to be zero for creep, and was known to be constant for purely elastic loading, and because condition (I.10) occurred in both instances and led to substantial mathematical simplification, attempts to incorporate these features in a composite solution did persist for some years, e.g.

- Soderberg (1941)
- Coffin *et al.* (1949)
- Johnson & Kahn (1963)
- Rabotnov (1969 translation of 1966 book)

It is, however, mathematically unsound to superimpose a non-linear creep solution on a linear elastic solution. The correct solution of the problem requires a numerical approach, and the real difficulty lies in formulating appropriate constitutive equations for creep.

I.4 Equivalence of stress systems

In designing a multi-axial stress system it is usually necessary to make use of uni-axial tensile test data because multi-axial data are sparse. The initial approach involves assumption of material isotropy, which may not always pertain in practice.

Given an appropriate tensile creep curve (time versus creep deformation at constant stress), to obtain creep rate over a short time interval it was found convenient to use a numerical chordal creep rate on the curve where

$$\text{Chordal tensile creep rate, } \dot{e} = \frac{e(t + \Delta t) - e(t)}{\Delta t} \tag{I.11}$$

Johnson (1960), which also references papers from 1948 to 1951, confirmed that primary creep curves for different stress levels were geometrically similar for aluminium, carbon steel, magnesium, and Nimonic 75. These findings suggested that time dependence of creep rate might be separated from stress dependence and that creep strain $\dot{e} = \phi(\sigma, t)$ could perhaps be written as $\dot{e} = f(\sigma)\varphi(t)$. This led to the general expression for multi-axial creep, involving stress deviators (temperature dependence is incorporated in the section on constitutive equations)

$$\dot{e}_{ij} = f(J_2')S_{ij}'\varphi(t) \tag{I.12}$$

For design purposes we require to know multi-axial creep rates, but generally only tensile creep data are available. Thus means of extracting multi-axial creep rates from tensile data are required. For equivalent complex stress creep rates in a multi-axial stress systems programme, Johnson (1962) showed that

$$\frac{\dot{e}_{ij}}{\dot{e}_{ii}} = \frac{f(J_2')S_{ij}'}{f(J_2')S_{ii}'} = \frac{S_{ij}'}{S_{ii}'} = \frac{3S_{ij}'}{2\sigma_{ii}} \tag{I.13}$$

viz.

$$\dot{e}_{ij} = \left(\frac{3}{2}\right)\frac{S_{ij}'e_{ii}}{\sigma_{ii}} \tag{I.14}$$

For multi-axial creep in a tube in which the stress and strain axes remain coincident, the multi-axial creep rates are then given by

$$\left.\begin{aligned}
\dot{e}_{ct} &= \left(\frac{3}{2}\right)(\sigma_t - \sigma_m)\left(\frac{\dot{e}_{tensile}}{\sigma_{tensile}}\right)\\
\dot{e}_{cr} &= \left(\frac{3}{2}\right)(\sigma_r - \sigma_m)\left(\frac{\dot{e}_{tensile}}{\sigma_{tensile}}\right)\\
\dot{e}_{ca} &= \left(\frac{3}{2}\right)(\sigma_a - \sigma_m)\left(\frac{\dot{e}_{tensile}}{\sigma_{tensile}}\right)
\end{aligned}\right\} \tag{I.15}$$

where

$$\sigma_m = \frac{\sigma_t + \sigma_r + \sigma_a}{3}$$

I.5 Fail-safe and safe-life

Both fail-safe and safe-life concepts have their place in design of structures. Fail-safe arises when one part of the structure fails and the remaining parts take up the load safely until repair is carried out. Safe-life is the permitted life of the structure before complete replacement is required. In thick cylinder design, safe-life is the appropriate criterion.

Several tensile (time versus creep rupture) parameters have been proposed, and using such parameters creep strain at rupture might be predicted. For Nimonic 90, of these parameters, the correlation due to Conrad (1959a,b) possessed the greatest similarity with expressions obtained by metallurgists working with dislocations.

Many empirical and theoretical models have been proposed for different alloys to explain different metallurgical mechanisms of creep. It is not proposed to survey the catalogue here, instead we concentrate on how the information is used in design.

I.6 Constitutive equations for creep

Uni-axial creep computation

When stress and temperature change with time, any attempt at numerical prediction of specimen behaviour is faced with the problem of deciding how creep rate in the next time interval is to be chosen.

Suppose the specimen is allowed to creep over a time interval Δt at steady values of stress and temperature until a creep strain of Δe has accumulated. Given that the rupture creep strain at these stress and temperature levels ($e_{rupture}$) is known, the *ductility fraction* ($\Delta e / e_{rupture}$) proposed by Goldhoff (1965) can be taken as a measure of useful creep life expended. If now creep continues at a new stress and temperature level, the point on the new creep curve where deformation is to begin is chosen to have the same value of ductility fraction as was reached at the end of the previous time interval. Over the new time interval the new chordal creep rate can now be calculated using equation (I.11) as before.

Under changing stress and temperature conditions, a running total of the ductility fraction can thus be kept, and computation proceeds until the ductility fraction reaches unity. At this point the safe-life of the specimen is deemed to have expired.

$$\sum \left(\frac{\Delta e}{e_{rupture}} \right) = 1 \qquad \text{(I.16)}$$

To obtain smooth values of creep strain at rupture for Nimonic 90 ($e_{rupture}$), the experimental creep rupture data of Walles (1959) were inserted into Clarke's (1966) representation of tensile creep curves to obtain values of time at creep rupture. Additional values of creep strain at rupture were obtained from the data of Betteridge (1958). These times were then used to obtain coefficients in Conrad's (1959a,b) correlation for creep rupture, which when used with Clarke's curves provided smooth values of creep strain at rupture. The empirical expressions of Clarke and of Conrad, are supported by expressions mathematically similar to those obtained by metallurgists working on dislocation theory.

Conrad's rupture parameter has the form

$$t_{rupture} = \exp\left[A + \frac{B}{\theta} + C(\theta)\sigma\right] \tag{I.17}$$

based on a dislocation model for creep proposed by Weertman (1957).

Complete creep curves can be represented by Clarke's (1966) method, and the ductility fraction used to obtain creep strain and chordal creep rate at any time.

Multi-axial creep computation

For multi-axial creep behaviour under time-dependent stresses and temperatures, a similar procedure to that for uni-axial creep is used for evaluating creep rates and expended safe-life in each time interval.

First, multi-axial stresses are reduced to equivalent uni-axial form using Maxwell's relation

$$\sqrt{(\sigma_t - \sigma_r)^2 + (\sigma_r - \sigma_a)^2 + (\sigma_a - \sigma_t)^2} = \sqrt{2}(\sigma_{11}) \tag{I.18}$$

The equivalent uni-axial creep rate found using of the ductility fraction concept as described above, and then equation (I.14) is invoked to find the appropriate multi-axial creep rates. Computation of the next time interval can now proceed.

During each time interval, if temperature changes are involved then equations for the temperature field are solved first (numerically). Physical properties can then be updated, new creep rates found to increment creep strains, and finally equations (I.1) to (I.5) for the stress field solved (numerically) to obtain new stress and strain values for the next time interval (Smith, 1964a, 1965b).

The size of the time interval is determined by the permitted change in stress and temperature levels in the computation.

Multi-axial creep-life expenditure

In a thick cylinder the stresses (σ_t, σ_r, σ_a) are evaluated at each time interval. During deformation both tensile and compressive values may exist at different times in different directions. Creep damage by void formation occurs only under tensile stress (Mohr condition), thus the ductility fraction summations have to be increased for each principal stress direction at stations across the radius, but only when the stress is tensile. This multi-axial creep-life summation is done in the same manner as was done to find creep rates, but plays no part in calculating the deformation and stress redistribution.

When the safe-life in one direction is reached, the tube is assumed to have completed its service. Although the tube may survive under a redistributed load after voids have coalesced, a predictable stress distribution no longer exists.

I.7 Clarke's creep curves

It is worth taking a more detailed look at Clarke's representation of creep-strain data because it points a way to possible further improvements in safe-life prediction. A typical creep strain versus time curve for Nimonic 90 presented by Clarke (1966) is shown in Fig. I.1.

Clarke re-plotted the data from Fig. I.1 in terms of natural logarithms, ln(strain) versus ln(time), and found that the shape of the curve was typical for all the alloys of the Nimonic series. He then proposed that it could adequately be represented by a hyperbola (Fig. I.2).

With this assumption, Clarke fitted a hyperbola for all the data at each test temperature using

$$3x^2 + 3y^2 - 10xy = 8A^2 \tag{I.19}$$

where

$$x = \ln\left(\frac{t}{\tau}\right) \quad \text{and} \quad t = \text{time},\ \tau = \exp[D(\theta) + E(\theta)\sigma]$$

$$y = \ln\left(\frac{e_c}{\lambda}\right) \quad \text{and} \quad e_c = \text{creep strain},\ \lambda = \exp[B + C\theta]$$

$$A, B, C, D(\theta), E(\theta) = \text{material constants}$$

Although Clarke claimed only that his data-fit was empirical, the form of his expressions did correspond to those anticipated from metallurgical considerations

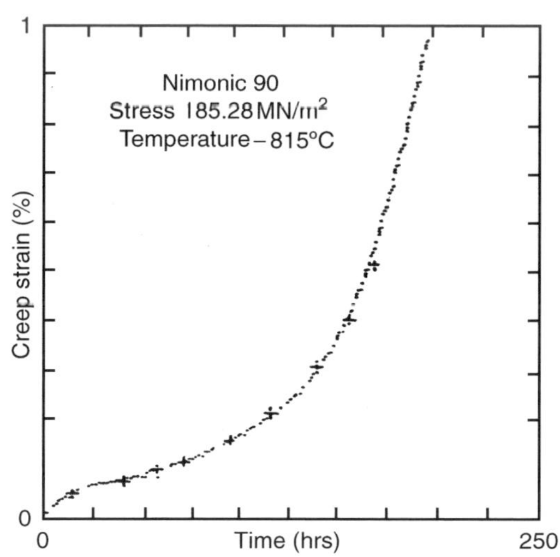

Fig.I.1 Typical uniaxial tensile creep curve for Nimonic 90

involving dislocations. Its terms also corresponded closely to those proposed by Conrad for his creep-rupture parameter.

This method of representing data also allowed explicit expressions for strain and strain rate or time and strain rate. However, such expressions are less appropriate with the more general form of equations (I.20) when a numerical approach is easier to apply.

Taking natural logarithms of raw creep data, viz. $x = \ln(\text{hours})$, $y = \ln(\text{creep strain})$, a more general hyperbola is first fitted to the data

$$Lx^2 + My^2 + Nxy + Qx + Ry + S = 0 \qquad \text{(I.20)}$$

The origin is then determined and the data then adjusted to the new origin to produce the simpler form of hyperbola

$$Lx^2 + My^2 + Nxy + S = 0 \qquad \text{(I.21)}$$

On the ln/ln plot of Fig. I.2 the point of minimum creep rate occurs before the circle defining the curve 'elbow'. The 'elbow' point on the ln/ln creep curve is of much greater interest than the point of minimum creep rate and its location can be found numerically.

Metallurgically, Ishida & McLean (1967) found that voids in creeping material occurred at right angles to the tensile stress, cavities being strung out along grain boundaries. Woodford (1969) found strong evidence that the number of voids was controlled by total strain rather than by time, and Dyson & McLean (1972) found a strong linear relationship between cavity density and strain. Davies *et al.* (1966) worked on Nimonic 80A and confirmed that annealing in the late secondary stage

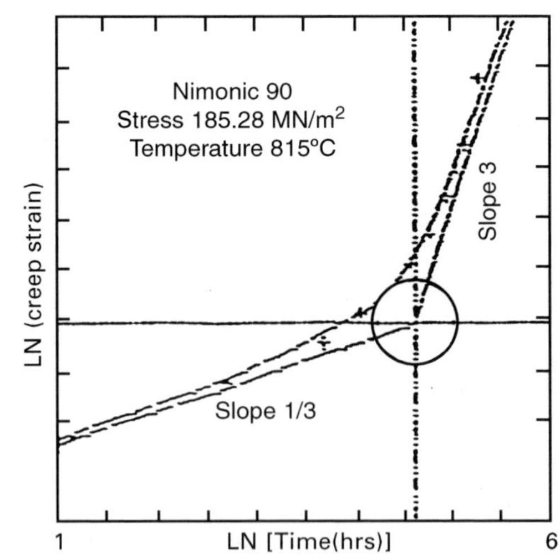

Fig.I.2 Re-plotting of data for Fig. I.1 as natural logarithms

of creep was more effective in extending life than annealing in the early tertiary stage of creep.

Such observations suggested that the end of safe-life for Nimonic 90 might be assumed when the 'elbow' of the ln/ln creep curve was reached. This is a well-defined point appropriate in design analysis of structures. The simpler criteria of strain-to-rupture is less precise by the way in which macroscopic cavities in the material coalesce into bigger cavities, making analysis of complex stress systems invalid in the final stages before failure.

A correlation which defines the 'elbow' point in the Clarke representation is required, and the form of Conrad's rupture parameter suggests itself. Following observations by metallurgists of void formation near the start of tertiary creep, the ductility fraction concept could then be expressed as $(\Delta e/e_{elbow})$.

I.8 Further and recent developments

The fundamental equations (I.1) to (I.4) derived from the basic axioms of physics, and their numerical solution, hold whatever constitutive equations may be injected into the thick tube problem. However, this problem is a special case of deformation in which initial directions of stress and strain tensors are maintained. This simplified the 1-space–1-time problem considerably.

Betten's (2001) extensive review of investigations into creep behaviour, which discusses 243 significant papers written over the past two or three decades, shows that mathematical representation of creep damage can now been extended to include complex stress situations in which the stress and strain tensors do not remain coincident during deformation. Betten's review is a most timely contribution to the subject, but two practical considerations remain, viz.:

1. What is the shortest time required to collect sufficient experimental data to permit creation of new constitutive equations?
2. How long will it take the fastest computer to compute the creep behaviour of real components (minimum two-space–one-time problems)?

I.9 Acknowledgements

The data for Nimonics used in computation were the extensive results obtained by Walles and Graham at the National Gas Turbine Establishment, RAE Farnborough. Henry Wiggin & Co., Hereford were equally helpful in providing data on Nimonics. Computing facilities were provided courtesy of Professor Ewan Page, Director of the Computing Laboratory at the University of Newcastle upon Tyne.

References

Anon. (1961) The Nimonic series of high temperature alloys. *Henry Wiggin Publication 2358.*

Anon. (1966) Nimonic alloys – physical and mechanical properties. *Henry Wiggin Publication 3270.*

Bailey, R.W. (1935) The utilisation of creep test data in engineering design. *Proc. Instn Mech. Engrs*, 131–349.

Betten, J. (2001) Mathematical modelling of material behaviour under creep conditions. *Appl. Mechanics Rev.*, **54**(2), March, 107–132.

Betteridge, W. (1958) The extrapolation of the stress rupture properties of the Nimonic alloys. *J. Inst. Metals*, **86**, 232–237.

Clarke, J. M. (1966) A convenient representation of creep strain data for problems involving time-varying stresses and temperatures. NGTE Report No. R.284, National Gas Turbine Establishment, Pyestock, Hants.

Coffin, L.F., **Shepler, P.R.**, and **Cherniak, G.S.** (1949) Primary creep in the design of internal pressure vessels. *Trans. ASME, J. Appl. Mechanics*, **16**, 229–241.

Conrad, H. (1959a) Correlation of high temperature creep and rupture data. *Trans. ASME, J. Basic Engng, Ser. D*, **81**, Paper 58-A-96.

Conrad, H. (1959b) Correlation of stress-rupture properties of Nimonic alloys. *J. Inst. Metals*, **87**(10), June, 347–349.

Davies, P.W., **Dennison, J.P.**, and **Evans, H.E.** (1966) Recovery properties of a nickel-base high temperature alloy after creep at 750°C. *J. Inst. Metals*, **94**, 270–275.

Dyson, B.F. and **McLean, D.** (1972) New method of predicting creep life. *Metal Sci. J.*, **6**, November, 220–223. (See also IMS Internal Report 44, National Physical Laboratory, UK.)

Ellison, E.G. and **Smith, E.M.** (1973) Predicting service life in a fatigue-creep environment. *Fatigue at Elevated Temperatures*, American Society for Testing Materials, ASTM STP 520, pp. 575–612.

Goldhoff, R.M. (1965) Uniaxial creep-rupture behaviour of low alloy steel under variable loading conditions. *Trans. ASME, J. Basic Engng, Ser. D*, **87**(2), June, 374–378.

Ishida, Y. and **McLean, D.** (1967) Formation and growth of cavities in creep. *Metal Sci. J.*, **1**, September, 171–172.

Johnson, A.E. (1960) Complex stress creep of metals (references to earlier work in 1948, 1949, 1951). *Metallurgical Rev.*, **5**(20), 447–506.

Johnson, A.E. (1962) *Complex Stress, Creep Relaxation and Fracture of Nimonic Alloys.* HMSO.

Johnson, A.E. and **Kahn, B.** (1963) Creep of metallic thick-walled cylindrical vessels subject to pressure and radial thermal gradient at elevated temperatures. In *Conference on Thermal Loading and Creep in Structures and Components, Proc. Instn Mech. Engrs*, **178**, *Part L(3)*, 29–42.

Johnson, A.E., **Henderson, R.J.**, and **Mathur, V.** (1958) Creep under changing complex stress systems. *Engineering*, Part 1, **206**(5350), 8 August, 209–210. Part 2, **206**(5351), 15 August, 251–257. Part 3, **206**(5350), 22 August, 287–291.

Rabotnov, Yu.N. (1969) *Creep Problems in Structural Members*, North-Holland Publishing Co, Holland. (Translation of 1966 Russian book, English version edited by F.A. Leckie.)

Smith, E.M. (1965a) Analysis of creep in cylinders, spheres and thin discs. *J. Mech. Engng Sci.*, **7**(1), March, 82–92.

Smith, E.M. (1965b) Estimation of the useful life and strain history of a thick tube creeping under non-steady conditions. *J. Strain Analysis*, **1**(1), 44–49.

Soderberg, C.R. (1941) Interpretation of creep tests on tubes. *Trans. ASME*, **63**, 737–748.

Walles, K.F.A. (1959) A quantitative presentation of the creep of Nimonic alloys (valid in the range 650 to 870 °C for stresses up to 541 MN/m^2). NGTE Note NT 386, National Gas Turbine Establishment, Pyestock, Hants.

Weertman, J. (1957) Steady-state creep through dislocation climb. *J. Appl. Physics*, **28**, 362–364. (See also pp. 1185–1189.)

Woodford, D.A. (1969) Density changes during creep in nickel. *Metal Sci. J.*, **3**, 234–240.

Bibliography

Glenny, R.J.E., **Howe, P.W.H.**, **Islip, L.**, and **Barnes, J.F.** (1967) Engineering in High Duty Materials. *Bulleid Memorial Lectures 1967*, vol. IV, University of Nottingham.

Smith, E.M. (1964a) Primary creep behaviour of thick tubes. In *Conference on Thermal Loading and Creep in Structures and Components, Proc. Instn Mech. Engrs*, **178**, *Part L(3)*, 135–141.

Smith, E.M. (1964b) Axial deformation in thick tubes creeping under internal pressure. *J. Mech. Engng Sci., Research Note*, **6**(4), 418–420.

Tilly, G.P. (1972) Relationships for tensile creep under transient stresses. (NGTE, data from K.F.A. Walles & A. Graham). *J. Strain Analysis*, **7**(1), 61–68.

Appendix J
Compact Surface Selection for Sizing Optimization

Search for improvement within constraints

J.1 Acceptable flow velocities

All notation in Appendix J follows that used in Chapter 4. The side of the two-stream exchanger with the lowest pressure level will usually require the lowest pressure loss. This determination is reinforced if the side with the lowest pressure loss also carries the higher-temperature fluid.

Since no published velocity constraint is specified with Kays & London (1964) plate–fin surface correlations, quite high velocities can arise in the core of a compact exchanger design, and this may not be discovered unless velocity values are evaluated – which is not always carried out. Values of 32.3 and 12.47 m/s, respectively, for hot and cold fluids were found in a design presented in Section 9.2 of the text by Shah & Sekulic (2003), corresponding to Reynolds numbers of 589 and 542.

A clue to selection of pressure loss in the text by Walsh & Fletcher (1998, Section 5.13.8) is to keep the Mach number at engine exhaust flange below a Mach number of 0.05 to minimize the dump pressure loss. For a conservative velocity value using the gas-side exchanger exhaust temperature of 564.4 K, the velocity value was found to be

$$u = \text{Ma} \times \sqrt{\gamma RT} = 0.05 \times \sqrt{1.4 \times 287.07 \times 564.4} = 23.81 \text{ m/s}$$

The above velocity values may be used in checking values found in exchanger design. When accurate fouling data become available then appropriate adjustments to the above velocity values can be made.

J.2 Overview of surface performance

It is convenient to represent flow-friction and heat-transfer correlations by procedure interpolating cubic spline-fits which automatically keep values f and j within the validity range of the correlations. Also surface geometries corresponding to the correlations are known and fixed. In preliminary investigations, hand calculations

Table J.1 Minimum/maximum range of ROSF surfaces for Manglik & Bergles correlations

Geometrical parameters	*Plate spacing, b (mm)*	*Cell width, c (mm)*	*Strip length, x (mm)*	*Fin thickness, tf (mm)*
Maximum	8.9660	2.127	12.70	0.152
Minimum	1.9050	0.940	2.540	0.1016

may be used, using cubic fits where data is smooth, preferably interpolating in the middle range of four points (see Appendix B.7).

Rectangular offset strip-fins

When universal correlations are employed, e.g. the Manglik & Bergles (1990) algebraic equations for f and j for rectangular offset strip-fins (ROSF) surfaces may be found in Appendix C.4 and these results were applied in deriving performance graphs given in Appendix C.3. When such universal correlations are employed then two additional constraints must be applied during computation, viz:

- maximum and minimum permitted values of Reynolds number
- range limits of surface geometries under consideration

On the low-pressure side of an exchanger we might reasonably expect to use ROSF surfaces, and the maximum cell geometry to be considered would therefore not exceed that shown in Table J.1.

Plain rectangular ducts

Figure 4.11 of Chapter 4 shows the relative performance of plain rectangular ducts against duct aspect ratio developed using *theoretical* results for performance of plain rectangular ducts given in Table J.2. The flow area parameters b, c for cells with zero fin thickness in that figure correspond to $(b - tf)$, $(c - tf)$ used in describing rectangular surface geometries[1].

In this work the author employed a specific performance parameter for unambiguous comparison of performance of heat exchangers, viz.

$$\dot{Q}_{spec} = \frac{\dot{Q}_{duty}}{V\Delta\theta_{mean}} = \frac{US_{ref}}{V} \tag{J.1}$$

and plotted fin efficiency ϕ, duct length L m, and specific performance parameter $\dot{Q}_{spec}$ kW/(m^3 K) against duct aspect ratio. The results revealed that square ducts gave the worst possible performance. The right-hand end of Fig. 4.11 approaches primary surfaces, but this probably has applications only for thin crossflow

[1]Figure 4.11 was constructed using an assumed constant value of flow area $A = 8.0\,\text{mm}^2$, with constant mass velocity of $G = 12.5\,\text{kg/(m}^2\,\text{s)}$, constant density $\rho = 0.550\,\text{kg/m}^3$, giving constant flow velocity of $u = G/\rho = 22.73\,\text{m/s}$.

Table J.2 Extract from Shah & London (1974) (fully developed forced laminar flow)

Duct aspect $(b - tf)/(c - tf)$	Nu_{H1}	$f \times Re$
8/1	6.490	20.585
6/1	6.049	19.702
4/1	5.331	18.233
2/1	4.123	15.548
1/1	3.608	14.227

exchangers (car radiators), while the left-hand of Fig 4.11 finds applications with block contraflow exchangers (gas turbine recuperators). However it is not desirable to go to very short exchangers as this results in greater longitudinal conduction, nor is it desirable to go to excessive duct heights as this leads to minimization of improvement.

Plate and fin material

Fin thickness, mm	$tf = 0.1524$
Plate thickness, mm	$tp = 0.3048$
Thermal conductivity, J/(m s K)	$\lambda_w = 20.77$
Density, kg/m^3	$\rho_w = 7030.0$

Note that the x-axis of Fig. 4.11 uses *LOG(duct base/duct height)*, while in this section we shall reverse the notation and use Duct Aspect = (*duct height/duct base*).

Optimization of block contraflow exchangers

Full optimization of plate-fin surfaces is possible using either the direct-sizing approach (Smith 1994, 1997–99), or by following the genetic algorithm approach (Cool *et al.*, 1999). Both methods are capable of exploring the complete envelope of possible surface geometries to arrive at a fully optimized exchanger core.

The direct-sizing approach is not completely optimized because only one cell parameter was allowed to vary its geometry over the permitted full range while the other parameters were maintained at some mean condition. Genetic algorithms avoid this constraint, and the search allows the whole geometry to vary one parameter at a time, simply selecting and following the best incremental improvement.

Results for direct-sizing with ROSF surfaces were presented by Smith (1997–99) as a series of four plots of trend curves, while Cool *et al.* (1999) presented results for genetic algorithms in the form of scatter diagrams covering the research area for plain rectangular ducts.

Trend curves

In 1994 the author used direct-sizing on an exchanger with a cold-side/hot-side pressure ratio of 6/1 to investigate the effect on performance of changing the geometry of ROSF surfaces. Manglik & Bergles (1993) universal heat-transfer and flow-friction correlations for both single-cell and double-cell geometries were employed, and only minor differences in the results between single and double cell configurations were found. The effects of fin thickness *tf*, separating plate thickness *tp*, and splitter thickness *ts* were negligible and did not affect the results. The range of rectangular offset strip-fin geometries used were

$$2.0 \le b \le 10.0\,\text{mm}$$
$$1.0 \le c \le 4.0\,\text{mm}$$
$$3.0 \le x \le 6.0\,\text{mm}$$

The value for $b = 1.0$ mm can be avoided as this would have led to a square cell, which was shown to be the worst choice – see Fig. 4.11.

A nominal surface geometry was selected, viz.

$$b = 5.0\,\text{mm} \quad c = 2.0\,\text{mm} \quad x = 6.0\,\text{mm}$$
$$tf = 0.15\,\text{mm} \quad ts = 0.15\,\text{mm} \quad tp = 0.20\,\text{mm}$$

and the dimensions *b, c, x* were varied about their nominal size, changing one dimension at a time while keeping the others at their nominal size[2]. Six possible core minimization targets were investigated, viz.

- minimum volume of block
- minimum mass of block (excluding fluids)
- minimum length of block
- minimum frontal area of block
- minimum plate surface area
- minimum total surface area

Six sets of four figures were generated, five sets of which were presented in the text by Smith (1997–99):

- minimum total and plate surface areas approximate to minimum core volume.
- minimum mass of the block is less frequently required.
- length of the block remains a free choice as it can be specified independently to match allowable pressure losses.
- minimum frontal area of a block may be approached by selecting minimum block volume and choosing a higher value for block length *L*.

From the figures generated using direct-sizing the case of minimum block volume showed consistency, in that each of four figures indicated that the best choice was to use rectangular cells of minimum width. Figures C.1 to C.4 in Appendix C provide just an outline for the more comprehensive list of figures presented by Smith (1997–99).

[2]In this respect the search is less thorough than that developed by Cool *et al.* (1999), which employs the genetic algorithm approach, but which produces scatter diagrams.

Table J.3 Input data for contra-flow exchanger ($\dot{Q}_{duty} = 5.4724$ MW, effectiveness = 0.842)

Flow stream parameter	*Hot LP gas*	*Cold HP air*
Specified parameters		
Mass flow rate, kg/s	$\dot{m}_h = 24.683$	$\dot{m}_c = 24.318$
Inlet absolute pressure, bar	$p_h = 1.02735$	$p_c = 9.1014$
Inlet temperature, K	$T_{h1} = 702.59$	$T_{c2} = 448.15$
Outlet temperature, K	$T_{h2} = 497.69$	$T_{c1} = 662.57$
Values by splinefits		
Specific heat at const. pressure, J/(kg K)	$C_h = 1084.5$	$C_c = 1051.90$
Absolute viscosity, kg/(m s)	$\eta_h = 0.00003015$	$\eta_c = 0.00002850$
Thermal conductivity, J/(m s K)	$\lambda_h = 0.048817$	$\lambda_c = 0.044744$
Density by gas law		
Gas constant, J/(kg K)	$R_h = 287.07$	$R_c = 287.07$
Density, kg/m^3	$\rho_h = 0.59618$	$\rho_c = 5.70994$

J.3 Design problem

Input data

Consider the contraflow recuperator shown in Table J.3, data adapted from the examples for crossflow and contraflow exchangers described in Chapter 4.

Here we take the same mean temperature difference in crossflow, and apply this to the corresponding contraflow example. We anticipate that the results will differ from those for the contraflow example described in Chapter 4 as now $\dot{Q}_{duty} = 5.4724$ MW. Also physical properties would change slightly to correspond with the new mean temperatures encountered, but we shall use the values quoted in Chapter 4. Using plain rectangular cell surfaces, let us now seek minimum block volume.

Hot low-pressure side

Figures of Appendix C apply to a design of ROSF fins with a cold/hot pressure ratio of 6/1 (the same conclusions will be valid for other pressure ratios). For minimum core block volume we first select the cell dimension with the greatest slope. Using Fig. C.1 (this text) and Figs C.1 to C.4 (Smith 1997–99), the smallest cell widths were found to give minimum block volume, and the choice made was to select duct widths $(c - tf) = 1.0$ mm for both hot and cold sides[3].

Examining flow requirements in Table J.4, the hot low-pressure side will obviously require the largest single-cell duct height, say $(b - tf)_h = 8.0$ mm, while the height for the cold high-pressure side $(b - tf)_c$ requires further consideration.

[3]This choice reduces design of the exchanger to the study of performance of opposite cells, however there is no constraint on using different cell widths.

Table J.4 Summary of results for hot low-pressure side. Duct width, $(c - tf)_h = 1.0$ mm

Duct aspect $\left(\frac{b-tf}{c-tf}\right)_h$	*Heat trans. coeff.* $\hat{\alpha}_h, J/(m^2\, s\, K)$	*Core press. loss* $\frac{\Delta p_h}{L}, N/m^3$
8/1	896.292	14.2232
6/1	825.946	14.7500
4/1	678.960	15.8037
2/1	435.036	18.9650

We plan to employ data for plain rectangular ducts in laminar forced flow convection using the *theoretical* results of Shah & London (1978). Results obtained by selecting data from both rectangular offset strip-fins and plain rectangular ducts should not be in too serious error for obtaining reasonable starting values for optimization.

Table J.1 provides the necessary values of Nusselt number Nu and flow-friction expression fRe for the hot low-pressure side, viz.

$$\mathrm{Nu} - 6.490 \qquad f\mathrm{Re} - 20.585$$

and we choose to remain in the laminar region by assuming that $\mathrm{Re}_h = 500.0$, then

$$\mathrm{Re}_h = \frac{d_h G_h}{\eta_h} \qquad \text{where } G_h = \frac{\dot{m}_h}{A_h} \quad \text{and} \quad d_h = \frac{4 \times \text{area for flow}}{\text{wetted perimeter}}$$

hydraulic diameter $\quad d_h = \dfrac{4(b - tf)(c - tf)}{2[(b - tf) + (c - tf)]} = 1.7778\,\mathrm{mm}$

mass velocity $\quad G_h = \dfrac{\mathrm{Re}_h \eta_h}{d_h} = 8.47958\,\mathrm{kg/(m^2\, s)}$

flow area of single duct $\quad A_h = (b - tf)_h (c - tf)_h = 0.000008\,\mathrm{m^2}$

mass flow rate in duct $\quad \dot{m}_h = G_h A_h = 0.000067837\,\mathrm{kg/s}$

flow velocity $\quad u_h = \dfrac{G_h}{\rho_h} = 14.2232\,\mathrm{m/s}$

heat transfer coefficient $\quad \alpha_h = \dfrac{\mathrm{Nu}_h \lambda_h}{d_h} = 178.210\ \mathrm{J/(m^2\, s\, K)}$

fin efficiency $\quad \hat{m}_h = \sqrt{\dfrac{2\alpha_h}{\lambda_f t_f}} = 237.277\,\mathrm{m^{-1}}$

$$\hat{m}_h Y_h = 1.89821$$

$$\phi_h = \frac{\tanh(\hat{m}_h Y_h)}{\hat{m}_h Y_h} = 0.503677$$

fin performance ratio $\phi_h\left(\dfrac{b-tf}{c-tf}\right) = 4.029414$

plate heat transfer coefficient $\hat{\alpha}_h = \alpha_h\left[1+\phi_h\left(\dfrac{b-tf}{c-tf}\right)_h\right] = 896.292\ \mathrm{J/(m^2\,s\,K)}$

friction factor $f_h = \dfrac{f_h\mathrm{Re}_h}{\mathrm{Re}_h} = \dfrac{20.585}{500.0} = 0.04117$

pressure loss ratio $\dfrac{\Delta p_h}{L} = \dfrac{4f_hG_h^2}{2\rho_h d_h} = 5585.98\ \mathrm{N/m^3}$

It is useful at this point to assess the dependence of heat transfer performance on cell aspect ratio. On repeating the above calculations over the permitted range of fin heights the results are summarized in Table J.4.

The choice for the low pressure side is clear, choose the highest value of plate referred heat transfer coefficient, which also requires the least pressure loss. Plotting the heat transfer coefficient versus duct aspect ratio shows that we are approaching minimal performance return as the duct aspect ratio increases above 8/1. Beyond this point it may be desirable to go to double-cell surfaces, say $(b-tf)_h =$ 12.0 mm, i.e. with a duct cell 6.0 mm high carrying half the mass flow rate.

Cold high-pressure side

The Reynolds number on the cold high-pressure side may be obtained as follows

$$\frac{\mathrm{Re}_h}{\mathrm{Re}_c} = \left(\frac{d_h}{d_c}\right)\left(\frac{G_h}{G_c}\right)\left(\frac{\eta_c}{\eta_h}\right) = \left(\frac{4A_h}{P_h}\times\frac{P_c}{4A_c}\right)\left(\frac{\dot{m}_h}{z_hA_h}\times\frac{z_cA_c}{\dot{m}_c}\right)\left(\frac{\eta_c}{\eta_h}\right)$$

$$= \left(\frac{P_c}{P_h}\right)\left(\frac{\dot{m}_h}{\dot{m}_c}\times\frac{c_h}{c_c}\right)\left(\frac{\eta_c}{\eta_h}\right),\ \text{making use of edge-length } E = z_hc_h = z_cc_c$$

With $\mathrm{Re}_h = 500.0$ (laminar flow)

$$\frac{500.0}{\mathrm{Re}_c} = \left[\frac{(b-tf)_c+1.0}{8.0+1.0}\right]\left(\frac{24.683}{24.318}\times\frac{1.000}{1.000}\right)\left(\frac{0.00002850}{0.00003015}\right)\ \text{then}$$

$$\mathrm{Re}_c = \frac{4690.13}{(b-tf)_c+1.0}$$

Table J.5 Summary of results for cold high-pressure side. Duct width, $(c-tf)_h = 1.0$ mm

Duct aspect $\left(\dfrac{b-tf}{c-tf}\right)_c$	*Heat trans. coeff.* $\hat{\alpha}_c$, *J/(m² s K)*	*Core press. loss* $\dfrac{\Delta p_c}{L}$, *N/m³*
8/1	845.414	630.664
6/1	774.046	1073.10
4/1	630.216	2394.10
2/1	400.014	6389.06

Following the same calculation route as for the hot low-pressure side, Table J.5 summarizes findings

Search of core design

The overall heat transfer coefficient, log mean temperature difference, total plate surface, (half height cell surfaces + plate thickness), and block volume were obtained using the following equations.

Overall plate heat transfer coefficient, U in $J/(m^2\ s\ K)$, is obtained from

$$\frac{1}{U} = \frac{1}{\hat{\alpha}_h} + \frac{tf}{\lambda_f} + \frac{tp}{\lambda_w} + \frac{1}{\hat{\alpha}_c}$$

Log mean temperature difference, $\Delta\theta_{lmtd}$ is obtained using

$$\Delta\theta_{lmtd} = \frac{\Delta\theta_1 - \Delta\theta_2}{\ln(\Delta\theta_1/\Delta\theta_2)} = 44.61 \text{ deg K}$$

Total plate surface, S_{plate} m^2, is obtained from

$$S_{plate} = \frac{\dot{Q}_{duty}}{U\Delta\theta_{lmtd}}$$

(Half height cell surfaces + plate thickness) are

$$\left[\frac{b_h}{2} + tp + \frac{b_c}{2}\right] \text{ converted to metres}$$

Block volume, V in m^3 is then found from

$$V = S_{plate}\left[\frac{b_h}{2} + tp + \frac{b_c}{2}\right]$$

Table J.6 shows that a minimum exists for core volume. Using four-point cubic interpolation (Appendix B.7), it was found that the minimum volume of $V = 2.162482\ m^3$ occurred at $(b - tf)_c = 4.370676$ mm. Using this data, the following values for the optimum cold side design were obtained, each value also being

Table J.6 Search of core design for minimum volume

Cold-side duct aspect	*Overall heat transfer coeff. U, $J/(m^2\ s\ K)$*	*Plate surface S_{plate}, m^2*	*Half-height cells + plate thickness, m*	*Block volume V, m^3*
8/1	429.570	286.227	0.0084572	2.420678
6/1	410.346	299.636	0.0074572	2.234446
4/1	366.057	335.889	0.0064572	2.168902
2/1	274.351	448.165	0.0054572	2.445725

obtained by four-point cubic interpolation:

$$\begin{cases} \mathrm{Nu}_c = 5.4947112 \\ f\mathrm{Re} = 18.57827 \end{cases}$$

Reynolds number $\mathrm{Re}_c = \dfrac{4690.13}{(b - tf)_c + 1.0} = 873.284$

Repeating the same performance calculation illustrated for the hot low-pressure side, the following values were obtained

hydraulic diameter $d_c = \dfrac{4(b - tf)(c - tf)}{2[(b - tf) + (c - tf)]} = 1.627607\,\mathrm{mm}$

mass velocity $G_c = \dfrac{\mathrm{Re}_c \eta_c}{d_c} = 15.29154\,\mathrm{kg/(m^2\,s)}$

flow area of single duct $A_c = (b - tf)_c (c - tf)_c = 0.000004370676\,\mathrm{m^2}$

mass flow rate in duct $\dot{m}_c = G_c A_c = 0.00006683436\,\mathrm{kg/s}$

flow velocity $u_c = \dfrac{G_c}{\rho_c} = 2.678056\,\mathrm{m/s}$

heat transfer coefficient $\alpha_c = \dfrac{\mathrm{Nu}_c \lambda_c}{d_c} = 151.05327\,\mathrm{J/(m^2\,s\,K)}$

fin efficiency $\hat{m}_c = \sqrt{\dfrac{2\alpha_c}{\lambda_f tf}} = 308.9365\,\mathrm{m^{-1}}$

$$\hat{m}_c Y_c = 1.3502617$$

$$\phi_c = \frac{\tanh(\hat{m}_c Y_c)}{\hat{m}_c Y_c} = 0.6473670$$

fin performance ratio $\phi_c \left(\dfrac{b - tf}{c - tf}\right)_c = 2.829431$

plate heat transfer coefficient $\hat{\alpha}_c = \alpha_c \left[1 + \phi_c \left(\dfrac{b - tf}{c - tf}\right)_c\right] = 578.4481\,\mathrm{J/(m^2\,s\,K)}$

friction factor $f_c = \dfrac{f_c \mathrm{Re}_c}{\mathrm{Re}_c} = \dfrac{18.57827}{873.284} = 0.021274$

pressure loss ratio $\dfrac{\Delta p_c}{L} = \dfrac{4 f_c G_c^2}{2\rho_c d_c} = 1070.5716\,\mathrm{N/m^2}$

Final exchanger core

overall heat transfer coefficient $\dfrac{1}{U} = \dfrac{1}{\hat{\alpha}_h} + \dfrac{tf}{\lambda_f} + \dfrac{tp}{\lambda_p} + \dfrac{1}{\hat{\alpha}_c}$

$$U = 347.9689\,\mathrm{J/(m^2\,s\,K)}$$

log mean temp. difference $\Delta\theta_{lmtd} = \dfrac{\Delta\theta_1 - \Delta\theta_2}{\ln(\Delta\theta_1/\Delta\theta_2)} = 44.864\,\mathrm{deg\,K}$

total plate surface	$S_{plate} = \dfrac{\dot{Q}_{duty}}{U\Delta\theta_{lmtd}} = 350.540\,\text{m}^2$
core block volume	$V = S_{plate}\left[\dfrac{b_h}{2} + tp + \dfrac{b_c}{2}\right] = 2.328685\,\text{m}^3$
core block length	$L = 0.723\,\text{m}$ (this free choice affects Δp's)
exchanger (breadth × width)	$B \times W = \dfrac{V}{L} = 3.22087\,\text{m}^2$
pressure losses	$\Delta p_h = \left(\dfrac{\Delta p_h}{L}\right)L = 0.0408866\,\text{bar}$; $\Delta p_c = \left(\dfrac{\Delta p_c}{L}\right)L = 0.0077402\,\text{bar}$
pressure levels become	atmospheric pressure $p_{atm} = 1.01325\,\text{bar}$; hot-side exit pressure $p_{h2} = 0.98646\,\text{bar}$; cold-side exit pressure $p_{c1} = 9.09365\,\text{bar}$

The hot-side exit pressure level is below atmospheric pressure, a partial legacy from the original Kay & London design case (2nd edn, 1964). This result points to the importance of specifying pressure losses as input data in direct-sizing.

An immediate reduction in pressure loss for the low pressure hot-side may be achieved by selecting $(c - tf)_h = 1.25\,\text{mm}$ while keeping the aspect ratio at 8/1. The resulting hot-side plate heat transfer coefficent becomes $\hat{\alpha} = 830.365\,\text{J}/(\text{m}^2\,\text{s K})$, whilst the pressure loss reduces to $(\Delta p_h/L) = (4f_h G_h^2)/(2\rho_h d_h) = 2860.215\,\text{N/m}^3$. However the author did not pursue this design route, as both heat transfer performance data and friction factor data used were based on Shah & London's 1978 theoretical predictions. Note however that performance of the high-pressure cold-side should be re-calculated using new values of Reynolds numbers. The use of accurate data for f- and j-correlations is discussed later.

Comparison with earlier designs

Table J.7 compares the results of designing by direct-sizing with the new approximate approach. The contraflow design is not ideal for comparison with the optimum design, as the mean temperature differences do not correspond.

The specific performance parameter $\dot{Q}_{duty}/V\Delta\theta_m = US/V$ (kW/m^3 K) shows the optimized exchanger design to be best, Table J.7 also revealing that volume of the contraflow exchanger with mean temperature difference of 69.224 deg K is greater than that of the optimized exchanger with mean temperature difference of 44.864 deg K.

Experimental performance data

Kays & London (1964) provide heat transfer and flow-friction curves for only two plain rectangular fin geometries, surfaces 6.2 and 11.11(a) respectively, giving values of $StPr^{2/3}$ and f over the full range of Nu. Suitably modified theoretical parameters Nu and f Re plotted on the same figures suggest that performance values

Table J.7 Comparison of exchanger performance (first two sets of results from Chapter 4)

Parameter	*Crossflow exchanger*	*Contraflow exchanger*	*Optimum contraflow*
Hot inlet temp., K	702.59	702.59	702.59
Cold inlet temp., K	448.15	448.15	448.15
Hot outlet temp., K	521.28	521.28	498.22
Cold outlet temp., K	637.94	637.94	662.08
Effectiveness ε	0.7459	0.7459	0.8408
Hot Reynolds no., Re_h	1366.07	1555.85	500.0
Cold Reynolds no., Re_c	4079.98	2324.57	873.28
$\Delta\theta_m$ deg K	44.854	69.224	44.864
Ntu_h	4.0371	2.6134	4.5932
Ntu_c	4.2261	2.7347	4.8066
Plate surface S, m^2	571.50	398.1308	350.54
Volume V, m^3	3.803227	2.656238	2.328685
Overall heat trans. coeff. U, J/(m^2 s K)	189.373	175.704	347.969
$\dot{Q}_{duty}$, MW	4.8548	4.8548	5.4724
$\dfrac{\dot{Q}_{duty}}{V\Delta\theta_m} = \dfrac{US}{V}\left(\dfrac{\text{kW}}{\text{m}^3\,\text{K}}\right)$	28.4558	26.3354	52.3802
Hot surface	K&L plain 1/4-11.0 single cell	K&L plain 1/4-11.0 single cell	Plain rect. duct $c - tf = 1.0$ mm $b - tf = 8.0$ mm
Cold surface	K&L louver 3/8-06.06 single cell	K&L louver 3/8-06.06 single cell	Plain rect. duct $c - tf = 1.0$ mm $b - tf = 4.37$ mm

using Shah & London (1974) data may not match too closely the actual performance of plain rectangular ducts, reported by Kays & London (1964). For the plain rectangular duct surface 11.11(a) values at specific Reynolds numbers are compared with the values used in the new computational approach, Table J.8.

Directions for improvement

As illustrated in Table J.8, the present method of calculation should not be pressed too far in final design, but it does provide a good starting point for optimization, allowing the range of duct sizes to be tested to be minimized. Adequate allowance must be made to accommodate fouling particles, and it may be necessary to adjust one or both values of cell width appropriately.

Reliable experimental data for plain rectangular ducts in laminar flow is required, from which interpolating cubic splinefitted f- and j-correlations may be prepared. This may require an experimental single-blow testing program. Assessment of

Table J.8 Performance comparison for heat-transfer and flow-friction input data

Surfaces	*Reynolds number*	*Aspect ratio*	Nu_{HI}	$f\,Re$
Hot low-pressure side	500.0	8.0000	6.490	20.585
K&L surface 11.11(a)	500.0	5.7561	4.571	19.00
Cold high-pressure side	873.3	4.3707	5.4947	18.578
K&L surface 11.11(a)	873.3	5.7561	5.0789	19.581

duct thicknesses will be necessary to ensure that the ducts can withstand applied internal pressures.

Direct-sizing approach

Direct-sizing design is recommended as this involves specification of pressure losses as input data. A Pascal program is required, with input data designed to obtain minimum core block volume. Interpolating cubic spline-fits of original heat-transfer and flow-friction data should be used, and the heat-transfer data may involve Reynolds and Prandtl numbers.

The direct-sizing design software is not developed here, and should be added to the recommendations for future work listed in Appendix L.

Overall heat transfer coefficient

At this point it is worth considering what happens to the overall heat transfer coefficient. If we assume hot and cold fluid heat transfer coefficients to be α_1, α_2, and neglect thermal resistance of the separating plate, then the overall heat transfer coefficient U is found from

$$\frac{1}{U} = \frac{1}{\alpha_1} + \frac{1}{\alpha_2}, \quad \text{and when } \alpha_1 = \alpha_2, \qquad U = \frac{\alpha_2}{2}$$

$$\alpha_1 = 4\alpha_2, \qquad U = \frac{4}{5}\alpha_2$$

$$\alpha_1 \Rightarrow \infty, \qquad U \Rightarrow \alpha_2$$

from which the overall coefficient is seen to lie in the range $\alpha_2/2 < U < \alpha_2$ where α_2 is the smaller of α_1, α_2. There is great incentive to seek increase in the smaller coefficient α_2, and good incentive to seek increase in the larger coefficient α_1.

This observation supports the choice made to increase the size of the hot-side rectangular duct as discussed under 'Final Exchanger Core' above.

J.4 Exchanger optimization

Optimization route

This optimization problem was made possible by the following observations

- performance trend curve Figs C.1 to C.4 presented by Smith (1997–99) show that minimizing cell base values $(c - tf) = 1.0\,\text{mm}$ for both sides led to minimum core volume
- the hot low-pressure side required the largest practicable value for cell height, here $(b - tf)_h = 10.0\,\text{mm}$
- fixing cell geometry on the hot side with the constraint that Reynolds number was 500.0, allowed generation of cold side Reynolds numbers which depend on duct height only
- varying the cold side cell geometry revealed that a minimum exists for core volume, evaluated using four-point cubic interpolation.

J.5 Possible surface geometries

Plain rectangular fins

For many low fouling applications there may only be one prime candidate surface geometry – namely plain rectangular fins with duct aspect ratios in the range between 3/1 to 10/1 and cell spacing (fin pitching) between 1.15 to 1.5 mm (see Fig. 4.11).

Because of the narrowness of the finned channels there is no longer scope for use of rectangular offset-strip fins, nor for the use of wings. In any case little advantage is evident in use of strip fins – see Appendix C, Figs C.1 to C.4 (Smith 1997–99). It is also known that some perforated fin surfaces can generate unwanted noise. However the use of wavy channels might still prove to be practicable.

Printed circuit geometries have a maximum depth of etching around 2.0 mm which is well below the required height of duct. Two etched plates placed face to face would just approach the lower limit required, but the width of etching may prove to be too great.

Round tubes as surface geometry

Webb (1994, Table 2.2) shows round geometries to have reasonable thermal performance values of $(\text{Nu}_H = 4.364, f\,\text{Re} = 16.00)$. To ensure good thermal contact each separating plate would need to have cusped circular profiles machined on one side to ensure snug fit with the round tubes. Very small tubes of under 1.0 mm external diameter may prove to be of interest in special applications.

References

Cool, T., **Stevens, A.**, and **Adderley, C.I.** (1999) *Heat exchanger optimisation using genetic algorithms*, 6th UK National Heat Transfer Conference, Institution of Mechanical Engineers, Edinburgh, September, pp. 27–32.

Kays, W.M. and **London, A.L.** (1964) *Compact Heat Exchangers*, 2nd edn. McGraw Hill, New York.

London, A.L. and **Shah, R.K.** (1968) Offset rectangular plate-fin surfaces – heat transfer and flow friction characteristics, *ASME, J. Engng. Power*, **90**, 218–228.

Manglik, R.M. and **Bergles, A.E.** (1990) The thermal hydraulic design of the rectangular offset strip-fin compact heat exchanger. In *Compact Heat Exchangers – a Festschrift* *for A.L. London* (Eds R.K. Shah, A.D. Kraus, and D. Metzger), Hemisphere Press, pp. 123–149.

Shah, R.K. and **Seculic, D.P.** (2003) *Fundamentals of Heat Exchanger Design*, John Wiley & Sons Inc., New York.

Smith, E.M. (1994) Direct thermal sizing of plate-fin heat exchangers, *10th International Heat Transfer Conference, Industrial Sessions Papers* (Ed R.J. Berryman), Paper I/2-CHE-2, pp. 55–66.

Smith, E.M. (1997–99) *Thermal Design of Heat Exchangers*, John Wiley & Sons Ltd, Chichester, 1st edn, 1997, reprinted with corrections 1999.

Walsh, P.P. and **Fletcher, P.** (1998) *Gas Turbine Performance*, Blackwell Science, Oxford.

Webb, R.L. (1994) *Principles of Extended Heat Transfer*, John Wiley & Sons Inc., New York.

Appendix K
Continuum Equations

Basis of equations in solid and fluid mechanics – for those who insist on knowing where to begin

K.1 Laws of continuum mechanics

Valid range of theory

Continuum theory introduces the concept of small volumes and small time intervals, but these must not become too small otherwise the smoothness of macroscopic property values is no longer guaranteed. Below continuum (macroscopic) level we would enter the statistical (microscopic) level.

For fluid flow in small channels, Kandlikar & Grande (2002) proposed the following classification based on hydraulic diameter (d_h):

Conventional channels	$d_h > 3\,\text{mm}$
Minichannels	$200\,\mu\text{m} \leq d_h \leq 3\,\text{mm}$
Microchannels	$10\,\mu\text{m} \leq d_h \leq 200\,\mu\text{m}$
Transitional channels	$0.1\,\mu\text{m} \leq d_h \leq 10\,\mu\text{m}$
Nanochannels	$d_h < 0.1\,\mu\text{m}$

The microchannel range is influenced by rarefaction effects for gases described by the Knusden number (Kn) where

$$Kn = \frac{\lambda}{d_h}$$

and λ is mean free path for a gas obtainable from

$$\lambda = \frac{\eta\sqrt{\pi}}{\rho\sqrt{2RT}}$$

with

$R =$ gas constant, J/(kg K)
$\eta =$ absolute viscosity, N/(m s)
$\rho =$ density, kg/m^3
$T =$ absolute temperature, K

For air at 300 K, $\lambda = 0.068\ \mu$m (data from Kandlikar & Grande). McAdams (1954) suggests the following regimes for gases

Continuum flow	$Kn < 0.001$ implying $d_h > 0.068$ mm
Slip flow	$0.001 < Kn < 2$
Free-molecule flow	$Kn > 2$

Thus continuum flow may extend some way into the minichannel range suggested by Kandlikar & Grande. However, it would be desirable to calculate the Knusden number for each small duct application below 3 mm hydraulic diameter.

Table 4.9 for fully developed flow in plain rectangular ducts shows that

$$\mathrm{Nu}_H = \frac{\alpha d_h}{\lambda} = K_1 \qquad \text{and} \qquad f \times \mathrm{Re} = f \times \frac{d_h G}{\eta} = K_2$$

where (K_1, K_2) are constants for a given duct geometry.

As the hydraulic diameter (d_h) approaches the lower limit for continuum flow, the heat-transfer coefficient (α) must increase. Similarly the friction factor (f) must also increase, which implies that the duct length must be shorter to meet a pressure loss constraint.

In the limit, this means that we may be working very much within the developing flow region for laminar flow. The work of Shah & London (1978) provides a substantial amount of information, but not for developing flow on rectangular ducts with heat transfer.

There is need to explore this region further for the ducts of interest. One high-performance exchanger is reported to employ tubes 0.8 mm o.d. and 0.05 mm wall at temperatures up to 700 °C – evidently still in the continuum region.

Development of continuum theory

The main thrust of work in continuum mechanics was carried out by a group of applied mathematicians at John Hopkins University under the leadership of Clifford Truesdell. The objective of the work was to establish the exact form of equations for predicting the behaviour of continua. There was required to be exactly the same number of governing equations as there were unknowns.

This considerable undertaking clarified the approach to many problems, covering solid mechanics, single-phase and two-phase fluid flow, and more exotic phenomena. Much of the work was published in the journal *Archive for Rational Mechanics and Analysis*, and several textbooks also appeared, making the subject matter more accessible to engineers. The texts by Truesdell (1966a,b) and by Coleman *et al.* (1966) were followed by Jaunzemis (1967), Malvern (1969) and by many others.

Prior study of *Tensor Calculus* is desirable – on the surface this is just a compact way of writing equations, but it also carries the benefit of allowing an observer to view the physical phenomena from different directions while the equations remain unchanged. In engineering it may not be beneficial to compact the equations too much, as there may remain the requirement to solve the problem on a computer, and equations have eventually to be expanded before algorithms can be written.

Potential solutions developed from continuum theory might not always be fully practicable, in that too much data might be required to produce an engineering solution, but the considerable benefit remains that when simplifications are made to obtain a solution, the extent of the error could be known. A perfect example of this is illustrated in Sections K.2 and K.3, viz. *coupled* and *de-coupled* theories.

Both theories require first the concept of *Kinematics* – namely if a body is imagined to be cut into a series of rectangular blocks, then after deformation, all these distorted blocks will fit together exactly, leaving no holes. This deformation may be regarded as strain or as rate of strain – a concept involving pure mathematics only, and no equations.

There are five *Physical Axioms*, namely concepts in physics which are taken to be true as a result of observation – no proof being possible, but the concepts seem to fit the facts and can be expressed mathematically. Four of the expressions are equations, the last expression is an equality which simply has to be satisfied, viz.:

- balance of mass — one equation
- balance of linear momentum — three equations
- balance of moment of momentum — not required for irrotational flow
- balance of energy — one equation
- growth of entropy — an inequality

There are found to be more unknowns present than there are equations to be solved. This then requires construction of a sufficient number of *Constitutive* equations which can only be validated by experimentation.

The seemingly exotic nature of the last two constitutive equations in coupled theory (Section K.2) is probably due to the fact that they are seldom invoked – engineers find that most problems can be tackled by first simplifying the couple theory.

De-coupled theory (Section K.3) involves study of the energy equation, and examination of its two sub-forms, namely mechanical energy and thermal energy. There appears a 'stress power' term in each equation, and these terms differ in sign so that they seemingly disappear when mechanical and thermal energy equations are summed to form the total energy equation.

The stress power term is only important under ballistic impact conditions where it generates a significant amount of thermal energy – and may be neglected (i.e. included or omitted in different equations as appropriate) for most engineering applications. The mechanical energy equation then becomes identical to the balance of linear momentum equation. Hence the solution of problems involving both stress and temperature may be de-coupled and solved sequentially. Only two constitutive equations are now required – the stress/strain (or strain-rate) law and Fourier's law.

In the last set of equations presented in Appendix A.2 the term [pressure field] contains both the shear stress/strain law and an expression for pressure gradient. The stress/strain law has been modified to become an expression involving the friction factor (f) – see discussion under 'Pressure gradient due to friction' in Appendix A.1. The expression for pressure gradient takes care of flow acceleration/deceleration effects.

Fourier's law is built into each of the energy equations.

Transient equations

The de-coupled equations for a single fluid provide a foundation for the transient equation discussed in Appendix A – only now we have to cater for two fluids and one separating wall. These equations are only for one dimension in space and one in time. Further simplifications become possible when it is found that the Rayleigh dissipation terms are very small and may be neglected in the fluid equations.

In no circumstances should such equations be used to describe the flow of a non-Newtonian fluid, e.g. a thixotropic paint. It is best practice always to go back to the beginning and introduce the proper constitutive equation at the fundamental level.

Solution of equations

Solution of these equations may now be broken in to two groups. The first group contains the balance of mass (density field) and balance of linear momentum equations (density × velocity field) – one set for each fluid. The second group contains the balance of energy equations (temperature field) for the wall and both fluids.

First group

In each time interval there is need to use values of velocity from the previous time interval to set up the equations for solution. There may also be need to iterate solution of the balance of linear momentum equations until zero gradient is obtained at the assumed outlet condition.

The time intervals are small, being constrained by the Courant–Friedrichs–Lewy condition, but if the need should arise it would be possible to use the newly found velocity value and iterate solution of the first group of equations until change in the value of velocities was found to be less than some arbitrary small value.

The need to make such corrections might be omitted in a first computer run.

The same algorithms can be used for each fluid, it only being necessary to reverse-number input values for the second fluid before solution, and reverse renumber the solution when it appears – which approach minimizes coding errors.

Second group

The solution process is made straightforward by constructing the matrix to be solved mid-way between values for fluid temperatures, as there are then exactly the same number of equations as unknowns.

This approach uses arithmetic mean values of fluid temperatures, and any errors so introduced would be minimized by increasing the number of space intervals. A minimum of 50 space intervals is recommended.

K.2 Coupled continuum theory

Kinematics – geometry of deformation

$$\varepsilon_{ij} = \frac{1}{2}\left(\frac{\partial u_i}{\partial x_j} + \frac{\partial u_j}{\partial x_i}\right)$$

Basic axioms – physical principles

Balance of mass

$$\frac{D\rho}{Dt} + \rho\frac{\partial \dot{u}_i}{\partial x_i} = 0$$

Balance of linear momentum

$$\rho\frac{D\dot{u}_i}{Dt} = \frac{\partial \sigma_{ij}}{\partial x_i} + \rho b_i$$

Balance of moment of momentum (non-polar materials only)

$$\varepsilon_{ijk}\sigma_{ki} = 0 \qquad \text{implying symmetry of stress tensor} * \sigma_{ij} = \sigma_{ji}$$

Balance of total energy (mechanical + thermal)

$$\rho\frac{De}{Dt} + \rho\dot{u}_i\frac{D\dot{u}_i}{Dt} = \frac{\partial}{\partial x_i}(\sigma_{ij}\dot{u}_i) + \sigma\dot{u}_i b_i + \hat{r} - \frac{\partial \dot{q}_i}{\partial x_i}$$

Growth of entropy – an inequality, not an equation

$$-\rho\left(s\frac{DT}{Dt} + \frac{DF}{Dt}\right) + \sigma_{ij}\frac{\partial \dot{u}_i}{\partial x_j} - \frac{\dot{q}_i}{T}\frac{\partial T}{\partial x_i} \geq 0$$

where the Helmholtz function, $F = e - Ts$, is introduced in place of e.

Equations			*Unknowns*	
Axiomatic:				
Mass	1	1	Density	ρ
Linear momentum	3	3	Velocity components	$\dot{u}_i$
Moment of momentum	*	6	Stress components *	σ_{ij}
Energy	1	1	Specific internal energy	e
Constitutive:		3	Heat flux vector components	$\dot{q}_i$
Stress/strain	6	1	Temperature	T
Fourier's law	3	1	Specific entropy	s
Energy	1			
Entropy	1			
Equations $\Longrightarrow$	16	16	$\Longleftarrow$ Unknowns	

K.3 De-coupling the balance of energy equation

Balance of total energy

$$\rho \frac{De}{Dt} + \rho \dot{u}_i \frac{D\dot{u}_i}{Dt} = \frac{\partial}{\partial x_i}(\sigma_{ij}\dot{u}_i) + \sigma \dot{u}_i b_i + \hat{r} - \frac{\partial \dot{q}_i}{\partial x_i}$$

Balance of mechanical energy (scalar product of $\dot{u}_i$ and balance of linear momentum equation)

$$\rho \dot{u}_i \frac{D\dot{u}_i}{Dt} = \frac{\partial}{\partial x_i}(\sigma_{ij}\dot{u}_i) + \rho \dot{u}_i b_i - \sigma_{ij}\frac{\partial \dot{u}_i}{\partial x_j}$$

Balance of thermal energy

$$\rho \frac{De}{Dt} = -\frac{\partial \dot{q}_i}{\partial x_i} + \hat{r} + \sigma_{ij}\frac{\partial \dot{u}_i}{\partial x_j}$$

Adding the mechanical energy and thermal energy equations causes the stress power term

$$\left(\sigma_{ij}\frac{\partial \dot{u}_i}{\partial x_j}\right)$$

to hide in the total (mechanical + thermal) energy balance equation.

The stress power term is only important under ballistic impact conditions, and it can be neglected for most engineering applications. Thus the mechanical energy equation is not required to solve stress-field equations, and the solution of problems involving both stress and temperature may be de-coupled and solved sequentially.

Stress-field equations		*Unknowns*		
Axiomatic:				
Mass†	1	1	Density†	ρ
Linear momentum	3	3	Velocity components	$\dot{u}_i$
Moment of momentum	*	6	Stress components	σ_{ij}
Constitutive:				
Stress/strain (rate)	6			
Equations $\Longrightarrow$	10	10	$\Longleftarrow$ Unknowns	

† In small deformation of solids, density is assumed to be a prescribed constant, thus the balance of mass equation need not be invoked, and density may be deleted as an unknown.

Temperature-field equations		*Unknowns*		
Axiomatic:				
Thermal energy	1	1	Temperature	T
		3	Heat flux vector components	$\dot{q}_i$
Constitutive:				
Fourier's law	3			
Equations $\Longrightarrow$	4	4	$\Longleftarrow$ Unknowns	

References

Coleman, B.D., **Markovitz, H.**, and **Noll, W.** (1966) *Viscometric Flows of Non-Newtonian Fluids*, Springer, Berlin.

Jaunzemis, W. (1967) *Continuum Mechanics*, Macmillan, New York.

Kandlikar, S.G. and **Grande, W.J.** (2002) Evolution of microchannel flow passages – thermohydraulic performance and fabrication technology. *ASME International Mechanical Engineering Congress and Exposition*, New Orleans, 17–22 November, Paper IMECE2002-320453.

McAdams, W.H. (1954) *Heat Transmission*, 3rd edn, McGraw-Hill, New York.

Malvern, L.E. (1969) *Introduction to the Mechanics of a Continuous Medium*, Prentice-Hall, New Jersey.

Truesdell, C. (1966a) *Six Lectures on Modern Natural Philosophy*, Springer-Verlag, Berlin.

Truesdell, C. (1966b) *The Elements of Continuum Mechanics*, Springer-Verlag, New York.

Shah, R.K. and **London, A.L.** (1978) *Laminar Force Flow Convection in Ducts*, Supplement to *Advances in Heat Transfer*, Academic Press, New York.

Bibliography

Aparecido, J.B. and **Cotta, R.M.** (1990) Thermally developing laminar flow inside rectangular ducts. *Int. J. Heat and Mass Transfer*, vol. 33, no. 2, pp. 341–347.

Hunter, S.C. (1983) *Mechanics of Continuous Media*, 2nd edn, Ellis Horwood, Chichester.

Muzychka, Y.S. and **Yovanovich, M.M.** (2002) Laminar flow friction and heat transfer in non-circular ducts. Part I – Hydrodynamic problem, pp. 123–130. Part II – Thermal problem, pp. 311–319. *Compact Heat Exchangers: A Festschrift on the 60th Birthday of Ramesh K. Shah* (Eds., G.P. Celeta, B. Thonon, A. Bontemps, and S. Kandlikar), Begell House.

Truesdell, C. (1969) *Rational Thermodynamics*, McGraw-Hill, New York. (2nd edition has Appendix by C.-C. Wang, plus additional material contributed by 23 colleague authors.)

Appendix L
Suggested Further Research

Recommended extensions

L.1 Sinusoidal–lenticular surfaces

It may be useful to investigate further the thermal performance of sinusoidal–enticular geometries (Fig. L.1) as this class of surface geometry may possess special features absent from other surface geometries, viz.

- blockage of one channel by debris is limited to the point where the blockage occurs;
- migration of flow across the main flow direction may produce exchangers with improved mass flow distribution for both contraflow and crossflow;
- plain sinusoidal ducts currently show the highest thermal performance for compact exchangers;
- sinusoidal–lenticular ducts enlarge and contract throughout the heat-transfer surface and the wider portions will show improved thermal performance;
- offset-lenticular fins help restart boundary layers, leading to high heat transfer coefficients.

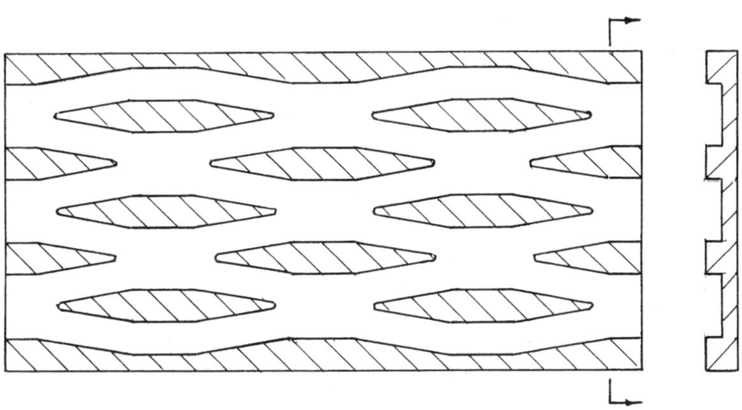

Fig.L.1 Sketch approximating to sinusoidal–lenticular surface geometry

L.2 Steady-state crossflow

In Chapter 3, examination of temperature and temperature-difference sheets of Figs 3.16 and 3.17 reveals that the pressure loss in parallel flow channels will not be the same. The core pressure loss is given by

$$\Delta p = \frac{4fG^2}{2\rho}\left(\frac{L}{d}\right) \qquad \text{where } f = \phi(\text{Re}) = \phi\left(\frac{dG}{\eta}\right) \qquad \text{and} \qquad G = \frac{\dot{m}}{A} = \rho u$$

thus $\Delta p = \varphi(\eta)$ where (η) is a function of temperature, and the temperature profiles of each flow path are different.

The assumption of equal mass flowrate across the inlet face of crossflow exchangers is made by nearly every researcher, and it now seems time that the validity of this assumption be explored, at least for the one-pass, unmixed–unmixed flow case.

For conventional surface geometries it is suggested that a programme of investigation might involve a number of sequential stages, viz.

1. Assume equal mass flowrates across the inlet face, and evaluate pressure loss profiles across the exchanger. It is possible that the solution approach may have to be by step-wise rating. This will provide a first measure of the extent of any potential problem.
2. Without changing the assumption of equal inlet mass flowrates, now apply Haselar's method of allowing for *cross-conduction* effects (Ch. 11, Section 11.6) to see if the problem is mitigated in any sense.
3. If the situation is serious enough, impose the condition of equal pressure loss for each flow channel and find the mass flowrate profiles across each inlet face.

This is not a trivial problem and may involve consideration of cell-by-cell heat transfer and flow friction over the complete exchanger.

L.3 Headers and flow distribution

Headers

Header design should be given as much attention as core design. In Chapter 8 the analytical solution of flow in a cylindrical gas burner with profiled insert, no dependence on inlet Reynolds number was required. From photographs of experimental work (see Dow's 1950 paper) the flow discharge along the header was level – but at different flame height – for different flowrates.

In a first approach to design of a header with varying rectangular cross-section (required for most heat exchangers), the numerical solution was less clear on this point.

For steady-state design it is recommended that a thorough analytical/experimental programme be initiated to provide a reliable design approach with confirmation by careful experimental investigation of the performance of Z-type and U-type header systems.

In later stages the work should be extended to investigate what happens during transients, with careful assessment of delays in entry to the exchanger core for both U-type and Z-type headers.

Flow distributors

The duty of a distributor is to connect the header to the core of the exchanger when necessary, and is not there primarily to exchange heat. Their presence may be unavoidable with certain contraflow designs.

The pressure loss needs to be investigated for tapering channels of rectangular cross-section for a range of rib angles in the range 0–15 degrees, and for varying mean widths of channel. The longer flow channels would have greater included angles and would need to be wider.

The target is the same equal minimum pressure loss for all inlet and outlet channels of both increasing and decreasing taper angles.

L.4 Transients in contraflow

The Crank–Nicholson finite-difference solution approach presented in Chapter 9, Appendix A and the supplement to Appendix B is slow in implementation. The reason for adopting Crank–Nicholson is that it permits a clear view of what is happening to each component of the governing equations, but it has not been fully computed, nor have its results been compared with those of an alternative approach.

The Method of Lines (Fletcher, 1991, vol. 1, Section 7.4) is a more efficient approach, and turns a single partial differential equation (PDE) into a series of ordinary differential equations (ODEs) at each station along the exchanger length, which are then solved using Runge–Kutta.

Thus

$$\frac{\partial \rho}{\partial t} + u\frac{\partial \rho}{\partial x} = -\rho\frac{\partial u}{\partial x}$$

may be discretized using backward differences to produce the ordinary differential equation to be solved by Runge–Kutta at every station

$$\left.\frac{d\rho}{dt}\right|_n^t = -u_n\left(\frac{\rho_n^t - \rho_n^t - 1}{\Delta x}\right) - \rho_n\left(\frac{u_n^t - u_n^{t-1}}{\Delta x}\right)$$

The neat geometrical description in geometrical terms of the Runge–Kutta method by Hamming (1962, Chapter 16) is reproduced below. An alternative description can be found in Press *et al.* (1989, Section 15.1). Given $\rho' = f(x, \rho)$ we compute in turn

$$k_1 = \Delta x f(x_n, \rho_n)$$

$$k_2 = \Delta x f\left(x_n + \frac{\Delta x}{2}, \rho_n + \frac{k_1}{2}\right)$$

$$k_3 = \Delta x f\left(x_n + \frac{\Delta x}{2}, \rho_n + \frac{k_2}{2}\right)$$

$$k_4 = \Delta x f(x_n + \Delta x, \rho_n + k_3)$$

from which

$$\Delta\rho = \frac{1}{6}(k_1 + 2k_2 + 2k_3 + k_4)$$

and

$$\rho_{n+1} = \rho_n + \Delta\rho$$

At the point (x_n, ρ_n) we compute the slope $(k_1/\Delta x)$ and with this value go one-half step forward and examine the slope there. Using this new slope $(k_2/\Delta x)$ we start again at (x_n, ρ_n) and go one-half step forward and again sample the slope. Using this latest slope $(k_3/\Delta x)$ we again start at (x_n, ρ_n) but now go a full step forward to examine the slope $(k_4/\Delta x)$. The four slopes are averaged using weights of $(1/6, 2/6, 2/6, 1/6)$, and using this average slope we make the final step from (x_n, ρ_n) to (x_{n+1}, ρ_{n+1}). Essentially Runge–Kutta simply integrates the curve between its two ends and adds the result to the initial value (x_n, ρ_n).

The reader may be struck by the similarity between the weights used by Runge–Kutta and the abscissae and weighting factors of the Gaussian four-point quadrature method using Legendre polynomials shown in Table E.4 of Appendix E. Hamming states that '... Gaussian methods of integration are effective in estimating the value of an integral from a few samples providing that the function can be accurately approximated by a polynomial' (Hamming, 1962, Section 12.5).

Formal mathematical derivation of the Runge–Kutta method from Taylor series is presented by Hildebrand (1976, Section 3.5). Spencer *et al.* (1977, vol. 2, Section 7.3) show that the fourth-order Runge-Kutta method is to be preferred, and that no higher-order method is viable. Implementation of the method can be found in the text by Press *et al.* (1989). Jeffrey (1989, Section 17.6) shows how the method of lines with Runge–Kutta may be readily extended to solve sets of simultaneous equations.

It is suggested that the relative performance of the Crank–Nicholson and method of lines approaches be computed and compared, using the same cubic-spline fitting of temperature-dependent physical data and an identical worked example.

Later consideration could be given to adding a diffusion term to the fluid energy equations as diffusion rates are more often driven by temperature than by turbulence alone. In the energy equation addition of diffusion terms can provide additional computational stability. The centred difference expression would be

$$D\left(\frac{\partial^2 t}{\partial x^2}\right) \Rightarrow D_n^t\left(\frac{T_{n+1}^t - 2T_n^t + T_{n-1}^t}{(\Delta x)^2}\right)$$

Rayleigh dissipation models shear stresses acting throughout the fluid which generate frictional heating, but this last effect is likely to be negligible for heat exchangers.

For fluids other than perfect gases, analytical equations of state in Bejan (1997), including Redlich–Kwong and Benedict–Webb–Rubin, might be used to extend the computational methods to other fluids, providing the restriction on equations (A.1), Appendix A, remains that the fluids retain Newtonian constitutive form, viz.

$$\sigma_{ij} = -pI + 2\eta\dot{e}_{ij} - \frac{2}{3}\eta\delta_{ij}\dot{e}_{kk}$$

References

Bejan, A. (1997) *Advanced Engineering Thermodynamics*, 2nd edn, John Wiley.

Fletcher, C.A.J. (1991) *Computational Techniques for Fluid Dynamics*, vols. 1 and 2, 2nd edn, Springer.

Hamming, R.W. (1962) *Numerical Methods for Scientists and Engineers*, McGraw-Hill, New York.

Hildebrand, F.B. (1976) *Advanced Calculus for Applications*, 2nd edn, Prentice-Hall, New Jersey.

Jeffrey, A. (1989) *Mathematics for Engineers and Scientists*, 4th edn, Van Nostrand.

Press, W.H., **Flannery, B.P.**, **Teukolsky, S.A.**, and **Vetterling, W.T.** (1989) *Numerical Recipies in Pascal*, Cambridge University Press, Cambridge.

Ramirez, W.F. (1997) *Computational Methods for Process Simulation*, Butterworth-Heinemann.

Spencer, A.J.M., **Parker, D.F.**, **Berry, D.S.**, **England, A.H.**, **Faulkner, T.R.**, **Green, W.A.**, **Holden, J.T.**, **Middleton, D.**, and **Rogers, T.G.** (1977) *Engineering Mathematics*, vols. 1 and 2, Van Nostrand.

Appendix M
Conversion Factors

Work in SI units

Length (L), m

1 inch = 25.4 mm

1 ft = 0.3048 m

1 mile = 5280 ft = 1.609 km

Area (cross-section A, surface S), m^2

1 in^2 = 645.2 mm^2

1 ft^2 = 0.0929 m^2

Volume (V), m^3

1 in^3 = 16 387 mm^2

1 ft^3 = 0.028 32 m^3

1 UK gallon = 0.004 564 m^3
= 4.546 litres

1 US gallon = 0.003 785 m^3
= 3.785 litres

Mass (M), kg

1 lbm = 0.4536 kg

1 ton = 2240 lbm = 1016 kg

1 tonne = 1000 kg

Density (ρ), kg/m^3

$1\ \text{lbm/in}^3 = 2.768 \times 10^4\ \text{kg/m}^3$

$1\ \text{lbm/ft}^3 = 16.0185\ \text{kg/m}^3$

$1\ \text{g/cm}^3 = 1000\ \text{kg/m}^3$

Pressure (p), N/m^2

$1\ \text{lbf/in}^2 = 6894.76\ \text{N/m}^2$

$1\ \text{kgf/cm}^2 = 98.0665\ \text{N/m}^2$

$1\ \text{tonf/in}^2 = 15.44\ \text{MN/m}^2$

$1\ \text{Pa} = 1\ \text{N/m}^2$

$1\ \text{bar} = 14.50\ \text{lbf/in}^2 = 10^5\ \text{N/m}^2$

1 atm. = 1.013 25 bar

1 atu = pressure over atmospheric (not desirable as a unit)

$1\ \text{mmHg} = 1\ \text{torr} = 133.322\ \text{N/m}^2$
($T = 0°\text{C}$, $P = 760$ mmHg)
($g = 9.806\ 65\ \text{m/s}^2$)

$1\ \text{mmH}_2\text{O} = 9.806\ 38\ \text{N/m}^2$
($T = 4°\text{C}$, $P = 760$ mmHg)
($g = 9.806\ 65\ \text{m/s}^2$)

Energy (W, Q)

1 ft pdl = 0.042 14 N m or J

1 ft lbf = 1.356 N m or J

1 Btu = 1055 N m or J

1 therm = 105.5 MN m or MJ

1 kWh = 3.600 MN m or MJ

Power, W

1 ft lbf/s = 1.356 N m/s or W

1 hp = 550 ft lbf/s = 745.6 N m/s or W

Force, N

1 pdl = 0.1383 N
1 lbf = 4.448 N
1 tonf = 9964 N
1 kgf = 9.807 N
1 dyne = 10^{-5} N

Torque, N m

1 lbf ft = 1.356 N m
1 tonf ft = 3037 N m

Velocity (u), m/s

1 ft/s = 0.3048 m/s
1 mile/h = 0.4470 m/s
1 knot = 0.514 444 m/s

Gas constant (R), J/(kg K)

1 ft lbf/(lbm *R*) = 5.380 95 J/(kg K)

Specific heat capacity (Cp), J/(kg K)

1 Btu/(lbm *R*) = 4186.8 J/(kg K)
1 kcal/kg K = 4186.8 J/(kg K)

Thermal conductivity (λ), J/(m s K)

1 Btu/(ft h *R*) = 1.730 73 J/(m s K)
1 cal/(cm s *C*) = 4186.8 J/(m s K)
1 kcal/(m h *C*) = 1.163 J/(m s K)

Thermal diffusivity [$\kappa = \lambda/(\rho \cdot Cp)$], m^2/s

1 ft^2/h = 0.000 025 806 m^2/s

Heat-transfer coefficient (α, U), J/(m^2 s K)

1 Btu/(ft^2 h R) = 5.678 26 J/(m^2 s K)
1 kcal/(m^2 h C) = 1.163 J/(m^2 s K)

Dynamic (absolute) viscosity (η), kg/(m s)

1 lbm/(ft h) = 0.000 413 kg/(m s)
1 poise = 0.1 kg/(m s)
1 centipoise = 0.001 kg/(m s)
1 (N s)/m^2 = 1 kg/(m s)
1 lbf/(ft s) = 1.488 16 kg/(m s)
1 (kgf s)/m^2 = 9.806 65 kg/(m s)
1 slug/(ft s) = 47.8802 kg/(m s)
1 (lbf s)/ft^2 = 47.8802 kg/(m s)
1 gm/(cm s) = 0.1 kg/(m s)
1 (dyne s)/cm^2 = 0.1 kg/(m s)

Kinematic viscosity ($\nu = \eta/\rho$) – convert to dynamic viscosity (η)

1 stoke = 10^{-4} m^2/s

Surface tension (σ), N/m

1 lbf/in = 175.127 N/m
1 dyne/cm = 10^{-3} N/m

Notation

SI units (preferred throughout)

Commentary

The new international standards for notation are followed, with some exceptions. Circumstances always arise where an awkward choice can be avoided and notation simplified, if there is departure from the standard. It was found that single-blow transients deserved such treatment, and the symbol for temperature was changed from T to θ, to allow the use of X, Y, T for dimensionless length and scaled time.

It was relatively easy to accept most of the new symbols, e.g.

- individual heat transfer coefficient (α for h)
- thermal conductivity (λ for κ)
- thermal diffusivity (κ for α)
- absolute viscosity (η for μ)

although in the last case the same symbol is now used for efficiency and absolute viscosity, while μ remains available, at least for single-species heat transfer.

While lengthy discussions to arrive at the final preferred list of international symbols must have occurred, this author will plead that the preferred list is for guidance of the experienced, and for observance by the novice. Most readers of this volume will fall into the first category, and will appreciate the problem of having too many subscripts. Where departure from the preferred convention has arisen, it has been solely to achieve clarity of presentation.

Examples of the important symbols used are

- surface area, S, associated with overall heat-transfer coefficient, U
- area of cross-section, A
- fluid mass flowrate, $\dot{m}$
- solid wall mass, M
- specific heat at constant pressure, C
- mass velocity of fluid, $G = \dot{m}/A$
- temperature, steady-state and transient, T
- temperature difference, $\Delta\theta$
- non-dimensional temperatures, θ
- time, dimensionless time, t, T

Dimensionless groups are treated at the end of Chapter 2, and will not be further listed in the tables of symbols. One or two of the less-used groups are explained where they arise.

Chapter 2 Fundamentals

Symbol	*Parameter*	*Units*
A	area of cross-section	m^2
C	specific heat at constant pressure	J/(kg K)
E	exergy	J/s
f	friction factor	
G	mass velocity, $\dot{m}/A$	$kg/(m^2 s)$
h	specific enthalpy	J/kg
ℓ	characteristic length	m
L	flow length	m
$\dot{m}$	mass flowrate	kg/s
N	number of overall transfer units, $U S/(\dot{m} C)$	
Ntu	larger value of N_h, N_c	
p	absolute pressure (bar $\times 10^5$)	N/m^2
q	heat flow	J/s
$\dot{Q}$	exchanger duty	W or J/s
R	ratio $M_w C_w/(\tilde{m} C)$, see Appendix A	
S	reference surface area	m^2
t	time	s
T	temperature	K
$Tspan$	temperature span of an exchanger	K
U	overall heat transfer coefficient	$J/(m^2 s K)$
W	ratio of water equivalents ($W < 1$)	
x,y,z	length	m
Greek symbols		
α	heat-transfer coefficient	$J/(m^2 s K)$
γ	ratio of specific heats (Cp/Cv)	
Γ	constant	
Δp	core pressure loss	N/m^2
$\Delta \theta$	temperature difference	K
ε	effectiveness	
θ	non-dimensional temperature	
κ	thermal diffusivity, $\lambda/(\rho C)$	m^2/s
λ	thermal conductivity	J/(m s K)
ξ	normalized length	
ρ	density	kg/m^3

Symbol	*Parameter*	*Units*
τ	residence time	s
Subscripts		
fg	latent heat	
h, c, w	hot, cold, wall	
m	mean	
lim	limiting	
lmtd	log mean temperature difference	
loss	loss	
1, 2	ends of exchanger	

Chapter 3 Steady-state temperature profiles

Symbol	*Parameter*	*Units*
A	area of cross-section	m^2
C	specific heat at constant pressure	J/(kg K)
f	friction factor	
G	mass velocity, $(\dot{m}/A)$	
H, C, W	finite-difference temperatures, (hot, cold, wall)	K
Kc, Ke	coefficients of contraction, expansion	
L	length of exchanger	m
$\dot{m}$	mass flowrate	kg/s
$\tilde{m}$	residence mass of fluid (constant velocity only)	kg
M	mass of solid wall	kg
n	number of local transfer units, $\alpha S/(\dot{m} C)$	
N	number of overall transfer units, $U S/(\dot{m} C)$	
p	absolute pressure (bar $\times\ 10^5$)	N/m^2
$\dot{Q}$	exchanger duty	W or J/s
R	ratio of thermal capacities, $(M_w C_w)/(\tilde{m}_h C_h)$ etc.	
s	curved length of an involute	m
S	surface area	m^2
t	angle in radians for an involute	
tp	plate thickness	m
T	temperature	K
U	overall heat transfer coefficient	$J/(m^2\ s\ K)$
v	specific volume	m^3/kg
V	volume	m^3
x, y	length	m
X, Y	normalized length, $(X = x/L_x,\ Y = y/L_y)$	

Symbol	*Parameter*	*Units*
Greek symbols		
α	heat-transfer coefficient	$J/(m^2\ s\ K)$
Δp	core pressure loss	N/m^2
ε	effectiveness	
η	absolute viscosity	kg/(m s)
θ	dimension less temperature	
κ	thermal diffusivity, $\lambda/(\rho C)$	m^2/s
λ	thermal conductivity	J/(m s K)
ξ	normalized length, (x/L)	
ρ	density	kg/m^3
σ	ratio $(A_{flow}/A_{frontal})$	
τ	residence time	s
Subscripts		
h, c, w	hot, cold, wall	
x, y	directions	
Local parameters		
A_0, A_1, A_2, A_3	defined in text	
p, q, r_2, r_3	defined in text	
a_{ij}	matrix coefficients	
$\beta_1, \beta_2, \rho, \mu$	defined in text	

Chapter 4 Direct-sizing of plate–fin exchangers

Symbol	*Parameter*	*Units*
a	individual cell flow areas	m^2
b	plate spacing	m
c	cell pitch	m
C	specific heat	J/(kg K)
D	cell hydraulic diameter	m
E	edge length	m
f	flow friction coefficient	
G	mass velocity	$kg/(m^2\ s)$
(h, l, s, t)	Manglic & Bergles parameters defined in text	
j	Colburn heat transfer coefficient	
L	flow length	m
$\dot{m}$	mass flowrate	kg/s
n	number of local transfer units, $\alpha S/(\dot{m} C)$	
N	number of overall transfer units, $U S/(\dot{m} C)$	
p	absolute pressure (bar $\times 10^5$)	N/m^2

Symbol	*Parameter*	*Units*
Per	cell perimeter	m
$\dot{Q}$	exchanger duty	W or J/s
R	gas constant	J/(kg K)
tf	fin thickness	m
tp	plate thickness	m
ts	splitter thickness	m
T	temperature	K
U	overall heat transfer coefficient	J/(m^2 s K)
x	strip length	m
z	number of cells	
Greek symbols		
α	heat-transfer coefficient	J/(m^2 s K)
α, δ, γ	Manglic & Bergles ratios defined in text	
Δp	core pressure loss	N/m^2
$\Delta\theta$	temperature difference	K
η	absolute viscosity	kg/(m s)
κ	thermal diffusivity	m^2/s
λ	thermal conductivity	J/(m s K)
ρ	density	kg/m^3
Subscripts		
h, c, w	hot, cold, wall	
lmtd	log mean temperature difference	
m	mean	
1, 2	ends of exchanger	
Surface parameters		
alpha	Stotal/Vexchr	1/m
beta	Stotal/Vtotal	1/m
gamma	Sfins/Stotal	
kappa	Stotal/Splate	
lambda	Sfins/Splate (*kappa* × *gamma*)	
sigma	Aflow/Aplate	

Chapter 5 Direct-sizing of helical-tube exchangers

Symbol	*Parameter*	*Units*
a	local area	m^2
A	total area	m^2
b	dimensionaless parameter	

Symbol	*Parameter*	*Units*
C	specific heat	J/(kg K)
d	tube diameter	m
D	mandrel, wrapper, mean coil diameters	m
f	friction factor	
G	mass velocity	kg/(m^2 s)
$K1, \ldots, K7$	factors defined in text	
ℓ	length of a single tube	m
ℓ_c	length of tubing in one longitudinal tube pitch	
ℓ_p	tubing in projected transverse cross-section	
L	length of tube bundle	m
m	integer number of tubes in outermost coil	
$\dot{m}$	mass flowrate	kg/s
n	integer number of tubes in innermost coil	
N	total number of tubes in the exchanger	
p	longitudinal tube pitch	m
p	absolute pressure (bar $\times 10^5$)	N/m^2
P_y	shell-side porosity	
$\dot{Q}$	exchanger duty	W or J/s
r	start factor (integer 1 to 6 only)	
S	reference surface area	m^2
t	transverse tube pitch	m
T	temperature	K
u	velocity	m/s
U	overall heat transfer coefficient	J/(m^2 s K)
V	volume	m^3
y	number of times shell-side fluid crosses a tube turn	
z	integer number of tubes in intermediate coil	
Greek symbols		
α	heat-transfer coefficient	J/(m^2 s K)
Δp	core pressure loss	N/m^2
$\Delta\theta_{lmtd}$	log mean temperature difference	K
η	absolute viscosity	kg/(m s)
λ	thermal conductivity	J/(m s K)
ρ	density	kg/m^3
ϕ	helix angle of coiling	
Subscripts		
a	annular	
i	inside	
max	maximum	
min	minimum	
s, t, w	shell-side, tube-side, wall	

Note: tube outside diameter (d) has no subscript, as this is the reference surface.

Chapter 6 Direct-sizing of bayonet-tube exchangers

Symbol	*Parameter*	*Units*
a, b	constants defined in equation (6.22)	
A, B	constants	
C	specific heat	J/(kg K)
d, D	diameter	m
ℓ	length of tube	m
L	length of exchanger	m
$\dot{m}$	mass flowrate	kg/s
N	number of overall transfer units, $N = U\,S/(\dot{m}\,C)$	
p	absolute pressure (bar $\times 10^5$)	N/m^2
P	perimeter transfer units, $P = N/L$	1/m
$\dot{Q}$	exchanger duty	W or J/s
s	spacing between two parallel flat plates	m
T	temperature	K
u	velocity	m/s
U	overall heat-transfer coefficient	$J/(m^2\ s\ K)$
x	distance	m
X	locus of minimum	m
Z	mean tube perimeter	m
Greek symbols		
α, β	parameters defined in the text	
Δp	pressure loss	N/m^2
ε	effectiveness	
η	absolute viscosity	kg/(m s)
θ	temperature for case of condensation	K
ϕ	function	
Subscripts		
b, e	bayonet, external	
i, o	inner, outer	
min	minimum	
1, 2, 3	defined in Figs 6.1, 6.4, 6.5, and 6.8	
Embellishments		
$\hat{}$	inner bayonet-tube fluid	
$\bar{}$	mean value	

Chapter 7 Direct-sizing of RODbaffle exchangers

Symbol	*Parameter*	*Units*
a	flow area per single tube	m^2
A	total flow area	m^2
B	number of RODbaffles	
d	diameter	m
D	shell diameter	m
f	friction factor	
G	mass velocity	$kg/(m^2\ s)$
k	baffle loss coefficient	
L	length	m
L_b	baffle spacing	m
$\dot{m}$	mass flowrate	kg/s
n	number of local transfer units, $\alpha S/(\dot{m}C)$	
N	number of overall transfer units, $US/(\dot{m}C)$	
p	tube pitch	m
p	absolute pressure (bar $\times 10^5$)	N/m^2
$\dot{Q}$	exchanger duty	W or J/s
r	baffle rod radius	m
T	temperature	K
u	velocity	m/s
U	overall heat-transfer coefficient	$J/(m^2\ s\ K)$
Z	number of tubes	
Greek symbols		
α	heat-transfer coefficient	$J/(m^2\ s\ K)$
Δp	core pressure loss	N/m^2
$\Delta\theta_{lmtd}$	log mean temperature difference	K
ε	surface roughness	m
η	absolute viscosity	kg/(m s)
λ	thermal conductivity	J/(m s K)
ρ	density	kg/m^3
Subscripts		
b, p	baffle, plain	
s, t	shell, tube	
Terms from paper by Gentry et al.		
C_L	coefficient in correlation $Nu = C_L(Re_h)^{0.6}\ (Pr)^{0.4}(\eta_b/\eta_w)^{0.14}$ where $C_L = (\xi_\ell)(C_\ell)$	
C_T	coefficient in correlation $Nu = C_T(Re_h)^{0.8}\ (Pr)^{0.4}(\eta_b/\eta_w)^{0.14}$ where $C_T = (\xi_t)(C_t)$	
C_1, C_2	coefficients in correlation $k_b = \phi(C_1 + C_2/Re_b)$	

Symbol	Parameter	Units
D_{bi}	exchanger baffle ring inner diameter	m
D_{bo}	exchanger baffle ring outer diameter	m
D_o	exchanger outer tube limit	m
D_s	shell inner diameter	m
ξ_ℓ, ξ_t	expressions defined in papers by Gentry *et al.*	

Chapter 8 Exergy loss and pressure loss

Symbol	Parameter	Units
a, b	constants in temperature ratios, and in friction factors	
A	area for flow	m^2
b	specific exergy, $b = h - T_0 s$	kJ/kg
$\dot{B}$	rate of exergy change, $\dot{B} = \dot{m}(b_{out} - b_{in})$	J/s
C	specific heat at constant pressure	J/(kg K)
D	hydraulic diameter	m
e	specific internal energy	J/kg
f	friction factor	
h	specific enthalpy	kJ/kg
$\dot{I}$	rate of irreversibility production	J/s
L	length of header	m
$\dot{m}$	mass flowrate	kg/s
$\dot{m}_0$	header inlet mass flowrate	kg/s
N_h, N_c	number of transfer units, $N_h = US/(\dot{m}C)_h$, $N_c = US/(\dot{m}C)_c$	
N_x	exergy loss number	
p	pressure	N/m^2
$\dot{q}$	specific heat flow	$J/(m^2\ s)$
$\dot{Q}$	heat flowrate	J/s
r_{hyd}	hydraulic radius	m
R	gas constant	J/(kg K)
s	specific entropy	J/(kg K)
S	reference surface area for heat transfer	m^2
$\dot{S}_{gen}$	entropy generation rate	J/(s K)
t	time	s
T	temperature	K
u	velocity	m/s
U	overall heat-transfer coefficient	$J/(m^2\ s\ K)$
v	specific volume	m^3
V	core volume	m^3

Symbol	Parameter	Units
W	work	N m
x, y	distance	m
Greek symbols		
γ	isentropic index	
$\Delta\dot{B}$	exergy change rate	J/s
Δp	pressure difference	N/m^2
$\Delta\theta$	local temperature difference	K
$\Delta\theta_{LM}$	log mean temperature difference	K
ε	effectiveness	
η	absolute viscosity	kg/(m s)
ρ	density	kg/m^3
$\phi()$	function of	
Subscripts		
c, h	cold, hot	
0	dead state	
1, 2	hot, cold end of exchanger	

Chapter 9 Transients in heat exchangers

Symbol	Parameter	Units
A	cross-sectional area	m^2
A, B, C	numerical coefficients in velocity-field algorithms	
C	specific heat	J/(kg K)
E, F, G, H	numerical coefficents in temperature-field algorithms	
f	friction factor	
G	mass velocity	$kg/(m^2\ s)$
L	length	m
m	number of space increments in exchanger length	
$\dot{m}$	mass rate of flow	kg/s
$\tilde{m}$	residence mass	kg
M	mass of exchanger core	kg
p	absolute pressure (bar $\times 10^5$)	N/m^2
S	reference surface area	m^2
t	time	s
T	temperature	K
u	velocity	m/s
U	overall heat-transfer coefficient	$J/(m^2\ s\ K)$

Symbol	Parameter	Units
W	flow work terms	K/s
x	distance	m
Greek symbols		
α	heat-transfer coefficient	J/(m^2 s K)
α, β	characteristic directions	
Δ	increment	
η	absolute viscosity	kg/(m s)
κ	thermal diffusivity	m^2/s
λ	thermal conductivity	J/(m s K)
ρ	density	kg/m^3
Subscripts		
h, c, w	hot, cold, wall	
j	subscript, indicating space station	
t	superscript, indicating time interval	

Chapter 10 Single-blow test methods

Symbol	Parameter	Units
a	arbitrary radius	m
a_0, a_1, b_1	numerical constants	
B	mean solid temperature excess $(\theta_b - \theta_i)$	K
$B^{\#}$	non-dimensional ratio (B_2/G_1)	
C	specific heat	J/(kg K)
D	non-dimensional inlet disturbance	
G	mean fluid temperature excess $(\theta_g - \theta_i)$	K
$G^{\#}$	non-dimensional ratio (G_2/G_1)	
k	numerical constant	
L	length of matrix	m
$\dot{m}$	mass flowrate of gas	kg/s
$\tilde{m}_g$	mass of gas in matrix	kg
M_b	mass of matrix	kg
Ntu	number of transfer units (one local value only)	
r	radius	m
R_{bg}	ratio $M_bC_b/(\tilde{m}_gC_g)$	
s	Laplace transform image of t	
S	surface area	m^2
t	time	s
t^*	time constant of inlet exponential temperature disturbance	

Symbol	*Parameter*	*Units*
T	temperature	K
u	gas velocity defined as $(\dot{m}_g L/\tilde{m}_g)$	m/s
V	volume of solid matrix	m^3
x	distance into matrix	m
Greek symbols		
α	heat-transfer coefficient	$J/(m^2\ s\ K)$
β	ratio (τ/Ntu)	
$\delta()$	delta function	
θ	temperature above reference state	K
κ	thermal diffusivity	m^2/s
ξ	non-dimensional scaling of length	
σ	dummy variable	
τ	non-dimensional time	
τ^*	non-dimensional time constant	
ω	rotational speed	1/s
ϖ	non-dimensional rotational speed	
Subscripts		
b, s	bulk, surface	
h, c	hot, cold	
g	gas	
i	initial isothermal reference state	
w	wall	
$1, 2$	inlet, outlet	

Chapter 11 Heat exchangers in cryogenic plant

Symbol	*Parameter*	*Units*
a, b	arbitrary limits	
c	sonic velocity	m/s
C	specific heat at constant pressure	J/(kg K)
h	specific enthalpy	J/kg
k	number of stages of compression	
p	absolute pressure (bar $\times 10^5$)	N/m^2
$\dot{Q}$	exchanger duty	W or J/s
r	compression ratio	
R	gas constant	J/(kg K)
S	entropy	J/(kg K)
T	temperature	K
W	work	W or J/s
x, y	fractions	

Symbol	Parameter	Units
Greek symbols		
α	blade angle, preferred notation for gas turbines	
γ	isentropic index, (C_p/C_v)	
η	efficiency	
θ	angle	
Subscripts		
e, n, o, p	equilibrium, normal, ortho-, para- (forms of hydrogen)	
fg	saturation field	
min	minimum	
s	isentropic	
0	dead state	
0, 1, 2, 3	stations in radial turbine analysis	
Embellishments		
$\bar{}$	mean value	

Chapter 12 Heat transfer and flow friction in two-phase flow

Symbol	Parameter	Units
a	numerical constant	
A	area for flow	m^2
B	numerical constant	
c	numerical constant	
C	numerical parameter depending on flow condition	
d	tube diameter	m
E, F, H	parameters in Friedel's correlation	
f	friction factor	
Fl	heat flux	W or J/s
g	acceleration due to gravity	m/s^2
G	mass velocity	$kg/(m^2\ s)$
ℓ	length	m
m	numerical constant	
$\dot{m}$	mass flowrate	kg/s
n	numerical constant	
p	absolute pressure (bar $\times\ 10^5$)	N/m^2
$\dot{q}$	heat flowrate	W or J/s
T	temperature	K
U	overall heat-transfer coefficient	$J/(m^2\ s\ K)$
x	dryness fraction	
X^2	ratio defined in text	

Symbol	*Parameter*	*Units*
Greek symbols		
α	heat-transfer coefficient	$J/(m^2\ s\ K)$
$\Delta\ell$	length increment	m
Δp	pressure loss	N/m^2
η	absolute viscosity	kg/(m s)
ρ	density	kg/m^3
σ	surface tension	N/m
ϕ	two-phase flow multiplier	
Subscripts		
crit	critical	
f	liquid	
fg	saturation	
g	vapour	
tp	two-phase	

Appendix A Transient equations with longitudinal conduction and wall thermal storage

Symbol	*Parameter*	*Units*
A	wall cross-section for longitudinal conduction	m^2
C	specific heat at constant pressure	J/(kg K)
e	specific internal energy	J/kg
$\dot{e}$	strain rate	
f	friction factor	
I	unit matrix	
L	length	m
$\dot{m}$	mass rate of flow	kg/s
M	mass of exchanger solid wall, ($M_w = \rho_w A_w L$)	kg
p	absolute pressure (bar $\times\ 10^5$)	N/m^2
$\dot{q}$	heat flow rate	$J/(m^2\ s)$
$\hat{r}$	radiation	$J/(m^3\ s)$
r_{hyd}	hydraulic radius	m
R	gas constant	J/(kg K)
S	reference surface area	m^2
t	time	s
T	temperature	K
u	velocity	m/s
V	total volume of exchanger solid wall	m^3

Symbol	*Parameter*	*Units*
W	dissipation terms	K/s
x,y	distance	m
Greek		
α	local heat transfer coefficient	J/(m^2 s K)
η	absolute viscosity	kg/(m s)
κ	thermal diffusivity $\lambda/(\rho C)$ – for re-defined thermal diffusivity (see below)	m^2/s
λ	thermal conductivity	J/(m s K)
ρ	density	kg/m^3
σ	stress	N/m^2
τ	shear stress	N/m^2
Φ	Rayleigh dissipation function	1/s^2
Subscripts		
c, h, w	cold, hot, wall	
i, j	tensor directions	
x, y	directions	

Re-defined thermal diffusivity
Where $\hat{A}_w$ = cross-sectional area for longitudinal conduction and A_w = cross-sectional area for working out volume of wall, these two areas may differ depending on how a compact surface is formed, then $\hat{\kappa}_w = (\hat{A}_w/A_w)\kappa_w$.

Appendix I Creep life of thick tubes

Symbol	*Parameter*	*Units*
$e_c, \dot{e}_c$	creep strain, creep strain rate	—, 1/s
E	Young's modulus	N/m^2
J_2'	second invariant of deviator stress tensor	(N/m^2)^2
ℓ, m, n	numerical coefficients	
r	radius	m
s_{ij}'	stress deviator $(\sigma_{ij} - \sigma_m)$	N/m^2
t	time	s
Greek		
α	coefficient of thermal expansion	1/K
θ	temperature	K
θ, ϕ	angles which asymptotes make with the x-axis	
v	Poisson's ratio	
σ	stress	N/m^2

Symbol	*Parameter*	*Units*
Subscripts		
a, r, t	axial, radial, tangential	
ea, er, et	elastic axial, elastic radial, elastic tangential	
ca, cr, ct	creep axial, creep radial, creep tangential	
$\theta a, \theta r, \theta t$	thermal axial, thermal radial, thermal tangential	
$1, 2$	inside, outside	

Index

Acceptable flow velocities (Mach number) 41
Air conditioning exchangers 340
Algorithms and schematic source listings 361
  Crank–Nicholson finite-difference formulation 383
  Extrapolation of data 376
  Finite-difference solution schemes for transients 377
    alternative aproaches 380
    Crank–Nicholson approach 377
  Geometries for rectangular offset strip fins 366
  Longitudinal conduction in contraflow 370
  Mean temperature distribution in one-pass unmixed crossflow 361
  Schematic source listing for direct-sizing:
    compact contraflow exchanger 365
    one-pass crossflow exchanger 364
  Spline-fitting of data 375
Annular mist flow 332
Annular no-mist flow 332
Applicability of dimensionless groups 56
Availability 232
Axial conduction – *see* longitudinal conduction 67, 83, 37

Baffles in heat exchangers 2, 208
Baffle-ring by-pass (RODbaffle exchanger) 414
Bayonet tube exchangers 8, 14
  Conclusions, isothermal and non-isothermal shell-sides 204
  Design illustrations 190
  Hurd number 190
  Isothermal shell-side conditions 177
    annulus temperature profile 180
    condensation 189
    evaporation 178
    inner temperature profile 181
    non-isothermal shell-side conditions 191
    results for cases A, B, C, D 182–190
  Non-isothermal shell-side conditions 191
    explicit solution 196
    general numerical solutions 199
    special explicit case 194
  Pressure loss
    bayonet-end pressure loss 201
    helical annular flow 203
    simple annular flow 201
Best of plain rectangular and triangular ducts 120
Best small plain rectangular duct 125
Boiling, nucleate 331
Buffer zone, or leakage plate 'sandwich' 130
By-pass control, part-load operation 174

Calculus of variations 426
Carnot efficiency above and below the dead state 43, 298
Catalysts and continuous conversion, ortho-para, para-ortho 302
Classification of exchangers 1
  Bayonet-tube 9, 14
  Helical-tube 3
  Helically-twisted flattened-tube 7
  Involute curved, plate-fin, tube-panel 11, 13
  Plate-fin 5
  Porous matrix heat exchangers 9
  RODbaffle 6
  Serpentine tube-panel 13
  Spirally wire-wrapped 8

Classification of exchangers (*Continued*)
  Wire-woven heat exchangers 9
Compact surface selection for sizing optimization 455
  Acceptable flow velocities (Mach number) 455
  Exchanger optimization using direct-sizing 466
  Formulae used to generate performance tables 459
  Overview of surface performance 455
  Plain rectangular ducts 127, 456
  Possible surface geometries 467
  Surface selection 464
Compact contraflow, schematic source listing 365
Compact crossflow, schematic source listing 364
Compactness and performance 42
Comparison of real exchangers by exergy loss 253
Condensation 340
Consistency in design method 132
Contact resistance 341
Continuum equations 349, 408
  Coupled continuum theory 473
  De-coupling the balance of energy equation 474
  Laws of continuum mechanics 469
Contraflow
  Concept of direct-sizing in contraflow 110
  Controlling pressure loss 41
  Dependence of exergy loss on absolute temperature 236
  Direct-sizing of plate-fin exchanger 113
  Direct-sizing of helical tube exchanger 114
  Direct-sizing of RODbaffle exchanger 207
  Optimum temperature profiles in contraflow 35, 426
  Optimum pressure losses in contraflow 40
  Required values of Ntu in cryogenics 42
Conversion factors 483
Creep life of thick tubes 443
  Acknowledgements 451
  Applications 443
  Clarke's creep curves 449
  Constitutive equations for creep 447
  Early work on thick tubes 445
  Equivalence of stress systems 446
  Fail-safe and safe-life 447
  Fundamental equations 443
  Further and recent developments 451
Cross-conduction 317
Crossflow
  Determined and undetermined 90
  Direct-sizing of unmixed crossflow plate-fin exchanger 106
  Governing equations for steady crossflow 74, 79
  Longitudinal conduction in one-pass unmixed crossflow 83
  Mean TD in one-pass unmixed crossflow 78
  Mean TD in two-pass unmixed crossflow 79
  One-pass unmixed crossflow 74
  Three-pass crossflow 268
  Two-pass unmixed crossflow 79
Cryogenic heat exchangers 14, 297
  Background 287
  Candidate refrigeration fluids 299
  Carnot efficiency above and below the dead state 298
  Catalysts and continuous conversion, ortho-para, para-ortho 302
  Commercial applications 321
    ceramic super conductors 321
    fuel cells 322
    liquid hydrocarbons 321
    liquid hydrogen in aerospace 322
    liquid nitrogen 321
    pressurized hydrogen gas 321
    methanol 321
    world hydropower potential 321
  Compressors 303
  Cryo-expanders 304
    optimum expansion ratios for minimum exergy loss 306
  Forms of hydrogen 299

equilibrium, normal, ortho, para hydrogen 299
Hydrogen liquefaction plant 303
Hydrogen molecule configurations 300
Liquefaction concepts and components 298
Liquefaction of hydrogen 313
Liquefaction of nitrogen 307
Minimum work of liquefaction 300
Mixtures of gasses 299
Nitrogen liquefaction plant 307
Optimization of multistream exchangers 321
Para-content versus temperature 300
Preliminary direct-sizing of multi-stream heat exchangers 314
estimate of mean temperature difference (ratio of mass flowrates) 315
splitting exchanger into two-fluid units (approx. direct-sizing) 315
stepwise rating of exchangers 315
Product and refrigerating streams 299
Rapid cooling with mixtures of gases (Paugh) 299
Required values of Ntu in cryogenics 42
Stepwise-rating of multistream heat exchangers 317
Haseler's allowance for cross-conduction effects 317
stacking patterns for multistream exchangers 320
Storage tank 'roll-over' 14, 340
Thermo-magnetic regenerators 298
Cryogenic heat exchanger design 298
Multi-stream exchangers 314
Cryogenic storage tanks 14
Bayonet-tube exchanger 14
'Roll-over' problem 14, 340
Cryo-expanders (inward radial flow turbines) 304
Effect of pressure ratio on cooling range 306
Monatomic and diatomic molecules 306
Cubic spline-fitting (interpolating) 375

Data fitting 375
Defrosting and frosting 342
Dehumidification 340
Dig deeper (to) 45
Dimensionless groups 47
Applicability of dimensionless groups 54
Approach via differential equations 47
Buckingham's π-theorem 47
Dimensionless groups in heat transfer and fluid flow 54
Rayleigh's method 47
Direct-sizing 1
Computer programs for direct-sizing 104
Concept of direct-sizing in contraflow plate fin exchangers 116
Contraflow direct-sizing – EDGEFIN program 116
Crossflow direct-sizing – KAYSFIN program 106
Direct-sizing of bayonet-tube exchangers 177
Direct-sizing of a contraflow exchanger 113
Direct-sizing of helical-tube exchangers 143
Direct-sizing of RODaffle exchangers 208
Direct-sizing of unmixed crossflow exchanger 106
Direct-sizing of plate-fin heat exchangers 99
Rating and direct-sizing design software 103
Directional headers, U-type & Z-type 249
Double-tube heat exchanger 333

Embedded heat exchangers 251
Energy balance equation 53
Effectiveness concept 46
Entropy, fixed loss due to temperature profiles 40
Evaporation 178, 326
Exclusions and extensions 1
Baffled exchanger cores 2
Lamella heat exchangers 3
Plate-frame designs 2
Porous metal developments 3

Exclusions and extensions (*Continued*)
Printed-circuit designs 3
Rapid prototyping 3
Single-spiral designs 2
Exchanger layup (compact) 99
Exchanger optimisation 460
Exergy destruction 94
Exergy loss number for heat exchangers 229
Allowing for fluid and heat leakage 240
Bejan's balanced counterflow exchanger 230
Commercial considerations 242
Contraflow exchangers 234
Dependence of exergy loss number on absolute temperature level 236
Destruction of exergy 94
Dimensionless exergy loss number 231
Discussion of earlier work 230
Effect of temperature level on exergy loss number 236
Exergy change for any flow process 231
Exergy loss for any heat exchanger 233
Grassmann and Kopp 236
Historical development 230
Instantaneous exergy loss 234
Multi-stream exchangers 234
Minimum entropy generation 230
Minimum exergy loss 231
Optimum temperature profiles 236
Performance of cryogenic plant 238
Reference temperature 231
Specific availability 232
Specific exergy difference 232
Experimental test rigs (contraflow, singleblow) 251, 275, 423
Exponential spline fitting 375
Extrapolation of data 376

Fine-tuning of compact surfaces 127
Flow-friction and heat-transfer correlations 129, 133, 135, 154, 212, 413, 456
Flow distributors 130, Appendix 1, p3
Flow mal-distribution 250
Fouling, detection, references 442
Friedel's two-phase pressure loss 338
Frosting and defrosting 342
Fundamentals of heat exchangers 19
Compactness and specific performance 42
performance comparison 42
specific performance 42
Comparison of LMTD–Ntu and ε–Ntu approaches 33
Condenser 19, 66
Crossflow, one- and two-pass 74, 79
Contraflow, parallel flow 59, 61
De-superheating feed heater 20
Dimensionless groups 47
comparison with analytical solution 51
convective heat transfer 53
fundamental approach via differential equations 47
Rayleigh's method and Buckingham's π-theorem 47
Directional headers, U-type & Z-type 249
similarity in transient thermal conduction 48
Effectiveness and number of transfer units 27
Effectiveness and Ntu plots 31
Evaporator 19, 66
Exergy loss minimization below the dead state 35
ε–Ntu sizing problem 32
Intermediate wall temperature 65
Link between Ntu values and LMTD 26
LMTD–Ntu rating problem 23
LMTD–Ntu sizing problem 25
Log mean temperature difference 21
Ntu depends on terminal temperatures 44
Optimum pressure losses in contraflow 40
controlling pressure loss 41
exergy approach 40
Mach number approach 41
Optimum temperature profiles in contraflow 35
Parallel flow 20, 61
Rating problems, LMTD–Ntu, ε–Ntu 23, 31
Required values of Ntu in cryogenics 42

Simple temperature distributions 19
Sizing problems, LMTD–Ntu, ε–Ntu 25, 32
Sizing when $\dot{Q}$ is not specified 34
Temperature cross-over 20
Theta methods 26
To dig deeper 45
the effectiveness concept 46
units in differential equations 46
Values of Ntu required in cryogenics 42

Gas turbine, recuperated 12, 94
Inter-cooler, recuperator 19
Gaussian quadrature 422
Geometry of ROSF surfaces 133, 364
Grassman and Kopp, optimum temperature profiles in contraflow 35, 229

Headers
Compact flow distribution 249
Control of flow distribution 243
Design for zero pressure loss 244
Directional headers 249
Dow's theory of header design 244
Exchanger aspect ratios 248
Headers of varying rectangular section 246
U-type, Z-type 249
Heat transfer correlations
Helical tube multi-start coil exchangers 154, 164
Manglik & Bergles universal ROSF for compact exchangers 135, 409
Plain rectangular ducts 129
RODbaffle exchangers 211, 411
Helical-tube multi-start coil exchangers 3, 144
Central duct 151
Completion of the design 160
Consistent geometry 145
Correlations and constraints 154
Cryogenic designs 4
Design for curved tubes 168
fine tuning with curved-tube correlations 168
heat transfer (referred to outside tube surface) 170
individual coil design 169
overall heat transfer coefficient 170
shell-side heat transfer coefficient 170
shell-side pressure loss 169
straight tube correlations 168
tube-side heat transfer coefficient 170
tube-side pressure loss (coiled) 171
variations in mass flowrate 171
Design window 163
Direct-sizing design framework 143
Discussion 172
Exchanger with central duct 151
Fine-tuning the design 163, 168
Flow-friction correlations 154, 163, 168
Heat transfer constraints 158
Heat transfer correlations 154, 163, 168
Helix angle of coil 146
Laminar flow friction-factor, heat-transfer 164
Length of tube bundle 146
Length of tubing in one longitudinal tube pitch 147
Mean diameter of the z-th coil 145
Nuclear designs 4
Number of times that shell-side fluid crosses a tube turn 147
Number of tubes in exchanger 146
Optimized design 173
Part-load operation with by-pass control 174
Pressure loss constraints 158
Shell-side constraints 156
Shell-side correlations 154
Shell-side minimum area for axial flow 147
Shell-side porosity 151
Shell-side to tube-side flow area ratio 151
Simplified geometry 151
Start factor 145
Thermal design 153
Thermal design results for $(t/d) = 1.346$ 162, 173
Transition Reynolds number 164
Tube-side area for flow 151
Tube-side constraints 155
Tube-side correlations 154

Helical-tube multi-start coil exchangers (*Continued*)
  Tubing in a projected transverse cross-section 147
  Turbulent flow friction-factor, heat-transfer 166, 167
  Velocity constraints 157
Helically baffled exchangers 223
Helically-twisted flattened-tube exchanger 7
Helixchanger 223
Hurd number 190
Hydraulic diameter 121, 128, 132
Hydrogen 299
  Catalysts and continuous conversion in liquefaction 302
  Equilibrium-hydrogen 299
    normal-hydrogen 299
    ortho-hydrogen 299
    para-hydrogen 299
    spins of protons 299

Ice harvesting 342
Icing 342
Intercooler 12
Intermediate wall temperature 65
Interpolating cubic spline-fit 375
Involute-curved plate-fin exchangers 11
Inward radial flow turbines 305

Kroeger's method 67
  Longitudinal conduction in balanced contraflow 68

Labelling of exchanger ends xiii
Laplace transforms 419
Leakage buffer zone 130
Leakage plate 'sandwich' 130
Liquefaction plant 298
  Catalysts and continuous conversion 302
  Compressors 303
  Concepts and components 298
  Cryo-expanders 304
  Hydrogen 299, 313
  Nitrogen 307
Lockhart–Martinelli two-phase pressure loss 327
Log mean temperature difference (LMTD) 21
  Comparison of (LMDT–Ntu) and (ε–Ntu) approaches 33
  Link between Ntu and LMTD 26
  (LMTD–Ntu) rating 23
  (LMTD–Ntu) sizing 25
  Reduction factor due to longitudinal conduction (balanced) 67
  Reduction factor due to longitudinal conduction (unbalanced) 72
  'Theta' methods 26
Longitudinal conduction in transient flow 263
Longitudinal conduction in contraflow (steady-state) 67, 370
Longitudinal conduction in one-pass unmixed crossflow (steady-state) 83

MacCormack finite-difference scheme 257, 380
Mach number 41, 455
Manglik & Bergles universal correlations 132, 135, 405
Mean temperature difference in one-pass unmixed crossflow 74, 362
Mean temperatue difference in two-pass unmixed crossflow 77
Mean temperature difference in complex arrangements 93
Method of characteristics 258
Mist flow 332
Multi-stream exchangers 130, 317
  cross-conduction effect 317
  three-fluid exchangers 94
Most efficient temperature distribution in contraflow 425
  Calculus of variations 425
  Optimum temperature profiles 426

Navier–Stokes equation 53
Newtonian constitutive equation 53
Nitrogen liquefaction 307
Notation 487
Ntu from terminal temperatures only 42
Nucleate boiling 331

Optimization of rectangular offset-strip plate-fin surfaces 405
  Fine-tuning of rectangular offset-strip fins 405
  Manglik & Bergles correlations 409
  Optimization graphs 408
  Trend curves 407

Optimum pressure losses in contraflow 40
Optimum temperature profiles in contraflow (Grassmann & Kopp) 35, 236, 426
Overview of surface performance 455

Part-load operation, by-pass control 174
Performance data for RODbaffle exchangers 411
  Baffle-ring bypass 414
  Further heat-transfer and flow friction data 411
Physical properties of materials and fluids 429
  Fluids 429
  Solids 431
  Sources of data 429
Pinch technology 92
Plain rectangular duct 120, 129
Plate-fin heat exchangers 5, 99
  Alternative contraflow design 120
  Best of plain rectangular and triangular ducts 120
  Best small plain rectangular duct 125
  Buffer zone or leakage late 'sandwich' 130
  Cautionary remark about core pressure loss 92
  Computer software for direct-sizing 104
  Concept of direct-sizing in contraflow 110
  Conclusions 138
  Consistency in design methods 132
  Contraflow exchanger – EDGEFIN program 115
  Crossflow exchanger – KAYSFIN program 106
  Direct-sizing of a contraflow exchanger 113
  Direct-sizing of an unmixed crossflow exchanger 106
  Exchanger layup 99
  Fine tuning of ROSF surfaces 127
  Flow-friction correlations 103
  Geometry of rectangular offset strip fins 133
  Headers, distribution 130
  Heat-transfer correlations 103
  Involute curved layup 11
  Longitudinal conduction losses using LOGMEAN 125
  Manglik & Bergles universal correlations 135, 409
  Multi-stream design 130
  Overview of surface performance 127
  Rating and direct sizing 103
  Specific performance comparison of plain rectangular ducts 129
  Surface geometries 103, 120, 125, 129, 133, 135
  Total pressure loss 105
  Universal ROSF correlations 135
Porous matrix heat exchangers 9
Pressure loss
  Cautionary remark concerning evaluation 92
  Compact flow distributors 249
  Control of flow distribution (temperature dependent fluid properties) 243
  Dow's theory of header design 244
  Exit loss (expansion) 93
  Flow acceleration 93
  Flow maldistribution (minimization) 250
  Friedel two-phase flow pressure loss 338
  Header design for zero pressure loss 244
  Headers of varying rectangular section 246
  Inlet loss (contraction) 93
  Kay's and London expression for losses 93
  Lockhart–Martinelli two-phase pressure loss 327
  Minimizing effects of flow maldistribution 250
  Pumping power 253

Pressure loss (*Continued*)
Test rig for transients in model heat exchanger 251
U-type and Z-type headers 249
Optimum pressure losses in contraflow 40
Primary surface heat exchanger 129
Propulsion systems 10
Intercoolers 12
Large recuperators 11
Liquid hydrogen propulsion 12
Small recuperators 11
Proving the single-blow test method – theory and experiment 420
Analytical approach using Laplace transforms 419
Experimental test equipment 423
Numerical evaluation of Laplace outlet response 420
Pumping power 253

Rayleigh dissipation function 53
Rating and direct-sizing software 103
Rectangular offset strip fins (ROSF), fine tuning 133, 405
Reduction factor for LMTD (due to longitudinal conduction) 67
Balanced contraflow 68
Unbalanced contraflow 72
Reduction in meanTD in one-pass unmixed crossflow 83
Refrigeration fluids 299
Regenerators 290
Roadmap, thermal design xxviii
RODbaffle exchangers 6
Approach to direct-sizing 208
Baffle-ring by-pass 411
Characteristic dimensions 209
Configuration of the RODbaffle exchanger 208
Design correlations 210
Design framework 207
Direct-sizing 215
Flow areas 209
Flow-friction correlations 213, 411
baffle-rings 214
shell-side 214
tube-side 213
Further flow-friction and heat-transfer data 411
Generalized correlations 220
shell-side baffle pressure loss 221
shell-side heat transfer 220
Heat-transfer correlations 211, 411
shell-side 211
tube-side 212
tube-wall 212
Other shell and tube designs 222
Phadke tube count 216, 217
Practical design 217
Recommendations 222
Reynolds numbers 211
Shell-by-pass flow 416
Tube-bundle diameter 217
'Roll-over' 14, 340

Schematic algorithms 361
Segmental baffles 2
Shell-and-tube exchangers 222
Conventionally baffled 222
Flattened and helically twisted tubes 223
Helically baffled 223
RODbaffled 208
Small tube inclinations 266
Similarity 48
Single-blow testing 275
Accuracy of outlet response curves in experimentation 284
curve matching, initial rise, maximum slope, phase angle & amplitude 284
Additional effects 287
axial and longitudinal conduction in the fluid 287
conduction into the solid interior 287
internal heat generation 287
longitudinal conduction in the solid 287
surface losses from matrix exterior 287
Analysis of coupled fluid and solid equations 278
Analytical and physical assumptions 277
Boundary conditions 280

Choice of theoretical model 276
Complete curve matching 284
Conclusions on test method 287
Coupled fluid and solid equations 278
Experimental test rig and equipment 275, 423
Exponential inlet disturbance 383
Features of test method 275
Generating theoretical response curves 286
Harmonic inlet disturbance 282
Initial rise method 284
Inlet disturbances 277
Inverse Laplace transforms 281
Laplace transforms 420
Longitudinal conduction 288
Mathematical assumptions & physical requirements 277
Maximum slope 284
Numerical evaluation of integrals 420
Practical considerations 288
full equations 288
longitudinal conduction 288
Phase angle and amplitude 285
Regenerators 290
Relative accuracy of outlet response curves in experimentation 284
Simple theory 278
Simplification 290
Solution of basic equations using Laplace transforms 280
Solution by finite-differences 286, 420
full computation 289
neglecting longitudinal conduction 290
Step inlet disturbance 284
Theoretical modelling 276
Theoretical outlet response curves 285
Single-pass crossflow 74
Sinusoidal–lenticular surfaces 477
Sizing when $\dot{Q}$ not specified 34
Solution of transient temperature fields in contraflow 379, 388, 399
Solution of transient velocity fields in contraflow 379, 384, 386
Source books on heat exchangers 433
Exchanger types not already covered 439
Fouling – some recent literature 442
Texts in chronological order 433
Single-spiral heat exchangers 2
Specific performance 42, 129, 139, 219
Spirally wire-wrapped exchanger 7
Spline-fitting of data 375
Cubic, exponential, taut, variable power 375
Steady-state temperature profiles 59
Cautionary remark about core pressure loss 92
Condensation 66
Contraflow 61
Determined and undetermined crossflow 90
Evaporation 66
Exergy destruction 94
Extension to two-pass unmixed crossflow 79
Involute-curved plate-fin exchangers 82
Linear temperature profiles in contraflow 59
Longitudinal conduction in contraflow 67
equal water equivalents 68
schematic temperature profiles 71
unequal water equivalents 72
Longitudinal conduction in one-pass unmixed crossflow 83
Mean temperature difference in complex arrangements 93
Mean temperature difference in unmixed crossflow 74
Parallel flow 61
Pinch technology 92
Possible optimization criteria 92
Three fluid exchangers 94
Wall temperatures 65
Stepwise rating of multistream exchangers 317
Stratified flow 331
Suggested further research 477
Header design 478
Steady-state crossflow 478
Transients in contraflow 479

Supplement to Appendix B – Transient algorithms 383
Balance of energy 388
Balance of linear momentum 386
Balance of mass 384
extrapolation 385
zero gradient 386
Coding of temperature matrix TMATRIX 397
Conclusions 404
Crank–Nicholson finite-difference formulation 383
Preparation of algorithms 383
TMATRIX 399

Taut spline fitting 375
Temperature crossover 20,80
Test rigs, contraflow, singleblow 251, 423
Thermal design roadmap xxviii
Thermal storage in wall 349
'Theta' methods 26
Three-fluid exchangers 94
Three-pass crossflow 268
Time constant 421
To dig deeper 45
Transient equations with longitudinal conduction and wall storage 349
Computational approach 355
change in sign of velocity 358
development of algorithms 359
energy equations 357
fluid flow equations 356
numerical considerations 355
potential problems with crossflow 358
pressure field terms 357
reflection of transients in contraflow 357
selection of time intervals 355
splitting the problem 355
transients travelling against the flow in contraflow 358
Mass flow and temperature transients in contraflow 349
alternative form of balance of linear momentum 351
constitutive equation for Stokes fluid 350
pressure gradient due to friction 350
Summarized development of transient equations for contraflow 352
cleaned up 354
expanded and rearranged 353
fundamental 352
simpified for computation 354
Transients in heat exchangers 257
Contraflow review of solution methods 257
characteristics, method of 258
direct finite-differences 257
Laplace transforms with numerical inversion 258
MacCormack's finite difference method 257, 380
method of characteristics 258
other approaches 258
Rayleigh dissipation function 258
Contraflow with finite-differences 259
convective mesh drift 262
disturbances, shape of 264
enginering applications – contraflow 266
extrapolation schemes 385
finite-difference solution schemes 383
flow-friction and pressure terms 262, 265
interpolating cubic splinefits 263
longitudinal conduction 263, 349
Mach numbers 263
mass flow and temperature transient equations 349, 352
mesh drift, convective 262
one dimensional plug flow 263
order of solution 264
phase-lag, cross-conduction and boundary conditions 265
physical properties 263
pressure terms and flow friction 262, 265
Rayleigh dissipation function neglected 260
results of computation (without pressure field equations) 265
selection of time intervals 260, 383

shape of disturbances 264
shell-and-tube exchangers with small tube inclinations 266
shell heat leakage 262
space and time intervals 260, 383
summarized development of transient equations 352
temperature difference across solid wall 263
time interval selection 260, 383
Crossflow review of solution methods 267
axial dispersion terms 259
engineering applications 268
summary of past work 267
solution methods 268
Supplement to Appendix B – Transient algorithms 383
TMATRIX coding 399
Transition in two-phase flow (annular mist to mist flow) 332
Trend curves for selection of ROSF surfaces in contraflow 407
Tubular heat exchangers 13
Involute tube panel 13
Serpentine tube panel 13
Tube-and-fin (fin-and-tube) heat exchangers 341, 439
Twisted-tube heat exchanger 7, 223
Two-pass unmixed crossflow 79
Two-phase flow 12-2
Aspects of air-conditioning 340
Condensation 343
Contact resistance 340
Fin-and-tube (tube-and-fin) heat exchangers 341
Friedel two-phase pressure loss correlation 338
Frosting and defrosting 342
Ice harvesting 342
Lockhart–Martinelli two-phase pressure loss correlation 327
Plate-fin surfaces 339
Rate processes 343
Supporting work 339
Two-phase design of a double-tube exchanger 333
Two-phase flow regimes 326
Two-phase heat transfer correlations 331
annular mist flow 332
annular (no-mist) flow 332
demarcation mass velocity 333
mist flow 332
nucleate boiling 331
stratified flow 321
transition from annular mist flow to mist flow 332
Two-phase pressure loss 327
Friedel 327, 338
Lockhart–Martinelli 327, 328
With and without phase change (two-phase flow) 325

Units
in differential equations 46
nomenclature 487

Variable power spline fitting 375

Wall temperature, intermediate 65
Wall thermal diffusivity 349
Wire-woven heat exchangers 9